CHEMICAL MUTAGENS

Principles and Methods for Their Detection

Volume 10

A Continuation Order Plan is available for this series. A continuation order will bring delivery of each new volume immediately upon publication. Volumes are billed only upon actual shipment. For further information please contact the publisher.

Sponsored by the Environmental Mutagen Society

CHEMICAL MUTAGENS

Principles and Methods for Their Detection

Volume 10

Edited by Frederick J. de Serres

National Institute of Environmental Health Sciences
Research Triangle Park, North Carolina

PLENUM PRESS • NEW YORK and LONDON

The Library of Congress cataloged the first volume of this title as follows:

Chemical mutagens. v. 1–
New York, Plenum 1971–

v. ill. 24 cm.

"Principles and methods for their detection."
Vols. 1– sponsored by the Environmental Mutagen Society.
Key title: Chemical mutagens, ISSN 0093-6855.

1. Chemical mutagenesis—Collected works. I. Environmental Mutagen Society.

QH465.C5C45	575.2′92	79-640909
Library of Congress	79	MARC-S

ISBN 0-306-42171-2

This book was edited by F. J. de Serres in his private capacity. No official support or endorsement of DHHS is intended or should be inferred.

A Division of Plenum Publishing Corporation
233 Spring Street, New York, N.Y. 10013

Printed in the United States of America

Contributors

Gerald M. Adair, Science Park Research Division, University of Texas System Cancer Center, Smithville, Texas 78957

Bruce M. Cattanach, Medical Research Council, Radiobiology Unit, Chilton, Didcot, Oxon OX11 ORD, England

Peter de Boer, Department of Genetics, Agricultural University, 6703 BM Wageningen, The Netherlands

Linda Dousman, Microbial Genetics Department, SRI International, Menlo Park, California 94025

Arthur Falek, Human and Behavioral Genetics Research Laboratory, Georgia Mental Health Institute, and Department of Psychiatry, Emory University School of Medicine, Atlanta, Georgia 30306

Polly A. Foureman, Department of Zoology and Physiology, Louisiana State University, Baton Rouge, Louisiana 70803

Hansjörg Frei, Institute of Toxicology, Swiss Federal Institute of Technology and University of Zurich, CH-8603 Schwerzenbach/Zurich, Switzerland

Hansruedi Glatt, Institute of Toxicology, University of Mainz, D-6500 Mainz, West Germany

Ulrich Graf, Institute of Toxicology, Swiss Federal Institute of Technology and University of Zurich, CH-8603 Schwerzenbach/Zurich, Switzerland

Leigh Henderson, Centre for Medical Research, School of Biological Sciences, University of Sussex, Falmer, Brighton BN1 9QG, England

Cynthia A. Hoy, Biomedical Sciences Division, Lawrence Livermore National Laboratory, University of California, Livermore, California, 94550. *Present address:* Department of Biological Sciences,

Stanford University, Stanford, California 94305–2493

Ronald M. Humphrey, Science Park Research Division, University of Texas System Cancer Center, Smithville, Texas 78957

T. C. Hsu, Department of Genetics, University of Texas System Cancer Center, M. D. Anderson Hospital and Tumor Institute at Houston, Houston, Texas 77030

Jan C. Liang, Department of Genetics, University of Texas System Cancer Center, M. D. Anderson Hospital and Tumor Institute at Houston, Houston, Texas 77030

Kenneth I. Mandelberg, Department of Mathematics and Computer Science, Emory University, Atlanta, Georgia 30322

Kristien E. Mortelmans, Microbial Genetics Department, SRI International, Menlo Park, California 94025

Franz Oesch, Institute of Toxicology, University of Mainz, D-6500 Mainz, West Germany

Kanithahalli L. Satya-Prakash, Department of Genetics, University of Texas System Cancer Center, M. D. Anderson Hospital and Tumor Institute at Houston, Houston, Texas 77030

Antony G. Searle, Medical Research Council, Radiobiology Unit, Chilton, Didcot, Oxon OX11 ORD, England

David A. Shafer, Human and Behavioral Genetics Research Laboratory, Georgia Mental Health Institute, and Department of Psychiatry, Emory University School of Medicine, Atlanta, Georgia 30306

Michael J. Siciliano, Department of Genetics, University of Texas System Cancer Center, M. D. Anderson Hospital and Tumor Institute at Houston, Houston, Texas 77030

Raymond L. Stallings, Department of Genetics, University of Texas System Cancer Center, M. D. Anderson Hospital and Tumor Insitute at Houston, Houston, Texas 77030

Charles Tease, Medical Research Council, Radiobiology Unit, Chilton, Didcot, Oxon OX11 ORD, England

Larry H. Thompson, Biomedical Sciences Division, Lawrence Livermore National Laboratory, University of California, Livermore, California 94550

Ekkehart W. Vogel, Department of Radiation Genetics and Chemical Mutagenesis (Sylvius Laboratories), State University of Leiden, Leiden, The Netherlands

Friedrich E. Würgler, Institute of Toxicology, Swiss Federal Institute of Technology and University of Zurich, CH-8603 Schwerzenbach/ Zurich, Switzerland

Preface

Volume 10 of *Chemical Mutagens* contains a series of papers on work aimed at facilitating the detection of the genetic effects of environmental agents. These include two new approaches with *Drosophila* (one to detect somatic mutations and another which makes use of new mutagen-sensitive strains), and the detection of gene mutations in both normal and repair-deficient strains of CHO cells in culture. Two chapters are devoted to the analysis of the genetic consequences of the induction of chromosome aberrations in mice. Three chapters deal with the detection of chemicals that can cause aneuploidy, in *Drosophila,* in mammalian cells in culture, and in the mouse. Another chapter deals with the effect of different plasmids on the process of mutation induction in *Salmonella.* The remaining chapters are concerned with structural factors that affect the mutagenicity of polycyclic aromatic hydrocarbons, the detection of transplacental genotoxic agents, and computer automation of cytogenetic analysis.

This volume also lists the Contents for the previous nine volumes in this series. This listing should greatly facilitate searches for particular chapters, especially in those cases where a volume may be missing from the set.

Research and development are stronger than ever in this field. There are a variety of tasks to be performed and different assay systems may be particularly useful for gathering information, for example, on basic mechanisms of mutation induction, but totally unsuitable for mass-screening programs. Progress is slow, but the objectives are clearly defined and the solutions will come with time and the design of appropriate assay systems and experiments. Many of the chapters in this volume describe exciting new approaches to the problem of detecting mutagenic activity with mammalian cells in culture and with the whole animal. The further development of these whole-animal systems is the only way that we can ensure a comprehensive evaluation of those chemicals that are positive in *in vitro* assays for mutagenicity and to evaluate their full potential for producing mutagenic/carcinogenic effects in those cases where there might be significant human exposure.

Frederick J. de Serres

Research Triangle Park, North Carolina

Contents of Earlier Volumes

Volume 1

Volume 2

Volume 3

Volume 4

Volume 5

Volume 6

Volume 7

Volume 8

Volume 9

Contents

Chapter 3

Chromosomal Mutations: The Genetic Approach **129**

Antony G. Searle

Chapter 4

Cytogenetic Assays for Mitotic Poisons Using Somatic Animal Cells 155

T. C. Hsu, Jan C. Liang, and Kanithahalli L. Satya-Prakash

Chapter 5

Detection of Aneuploidy in *Drosophila* 183

Polly A. Foureman

CHAPTER 1

In Vivo Mutagenicity Testing Using Somatic Cells of *Drosophila melanogaster*

Friedrich E. Würgler and Ekkehart W. Vogel

1. Introduction

In mutagenicity screening programs microbial tests have to be supplemented by tests with eukaryotic cells because certain kinds of chemicals

Abbreviations used in this chapter: A-α-C, 2-amino-α-carboline; ara-C, cytosine arabinoside; AF-2, furylfuramide; DA, 9,10-dimethylanthracene; DEB, diepoxybutane; DMBA, 7,12-dimethylbenz(*a*)anthracene; DMSO, dimethylsulfoxide; EMS, ethyl methanesulfonate; Glu-P-1, 2-amino-6-methyldipyrido(1m2:3′,2′-d) imidazole; Glu-P-2, 2-aminodipyrido(lm2-a:3′,2′-d) imidazole; IPCS, International Program on Chemical Safety; iPMS, isopropyl methanesulfonate; IQ, 2-amino-3-methylimidazo(4,5-f)quinoline; IS, insertion sequence; kb, kilobases; L1, first larval instar; L2, second larval instar; L3, third larval instar; MeA-α-C, 2-amino-3-methyl-α-carboline; MeIQ, 2-amino-3,4-dimethylimidazo(4,5-f)quinoline; MeIQx, 2-amino-3,8-dimethylimidazo(4,5-f)quinoxaline; ML, mosaic light spots in the eye; MMS, methyl methanesulfonate; MNNG, *N*-methyl-*N*′-nitro-*N*-nitrosoguanidine; SLRL, sex-linked recessive lethals; TEM, triethylenemelamine; Trp-P-1, 3-amino-1,4-dimethyl-5*H*-pyrido(4,3-b)indole; Trp-P-2, 3-amino-1-methyl-5*H*-pyrido(3,4-b)indole; TS, twin spot; UZ, unstable *white–zeste* eye mosaic system; //, symbol used to separate the genetic symbols of phenotypically different cell clones in twin spots.

Friedrich E. Würgler • Institute of Toxicology, Swiss Federal Institute of Technology and University of Zurich, CH-8603 Schwerzenbach/Zurich, Switzerland. **Ekkehart W. Vogel** • Department of Radiation Genetics and Chemical Mutagenesis (Sylvius Laboratories), State University of Leiden, Leiden, The Netherlands.

are undetectable or hardly detectable as mutagens by bacterial test systems. Fundamentally, the consequences of mutagen-induced changes in all genetically relevant cell components (in particular those organized differently in eukaryotes and in prokaryotes, such as nucleosomal DNA and tubulin-containing spindle fibers) must be detectable with appropriate screening schemes. Such changes can lead to mutagenic events in eukaryotes that are absent in bacteria (e.g., chromosome aberrations, sister chromatid exchange, mitotic recombination, and nondisjunction). In addition, the same chemical agent may induce different types of gene mutation when tested in bacteria or eukaryotic cells. For example, the bacterial frameshift mutagens benzo(*a*)pyrene diol epoxide and *N*-acetoxy-2-acetylaminofluorene yield predominantly base substitutions in human cells.[9] This calls for an analysis of potentially mutagenic chemicals in at least one eukaryotic assay system. As far as qualitative studies are concerned (detection of mutagenic potential of compounds), submammalian systems should be used extensively to avoid the unnecessary use of experimental warm-blooded animals. Another possibility is to use mammalian cells in culture, but with respect to xenobiotic metabolism, most of these systems pose the same problems as the bacterial assays and need an exogenous metabolizing system, such as an S9 liver extract.

Drosophila, with its intensively studied genetics, is a promising candiate for submammalian mutagenicity screening assays. An appreciable data base is available due to decades of research with *Drosophila* in radiation genetics[175] as well as chemical mutagenesis.[112,203,211] At present, the best-validated *Drosophila* screening test is the sex-linked recessive lethal (SLRL) assay. Of 421 compounds tested, 198 gave a positive mutagenic response, 46 were classified as nonmutagens, and 177 gave inconclusive results for different reasons, most frequently because of insufficient sample size.[112] The correlation between carcinogenic activity and mutagenic activity in the SLRL assay is excellent: between 83% (85/102) and 91% (50/55) of the classical carcinogens tested are mutagens in *Drosophila*.[112,211] Unfortunately, the SLRL assay is a relatively tedious and time-consuming method compared with the fast systems developed with bacteria and some lower eukaryotes. The flexibility one would wish to reach in theory in routine testing (e.g., fast repetition of tests under modified conditions) is, in practice, applicable for only a small number of chemicals.[215] Moreover, the possibility of elimination during meiosis of cells carrying a recessive lethal mutation can be a source of underestimation in the screening of chemicals for the induction of SLRL in spermatogonia, especially when larvae are exposed. In view of these intrinsic limitations of the SLRL method for the screening of large numbers of chemicals, other experimental approaches using somatic cells may improve the performance of *Drosophila* in screening programs. Homozygous mutant mosaic spots (clones) have long been known to form on the cuticle of adults after exposure of heterozygous larvae to mutagens, such as mustard gas,(8) X rays,(14) methyl methanesulfonate,(131) and triethylenemelamine.(17) Based on these observations, a method relying on the detection of mosaic spots was recommended for the screening of potential mutagens.(129,131) Based on new experimental data, several investigators have recently drawn attention to the potential of using somatic recombination and mutation assays.(63,78,80,160,196,210,211)

In general, the methods are based on the production of larvae with a genotype such that a mutational change in a cell leads to a genotypic change that is phenotypically manifested at a later stage in all the progenitor cells in the ensuing clone. For the most part manifestation occurs after the cell clone originating from the genetically altered larval cell has passed through metamorphosis in the pupal stage and has become part of the differentiated body of the adult fly. If suitable markers are used, the clones form spots on the external surface of the fly and allow easy detection of genetic changes. The detectable genetic alterations may be caused by mitotic recombination, somatic mutation, chromosome breakage, or nondisjunction and can be spontaneous or induced by genotoxic agents.

A number of genetic markers are suitable for clonal analysis in *Drosophila*.(18,67,151) They have been used for developmental studies of various organs and tissues. A prerequisite for mutagenicity testing is easy and unquestionable scoring after a minimal preparation of the body part to be analyzed. For this reason, only three systems have been explored: (1) the wing mosaic system, (2) the white/white-coral eye mosaic system, and (3) the unstable white–zeste eye mosaic system. The three assays will be described individually and then, based on the experimental data, some general conclusions will be drawn with regard to the present state of evaluation of the different *Drosophila* somatic recombination and mutation assays.

2. Basic Developmental Biology of *Drosophila*

Drosophila melanogaster is a holometabolic insect which undergoes complete metamorphosis. Under standard culture conditions (25°C and 60% relative humidity), the succession and approximate duration of the different stages in the life cycle are as follows: embryonic development, 1 day; first larval instar (L1), 1 day; second larval instar (L2), 1 day; third larval instar (L3), 2 days; prepupa, 4 hr; pupa, about 4.5 days; adult stage (imago), up to 40 days. An introductory description of the life cycle and some basic techniques are given in *A Handbook of Drosophila Development*[154] and in *The Genetics and Biology of Drosophila*, Vol. 2a.[6]

During development, groups of cells are set apart and are not involved in the formation of the larval body. These cell groups are frequently found as distinct small bodies within the larva. Most of them are discs in the anterior part of the larva. During the larval period the undifferentiated cells proliferate. The imaginal discs grow until they enter into differentiation at the pupal stage. They form distinct parts of the body of the adult fly (e.g., an eye is formed by an eye imaginal disc, a wing by a wing disc, legs by the leg discs). The histoblasts that form the abdominal cuticle of the adult fly are an exception because they represent cell groups that do not divide during the larval period. Detailed descriptions of the various developmental aspects of imaginal discs and cell nests can be found in review chapters included in *The Genetics and Biology of Drosophila*, Vol. 2c.[7]

Figure 1A shows cell numbers during imaginal disc growth. The specific dynamics of cell proliferation within the different cell groups during embryonic, larval, and pupal development are reflected by the different shapes of the curves.

Cell numbers occasionally have been determined from histological cell counts,[14] but usually are derived from analyses of X-ray-induced, genetically marked cell clones.[151] In the first study of X-ray-induced somatic recombination, Patterson[148] pointed out that the size of the marked clone is governed more by the animal's age at the time of radiation than by the X-ray dose. He reasoned that the size of a clone will be inversely proportional to the number of existing cells, but positively correlated with the number of cell divisions that occur after the genetic exchange. Accordingly, clone size can be used to estimate the cell number at different points of time during development.

The duration of a mitotic cycle during exponential growth of imaginal discs and histoblasts has been estimated from indirectly established growth curves.[30] The values are 11.9 hr for the eye disc, between 8.0 and 10.6 hr for the wing disc, and 2.8 hr for the hemitergite primordia of segments 2–6.[151]

Cell proliferation during development leads to an increase in the number of target cells, and an increased clone induction frequency is expected with increasing age of the mutagen-exposed larvae. In X-ray studies, the clone frequencies per disc depend on the developmental

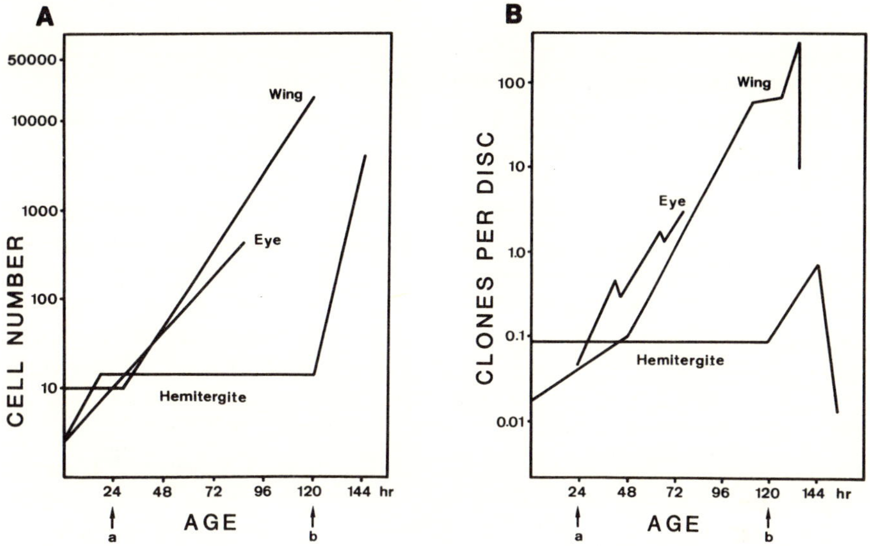

FIGURE 1. (A) Imaginal disc growth curves for the wing and eye discs and the growth curve for the histoblast nests forming the hemitergites. Cell numbers in various imaginal discs are estimated for different developmental stages from cell lineage data. (B) The clone frequencies per disc observed after 1000 R of X rays. *a*, Hatching of larvae; *b*, pupation. [Adapted from Postlethwait.[151]]

stage at the time of irradiation (Figure 1B). During much of the midlarval life period, the number of clones per disc increases geometrically for the anterior discs, but stays constant for the tergite primordia. Figure 1B also demonstrates that detectable clones in the eye are induced during the first and second instars, whereas in the wing disc, such clones are obtained after induction in any of the three larval instars. This difference between eye and wing results primarily from the fact that in the wing small spots affecting fewer than five cells are detectable; in the eye alterations in fewer than five cells are not easily detectable.

From the data on clone size and frequencies, one can estimate cell sensitivity to X-ray-induced somatic recombination. The frequency of somatic recombination induced in the left arm of chromosome 3 by 1000 R of X rays appears to increase somewhat during larval development from one in 500 cells in the second instar to one in 250 cells in the late third instar.[68] It is not known whether this apparent change in cell sensitivity is related to regular changes in sensitivity within the cell cycle, but there is evidence consistent with that suggestion.[151]

3. The Wing Mosaic System

3.1. The Genetic Basis of the Wing System

The most suitable genetic wing markers are those resulting in an altered phenotype of the trichomes (hairs) on the wing blade cells. One of the markers, multiple wing hairs (*mwh*, 3-0.0), is a homozygous viable recessive mutation which produces multiple trichomes per cell instead of the normally unique trichome. It is also expressed on other body parts, not only in the cells of the wing blade.[115,149] A second marker, flare (*flr*, 3-39.0), was isolated by Garcia-Bellido and Dapena.[67] This mutation is a zygotic recessive lethal, but homozygous (and possibly hemizygous) cells in the wing imaginal disc survive and lead to wing blade cells with short, thickened, and misshaped trichomes. The mutation has a somewhat variable expression,[80] but is usually clearly distinguishable from wild-type. In stock cultures, the *flr* alleles have to be kept over a multiply inverted balancer chromosome such as *TM1* or *TM2*.

The important feature of the progeny from a cross of *mwh*/*mwh* females to *flr*/+ males is that one-half are transheterozygous for the two markers on the left arm of chromosome 3. They are *mwh* +/+ *flr*. In this configuration somatic recombination in the chromosome region between the centromere and *flr* leads to twin spots (*mwh*//*flr*) consisting of two adjacent cell clones showing the multiple wing hairs and the flare phenotype, respectively. Somatic recombination between the markers *mwh* and *flr* leads to *mwh* single spots. Single spots for *mwh* and for *flr* may also arise from mutation or deletion of the respective wild-type alleles in the double heterozygote.

3.1.1. Repair-Proficient Larvae

In these crosses larvae are used that are proficient with respect to all DNA repair systems known in *Drosophila*,[26] in particular those modifying chemical mutagenesis (Würgler *et al.*, this volume, Chapter 10).

3.1.1a. The mwh–flr Cross. In the *mwh–flr* test cross, females of the genotype *y*; *mwh jv* are mated to males of the genotype *y Dp* sc^{J4}, *flr*/*TM1*, *Me ri* sbd^2. Detailed information on the genetic markers is found in Lindsley and Grell.[115] In the male stock, the *flr* marker is combined with a small terminal transposition containing the wild-type allele of the X-chromosomal body color marker yellow (*y*, 1-0.0). The transposition is attached to the tip of the left arm of chromosome 3. On a yellow background (if all X chromosomes of the strains used carry a mutant yellow allele), the progeny heterozygous for *flr* are recognized as nonyellow flies, while the progeny carrying the balancer chromosome are mutant yellow.

The standard *mwh–flr* cross used so far has the disadvantage of producing a relatively high and somewhat variable frequency of spontaneous small spots. Double-haired cells are often seen, which we do not record as multiple wing hairs cells. U. Graf (unpublished results) has found that the background frequency of these double-haired cells is temperature dependent. Elevated temperature (28–29°C) drastically increases the frequency of these aberrant cells. The duplication present on the *flr*-carrying chromosome seems to render the system particularly sensitive to this effect of increased temperature. To keep errors due to temperature sensitivity at a minimum, rigorous temperature control is indispensable.

3.1.1b. The mwh–flr^3 Cross. U. Graf (unpublished results) tested a new cross of *mwh* females to *flr^3/TM2, Ser* males. This cross leads to a lower spontaneous frequency of small single spots and a very low frequency of double-haired cells. This cross is only slightly temperature sensitive. Because the chromosome carrying the *flr^3* allele does not carry an additional marker gene, progeny not carrying *flr^3* are identified based on the *Ser* marker present on the balancer chromosome. *Ser(+)* individuals have the required double heterozygous constitution for the autosomal wing cell markers.

3.1.1c. The mwh–f Cross. An additional set of wing tester strains was constructed (J. Szabad, unpublished results) to study genetic events induced on the X chromosome. In a multiple wing hairs background, a transposition of the *mwh(+)* locus on an X chromosome[194] is used. In addition, the X chromosome carries a strong allele of forked (*f*, 1-56.7). The assay system can be used in a cross between *X/X; mwh* females to *X, mwh(+)f; mwh* males. The female progeny have the genotype *mwh(+) f/f(+); mwh/mwh*, and the wings can be scored for *mwh//f* twin spots as well as *mwh* and *f* single spots. The mutagenic and recombinogenic effects of two nitroso compounds (*N*-nitrosomorpholine and dinitrosopiperazine) have been studied[189,190] with this marker combination.

3.1.2. Excision Repair-Deficient Larvae

In order to study the influence of an excision repair deficiency on clone induction by chemical mutagens in the wing assay, U. Graf (unpublished results) introduced the X-chromosome mutation meiotic-

9 (i.e., the allele *mei-9*^{L1}[77,207])into one of the tester stocks (U. Graf, unpublished results; Refs. 76, 79, 176) Virgin *y mei-9*^{L1} *cv; mwh jv* females are crossed to males of the genotype *flr*^{3}*/TM2, Ser*. From this cross two types of progeny result: repair-proficient heterozygous females *(+/mei-9)* and excision repair-deficient males *(mei-9)*. In the males, the wings show the phenotype of the linked marker gene crossveinless (*cv*, 1-13.7). This is convenient for identifying *mei-9* wings even after mounting.

3.2. Scoring

To obtain permanent preparations, the wings are embedded in Faure's solution (gum arabic 30g, glycerol 20 ml, chloral hydrate 50 g, water 50 ml).[200] Fresh flies are placed in a drop of Faure's solution. Flies from storage in 70% ethanol are washed in water and then transferred into Faure's solution. The wings are removed, aligned with forceps on a clean slide, and allowed to spread. The wings arranged on the slide are kept in a dust-free atmosphere for at least 24 hr to become firmly glued to the slide. A droplet of Faure's solution is put on a coverslip, which, with the hanging drop is lowered onto the wings. The coverslip is weighted for at least 24 hr with several metal cubes (about 40 g each) while the preparation is drying and hardening. A permanent preparation is obtained by sealing the coverslip with nail polish. Screening of wings is carried out with a compound microscope, preferably at a magnification of 400×.

A faster mounting procedure, which permits analysis of the wings the same day the flies hatch, was described by Graf *et al.*[80] This method requires careful handling.

In the course of test development, it appeared that the correct identification of the mutant phenotypes, especially of the small *mwh* spots, could pose some problems. To be safe, it was decided to record as mutant only those spots that contain at least one cell with three or more hairlike (trichomal) elements.[80] The *flr* phenotype, also quite variable in appearance (from pointed, shortened, and thickened hairs to amorphic, sometimes balloonlike extrusions of melanotic chitinous material), usually poses no problems, except that in clones of fewer than four cells this mutant is practically never expressed.[196]

Because of the cell-autonomous expression of the phenotypes and the simple arrangement of all cells in two flat layers, it is not only possible to score large spots but, in the case of *mwh*, also very small ones consisting of just one or two mutant cells.

In our experience, the larger clones are detected without difficulty.

In contrast to Garcia-Bellido and Merriam,[68] but in accordance with Baker *et al.*,[13] we observed only a few split clones. We adopted the rule to count neighboring spots as separate clones if they are separated by more than two rows of normal hairs.

It is often not possible to count precisely the number of cells in very large spots. This does not introduce an important systematic error in spot size classification, because with large spots the limits of the size classes are very wide (higher powers of 2).

The variations of the mean spot size and the mean number of spots per wing were studied after acute exposure of larvae of different age to the unstable alkylating agent MMS (U. Graf, unpublished results) and the promutagen procarbazine (H. Juon, unpublished results). As expected, very large spots are observed after treatment of very young larvae. Due to the small number of target cells present in the wing disc of L1 larvae, only a few such spots are observed. This results in a low detection capacity; e.g., even with 100 wings analyzed, a high concentration of MMS (11.8 mM fed for 2 hr) did not yield statistically significant results (U. Graf, unpublished results). The same is true for 50 wings analyzed after an exposure of 36-hr-old larvae with 300 mM procarbazine for 6 hr. At the other extreme, exposure of 96-hr- and 108-hr-old larvae resulted in a large number of small single spots, but large spots were absent. Particularly important are the results for the twin spots, which are indicative of recombinogenic activity of a compound. Using standard exposures (11.8 mM MMS for 2 hr; 300 mM procarbazine for 6 hr), it was possible to obtain significant frequencies of twin spots if larvae of the age between 60 and 96 hr were exposed. After exposure of younger larvae no twin spots are observed, because the large *flr* parts of the induced twin spots are nonviable.[67] In this case only the *mwh* part is observed as a large single spot. With larvae older than about 96 hr only small spots are expected and here the *flr* part of small twin spots of four or fewer cells will not be phenotypically expressed due to the perdurance of flare.[68]

3.3. Data Analysis

Computer-assisted data analysis is carried out with the program SMART (somatic mutation and recombination test; F. E. Würgler, unpublished results). For each spot type, the spots are classified according to size: one cell, two cells, three to four cells, five to eight cells, etc., taking as class limits the powers of 2. This results in size classes reflecting the estimated minimal number of cell divisions between clone induction and differentiation. The following types of spots are analyzed

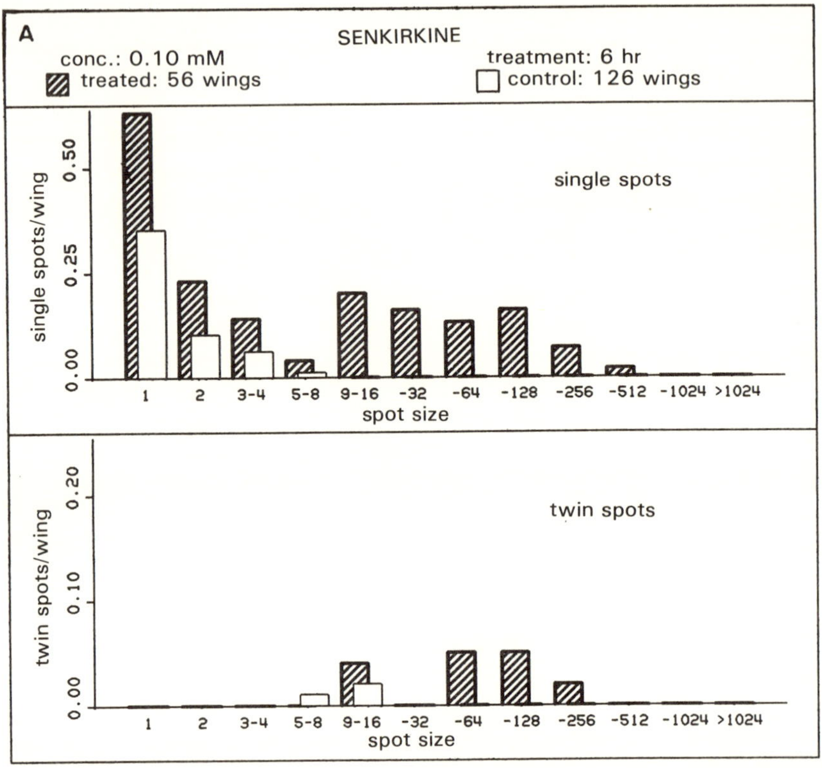

FIGURE 2. (A) Wing spots observed after feeding of 72-hr-old larvae for 6 hr with 0.10 mM senkirkine (a pyrrolizidine alkaloid). The spots are classified according to the number of cells affected, so that the size classes approximately reflect the number of cell divisions that occurred between clone induction and differentiation at metamorphosis. (B) Spots observed after a chronic exposure (for 48 hr starting with 72-hr-old larvae) with 0.05 mM. [All data from H. Frei (unpublished results).]

separately: small single spots (one or two cells in size), large single spots (more than two cells), total single spots, twin spots, and total spots.

The clone size distribution shows the difference between the spontaneous spots and the induced ones (Figure 2). A statistical comparison between control and treated groups can be based on the spot density distribution obtained from the recorded spot frequencies per wing analyzed. For these statistical analyses, we use the χ^2 test and the Kolmogorov–Smirnov test.[80] An overall statistical assessment of the experimental results can be obtained by applying to the different types of spots the analysis proposed by Selby and Olson.[180] This multiple-

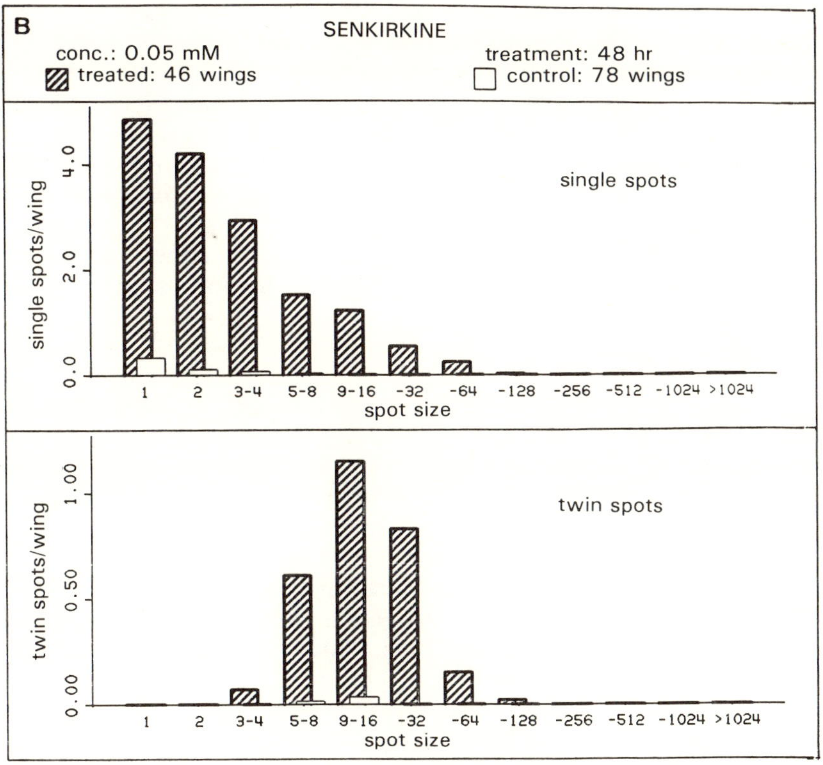

FIGURE 2. *(continued)*

decision procedure allows classification of results as positive, positive but weak, inconclusive, or negative.

X-ray studies of wing disc development have shown that a relationship exists between clone size and clone frequency with the time of clone induction during development.[68,89] Clones induced early are relatively large because they pass through several rounds of cell division in the period between induction and metamorphosis. They are not particularly numerous, because of the restricted number of target cells present at the time of induction. Clones induced later during development are smaller due to the limited number of cell divisions between induction and metamorphosis. They become progressively more frequent as the number of target cells increases in the proliferating wing primordium. Clones induced late in development are small.

The inverse relationship between clone size and clone frequency could be demonstrated by irradiation of larvae[68,89] or by acute chemical treatment with MMS (U. Graf, unpublished results) or with procarbazine (H. Juon, unpublished results) at different stages of development. However, in mutagenicity testing, the preferred treatment is to expose larvae continuously rather than acutely. For clones induced by chronic treatment, the time of induction is not known, but it has been shown to be possible to obtain an estimate of the overall per cell frequency of clone induction.

Szabad *et al.*[196] determined that in chronic treatment the average induction frequency per cell and per cell division is given by the relationship

$$f = n/NC$$

if only mwh clones are considered. In the equation, f represents the average frequency of *mwh* clone induction, n the total number of *mwh* clones recovered, N the number of wings analyzed, and C the number of scorable wing cells on a wing (according to Garcia-Bellido and Merriam).[68] We assume C to be about 24,000 in our tests because we do not score all of the approximately 30,000 cells on the wing, but consider only those in the distal compartments.

Determination of the clone induction frequency is best based on *mwh* clones alone. Since the *mwh* gene is located near the tip of the left arm of chromosome 3, *mwh* clone induction frequency accounts for somatic recombination occurring anywhere on this chromosome arm. The more proximally located marker *flr* would be less satisfactory in this respect, since it only allows detection of recombination events taking place in the proximal part of the chromosome arm. The *flr* marker is not suitable to determine an overall clone induction frequency, because small *flr* clones cannot be detected due to lack of phenotypic expression.

It remains to be determined how well the mean clone induction frequency can be used as the sole criterion in measuring mutagenic effects. Detailed studies are needed to determine the variation of the induction frequency during disc development. Estimates of the mean clone induction frequency may be influenced by factors such as the stability of a mutagen, possible variations in metabolic activation (e.g., induction) and/or deactivation, induction of DNA repair, and changes in proliferation patterns (e.g., in connection with cell death and subsequent regeneration).

3.4. Spontaneous Frequencies of Spots in Different Genetic Backgrounds

Graf *et al.*[80] have compiled sets of control data from different laboratories all using the *mwh–flr* cross and showed that, based on clone size distributions, the small spots (one or two cells in size) predominate in all control data. They have relatively high frequencies (0.22–0.47 spots per wing), which also may show some variation. By contrast, the rare large single spots (three or more cells in size) and the twin spots appear to have more constant frequencies, usually of the order of 0.02 spots per wing.

With the recently developed *mwh–flr*^{3} cross, low frequencies of spontaneous spots were obtained: 0.20 spots per wing for the small single spots, 0.03 for the large single spots, and 0.02 for the twin spots (U. Graf, F. E. Würgler, H. Frei, and A. Kägi, unpublished results). The finding that the background level of spontaneous appearance of spots depends on the cross used corroborates earlier studies reporting that the frequencies of spontaneous clones vary considerably for different stocks.[13,28,219] Frequencies can be quite different even for the same marker and the same chromosome arm.[183]

4. The White/White-Coral Eye Mosaic System

4.1. The Genetic Basis of the White/White-Coral System

The white/white-coral system was initially developed by Becker[14] in an attempt to study the developmental processes in the eye imaginal discs of *Drosophila*.[18] Several features made it attractive as a potential mutagenicity screening assay.[131] The system has been evaluated in detail by E. W. Vogel and J. A. Zijlstra (unpublished results).

Advantage is taken of the ability to distinguish the expression of two different alleles at the white eye locus on a sepia background. For the assay, larvae from a mating of w^{co} sn^2; *se h* virgin females to *w*; *se h* males are used. The sepia (*se*, 3-26.0) mutation gives a sepia background color of the eye and facilitates the distinction of the different allelic constitutions at the white locus in mosaic eyes. The other autosomal marker, hairy (*h*; 3-26.5), was helpful during the construction of the tester strains, but is irrelevant to the mutagenicity assay. The critical markers are the two alleles at the white locus (*w*, white and w^{co}, white-coral; 1-1.5) near the tip of the X chromosome. The distal position of the gene far from the centromere (approximately 70 cM) optimizes the possibility of recovery of spots due to mutagen-induced somatic recombination in XX females. Such an exchange leads to one homozygous w^{co} daughter cell and to another homozygous for *w*. The adjacent clones derived from these two cells will form a $w//w^{co}$ twin spot in the differentiated eye, consisting of an area of ommatidia with dark color and an adjacent area with white color, contrasting with the intermediate color of the heterozygous cells in the rest of the eye.

4.2. Scoring

The standard procedure (E. W. Vogel, unpublished results) is to transfer newly hatched females to fresh vials and score them 1 or 2 days later. The scoring of etherized flies is carried out in a liquid consisting of 90 parts ethanol, one part Tween (60 or 80), and nine parts water. The use of ethanol concentrations higher than 90% can cause changes (artifacts) in eye color pigmentation and should be avoided. Optimal illumination is an absolute necessity for accurate scoring. A ring-illuminator system with four light point emersions (Volpi Intralux cold light source, Urdorf/Zurich, Switzerland) fulfills this requirement when combined with a single-arm light guide.

The eyes of females that have developed from treated or untreated larvae are inspected for $w^{co}//w$ mosaic twin spots (TS) as well as mosaic light (or white) spots (LS) at a magnification of 50–75 ×. Regular scoring of dark (w^{co}) single spots is not recommended, because in calibration studies with some monofunctional alkylating agents, the frequencies of the dark single spots seemed disturbingly high compared with the rates obtained for the other two classes of spots (E. W. Vogel, unpublished results). Moreover, the majority of those dark single spots did not follow the characteristic pattern in which the developing eye tissue differentiates.[14,15] The conclusion was that mosaic dark spots were mostly variegation effects that may have nothing to do with chemically induced chromosomal changes. This conclusion was also put forward by Becker[14,18] to explain surprisingly high frequencies of dark ("w^{co}") single spots after X-ray treatment.

Another aspect concerns the occurrence of more than one mosaic spot per eye. Because it is frequently difficult to decide whether multiple spots are independent events or not, the conservation approach used (E. W. Vogel, unpublished results) is to count not more than one event per eye. If there is one TS and one LS per eye, only the mosaic twin spot is recorded. In our experience, this procedure will lead to an underestimate of the actual mutation rate only when strong mutagens at high doses are tested.

The white/white-coral mosaic method distinguishes between TS and LS; single spots may result from gene mutations or deletions affecting the white locus. A possible disadvantage of the method is the difficulty of detecting small clones with fewer than five to six ommatidia.

This technical problem can be overcome partially by using females in which w^{co} is replaced by the wild-type allele w^{+}. In this case somatic mutations, mitotic recombination, and other types of chromosomal rearrangements will lead to just one class of mosaic (i.e., light or white spots in the sepia eyes of the heterozygous w^{+}/w; *se h* females). The dark sepia eye color facilitates the accurate recognition of small white spots. The simplified technique may be attractive for mutagen screening because in this case the primary objective is to find out whether or not the chemical under test is genotoxic. In case of a positive result, additional studies are needed to decide whether mutagenic or recombinogenic activity is involved.

4.3. Data Analysis

The light single spots in the females are interpreted as somatic mutations of the white gene, leading to a lighter pigmentation or a white spot. In some cases, such light single spots could be the surviving light partner of an original twin spot, the dark (sister clone) having failed to develop.[18] Another problem inherent with the white/white-coral assay concerns the correct classification into TS and LS of small spots. A solution to these problems is to put the two classes into one category for statistical analyses, and to regard a compound as positive only if it gives at least a "doubling effect" in separate experiments. This procedure should prevent misclassification resulting from errors in the judgment of small spots.[209]

For statistical evaluation of results, Fisher's exact test can be applied to discriminate among the following three categories of chemicals:

1. +, Mutagenic. p values <0.01 are obtained when the pooled data from the treated groups (separate experiments) are compared with the pooled data from contemporary control.
2. M, Weakly mutagenic. Small (M = marginal), but reproducible effects (about a doubling of the spontaneous mutation frequency) are found in separate experiments, with $0.01 < p < 0.05$.
3. −, Nonmutagenic. No response is detectable in three to four separate experiments for the exposure tested.[209]

In the IPCS collaborative study on ten chemicals, a test substance was not considered satisfactorily negative in the white/white-coral system until 2000 eyes had been scored for spots in single-dose experiments and the substance found repeatedly ineffective in at least three separate experiments.[209] An alternative procedure would be to test two to three different exposure levels and to reduce the number of eyes tested to about 1000 per exposure.

4.4. Background Frequencies of Spots

The pooled spontaneous mutation rates from 25 separate experiments with the white/white-coral mosaic assay were 0.07% for twin spots and 0.33% for light singles, respectively, at a sample size of 8690 tested eyes.[209] For testing purposes, the best strategy seems to be the performance of concurrent control experiments with several (three to five) test groups. For efficient statistical analysis, one scores about equal numbers of eyes in the control and treated groups.

5. The Unstable White–Zeste Eye Mosaic System

5.1. The Genetic Basis of the Unstable White–Zeste System

The test system is based on the recessive zeste mutation (*z*, 1-1.0) and the unstable white locus (w^+, 1-1.5) described by Rasmuson and Green.[158] Normally, the zeste mutation manifests its phenotype of lemon-yellow eye color when the normal white allele is present in two copies as in normal females or in males with a duplication of the entire w^+ locus or at least the proximal part of it. In the case where the white allele is present in a single copy as in normal males or in females heterozygous for a deletion of the white locus, a red eye color is produced.[96] This *z–w* interaction reveals that the unstable white locus can occur in two different states: in males it can either suppress zeste, and thus act as a normal single w^+ allele, or it can let the zeste phenotype appear, in which case it acts as a duplication of w^+. Shifts between the two states occur spontaneously with a certain frequency. This phenomenon seems to be associated with a 1-kb insertion at the proximal end of the white region.[161] A piece of DNA seems to be integrated immediately to the right of the *w(sp)* site in the white locus.[157] This IS affects the regulatory part of the w^+ gene and interferes with the *z–w* interaction system. Together with the zeste mutation, the eye pigmentation phenotype in hemizygous males is zeste, but the system frequently reverts to the wild-type phenotype. Possibly, parts of the original duplication[158] remain and are responsible for the zeste phenotype. The nature of the instability is still obscure, but recent studies indicate that the change from zeste to red eye color is associated with an acquisition of a small IS element (J. Nygren, personal communication).

This is quite opposite to other white mutants analyzed by Zachar and Bingham[229] or Levis *et al.*[113] where large insertions in the proximal part of the white locus cause mutant phenotypes.

Mutations in the other direction, i.e., from red to zeste, when the IS is lost, are also frequent, as well as deletions and transpositions of the white locus.

A possible explanation for this instability is that the inserted element consists partly of foldback sequences that form loops; the element can either change orientation or leave its position, so that the white locus achieves different levels of activity. This would mean that DNA breakage and repair are necessary factors for the shifts to occur.[156]

For mutagenicity tests the unstable system can be used in either state, as wild type, *z* + (red); or as zeste *z* + (zeste). Mutations in somatic cells are registered in flies whose compound eyes have an aberrant pigmentation, zeste or unpigmented (white) in the *z* + (red) stock, red or unpigmented in the *z* + (zeste) stock.

The *z* + (red) and the *z* + (zeste) stocks also exist with a stable white locus designated as $Dp(1,1)w^{+R61e19}$. These stocks may be used for comparing induced mutation frequencies at the stable and the unstable white loci. The unstable stock of *z* + (zeste) as compared with the stable stock of *z* + (zeste) shows several times higher sensitivity to induction of the somatic red spots by X rays, EMS, and other chemical mutagens.[22,160,173]

5.1.1. Repair-Proficient Larvae

It is most convenient to use stocks in which the unstable X chromosome, *z* + (red) or *z* + (zeste), is kept in males crossed to females with attached-X chromosomes, *C(1)DX, y w f* or *C(1)DX, y f*. The former have a more conspicuous phenotype and are more easily separated from the male offspring to be scored, but *C(1)DX, y f* females usually are more fertile.[162]

5.1.2. Excision Repair-Deficient Larvae

Fujikawa *et al.*[64] constructed a stock of the genotype *C(1)DX, y f/ Y/sc z* w^+ (zeste) *mei-9*^{a}. In this stock the excision repair-deficient *mei-9*^{a} mutant is kept together with the somatic eye color mutation system on the X chromosome of the males. Stocks containing other mutagen-sensitive mutants, such as *mei-41*, have been constructed and have been used for the study of chemical mutagenesis.[60,62]

5.2. Scoring

Adult males are scored under a dissecting microscope for aberrant sectors in both eyes. When *z* + (red) larvae are treated, the sectors are yellow or white on a red background; when the larvae are *z* + (zeste), there are red and white sectors on a yellow background. The most sensitive and easily detectable indicator of somatic change are the red spots in the unstable *z* + (zeste) stock (K. Fujikawa, unpublished results; refs. 46, 47). The visible mosaics may comprise a few or more than 100 ommatidia. Since the ommatidia are not strictly of clonal origin,[(20)] very small spots of fewer than four ommatidia affected cannot be scored with reproducible certainty and are usually not included in the analysis.[(64)]

Recently, Rasmuson[(157)] studied the spot size after MMS and X-ray treatment of UZ male larvae (*sc z* w^+ *f*, unstable zeste) of different ages. The results show that the mean size of red spots decreases with increasing age of larvae at treatment, while the mutation frequency increases because of the multiplication of the mutagen-susceptible cells in the eye anlage. This is contradictory to the hypothesis maintained by Fahmy and Fahmy[(47)] that the somatic shifts are not mutagenic but epigenetic events, due to altered regulation of gene expression.

Red spots induced with MMS are smaller than X-ray-induced red spots, indicating a delay in the establishment of mutations from chemically induced lesions compared to irradiation damage. White spots, on the other hand, were equally large, irrespective of inducing agent and about twice the size of the chemically induced red spots, implying a faster and more direct action for fixation of deletions than for the production of MMS-induced shifts in eye color from zeste to red.

5.3. Data Analysis

The number of sectors (or animals with sectors, because multiple cases are rare) measures the frequency of mutations.[(162)] For statistical analysis of test results, either Fisher's exact test or the Kastenbaum–Bowman test[(101)] may be applied. Sample sizes necessary for drawing a positive conclusion at a given level of the mutation frequency in the treated and the nontreated series can be found in the sample size table prepared by Würgler *et al.*[(225)] In the IPCS collaborative study, dose–response relationships were taken into consideration together with the results of Fisher's exact test. Chemicals that caused a significant increase of the frequency of red sectors over the control levels in a dose-related manner were registered as positive.[(64)]

5.4. Background Frequencies of Spots

For the unstable *z* + (zeste) strain Rasmuson *et al.*[162] reported a spontaneous frequency of 43 sectors in 48,918 males scored. This frequency of 0.088% was obtained by pooling published data. A significantly lower frequency of 0.034% (5/14,526) is found with a stable system, $z\ Dp(1,1)w^{+R61e19}$.

5.5. Mechanism

The exact mechanism of the inducibility of the instability in the UZ system is not known. Although the instability seems to be in some way connected with the 1-kb insert (IS) in the white locus, simple integration and excision is not sufficient to explain the observed phenomena.[156] Modolell *et al.*[128] reported that certain spontaneous mutations, suppressible by *su(Hw)*, all were due to an insertion of a mobile element called "gypsy." Mutagen-induced mutations were not suppressible by *su(Hw)* and not associated with such insertions. Hence, the somatic mutations induced by mutagens in the UZ system are not caused by insertions, as the spontaneous germ line shifts, but are most probably due to other mechanisms.[157] There seems to be a fundamental difference in the induced transposition process between bacterial IS sequences and the *Drosophila* IS sequences. The *Drosophila* sequences respond to a wide variety of chemical mutagens (34 compounds listed in Table 1) and it is difficult to find "transposagens" active in *Escherichia coli*. Datta *et al.*[41] reported that among more than 100 compounds tested to stimulate the Tn9 transposition in *E. coli*, only seven were active: acetic acid, ammonium acetate, magnesium acetate, sodium acetate, dimethylsulfoxide, and the two detergents Brij 58 and octylphenoxypolyethoxyethanol (Nonidet P40). The nature of these substances suggests an involvement of lipid or membrane in the transposition process.[41] Among the negative compounds are such mutagens as mitomycin C, phleomycin, acridine orange, acriflavine HCl, hydroxylamine HCl, MMS, MNNG, 5-bromouracil, and 5-fluorouracil.[41] The mechanism(s) underlying the induction of spots in the UZ system of *Drosophila melanogaster* needs additional molecular studies.

6. Exposure Techniques

With *Drosophila* several routes of application are possible (feeding, inhalation, injection). We describe them in the order of practicality, considering particularly the treatment of larvae. If adult flies are not allowed to deposit their eggs directly on the medium containing the test compound, either eggs or larvae have to be collected and transferred into separately prepared treatment units. The exposure techniques require collection of eggs or larvae during a restricted period of time to obtain populations that are reasonably homogeneous in age.

6.1. Egg Collection

Flies may deposit their eggs on standard medium (usually containing live yeast) and, after a given time, pieces of food of a standard size are placed in treatment vials, which will then contain comparable population densities (P. Schweizer, personal communication).

Eggs deposited on black paper soaked with 1% acetic acid can be collected easily (after a few hours of egg deposition) with a fine brush and transferred in small batches into new culture units. If survival frequencies have to be determined, a known number of eggs or hatched larvae (100 or 200) is transferred into each test vial containing standard food[162] with or without the test compound.

6.2. Collection of Larvae

Larvae may be collected by washing them out of mass cultures with a concentrated aqueous solution of sucrose[136] or sodium chloride.[80] The suspension of larvae and food debris is poured into a separatory funnel. Because of a difference in specific weight, larvae float on the surface, while the food debris sinks to the bottom and can be released through a valve. After addition of tap water, the larvae sink to the bottom and can be released onto a nylon gauze, washed if desired, and placed in test vials.

6.3. Feeding

Feeding of larvae is an easy route of application of test compounds and has also been used for the treatment of larval germ cells.[106,108,162,214] Depending on the duration of the feeding period, one can distinguish chronic and acute treatments.

If fed, the chemicals enter the larva through the gut. The chemicals can be absorbed in different regions of the digestive tract, which have different pH, as in higher animals. Chemicals have to be transported to the target organs, the imaginal discs, via the hemolymph.

6.3.1. Chronic Larval Feeding

The term "chronic feeding" is used here for treatments that extend over 12 hr or more. Usually larvae are exposed to the test chemical during one or several of the larval instars (L3 or L2 + L3 or L1 + L2 + L3). The duration of exposure under feeding conditions can be given as an approximation only, because certain treatments retard development. Feeding will "automatically" be discontinued when the larvae stop feeding, leave the food, and pupate—although the chemical may still persist within the body.

The technically simplest way to feed larvae is to use vials with medium containing the test chemical. Adult flies from the appropriate stocks are allowed to deposit eggs for 24 hr. At the end of this period the adults are removed from the vials (they may be used for another egg-laying period or be discarded), and the progeny are allowed to complete their development in the vials. The progeny surviving to the adult stage are then analyzed for induced somatic spots.

To safeguard against mutagenic and/or sterilizing effects of the test chemical through effects on the parental germ cells, eggs may be collected from nonexposed parents in separate egg collection units.[162] Szabad *et al.*[196] collected eggs for 4 hr on standard medium, collected second-instar larvae at different intervals between 28 and 46 hr of age, and transferred them to freshly prepared food containing the test chemicals.

An alternative used by Fahmy and Fahmy[47,49] is a topical application to late embryos and newly hatched larvae. Parents were aged separately for 4 days and then 50 females and 20 males were introduced into each culture bottle of 250 ml capacity, containing 40 ml of normal agar food. These cultures were incubated at 25°C for 48 hr, to allow females to mate and oviposit, after which the adults were discarded. The laid eggs and the newly hatched larvae are normally confined to the surface of the food gel, thus permitting their treatment by covering them *in situ* by a thin film of the required molar concentration of the test compound in solution. This was invariably undertaken in 1% (v/v) aqueous dimethylformamide.

Fujikawa *et al.*[64] allowed mass-mated flies to lay eggs for 24 hr in standard culture bottles (180 ml in volume with 40 ml medium). After oviposition, the parental flies were discarded and the eggs allowed to develop for 24 hr. The culture bottles containing the newly hatched larvae were divided into two to five groups. One group served as the control and the rest were treated with the test chemicals. At the appropriate time, 1 ml of a test solution containing the desired concentration of the mutagen was pipetted onto the surface (18 cm^2) of the culture medium. Based on the experience with the ten chemicals tested for the IPCS study, this method seems to have limitations when chemicals with repellent effects are to be tested (e.g., *o*-toluidine[64,215]). It seems that if the medium is not homogeneously mixed with the chemical, some larvae can avoid its uptake. This can lead to substantial under-estimates of mutagenic effects.

The current method to add the test compound to the food consists in mixing the required amounts of substance, if possible in dissolved form, to standard corn-meal medium that has been cooked and cooled to a temperature that still allows homogeneous mixing.

As an alternative, commercially available instant media (e.g., "*Drosophila* Instant Medium" of Carolina Biological Supply Company, Burlington, North Carolina) offer the advantage that no cooking is necessary. The risk of decomposition of heat-labile chemicals or evaporation of relatively volatile compounds is avoided. Normally, dry medium is rehydrated in the test vials according to the instructions of the manufacturer with the appropriate volume of water containing the compound under test in the desired final concentration. For compounds with low water solubility, emulsions in water, buffer solutions, acidified solutions, solvent additives such as ethanol, Tween 80, Tween 60, or DMSO at low final concentrations may be used. Instant medium is sufficiently rich to support normal development from the egg to the pupal stage without addition of live yeast. In the wing mosaic assay, instant medium has been routinely used for chronic feeding mainly for 48 hr, but also for 72 and 96 hr.[80,226] Because of easy handling, it was also used as the routine mutagen-free medium following acute feeding, to raise the animals after completion of the mutagen exposure.

6.3.2. Acute Larval Feeding

The term "acute feeding" means that larvae are exposed to the chemical for just a few hours.[80] Larvae collected from mass cultures are kept in special feeding units. These units consist of a glass beaker into which a plexiglass tube containing the larvae is placed. The lower end of the plexiglass tube is covered with nylon gauze fixed to the wall of the tube by acetone. As a result the nylon gauze is glued to partly dissolved plexiglass. The beaker may contain powdered cellulose soaked with the test solution, which enters the tube through the gauze. At the end of the feeding period, the treated larvae are removed, washed, and transferred to vials with standard food. Acute feeding is useful for testing relatively unstable compounds such as β-propiolactone[80] or isopropyl methanesulfonate.[102] it is also useful for studying the dose response of potent chemicals by varying the duration of the feeding period.[80]

Testing of repellent compounds[1,10,74] is particularly critical if larvae 72 hr or older are treated. Such larvae can survive starvation and pupate without further feeding.[183] To cope with this problem, J. Szabad (personal communication) keeps the larvae floating in the feeding solution, which covers the bottom of the feeding unit to such a level that the larvae can reach the bottom with the mouth parts and still have the posterior spiracles above the surface of the solution to guarantee respiration. In this situation the larvae are unable to escape the feeding solution and can be treated for several hours.

Figure 2 demonstrates how the spot frequency distributions for single spots obtained after acute treatment differ from those seen after chronic exposure. The chronic exposure shows a more or less continuous distribution from relatively large to very small spots, and the acute treatment yields mainly two classes, relatively large spots with a mean size (depending on the period of treatment) of about 64 cells affected and very small spots with one to four cells affected. The latter may represent slowly growing aneuploid clones.

The acute treatment yielded a few relatively large twin spots. The mean size class is 33–64 cells. Chronic exposure is more efficient in inducing twin spots. The frequency of twins is higher, but they are on an average smaller (mean size class, 9–16 cells) than those induced by an acute treatment. The reduced size indicates that the majority of the spots are induced sometime (about two to three cell cycles) after the treatment was started.

6.3.3. Exposure of Axenic Larvae

The methods described for larval feeding may also be useful in studies with axenic larvae. This allows study of the potential role in activation and/or deactivation of test chemical by microorganisms in the medium and/or the intestinal flora. In contrast to adult feeding,[23,106] no experiments with somatic cell assays have been published.

6.4. Inhalation

Inhalation is expected to be an efficient route of application because the tracheae and tracheoles transport the gaseous compounds directly to the target tissues, the imaginal discs. For inhalation studies only acute gas exposures have been used.[80,225] The technique is similar to the inhalation procedure used by Verburgt and Vogel[205] and Kramers and Burm[107] to treat *Drosophila* with vinyl chloride and halothane, respectively. In addition, sophisticated experimental equipment with adjusted flow rates have been designed for experiments with larvae,[99,100] but have not yet been employed in studies with somatic cells.

Techniques for aerosol exposure of adults were developed by von Borstel,[217] Sega and Lee,[179] and Aaron *et al.*[2] Potentially, aerosol exposure is useful for somatic assays with *Drosophila* larvae.

6.5. Injection

In experiments with adult males, compounds have been found that are exceptional because they are only mutagenic after injection but not after feeding (e.g. epichlorohydrin).[212] In principle the injection techniques developed for transplantation experiments in developmental biology research with *Drosophila*[155,201] could be adapted for the injection of chemicals into larvae. It remains to be determined, however, whether injection is practical for genotoxicity testing with somatic cells in larvae. It may be a drawback that one larva has to be injected for every pair of discs. Mortality due to toxic effects may require that even more than the minimum number of injections be carried out.

6.6. Solvents

Water-soluble compounds may be tested without further difficulty. Solvents have to be used for chemicals not soluble in water. The widely used solvent dimethylsulfoxide (DMSO) should be avoided because it can interfere with the inducibility of xenobiotic metabolism.[120] Several authors[209,214,226] have shown that various mixtures of ethanol and Tween 80 (or Tween 60) are suitable as solvents. The actual composition of the mixtures may be varied depending on the extent of toxic effects resulting from the solvents and/or the test compounds. For certain cases (e.g., polycyclic hydrocarbons such as DMBA) more complex solvent mixtures have been used,[214,230] sometimes together with a low content of DMSO.

An advantage of using ethanol as solvent is that larvae and flies readily metabolize ethanol by means of a potent alcohol dehydrogenase enzyme system.[198] Sometimes problems may arise from the use of ethanol as a solvent. The excision repair-deficient larvae used for the *mei-9* wing spot test showed an increased frequency of spots when exposed to ethanol (U. Graf, unpublished results; ref. 226). This increase might have resulted from a greater sensitivity of the excision repair-deficient imaginal disc cells to the mutagenic effects of the ethanol metabolite acetaldehyde.[140] U. Graf (unpublished results) has recently found that acetaldehyde induces somatic spots in the wing assay.

To avoid problems related to ethanol, Fujikawa *et al.*[63] added the test compounds dissolved in ethanol to hot medium. The mixtures were then heated to 80°C and stirred for 5–10 min to allow the ethanol to evaporate. Obviously, this technique can be applied only with heat-stable compounds that have low volatility.

6.7. Dosimetry

In all reported experiments with somatic cells of *Drosophila*, only exposure doses[45] and no pharmacological or molecular doses were determined. It is desirable for the wing as well as for the eye imaginal discs that methods become available that allow the determination of tissue dose and/or molecular dose. The latter might be obtained from application of sensitive methods that allow measurement of the formation of specific types of lesions (single-strand breaks, DNA adducts, etc.).[3] Potentially, the detection of specific lesions by use of monoclonal antibodies seems to be a promising approach that should be explored at least for important chemicals. Molecular dosimetry probably will not be applied routinely, but it will be important for the exploration of assays with somatic cells and may become important in connection with studies of tissue specificities, repair processes, and mechanisms of mutation fixation.

7. DISCUSSION

7.1. Genetic Mechanisms

A number of different genetic mechanisms are involved in the formation of somatic mosaic spots. We discuss here in detail the white/white-coral test and the wing system, because the unstable white–zeste system, as a special case measuring the specific effects on an insertion element, has already been described in Sections 5.1 and 5.5.

7.1.1. Recombination

In the light of recent studies, the detection of recombinogenic activity in addition to mutagenic activity of chemicals becomes increasingly important. It has been suggested that somatic recombination is, in addition to other types of genetic changes, involved in certain steps of carcinogenesis, such as promotion.[32,50,104,145,152] Somatic recombination has long been assumed to occur in mammals,[82,93] including man.[216] A cytological demonstration of mitotic crossing-over in human lymphocyte metaphases from a patient with Fanconi anemia was reported by Therman and Kuhn.[197] Recently experimental evidence has been presented of molecular DNA recombination mechanisms in a Chinese hamster cell line[168] and in cultured mouse L cells[185] as well as in somatic cells of the mouse *in vivo*.[50] The *Drosophila* assays with somatic cells *in vivo*, which detect mutations as well as somatic recombination, are capable of detecting genotoxic agents, which appear to be of importance with respect to tumor initiation as well as to tumor promotion.

Phenomenologically, recombination between homologous chromatids in somatic cells has consequences comparable with those of meiotic crossing-over in germ cells. The segregation pattern of the genetic markers is changed. Two different arrangements of genetic markers may be used to detect somatic recombination. The first is a heteroallelic combination at one locus with alleles that in homozygous condition will lead to phenotypes distinguishable from the phenotype realized in the heteroallelic condition (e.g., white/white-coral). The second is a transheterozygous configuration at two different loci on two homologous chromosome arms. A suitable variety of marker combinations is available.[18] In the wing assay such a marker combination on the left arm of chromosome 3 is used.

Somatic chromosome pairing, a prerequisite for high frequencies of spontaneous and induced somatic recombination, was studied by Merriam *et al.*[126] Mitotic recombination can probably be induced by treatment at any stage of the cell cycle, but there is no possibility of distinguishing between events taking place in the two-strand or in the four-strand configuration,[18] and so it is not possible to discriminate between S-phase-dependent and S-phase-independent mutagens.

In the case of the heteroallelic white locus in the white/white-coral assay, mitotic recombination between the white locus and the centromere leads to the formation of one homoallelic *w*/*w* and one homoallelic w^{co}/w^{co} daughter cell. This pair of cells leads to two clones with differentially colored eye pigment cells. In the wing test, recombination in the chromosome region between the *flr* locus and the centromere leads to two daughter cells, one homozygous for *flr*, the other homozygous for *mwh*. Cell clones originating from these two daughter cells usually develop in adjacent wing areas, which express the phenotypes flare and multiple wing hairs, respectively. This represents a twin spot.

Recombination between two markers on the same chromosome arm (e.g., between *mwh* and *flr* in the case of the wing assay), results in homozygosity for the distal marker only and leads in this particular case to a *mwh* single spot.[67]

In the wing assay twin spots are in general of medium size, 5–64 cells per spot. Very small (up to four cells) and very large (more than 64 cells) twin spots are seen only occasionally. This large-spot phenomenon seems to be related to particular characteristics of the *flr* marker. The reduced viability of flare clones in the wing has already been described by Garcia-Bellido and Dapena.[67] Clones larger than 400 cells were not found, although they were expected in a given frequency in twin spots with *mwh*. The nonviability of the large *flr* clones is probably due to the same lethal effect that brings about the zygotic lethality in *flr* homozygotes. For clones larger than 20 cells, the size of the *flr* clones decreased relatively to that of their *mwh* counterparts.[67] With respect to the absence of very small twin spots, Szabad *et al.*[196] stated that the *flr* allele is not fully expressed in clones smaller than four cells. This is most probably due to the effect of perdurance,[70] which is the persistence of the flr^+ gene product in cells that have become *flr*/*flr* in consequence of mitotic recombination. For *flr* the perdurance seems to be effective for one to two cell cycles, that is, for 8.5–17 hr.[68] In small twin spots only the *mwh* part will be phenotypically expressed, and therefore the clones that correspond to small twin spots are manifested only as small *mwh* single spots.

7.1.2. Gene Mutations

Gene mutations at a marker locus (e.g., the induction of a noncomplementing new mutant allele at the site of the wild-type allele of a marker locus) will result in a detectable single spot. At present, we have no direct evidence concerning the mutational events that lead to the formation of somatic single spots, but extrapolation from what is known for mutational events occurring in germ cells of *Drosophila* seems legitimate. There is good evidence for induced gene mutations in germ cells exposed to EMS.[114] In a complementation assay, the majority of EMS-induced sex-linked recessive lethals was found to map as nonoverlapping mutants.[114] The detection of about 10% of temperature-sensitive induced mutations after EMS treatment has been taken as evidence for the presence of base-pair substitutions of the missense type.[111,191,192] Kramers *et al.*[109] discovered a high proportion of multilocus deletions among hycanthone-induced recessive lethals. Based on genetic assays, by complementation studies and some cytological confirmations, it has been demonstrated that EMS also induces a small number of deletions.[114,177,181,223] A minimum of one-third of the diepoxybutane-generated mutants are deletions.[146] Among sex-linked recessive lethals induced by MMS in unstored mature spermatozoa, only four out of 55 were deletions; after storage, the proportion of deletions was increased drastically to 47.1%.[147] With DNA sequencing techniques Benyajati *et al.*[19] showed that formaldehyde-induced gene mutations are often small deletions. This type of change at the gene level will occur not only in genes mutating to recessive lethals, but also in genes mutating to new visible alleles. Such a mutation in a wild-type allele of the markers used in the wing assay or in one of the white alleles used in the white/white-coral test will be expressed as a single spot.

Theoretically, twin spots could also arise in a similar way if two gene mutations occurred simultaneously in the two relevant alleles used in the marker system. Haendle[84] has shown that in radiation experiments with the white/white-coral system, the expected frequency of such double mutants is about 10,000 times smaller than the actually observed frequency of twin spots. Therefore, gene mutations are irrelevant as a basis for X-ray-induced twin spots. The same might be true for chemically induced twin spots, but the experimental proof might become more difficult than with X rays, especially if nonrandom uptake or variable distribution of a chemical does affect the relative frequency of twin spots.

Backmutations, which have been reported for alleles at the white locus,[81] would in theory also be recognized as single spots. Their contribution is expected to be small, because backmutations are generally rare events in *Drosophila*. In the wing assay backmutations are not recognizable.

In the wing assay the presence of single spots *per se* is no proof for the existence of gene mutation (and/or deletion) because, as shown above, recombination between the markers as well as small twin spots are recovered as single spots. It is possible to examine the *mwh*-heterozygous, inversion-heterozygous progeny (*mwh/TM1, Me ri sbd*^{2}, or *mwh/TM2, Ser*) for the presence of *mwh* single spots. Since recombination events within the multiple inversions of *TM1* or *TM2* result in nonviable daughter cells,[126] only mutational events in the immediate vicinity of the *mwh* locus are expected to lead to *mwh* single spots. A few small spots may be formed by aneuploid clones.[125] The probability of obtaining single spots due to recombination near the *mwh* locus is much lower with the *TM2* balancer because the leftmost breakpoint in this chromosome is practically at the *mwh* locus (A. Garcia-Bellido, personal communication). Although only few data are available, the observations of *mwh* spots in inversion heterozygotes treated with DMN,[103] iPMS,[102] senkirkine (H. Frei, unpublished results), *N*-nitrosomorpholine, *N*-methyl-*N*-nitroso-*p*-toluenesulfonamide, and *N,N'*-dinitroso pyridinol carbamate[190] indicate that gene mutations add/or deletions may be induced by these agents in somatic cells of *Drosophila*.

7.1.3. Chromosome Aberrations

Based on complementation studies and/or cytogenetic analyses of mutants induced in *Drosophila* germ cells by X rays, EMS, MMS, MNNG, TEM, formaldehyde, ethyleneimine, and hycanthone,[147] we infer that most chemical mutagens induce deletions. In somatic cells we also expect an array of changes of different sizes, ranging from those affecting one or a few nucleotides (frameshift mutations) to large multilocus deletions. If in a heterozygous condition the wild-type allele of a marker gene is deleted (rendering the mutant allele hemizygous), the mutant will be expressed phenotypically.

In the case of the wing assay the *mwh* marker may be uncovered by a terminal deletion. In the white/white-coral test either of the two mutant alleles may be lost by a terminal deletion, leading to the expression of the remaining hemizygous allele on the nondeleted X chromosome. It is still not clear whether complete terminal deletions including the loss of the telomere can occur or whether only subterminal deletions survive which have the broken chromosome capped by a telomere.[121,234] Roberts[169,170] concludes that in repair-proficient cells of *Drosophila* the probability of "healing" of a broken chromosome end to form a stable terminal deletion is vanishingly small. Mason *et al.*[121] describe an X-ray-dependent mutator gene in *Drosophila* which specifically increases the recovery of deletions extending to the very end of the chromosome. They propose that the mutator blocks the repair of DNA double-strand breaks and allows the formation of new telomeres *de novo*. If in repair-competent cells terminal deletions result in breakage–fusion–bridge cycles, the chromosome involved might get lost and a resulting small spot may be expected.

Segmental aneuploid cells can arise from unequal recombination between homologous chromosomes,[17,84] from somatic exchange between nonhomologous chromosomes (formally a translocation),[177] or from deletion.[13] By means of mitotic recombination, Ripoll[166] generated somatic cells bearing duplications and deletions of known terminal chromosome segments. He found that the presence of large duplications in epidermal cells does not seem to affect cell viability.

The recovery of deficiency-bearing cells in clones and their proliferation capacity decrease as a function of the size of the deleted fragment. Therefore, small clones induced in relatively young larvae may represent cells with segmental aneuploidy.[13] Monosomy and trisomy are the limiting cases as segmental aneuploidy increases. In the systems studied by Ripoll[166] indications were obtained that monosomic cell clones and possibly also trisomic ones may not be recoverable. Since viability and proliferation capacity strongly depend on the genetic background (as exemplified by the Minute mutations[183,184]) the question of the recoverability of monosomic and trisomic somatic cell clones should be studied in specific experiments.

7.1.4. Gene Conversion

Gene conversion as a basis for the production of a single spot remains a theoretical but unsubstantiated possibility. Gene conversion has been shown to occur in the germ line of *Drosophila melanogaster*.[51] Observations at different loci (rosy and maroonlike,[36] rudimentary,[34] garnet,[90] and shibire[98]) indicate that nonreciprocal intragenic recombination is a general phenomenon in *Drosophila*. Kahn and Sick[98] tried to promote gene conversion in germ cells by EMS treatment. The effect was not sufficiently large or reproducible to prove that gene conversion had been induced. It remains to be determined how the situation in somatic cells compares with that in germ cells. That mitotic gene conversion is a realistic possibility is substantiated by the fact that chemically induced gene conversion can be studied experimentally in fungi (e.g., in yeast).[137,235]

7.1.5. Nondisjunction and Chromosome Loss

Missegregation of autosomes (e.g., as a consequence of nondisjunction) leads to two aneuploid daughter cells: one trisomic, the other monosomic for the autosome involved. In the wing assay, only the monosomic cell line will have a genetic marker that is expressed phenotypically. A single spot is expected in this case. A twin spot may be formed if the monosomic cell lineage survives and if in the trisomic line the two nondisjunctional chromosomes are maintained, whereas the third one is lost. Such events have been observed in *Aspergillus*.[97]

In the white/white-coral assay, the monosomic cell line would be XO. Because of interference with the sex-determination state of the cells, such clones usually will not survive.[174]

A supposedly rare but possible event leading to a twin spot would be simultaneous double nondisjunction of homologous chromosomes.[13] In this way two euploid cell clones would be formed, which should, with high probability, survive and proliferate normally.

Monosomic cells can also result from chromosome loss (e.g., as a consequence of chromosome breakage). In this case a single spot would be formed if the hypoploid clone survives.

Spindle damage is assumed to be a major cause of chemically induced nondisjunction. Oenfelt and Ramel[141] concluded that extensive chromosome contraction and spindle disturbances (c-mitosis) can be induced by almost any chemical if given in a dose related to its water solubility. They suggested the existence of a common mechanism by which most chemicals act. This mechanism may act via the nuclear membrane and/or the cell membrane. Some compounds induce c-mitosis at much lower concentrations than would be expected on the basis of their water solubility. These chemicals act on specific targets (e.g., colchicine binding to tubulin is a specific inhibitor of microtubule assembly and affects spindle formation and function). Changes in the levels of protein sulfhydryl groups and reduced glutathione are important biochemical factors in connection with the induction of nondisjunction.

Chemically induced nondisjunction is well known in germ cells of *Drosophila*.[153,203] In the wing assay, the cytostatics vinblastine[80] and vincristine (H. Juon, unpublished results) were studied because they are known spindle poisons.[186,221] The induction of significant frequencies of single spots may be taken to indicate that nondisjunction can be induced by these agents. These cytostatic agents may have other effects in the treated cells that could indirectly give rise to somatic mosaics (e.g., via disturbances of the nucleotide pools).[110,224] Vinblastine sulfate is known to interfere with the amino acid pathways of glutaminic acid leading to the Krebs cycle and to urea.[135] Whether or not such metabolic disturbances can lead to induced somatic spots is unknown.

Definite proof of chromosome loss, which can also be a consequence of somatic nondisjunction, will become possible only when the effects of suspected chemicals can be studied with a *mwh* chromosome containing a suitable second marker on the right arm.

7.2. Differences between Genotypes

7.2.1. Sex Differences

Drosophila larvae of opposite sex may differ in their mutagen sensitivity. For example, Gatti *et al.*[71] reported that MMS-induced chromatid aberrations were three times more frequent in neural ganglia cells of females than males.

In the wing assay in which autosomal markers are used, wings of males and females are equally suitable for scoring, although they are mounted routinely as separate groups on the slides and recorded accordingly. The computer program SMART performs all calculations for male, female, and total wings separately. With the large majority of compounds studied so far, no substantial differences between spot frequencies on male and female wings were detected. This agrees with the observation of Becker[16] that X-ray-induced mitotic recombination on the left arm of chromosome 3 does not significantly differ between males and females, neither on the basis of twin spots nor on the number of total spots.

Only with caprolactam* were there more spots on the wings of females than on those of males. The observation was reproducible with chronic as well as acute exposures. Although the basis for the differential response of the two sexes is not known, it points to the possibility that the two sexes might react differentially to the mutagenic activity of certain chemicals.

7.2.2. Strain Differences

Detailed comparisons between the different assay systems using somatic cells in *Drosophila* are made problematic by the limited number of compounds tested in several assays. In addition, not only may larvae from different crosses differ with respect to the genetic markers used for the assay, but many other important characteristics may differ to variable extents. Larval feeding behavior measured as the number of cephalopharyngeal retractions per time period can vary between lines of isofemales.[144] Lines can be selected for fast or slow larval feeding rate.[31] This rate is affected by genes on all three major chromosomes.

Of special concern is the fact that the complex metabolic factors modulating the mutagenic response of *Drosophila* are under genetic control. A number of studies from several laboratories have demonstrated strain differences for activities in xenobiotic metabolism.[120,218,232]

For example, the activities of the mixed-function oxidases that metabolize dimethylnitrosamine, *p*-nitroanisole, aminopyrine, benzphet-

* U. Graf and F. E. Würgler (unpublished results); under further investigation for the IPCS Collaborative Study on Short-Term Tests for Genotoxicity and Carcinogenicity (CSSTT), Part II.

amine, 7-ethoxycoumarin, and benzo(*a*)pyrene depend on genotype, age, and developmental stage.

Also, the step of mutation fixation and the repair of premutational lesions may differ due to genetic differences between strains. So far, these factors have been studied only with strong repair-defective alleles,[26] but weaker alleles might also modulate mutagen sensitivity[38] and possibly also mutation fixation.

7.3. Toxic Effects and Cell Death

7.3.1. Preimaginal Lethality

Larval death is the most obvious toxic effect that can be observed after treatment of larval populations with noxious substances. Induced lethality can result either from lethal damage to larval structures and tissues, causing larval death, or from lethal damage to imaginal disc cells, expected to lead to pupal death. At a given concentration, some genotoxic chemicals act primarily on dividing cells and may predominantly damage the mitotic cells of the imaginal discs rather than nondividing larval cells, so that the exposed individuals die at metamorphosis in the pupal stage. Pupal death can be recorded by counting the number of dead pupae.

A rough impression of strong toxic effects is obtained if the number of progeny from control and exposure vials are compared, provided all vials contain approximately the same number of larvae at the beginning of the experiment. This is feasible if parents were allowed to lay eggs in the vials for a defined period of time (e.g., 24 hr[64]) or if approximately equal batches of larvae are transferred into the vials.[80] In the wing assay,[80] toxic effects are usually recorded qualitatively only (strong reduction in the number of progeny after mutagen treatment).

A precise determination of survival frequencies is more laborious, since it requires the counting of larvae.[156,166] In some studies chemically induced larval and pupal death (preimaginal lethality) was determined in this way.[64,166] Quantification of preimaginal lethality should not be overemphasized, since effects other than genotoxic damage (e.g., membrane damage) may contribute substantially to the lethal effects. However, toxicity may be taken as an indication that the chemical is being taken up.

7.3.2. Cell Death in Imaginal Discs

Programmed cell death appears to be a normal feature of developing imaginal discs. In certain mutants, e.g., vestigial and scalloped, cell death in the wing discs occurs throughout the third larval instar, and a small amount of cell death was also detectable in a similar location in wild-type discs at the beginning of the third larval instar. This led to the conclusion that these mutations act by increasing the extent of a degeneration event that occurred during normal development.[72]

If, following a mutagenic treatment, only part of the disc cells are killed, the disc may survive, and regeneration may restore the normal cell number. X-ray-induced cell death in wing discs was documented by Haynie and Bryand.[89] Regeneration in X-rayed discs was discovered by Schweizer.[178] He found that a relation exists between X-ray-induced cell death and the size of clones induced by mitotic recombination in the imaginal genital disc of *Drosophila.* A gradual increase in the size of induced clones with increasing X-ray doses was observed. This was presumably due to the elimination of a part of the disc cells and the compensatory growth of the remaining cells, including those in the clone. Haynie and Bryant[89] confirmed these results and showed an increase in average clone size on the wing blade with increasing X-ray dose. They observed an increase in the relative frequency of large clones and a decrease of smaller clones.

Direct evidence for X-ray-induced death of imaginal disc cells (increased frequency of pycnotic cells) has been presented by Haynie and Bryant[89] for *Drosophila melanogaster* and by Spreij[187] for *Calliphora erythrocephala.* Although experimental proof is lacking, chemicals at certain concentrations also are expected to induce cell death.

After irradiation of larvae, adult organs (eye, wing, etc.) with substantially altered structures are rare. Normal pattern formation is the most common result even after larval treatments that lead to killing or mitotic inhibition of at least 75% of primordial cells.[89] The generally successful pattern formation following extensive and presumably scattered cell death indicates that prior to differentiation, the restitution of a nearly normal cell number is accompanied by a regulation of the normal determination pattern. The success of this regulative response to cell death can be understood as the result of intercalary regeneration, which has been shown experimentally to take place in transplants in which disc fragments from opposite locations were combined.[88] With respect to the direction of division, the pattern of mitoses during intercalary regneration appears to be similar to that occurring during normal development. This is demonstrated by clones induced at low exposures. Clones take on the characteristic orientation in which they are aligned in the direction of the proximodistal axis of the wing blade. The rate of cell division during the extra growth need not be markedly increased, since pupation is often delayed in mutagen-exposed animals.

7.4. Status of Validation

Validation of the somatic assays used in *Drosophila* includes a determination of the detectability of mutagens and promutagens from many different chemical classes (see Table 1) and a comparison of the test performance with other genotoxicity assays, in particular with yeast, *Drosophila*, and the mouse. In a second step the correlation between mutagenicity in somatic cells of *Drosophila* and carcinogenicity in mammals has to be determined.

7.4.1. DNA Repair-Proficient Larvae

Table 1 has been compiled to demonstrate the status of validation reached with the different somatic cell assays in *Drosophila*. The table contains the published information available for all three assays as well as some unpublished data. Included in the table are only compounds that have been tested in at least one somatic assay with *D. melanogaster*.

The three somatic assays discussed in this review have reached different levels of validation. The number of compounds tested are 46 for the unstable white–zeste, 57 for the white/white-coral, and 92 for the wing test.

A critical part in the validation of the somatic assay systems is the study of promutagens. It has been shown that subcellular preparations from *Drosophila* larvae of a series of wild-type strains contain enzymatic activities that play key roles in xenobiotic metabolism.[11,53,54,86,91,150,231] Since major parts of the activities most probably occur in the fat bodies, Malpighian tubules, and various parts of the digestive tract,[35] it is important to examine whether promutagens are active in cells of the wing and/or eye imaginal disc. Reactive metabolites either may be produced within the target cell or may diffuse to the target cell from a more or less distant site of activation (e.g., from another tissue). The high mutagenic effects observed after treatment with the procarcinogens DMN and DEN, agents known to form short-lived intermediates, can be taken to indicate that imaginal disc cells have at least certain intracellular activating enzymes.

Although the cytoplasm of larval disc cells is typical of undifferentiated cells, containing little endoplasmatic reticulum, some membrane-bound P-450-dependent enzyme activities can be expected to be present in the disc cells.[57,202] For definitive proof additional detailed studies are necessary.

The supposition that disc cells are capable of at least some xenobiotic metabolism is further substantiated by the fact that the somatic assays

TABLE 1. Chemical and Physical Agents Tested for Mutagenic and Recombinogenic Activity in at Least One Somatic Assay in *Drosophila melanogaster*[a]

Chemical Compound (CAS registry no.)	Wing *mwh/flr*						Eye *w/w^co*			Eye UZ				Effect[b]							
	exr⁺				exr⁻		exr⁺			exr⁺		exr⁻									
	S	L	T	Ref.	Sp	Ref.	S	T	Ref.	Sp	Ref.	Sp	Ref.	S	T	C	N	Y	M	Ca	MA
1. *N*-Nitroso compounds																					
α-Acetoxymethyl-*N*-methyl-nitrosamine (AcO-DMN) (65986-79-0)	·	·	·		·		·	·		+	46, 47	·		·	·	·	·	·	·	·	*
Diethylnitrosamine (DEN) (55-18-5)	+	+	+	78, 80	·		+	+	208, 213, 215	+	138	·		+	i	+	·	+	−	+	*
Dimethylnitrosamine (DMN) (62-75-9)	+	+	+	80	·		+	+	208, 215	+ +	46, 47 138	·		+	+	+	·	+	i	+	*
Dinitrosopentamethylene-tetramine (DNPT) (101-25-7)	·	·	·		·		+	+	208, 215	·		·		·	·	·	·	·	·	·	*
N,N′-Dinitrosopiperazine (140-79-4)	p	o	s	189, 190	·		·	·		·		·		+	+	·	·	·	·	+	*
N,N′-Dinitroso pyridinol carbamate (71799-98-9)	p	o	s	189, 190	·		·	·		·		·		·	·	·	·	·	·	·	·
Dipropylnitrosamine (DPN) (621-64-7)	·	·	·		·		+	+	208, 215	·		·		·	·	·	·	·	·	+	*
N-Ethyl-*N′*-nitro-*N*-nitroso-guanidine (ENNG) (4246-77-6)	·	·	·		·		+	+	208, 213, 215	+	61	+	61	+	·	·	·	·	·	+	*
N-Ethyl-*N*-nitrosourea (ENU) (759-73-9)	+	+	+	225	·		+	+	208, 213, 215	+	173	·		+	−	+	·	+	+	+	
Methylethylnitrosamine (MEN) (10595-95-6)	·	·	·		·		+	+	208, 215	·		·		·	·	·	·	·	·	+	*

N-Methyl-*N'*-nitro-*N*-nitroso-guanidine (MNNG) (70-25-7)	+ + +	78[d]	·	+ +	208, 215	+	59, 61	+	61	+ + + + + + +	*
N-methyl-*N*-nitrosourea (MNU) (684-93-5)	+ + +	224	·	+ +	208, 215	+	173	·		+ + + · · – +	
N-Methyl-*N*-nitroso-*p*-toluenesulfonamide (80-11-5)	p o s	189, 190	·	· ·		·		·		· · · · · · +	
N-Nitrosodiethanolamine (1116-54-7)	· · ·		·	· ·		+	49	·		+ · · · · · +f	
N-nitrosomorpholine (59-89-2)	p o s	189, 190	·	· ·		·		·		+ + · · – · +	*
N-Nitrosopiperidine (100-75-4)	+ + i	228	·	· ·		·		·		+ + + · · · +	
Streptozotocin (18883-66-4)	+ + +	[e]	·	· ·		·		·		· · · · · · +	
2. Aryldialkyltriazenes											
Phenyldimethyltriazene (PDMT) (7227-91-0)	i + +	[f]	·	+ +	214	·		·		+ + + · + · +	*
1-(2,4,6-trichlorophenyl)-3,3-dimethyltriazene (tri-Cl-PDMT) (50355-74-3)	· · ·		·	+ + + +	215 130	·		·		+ · i · · · ·	*
3. Pyrrolizidine alkaloids											
7-Acetylintermedine (74243-01-9)	+ + +	55[g]	·	· ·		·		·		· · · · · · ·	*
7-Acetyllycopsamine (73544-48-6)	+ + +	55[g]	·	· ·		·		·		· · · · · · ·	*
Indicine (480-82-0)	+ + i	[e]	·	· ·		·		·		· · · · · · ·	*
Indicine-*N*-oxide (41708-76-3)	+ + +	[e]	·	· ·		·		·		· · · · · · ·	*
Intermedine (10285-06-1)	+ + +	[e]	·	· ·		·		·		· · · · · · ·	*

(continued)

TABLE 1. *(continued)*

Chemical compound (CAS registry no.)	Wing *mwh/flr* exr⁺ S	L	T	Ref.	Wing exr⁻ Sp	Ref.	Eye *w/w^co* exr⁺ S	T	Ref.	Eye UZ exr⁺ Sp	Ref.	Eye UZ exr⁻ Sp	Ref.	Effect[b] S	T	C	N	Y	M	Ca	MA
Monocrotaline (315-22-0)	+	+	+	[e]	·		·	·		·		·		+	+	·	·	·	·	+	*
Retrorsine (480-54-6)	+	+	+	[e]	·		·	·		·		·		+	·	·	·	·	·	+c	*
Senecionine (130-01-8)	+	+	+	[e]	·		·	·		·		·		+	·	·	·	·	·	·	*
Seneciphylline (480-81-9)	+	+	+	55, 225[g]	·		·	·		·		·		+	·	·	·	·	·	·	*
Senkirkine (2318-18-5)	+	+	+	55[g]	·		·	·		·		·		+	·	·	·	·	·	+c	*
Supinine (551-58-6)	−	−	i	[e]	·		·	·		·		·		+	·	·	·	·	·	·	*
Symlandine (74410-74-5)	+	+	+	[e]	·		·	·		·		·		·	·	·	·	·	·	·	*
Symphytine (22571-95-5)	+	+	+	228	·		·	·		·		·		·	·	·	·	·	·	+c	*
4. Hydrazines																					
Acethydrazide (1068-57-1)	+	+	+	[h]	·		·	·		·		·		·	·	·	·	·	·	·	·
1,2-Dimethylhydrazine (540-73-8)	−	+	+	[h]	·		·	·		·		·		·	·	·	·	+	·	+	*
Hydralazine (86-54-4)	i	+	+	78[h]	·		·	·		·		·		−	·	·	·	·	·	·	*
Hydrazine (302-01-2)	+	+	+	[h]	·		·	·		·		·		i	·	·	·	+	·	+	
Iproniazid (305-33-9)	+	+	i	[h]	·		·	·		·		·		·	·	·	·	·	·	·	·
Isoniazid (54-85-3)	+	+	+	[h]	·		·	·		·		·		·	·	·	·	·	+	+L	
N-Methyl-*N*-formyl hydrazine (758-17-8)	+	+	+	[h]	·		·	·		·		·		·	·	·	·	·	·	+	*
Phenylhydrazine (100-63-0)	−	+	+	[h]	·		·	·		·		·		·	·	·	·	·	·	·	
Procarbazine (Natulan) (366-70-1)	+	+	+	78, 80	·		+	+	208, 215	+	127	·		+	−	+	·	+	+	+	*
5. Benzodioxole																					
Safrole (94-59-7)	+	+	−	226	+	226	+	i	215			−	64	−	·	·	·	+	·	+	*

6. Acridines											
Acridine orange (65-61-2)	− + −	[d]	+	[d]	· ·		+	227	·	+ · + · + + ·	
Acriflavine (8048-52-0)	· · ·		·		· ·		+	127, 227	·	+ · + · + · ·	
5-Aminoacridine (578-07-4)	+ + −	78, 80	·		· ·		·		·	· · · · · · ·	
Proflavine (92-62-6)	· · ·		·		· ·		+	[i]	·	· · i + · · *	
7. Aziridines											
Tepa (545-55-1)	+ + +	[f]	·		· ·		·		·	+ + · · · · ·	
Thio-TEPA (52-24-4)	+ + +	[f]	·		+ +	[j]	·		·	+ · · · · · +	
Trenimon (68-76-8)	+ + +	78, 80	·		+ +	130	·		·	+ + + · · − +	
Triethylenemelamine (TEM) (51-18-3)	+ + +	[f]	·		+ +	17	+	[i]	·	+ + + · + + +L	
8. Oxirane, thiirane											
Diepoxybutane (1464-53-5)	− + +	78, 80	·		· ·		+	[i]	·	+ + + · + · +	
1,2;7,8-Diepoxyoctane (2436-07-5)	+ + +	[f]	·		· ·		·		·	· · · · · · ·	
Epichlorohydrin (106-89-8)	+ + +	[d,f]	+	[d]	+ +	[j]	·		·	+ · · · · · +	
9. Lactone											
β-Propiolactone (57-57-8)	+ + +	80	·		+ +	[j]	·		·	+ + + · + · +	
10. Heteroaromatic											
Aflatoxin B1 (1162-65-8)	− + +	79, 80	·		· ·		+	46, 47, 173	·	+ i · · + · +	*
Aflatoxin B1-2,3-dichloride (AFB1-C12) (58209-98-6)	· · ·		·		· ·		+	46, 47	·	· · · · · · ·	
Hycanthone methanesulfonate (23255-93-8)	i + i	[k]	·		· ·		·		·	+ − + · + i ·	
11. Haloalkanes, haloolefins											
1,2-Dibromo-3-chloropropane (96-12-8)	· · ·		·		+ +	215	·		·	· + + · · · +	*, 0

(continued)

TABLE 1. *(continued)*

Chemical compound (CAS registry no.)	Wing *mwh/flr* exr⁺ S	L	T	Ref.	Wing exr⁻ Sp	Ref.	Eye *w/w^co* exr⁺ S	T	Ref.	Eye UZ exr⁺ Sp	Ref.	Eye UZ exr⁻ Sp	Ref.	Effect[b] S	T	C	N	Y	M	Ca	MA
1,2-Dibromoethane (106-93-4)	−	+	+	80	·		+	+	208, 215[f]	·		·		+	·	·	·	·	·	+	*, 0
2,3-Dibromo-1-propanol (96-13-9)	·	·	·		·		+	+	215	·		·		·	+	+	·	·	·	·	*, 0
1,2-Dichloroethane (107-06-2)	·	·	·		·		+	+	[j]	+	138, 139	·		+	·	i	+	·	+	+	*, 0
Methylbromide (74-83-9)	i	+	+	[f]	·		·	·		·		·		·	·	·	·	·	·	·	
1,2,3-Tribromopropane (96-11-7)	·	·	·		·		+	+	215	·		·		·	·	·	·	·	·	·	*, 0
Tris(2,3-dibromopropyl)phosphate (126-72-7)	·	·	·		·		+	+	215	·		·		+	+	·	·	·	·	+	*, 0
12. Sulfates, sulfatones, sultones																					
Busulfan (Myleran) (55-98-1)	+	+	+	[f]	·		·	·		·		·		+	·	·	·	·	·	?	
Diethylsulfate[c] (64-67-5)	·	·	·		·		+	+	208, 213	+	204	·		+	+	+	·	+	−	+	
2-Chloroethyl methanesulfonate (3570-58-9)	+	+	+	[l]	·		·	·		·		·		+	+	+	·	·	·	·	
Ethyl methanesulfonate (EMS) (62-50-0)	+	+	+	78, 80	+	[d]	+	+	208, 213, 215	+	64, 170	+	64	+	+	+	·	+	+	+	
Methyl methanesulfonate (MMS) (66-27-3)	+	+	+	[d]	+	[d]	+	+	208, 213,	+	139	+	167	+	+	+	−	+	+	+	
							+	+	215, 131	+	64	+	64						−		
iso-Propylmethanesulfonate (iPMS) (926-06-7)	+	+	+	102[f]	·		·	·		·		·		+	+	+	·	·	·	·	
13. Aromatic amines																					

2-Acetylaminofluorene (2-AAF) (53-96-3)	– i –	m	i	m	– M	208, 211	– –	44 66	+	66	i · · · + + +	*
4-Acetylaminofluorene (4-AAF) (28322-02-3)	i i –	n	–	m	· ·		–	66	–	66	· · · · i · ·	
2-Aminofluorene (153-78-6)	+ + +	d	–	d	· ·		·		·		· · · · · · +	*
Benzidine (92-87-5)	· · ·		·		+ M	214, 215	·		·		· · · · · · +	*
4-Dimethylaminoazobenzene (4-DAB) (60-11-7)	· · ·		·		M M	214, 215	·		·		i · · · – · +	*
2-Naphtylamine (91-59-8)	· · ·		·		M M	208	·		·		i · · · + · +	*
14. Polycyclic hydrocarbons												
Benz(*a*)anthracene (56-55-3)	· · ·		·		· ·		–	46, 47	·		i · · · – · +	*
Benzo(*a*)pyrene (50-32-8)	+ + –	m •	·		– +	211[j]	+ +	46, 47 65	+	65	+ · · – + + +	*
9,10-Dimethylanthracene (781-43-1)	· · ·		·		+ +	208, 211, 214, 215[a]	·		·		· · · · + · ·	*
7,12-Dimethylbenz(*a*)anthracene (DMBA) (57-97-6)	+ i i	d,n,o	·		+ +	208, 211	+	46, 47	·		+ · · · – · +	*
3-Methylcholanthrene (3-MC) (56-49-5)	– – i	k	·		– +	208, 211[j]	·		·		i · · · · · +	*
Pyrene (129-00-0)	– – –	m	·		– –	j	–	65	–	65	· · · · + · ·	
15. N-, S-, or O-Mustards												
Cyclophosphamide (50-18-0)	+ + +	78, 225	·		+ +	208, 215	+	127	·		+ + + · + + +	*
ICR-170 (146-59-8)	+ + +	80[f]	·		· ·		·		·		+ + + · + · +L	
Mitomycin C (50-07-7)	+ + +	78, 80	·		· ·		·		·		+ · + – + + +	
Nitrogen mustard (HN2) (51-75-2)	+ + +	78	·		· ·		·		·		+ + + · · · +	
16. Antimetabolites, DNA-synthesis inhibitors												

(continued)

TABLE 1. *(continued)*

Chemical compound (CAS registry no.)	Wing *mwh/flr* exr⁺ S	L	T	Ref.	Wing exr⁻ Sp	Ref.	Eye *w/w^co* exr⁺ S	T	Ref.	Eye UZ exr⁺ Sp	Ref.	Eye UZ exr⁻ Sp	Ref.	Effect[b] S	T	C	N	Y	M	Ca	MA
Actinomycin C (8052-16-2)	p	o	s	193	·		·	·		·		·		·	·	·	·	·	·	i	
Adriamycin (23214-92-8)	+	+	+	56	·		+	+	215	·		·		+	·	·	·	·	·	+	
5-Azacytidine (320-67-2)	+	+	+	[f]	·		·	·		·		·		·	·	·	·	·	·	+L	
Azaserine (115-02-5)	+	+	+	[c]	·		·	·		·		·		·	·	·	·	·	·	+	
6-Azauridine (54-25-1)	n	e	g	193	·		·	·		·		·		i	·	·	·	·	·	·	
Bleomycin (11056-06-7)	+	+	i	78, 80[p]	·		+	+	208, 215	+	[p]	+	[p]	·	·	·	·	+	·	·	
5-Bromodeoxyuridine (59-14-3)	·	·	·		·		·	·		+	[i]	·		+	i	·	·	−	·	·	
Cytosine arabinoside (ara-C) (147-94-4)	+	+	+	[f]	·		+	+	215[j]	·		·		−	·	·	·	·	·	·	
Daunomycin (Daunorubicin) (20830-81-3)	+	+	+	[n,p]	·		+	+	215	+	[p]	·		·	·	·	·	·	·	+	
5-Fluorouracil (500-91-9)	+	+	+	[f]	·		·	·		·		·		i	·	i	+	+	·	i	
Methotrexate (amethopterin) (59-05-2)	+	+	+	224[p]	·		+	+	[j]	·		·		−	·	·	·	+	·	−	
cis-Platinum diamine dichloride (*cis*-PDD) (15663-27-1)	+	+	+	[f]	·		+	+	[j]	·		·		+	·	·	·	·	·	+L	
Vinblastine sulfate (143-67-9)	+	+	−	80[h]	·		+	+	[j]	·		·		·	·	·	·	·	·	·	
17. Pyrolysis products																					
A-α-C (26148-68-5)	+	+	+	228	·		·	·		·		·		·	·	·	·	·	·	+h	*
Glu-P-1 (67730-11-4)	+	+	i	228	·		·	·		·		·		·	·	·	·	·	·	+h	*
Glu-P-2 (67730-10-3)	+	+	+	228	·		·	·		·		·		·	·	·	·	·	·	+h	*

IQ (76180-96-6)	+ + +	228	·		· ·		·		·		· · · · · · +i	*
MeA-α-C (68006-83-7)	+ i +	228	·		· ·		·		·		· · · · · · +g	*
MeIQ (no CASRN)	+ + +	228	·		· ·		·		·		· · · · · · ·	*
MeIQx (no CASRN)	+ i +	228	·		· ·		·		·		· · · · · · ·	*
Trp-P-1 (62450-06-0)	+ + +	228[p]	·		· ·		+	63	i	[p]	M · · · · · +a	*
Trp-P-2 (62450-07-1)	+ + +	228[p]	·		· ·		+	63	+	[p]	M · · · · · +b	*
18. Phenyls and biphenyls												
Benzene (71-43-2)	i − −	226	+	226	M −	209	−	139	−	64	· · · · · · +L	
Benzyl chloride (BC) (100-44-7)	· · ·		·		· ·		+	48	·		− · · · + · +L	
4-Chloromethyl biphenyl (4CMB) (1667-11-4)	· · ·		·		· ·		+	48	·		− · · · · · ·	
4-Hydroxymethyl biphenyl (4HMB) (3597-91-9)	· · ·		·		· ·		+	48	·		· · · · · · ·	
19. Food colors												
Brown FK (8062-14-4)	· · ·		·		· ·		−	44	·		· · · · · · ·	
Red 2G (3734-67-6)	· · ·		·		· ·		−	44	·		· · · · · · ·	
Red 10B (3567-66-6)	· · ·		·		· ·		−	44	·		· · · · · · ·	
20. Nitroheterocyclics												
Moxnidazol (52279-59-1)	− + i	[k]	·		· ·		·		·		+ · · · · + ·	*
Nitrofurantoin (67-20-9)	− i −	[k]	·		· ·		·		·		i · · · · · ·	*
4-Nitroquinoline 1-oxide (4-NQO) (56-57-5)	− + +	[d,g]	·		· ·		·		·		+ · · · + + +	*
21. Aldehydes												
Acetaldehyde (75-07-0)	+ + +	[d]	·		· ·		·		·		· · · · · · ·	
Formaldehyde (50-00-0)	i + i	[d]	+	[d]	· ·		+	[i]	·		+ + · · · · +	
	p o s	196			+ +	196						
Malonaldehyde (542-78-9)	p o s	196	·		+ +	196	·		·		· · · · · · ·	

(*continued*)

TABLE 1. *(continued)*

Chemical compound (CAS registry no.)	Wing *mwh/flr* exr⁺ S	L	T	Ref.	exr⁻ Sp	Ref.	Eye *w/w^co* exr⁺ S	T	Ref.	Eye UZ exr⁺ Sp	Ref.	exr⁻ Sp	Ref.	Effect[b] S	T	C	N	Y	M	Ca	MA
22. Phosphates																					
Trimethylphosphate (512-56-1)	+	+	+	[f]	·		·	·		·		·		+	·	·	·	·	·	+	
23. Flavonoids																					
Rutin (water soluble) (153-28-4)	+	−	−	78	·		·	·		·		·		·	·	·	·	·	·	·	
24. Stilbenediols																					
Diethylstilbestrol (56-53-1)	−	−	−	226	·		M	M	209	·		−	64	+	·	·	·	−	·	+	
25. Inorganic																					
Cadmium chloride (10108-64-2)	·	·	·		·		−	i	[j]	·		·		·	·	+	·	+	·	+	
26. Miscellaneous																					
Acetamide (60-35-5)	·	·	·		·		−	−	[j]	−	127	·		·	·	·	·	−	·	+	
Acrylonitrile (107-13-1)	+	+	−	226	+	226	+	i	209	·		+	64	i	·	·	·	·	·	+	*
o-Aminoazotoluene (97-56-3)	·	·	·		·		·	·		−	44	·		·	·	·	·	·	·	+	
Aristolochic acid (12770-90-0)	+	+	+	56	·		·	·		·		·		+	·	+	·	·	·	+e	*
Auramine (492-80-8)	·	·	·		·		−	−	[j]	·		·		·	·	·	·	+	·	+d	
Benzoin (119-53-9)	−	−	−	226	·		−	−	209	·		−	64	·	·	·	·	·	·	·	
Caprolactam (105-60-2)	+	+	i	226	·		+	i	209[j]	·		+	64	·	·	·	·	·	·	·	
Captan (133-06-2)	·	·	·		·		·	·		−				+	−	·	·	+	·	+L	

p-Chloroaniline (106-47-8)	+	+	+	[d]	+	[d]	·	·		·		·		·	·	·	·	·	·	·	
Chloroquine (54-05-7)	·	·	·		·		·	·		+	227	·		·	·	·	·	·	·	·	
2,4-Dichlorophenoxyacetic acid (2,4-D) (94-75-7)	·	·	·		·		·	·		+	159	·		−	·	i	−	−	·	·	
Di-(2-ethylhexyl)phthalate (117-81-7)	−	−	−	226	−	226	M	−	209	·		M	64	·	·	·	·	·	·	+	
Dimilin (35367-38-5)	n	e	g	193	·		·	·		·		·		·	·	·	·	·	·	·	
Ethidium bromide (1239-45-8)	+	−	i	[d]	+	[d]	·	·		+	227[i]	·		+	·	·	·	−	·	·	*
Furylfuramide (AF-2) (3688-53-7)	−	+	i	[p]	·		·	·		+	59, 60	·		i	·	·	·	+	·	+	*
Hexamethylphosphoramide (HMPA) (680-31-9)	+	+	+	193, 226	+	84	+	+	208, 209	·		+	64	+	+	·	·	+	+	+	*
Hydroxylamine-HCl (5470-11-1)	i	+	+	78[g,h]	·		·	·		·		·		i	·	−	i	+	·	·	
Methoprene (40596-69-8)	n	e	g	193	·		·	·		·		·		·	·	·	·	·	·	·	
Nitroprusside (15078-28-1)	−	−	−	25	·		·	·		·		·		·	·	·	·	·	·	·	
Petrol (commercial) (no CASRN)		·	·		·		·	·		+	139	·		·	·	·	·	·	·	·	
Phenobarbital (50-06-6)	−	−	−	226	·		−	−	209	·		−	64	·	·	·	·	·	·	·	
Quinacrine (83-89-6)	·	·	·		·		·	·		−	227[i]	·		·	·	·	·	·	·	·	
Sodium nitrite (7632-00-0)	+	−	−	[d]	·		·	·		·		·		+	·	·	·	+	·	·	
Strychnine (60-41-3)	+	+	+	[n]	·		·	·		·		·		−	·	−	·	+	·	·	
Tetracycline hydrochloride (64-75-5)	·	·	·		·		·	·		−	127	·		·	·	·	·	·	·	·	
o-Toluidine (95-53-4)	+	i	i	226[d]	i	226[d]	+	i	209	·		−	64	·	·	·	·	−	·	+	
2,4,5-Trichlorophenoxyacetic acid (2,4,5-T) (93-76-5)	·	·	·		·		·	·		−	159	·		+	·	i	−	·	·	·	

(*continued*)

TABLE 1. *(continued)*

Chemical compound (CAS registry no.)	Wing *mwh/flr* exr⁺ S	L	T	Ref.	exr⁻ Sp	Ref.	Eye w/w^{co} exr⁺ S	T	Ref.	Eye UZ exr⁺ Sp	Ref.	exr⁻ Sp	Ref.	Effect[b] S	T	C	N	Y	M	Ca	MA
27. Physical agents																					
Microwave radiation (2460 MHz) (no CASRN)	·	·	·		·		·	·		–	87	·		·	·	·	·	·	·	·	·
Pions (no CASRN)	+	+	+	22	·		·	·		·		·		·	·	·	·	·	·	·	·
X rays (no CASRN)	+	+	+	69	·		+	+	131	+	160, 173	+	p	+	+	+	+	+	+	+	

[a] Where available, the response in other *Drosophila* tests, in the yeast homozygosity test for genetic recombination and/or gene conversion, in the mouse spot test, and in the rodent carcinogenesis bioassays are given. Compounds known to lead to mutagenic metabolites are marked with asterisks. Abbreviations: Wing *mwh/flr*: Somatic mutation and recombination assay using the wing hair markers *mwh* (multiple wing hairs) and *flr* (flare). exr⁺, Excision-repair-proficient larvae exposed; exr⁻, excision-repair-deficient (*mei-9*) larvae exposed; S, small single spots; L, large single spots; T, twin spots; SP, spots; Ref., references. Eye w/w^{co}: Somatic mutation and recombination assay using the two eye color markers at the white locus, *w* (white) and w^{co} (white coral). Eye UZ: Somatic mutation assay using the unstable white–zeste system. Symbols: positive, +; weak positive, W; negative, –; inconclusive, i; marginal positive, M; no data available, ·. For formaldehyde, malonaldehyde,[196] *N,N'*-dinitrosopiperazine, *N,N'*-dinitroso pyridinol carbamate, *N*-methyl-*N*-nitroso-*p*-toluenesulfonamide, *N*-nitrosomorpholine,[189,190] 6-azauridine, dimilin, methoprene,[193] and nitroprusside[25] the diagnosis for the wing test is simply given as "pos" or "neg" because the frequencies of twin spots and large single spots are not given explicitly in the papers.

[b] S, Sex-linked recessive lethals induced in *Drosophila* germ cells: All from Lee *et al.*,[112] with the exception of: adriamycin,[37] aristolochic acid,[56] diethylstilbestrol,[39] DMBA,[52] ethylene dibromide,[99] *cis*-platinum,[222] benzyl chloride, 4-chloromethyl biphenyl,[117] methotrexate (amethopterin), cytosine arabinoside,[199] seneciphylline, senkirkine,[33] Trp-P-1, Trp-P-2,[63] ethidium bromide,[116] moxnidazol, 4-NQO, nitrofurantoin,[106] Hydralazin (H. Juon, unpublished results), procarbazine,[21] strychnine (F. E. Würgler, U. Graf, and A. Kägi, unpublished results), 2-chloroethyl methanesulfonate (F. E. Würgler and U. Graf, unpublished results; ref. 112), *N*-nitrosodiethanolamine.[49]
T, Heritable reciprocal translocations and C, clastogenic effects, induced in *Drosophila* germ cells: All from Valencia *et al.*,[203] with the exception of: DEN, DMN, *N*-nitrosopiperidine, diepoxybutane, β-propiolactone, hycanthone menthansulfonate, 1,2-dibromo-3-chloropropane, 2,3-dibromo-1-propanol, diethylsulfate, cyclophosphamide, nitrogen mustard,[75,233] mitomycin C, acridine orange, acriflavin,[1] strychine (F. E. Würgler, U. Graf, and A. Kägi, unpublished results), 2-chloroethyl methanesulfonate (F. E. Wurgler and U. Graf, unpublished results; ref. 203), aristolochic acid.[56]
N, Nondisjunction induced in *Drosophila* germ cells.[203]

Y, Mitotic recombination and/or gene conversion induced in *Saccharomyces cerevisiae*: All from Zimmermann *et al.*,[236] with the exception of: HMPA, pyrene, benzo(*a*)pyrene, 9,10-dimethylanthracene, 2-AAF, 4-AAF, *N*-nitrosomorpholine,[13] DMN, EMS, ICR-170,[29] hydrazine,[201] strychnine (R. Fahrig, unpublished results).

M, Somatic spots induced in the mouse spot test: All from Russel *et al.*,[172] with the exception of: *N*-methyl-*N*-nitrosourea, trenimon, diethylsulfate, MMS,[27] 1,2-dichloroethane, 2-AAF, nitrofurantoin.[73]

Ca, Carcinogenicity in mammals and/or man: +, positive (= carcinogenic)[131]; +L, limited evidence, limited positive[131]; ?, data equivocal[131]; + a, Ishikawa *et al.*[95]; +b, Hosaka *et al.*[92]; +c, Culvenor[10]; +d, probably carcinogenic for humans[91]; +e, Mengs *et al.*[122,123]; +f, Fahmy and Fahmy[19]; +g, Nagao[133]; +h, Ohgaki *et al.*[142]; +i, Ohgaki *et al.*[143]

MA, Metabolic activation needed (=); two possibilities (*, 0).

[c] In the paper of Rasmuson *et al.*[162] diethylsulfonate should read diethylsulfate (B. Rasmuson, personal commmunication).

[d] U. Graf (unpublished results).

[e] E. Frei (unpublished results).

[f] A. J. Katz (unpublished results).

[g] H. Frei, J. Brauchli, J. Lüthy, and J. Zweifel, (unpublished results).

[h] H. Juon (unpublished results).

[i] H. Svalin (unpublished results, cited in ref. 162).

[j] E. Vogel and J. A. Zijlstra (unpublished results).

[k] F. E. Würgler, U. Graf, H. Frei, H. Juon, and A. Kägi (unpublished results).

[l] F. E. Würgler and U. Graf (unpublished results).

[m] U. Graf and F. E. Würgler (unpublished results); see text, footnote to p. 32.

[n] A. Kägi (unpublished results).

[o] Under further investigation.

[p] K. Fujikawa (unpublished results).

[q] Under further investigation by S. Kondo.

are capable of detecting mutagenic activity of a large variety of promutagens with diverse metabolic activation pathways (see Table 1). For instance, several nitrosamines, certain triazenes, practically all pyrrolizidine alkaloids, procarbazine, aflatoxin B1, many haloolefins and haloalkanes, and some aromatic amines and polycyclic hydrocarbons give positive results.

In experiments designed to assess the general performance of the white/white-coral system, 57 chemicals belonging to a variety of structurally unrelated classes of chemicals have been tested. Among them were 22 mutagens that depend on enzymatic activation by the mixed-function oxygenase system. From their mutagenic effects on eye imaginal disc cells it is evident that *Drosophila* larvae, similar to adult flies, are capable of activating different groups of promutagens (Table 1). Because this study is still in progress (E. W. Vogel, unpublished results), no attempt will be made to discuss in more detail the actual experimental results.

Detailed comparisons of the three assays are not yet possible because only a limited number of (frequently different) compounds have been studied in these assays and because more quantitative data are needed. Nevertheless, some general conclusions can be drawn from Table 1: (1) Powerful reference mutagens, such as the direct-acting mono- and polyfunctional alkylating agents, are detected without difficulty in somatic assays. (2) Unstable, direct-acting mutagens are detected in the wing assay as well as in the white/white-coral assay using simple feeding procedures. A good example is β-propiolactone, with a half-life of about 3.5 hr in aqueous solution.[105] (3) The detection of intercalating agents, several of which have been studied in the wing test, seems to be more difficult. The limited experience gained indicates that intercalators of the acridine type induce only small single spots, if any. Acridine orange has been shown to induce homozygosity (genetic recombination and gene conversion) in the yeast *Saccharomyces cerevisiae*.[182] The question of whether it is possible to improve the experimental conditions for their detection in *Drosophila* is currently under investigation by P. Alba and R. Marcos.

In the wing assay, ICR-170, a molecule with an intercalating part and an alkylating side chain, is capable of inducing all three types of spots (A. J. Katz, unpublished results; ref. 80). The antibiotics adriamycin and daunomycin, which were active in both the wing and the white/white-coral test, are also thought to act through an intercalation mechanism similar to that described for acridines.[206]

It is of particular practical implication that mutagens that do not directly interact with DNA, such as methotrexate (amethopterin), an antifolate which leads to disturbances in the nucleotide pools, are detected in the somatic assays testing proliferating cells. This is true for the white/white-coral as well as for the wing assay.

One member of the class of structurally varied compounds that induce a volumetric increase of peroxysomes in hepatocytes following subchronic administration to rodents, di-(2-ethylhexyl)phthalate, was tested in all somatic assays. These compounds lead to an increased incidence of hepatocellular carcinoma in rats and mice.[21,163] Such carcinogenic peroxisome proliferators do not appear to interact with DNA, although Reddy *et al.*[163,164] proposed that they may act indirectly upon the genome through production of excess H_2O_2 and the associated oxidative damage. di-(2-Ethylhexyl)phthalate gave negative results in the wing assays and negative or marginal results in the eye assays. Preliminary feeding studies have also shown that clofibrate (a lipid-lowering agent and peroxysome proliferator) appears to be negative in the wing assay (H. Frei, unpublished results). It remains to be determined whether this class of compounds is important based on interactions with other mutagens and promutagens.

7.4.2. Excision Repair-Deficient Larvae

In the wing assay 16 compounds (see Table 1) were tested with DNA repair-proficient as well as with excision repair-deficient larvae. In general the results obtained with the two types of larvae agree. Only one compound, benzene, gave a positive result with repair-deficient larvae, while the result of the standard cross was inconclusive. With the set of compounds studied so far, there was no indication that inclusion of a *mei-9* mutation increases the detection capacity of the assay. In contrast, the use of the *mei-9* strain resulted in additional problems. Often the excision repair-deficient larvae exhibited an increased lethality resulting from an increased senstivity to toxic effects of the chemical under test. In these cases, therefore, only exposure concentrations could be tested that were lower than those used with repair-proficient larvae. A major drawback of this system is the limited size of the spots. In *mei-9* hemizygous individuals, small single spots (one or two cells in size) are found exclusively, even after exposure to MMS and EMS. It seems, therefore, that affected cells do not proliferate normally in *mei-9* wing imaginal discs. This basic bias, together with the poor fertility of *mei-9* females and the fact that scoring of yellow wings is not as easy as scoring wild-type wings, led to the conclusion to discontinue using this test system (U. Graf, unpublished results).

In the UZ system Rasmuson[156] found that strains containing the *mei-9*^{b} or *mei-41*^{D5} mutation had an increased spontaneous germ-line mutation rate. It is assumed that this results from a high frequency of unrepaired spontaneous chromosome breaks occurring in these chromosome-destabilizing mutants. On the other hand, the two mutations did not influence the spontaneous frequency of somatic spots in the UZ strain. This is an interesting contrast to the effects reported for the assays with autosomal markers. Baker *et al.*[(12)] found that *mei-41* has no extra twin spots, but has a 20 times higher frequency of spontaneous somatic single spots expressing chromosome 3 somatic cell markers than the control. In *mei-9*^{b}, they found high frequencies of both twin spots and single spots.

Strains containing a meiotic mutation in combination with the UZ system were exposed to MMS. The *mei-9* and *mei-41* strains were markedly sensitive to MMS treatment, while *mei-218* (which does not affect DNA repair) did not affect survival. Only the mutant *mei-41*, which is lacking postreplication repair, affected the somatic mutation rate of the UZ system, decreasing it by 50%.[(156)]

The usefulness of the excision repair-deficient strains with the UZ system is still under test by Fujikawa *et al.* (unpublished results) in connection with the collaborative study for the IPCS. Conclusions with regard to the future use of this assay have to be postponed until the termination of the validation studies still being conducted.

7.5. Comparison of Germ Cells versus Somatic Cells

Comparisons between germ cells and somatic cells may be either indirect or direct.

7.5.1. Indirect Comparison

We may compare the data from somatic assays with the published results of the sex-linked recessive lethal (SLRL) assay, heritable translocation test, etc. Table 1 permits such a comparison, although at present only for a limited number of compounds. In general, there is good agreement between the SLRL assay results and the results of the somatic assays. So far, no group of chemicals has been discovered that would be detected in one assay but not in the other. Having studied in detail the mutagenic activity of selected aromatic amines and polycyclic hydrocarbons, Vogel *et al.*[(214)] came to the conclusion that somatic assays are particularly suitable to complement the germ-line-specific SLRL test.

The comparison between results of somatic assays and those of the

SLRL assay in male germ cells is based on 80 compounds. Of these, 55 (69%) are positive in both assays. Only two compounds (captan and 2,4,5-trichlorophenoxyacetic acid) have tested positive in the SLRL assay but negative in the somatic test. These two compounds were tested only with the unstable white–zeste system and should be retested in larvae using the white/white-coral and/or the wing assay. Eight compounds (ara-C, benzyl chloride, 4-chloromethylbiphenyl, 2,4-dichlorophenoxyacetic acid, methotrexate, safrole, strychnine, vincristine) are positive in the somatic cells but negative in the SLRL germ cell assay. Another seven compounds (hydrazine, Trp-P-1, Trp-P-2, acrylonitrile, AF-2, hydroxylamine, and 5-fluorouracil) gave positive results in the somatic assays, but inconclusive results with the SLRL assay. Many antimetabolites are found among these compounds. A possible explanation is that in the SLRL studies the germ cells samples were predominantly postmeiotic stages, and therefore the strongly S-phase-dependent mutagenic activity could not be detected. Tests for the induction of autosomal recessive lethals in spermatogonia or X-linked lethals in oogonia might clarify this apparent discrepancy in response between somatic cells and germ cells. Among the compounds presently judged inconclusive or marginally positive in the somatic assays, two were found positive in the SLRL assay (supinine and diethylstilbestrol), and four others were difficult to study in the germ cell assay and yielded an inconclusive result (2-acetylaminofluorene, 4-dimethylaminoazobenzene, 3-methylcholanthrene, and nitrofurantoin). Two compounds, benz(*a*)anthracene and 6-azauridine, were found negative in the somatic assays and inconclusive in the SLRL assay. Benz(*a*)anthracene was also negative in the yeast homozygosity tests for mitotic recombination and gene conversion.

For a comparison between somatic assays and germ cell assays for chromosome breakage (heritable translocations and clastogenicity) in *Drosophila*, results for 46 compounds are available. Of these, 35 are positive in at least one somatic assay and one chromosome breakage assay in germ cells. For seven compounds the somatic assays are positive, but the germ cell tests gave inconclusive results. Two compounds (hydroxylamine and strychnine) were positive in somatic cells, but negative in the chromosome breakage test. Among the two negative compounds in the somatic assays, one was inconclusive (2,4,5-trichlorophenoxyacetic acid) and one was negative (captan) in the germ cell chromosome breakage assays.

The set of chemicals tested in both the somatic assays and the nondisjunction assays is too small for a comparison.

TABLE 2. Chemically Induced X-Linked Recessive Lethals and Eye Spots (in the White/White-Coral Test) after Treatment of *Drosophila* Larvae[a]

			Eye spots			Recessive lethals		
Chemical	Method	Concentration	TS, %	LS, %	Activity	Brood, days	Females, %	Males, %
MMS	B	5.0 mM	5.27	6.10	+	0–2	0.41 (+)	0.24 i
						2–4	0.87 +	0.15 i
ENU	B	1.0 mM	1.19	3.79	+	0–2	7.20 +	10.80 +
						2–4	6.70 +	6.20 +
Bleomycin	B	0.005 %	2.35	0.85	+	0–2	0.50 (+)	0 −
						2–4	0.47 (+)	0 −
DMBA	A	1.0/2.0 mM	0.84	3.57	+	0–2	0.25 i	1.60 +
						2–4	0.40 (+)?	1.00 +
DA	A	1.0/4.0 mM	0.79	1.37	+	0–2	1.60 +	0.16 i
						2–4	0.55 (+)	0.30 i

[a] Method A, food supplemented with test chemical; method B, dropping 0.2 ml of mutagen solution on the surface of the food; +, mutagenic; (+), weakly mutagenic; i, inconclusive; −, no activity detectable; TS, mosaic twin spots; LS, mosaic light spots. For details see Vogel.[208]

7.5.2. Direct Comparison

The UZ system allows a direct comparison between effects in germ cells and those in somatic cells.[166] The advantage of this approach is the possibility to monitor identical genetic events in somatic cells and germ cells using the same set of tester strains. Only a few compounds have been studied using this approach. This elegant possibility of recording identical genetic events in the two cell types has an important potential for studies in which the quantitative extrapolation from somatic to germ cells requires experimental proof.

A similar comparison between somatic cells and mitotically dividing cells of the germ line is possible by opposing somatic recombination to mitotic recombination in spermatogonial stem cells.[5,203,220] This approach would be more laborious than studies with the UZ system.

For a number of model mutagens a direct comparison between the genotoxic effectiveness in male and female germ cells with their activity in the somatic tissue of females was determined.[208] In germ cells the frequency of sex-linked recessive lethals was recorded. The effect in somatic cells was studied with the white/white-coral test. Considering first the results on recessive lethals, it can be seen in Table 2 that ENU, showing strong mutagenic effectiveness, was the only reference mutagen readily detectable in both sexes, whereas the identification as mutagens of MMS and bleomycin caused severe problems. Both mutagens were either ineffective or only showed marginal activity at rather high dosages. DMBA was more active in larval testis, DA more so in the female germ line. Except for ENU, which predominantly induces point mutations, and with DA, for which a few clusters were found, no large clusters of mutants of common origin were observed.

This pattern of effects described for the recessive lethal assay is quite different from the markedly high response of somatic tissue to the five chemicals studied. The most striking difference between both assays is the high response of somatic tissue to bleomycin and MMS. Thus, a drastic difference in detection capacity will be noted between germ-line assay and somatic mutation test. However, quantitative comparisons cannot be made, because the eye disc at the time of response consisted of at least 20 cells per female. This somewhat reduces the difference between germinal and somatic changes. Another interesting aspect of this analysis concerns the proportion of mosaic twin spots relative to mosaic single light spots. Twin spots are assumed to reflect somatic recombination events and chromosomal aberrations; single light spots may predominantly be due to mutations or deletions.[15] It is seen in Table 2 that the proportion of twin spots versus light spots varied considerably as a function of the type of DNA damage introduced. Bleomycin and MMS, reagents that are known to be efficient inducers of chromosome breaks, produced the highest twin spot to light spot ratios. Thus, we are inclined to the view that strong segregational elimination during meiosis may be responsible for the low yield of heritable mutations seen with these reagents.

7.6. Comparison of *Drosophila* Tests with Other Assay Systems

7.6.1. Comparison with Yeast and Mouse Assays

In a comparison of the somatic assays in *Drosophila* with the homozygosity assays in yeast to detect mitotic recombination and gene conversion, we found that 75% of the chemicals tested in both systems (36/48) give consistent results. For eight compounds tested the results disagree: six are positive in *Drosophila* but negative in yeast, and two differ in the opposite direction. One positive (2-acetylaminofluorene) and three compounds negative in yeast (4-dimethylaminoazobenzene, diethylstilbestrol, and acetamide) gave inconclusive or marginally positive results in *Drosophila* They should be tested further in *Drosophila* to obtain conclusive decisions about their activity in the somatic assays.

Twenty-three compounds were tested in both somatic assays in

Drosophila and the mouse spot test. Among this group, 17 chemicals are positive in both assays. This includes MMS, which has been classified as positive by the Gene-Tox committee,[172] but was recently reported as negative in the mouse spot test by Braun *et al.*[27] Four compounds (diethylnitrosamine, methylnitrosourea, trenimon, and diethylsulfate) are positive in *Drosophila* somatic assays but negative in the mouse spot test. Two chemicals (DMN and hycathone methanesulfonate) that are inconclusive in the mouse spot test have been found positive in *Drosophila*.

7.6.2. Comparison with *in Vivo* Carcinogenesis Data

It was found that for the compounds tested so far the correlation between carcinogenic activity in rodents and the *Drosophila* tests is high. Vogel *et al.*[211] found that among 102 carcinogens tested adequately in the *Drosophila* sex-linked recessive lethal test, 85 compounds (83%) were mutagenic. The remaining 17 chemicals did not induce lethals in *Drosophila*.

Among the 76 carcinogens tested in somatic assays (Table 1), 67 compounds (88%) were found to be mutagenic in at least one somatic assay. Three gave marginal effects and six (6/76 = 8%) were not detected by the somatic assays.

Of particular interest are the studies of Kondo and co-workers.[58,228] They studied the mutagenic potency of amino acid pyrolyzates in the *Drosophila* wing spot test and its correlation with carcinogenic potency. Nine amino acid pyrolyzates, which require metabolic activation for mutagenicity, were fed to larvae. All the compounds (Trp-P-1, Trp-P-2, Glu-P-1, Glu-P-2, IQ, MeIQ, MeIQX, A-α-C, MeA-α-C) did induce, to variable extents, *mwh* spots on the wings.

Based on an attempt to interpret the data in a quantitative manner, Kondo and his co-workers[228] reached the following conclusion: The reported values of carcinogenic potency in the mouse assay for seven of the nine compounds are in excellent correlation with mutagenic potency values obtained in the *Drosophila* wing assay (Figure 3), but not so with those obtained in the *Salmonella* assay.[188]

7.7. Future Developments

The list of compounds positive in calibration studies not only covers strong reference mutagens but also includes some chemicals that are difficult or impossible to detect as mutagens in other assays, such as the *Drosophila* SLRL or the *Salmonella* assay. It is felt that differences in

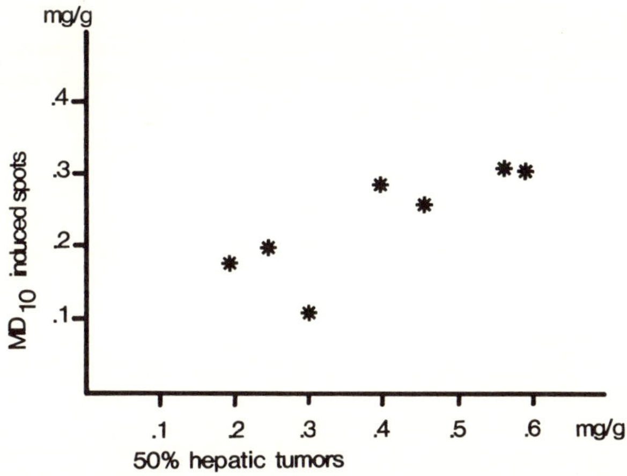

FIGURE 3. Quantitative comparison of mutagenic potency values in the *Drosophila* wing spot test and the carcinogenic potency values in the mouse *in vivo* assay estimated for amino acid pyrolyzates and heterocyclic amines. The dose in mg mutagen per g medium needed in the wing assay to induce 10% small single spots (MD_{10}) is plotted against the dose in mg carcinogen per g food resulting in a 50% incidence of hepatic tumors in mice. The correlation is based on the following compounds (from left to right): Trp-P-2, Trp-P-1, IQ, Glu-P-2, Glu-P-1, MeA-α-C, and A-α-C. [Based on data presented by Yoo *et al.*[228]]

sensitivity among different tissues (gonads, different imaginal discs) could be a source of variation. For instance, variation in mutagen sensitivity of different cell stages is a common observation made both for the male and female germ cell cycle in *Drosophila*.[213]

The mutagenicity assays using somatic cells of *Drosophila* have proved their usefulness for the detection of mutagenic activity. They appear to be of particular interest with respect to mutagenicity screening, because they are faster than the presently used *Drosophila* germ cell assays. Validation should include the technical aspects of the testing procedures as well as a more detailed study of the detectability of promutagens. The possible optimization of *in vivo* metabolic activation should be explored. Validation studies should also concentrate on compounds for which mammalian data exist so that sound comparisons of *Drosophila* and mammalian assays will become possible. In particular those assays that allow for the registration of recombination as well as mutation should be validated with high priority.

7.7.1. Evaluation of Combined Assays

A possible limitation of the eye system may be the correct identification of chemicals as mutagens that specifically act late during larval development, because only clones of small size will be produced in this case. In such a situation, the wing assay could complement the eye assay because spots comprising only one or two cells can be detected on the wing.

The data presented in Table 1 allow a first qualitative comparison of the assay systems; quantitative comparisons require additional data. With the information available, discrepancies are difficult to explain, as will be shown for the case of bleomycin. This compound is a potent inducer of twin spots in the white/white-coral assay, but hardly induces twin spots in the wing. In this same organ it is capable of inducing significant frequencies of single spots.[80] and produces somatic recombination in abdominal histoblasts and in mitotically dividing male germ cells.[42] Whether this discrepancy results from strain differences (uptake, distribution, metabolism, excretion, etc.) or is a result of a differential response of different developing tissues to the same genotoxic damage can only be decided after several systems have been studied together in the same treated larvae. A combined system in which the eye, the wing, and the chorion-producing cells can be tested simultaneously has been constructed by Szabad *et al.*[196] and was used to study the genotoxic activity of formaldehyde and malondialdehyde. Other combined systems with special advantages (easily scorable phenotypes, increased sensitivity, etc.) are in preparation (A. Kägi, unpublished results; E. W. Vogel, unpublished results).

Since one of the tester strains presently used for the white/white-coral test contains a singed mutation (sn^3; 1-21.0), the female larvae are also heterozygous for this bristle marker. This allows following clones that may extend from the eye into the cuticule around the eye where they could be recognized as a group of cells with singed bristles.

To account for possible cases of tissue-specific mutagen sensitivity, E. W. Vogel C. A. Raaymaker Jansen-Verplanke, and J. A. Zijlstra (unpublished results) constructed a set of strains which carry the markers for the white/white-coral test and the markers yellow and forked, enabling the simultaneous scoring of genetic changes affecting the eyes and/or the bristles on the cuticule.

Most recently J. Szabad (unpublished results; ref. 190) constructed sophisticated multipurpose stocks containing the wing system (*mwh/f*), the eye system (w/w^{co}) and the female germ-line dependent system (fs^1 *K10*). These stocks are used in a cross of fs^1 *K10 w/C1B*; *mwh se e/mwh se e* females to mwh^+ w^{co} *f/Y*; *mwh se/mwh se* males.

7.7.2. Metabolic Activation

In *Drosophila* a number of genetically determined strain specificities with respect to the mutagenicity of promutagens are known.[52,232] With the wing assay, such differences also seem to be important. Mutagenic activity of B(*a*)P* and of DMBA (A. Kägi, unpublished results) was detectable only with the *mwh–flr*^{3} cross, but not with the *mwh–flr* cross. Before any particular assay can be recommended for routine screening, the larvae from the particular cross should be checked for their ability to metabolize a set of promutagens with different activation pathways.

7.7.3. Effects in Nonproliferating Cells

All the assays discussed above study effects on rapidly proliferating imaginal disc cells. It remains to be determined whether the detection capacity can be improved if nonproliferating cells such as histoblasts are studied. Histoblasts, once they are developmentally determined, undergo one division during embryogenesis, but then do not divide until puparium formation. At this stage histoblast proliferation resumes and a series of rapid cell divisions takes place.(69,83,119,195) It has been calculated that only about eight division cycles are required to form the adult abdominal epidermis.(118,177)

Histoblasts are sensitive to X-ray-induced mitotic recombination, suggesting that they are in the G2 stage of the cell cycle.(66) The study of repair following fractionated irradiation suggested that the most important part of induced mitotic recombination occurs at the time of X-irradiation.(124) Thus, the histoblasts probably represent a homogeneous population of cells in interphase.

The histoblasts are also special in other respects. A pronounced increase in cell volume occurring during larval development probably allows the accumulation of reserves, enabling the initial, cleavagelike divisions to take place at the time the animal stops feeding. Cell competition, which is very effective in imaginal discs, is practically absent among histoblasts (no Minute effect).(132,165,183,184) Clones homozygous for many lethal mutations that are inviable in the thoracic discs form normal structures in the tergites developing from histoblasts.(132) It is also possible that histoblasts tolerate aneuploidy better than do cells of imaginal discs. It is possible that the XO clones induced in females survive in the tergites although their X-chromosomal genetic information is only transcribed at the low rate typical of XX female cells.(177)

Differences in radiosensitivity have been observed between disc cells and histoblasts. Spots of the mutant *mwh* appear following 1000-R X-irradiation to larvae 24–48 hr before puparium formation with a

* See footnote on p. 32.

frequency of 3 per wing and 1.4 per abdomen.[68–70] Postlethwait[151] analyzed the sensitivity of cells in discs and histoblasts with respect to somatic recombination in the left arm of chromosome 3 in different discs following irradiation with 1000-R X rays. He found that during the larval period the imaginal disc cells undergo 0.1–0.3 recombinations per 100 cells in the logarithmic phase of their growth, compared with 1.5 events in the mitotically inactive histoblasts. for details of data and calculations see Postlethwait.[151]

Differences between disc cells and histoblasts with respect to the mutagenic response to chemicals may also be expected. During the larval stages DNA repair processes may play an important role in mitotically inactive histoblasts exposed to mutagens. The involvement of DNA repair phenomena may be studied using mutants deficient in excision repair (e.g., *mei-9, mus-201*, or *mus-302*) or postreplication repair (*mei-41*). (Würgler *et al.*, this volume, Chapter 10).

7.8. Test Performance

For all three assays test performance has been well documented for strong reference mutagens. Weak mutagens and nonmutagens have been tested only to a limited extent, although some data are available from the IPCS studies.[215] More studies with this group of compounds are needed to provide a solid data base regarding the advantages and limitations of the various assays, in particular with regard to the reproducibility of weak effects. This information will be helpful in the reexamination of the criteria set up to define a nonmutagen in these somatic assays. Additional efforts are necessary to establish dose–response relationships, cell proliferation dynamics, regeneration phenomena, and molecular dosimetry techniques.

8. Conclusions

The studies conducted with somatic cells of *Drosophila melanogaster* clearly demonstrate that these assays are well suited for mutagenicity testing. The tests designed for this purpose are technically simple and need, besides a microscope, no expensive equipment or laboratory facilities. Most of the problems connected with the construction of appropriate tester strains, the exposure techniques (feeding and inhalation), the registration of the genetic effects, and the statistical analysis and interpretation of data have been solved. The identification of mutagenic agents using somatic assays is clearly faster than the *Drosophila* germ cell assays. In addition, the simple and fast techniques give the assays the flexibility one would like for routine screening assays. There is good evidence from the mutagenic activity of promutagens in the *Drosophila* recombination and mutation assays as well as from direct biochemical studies that *Drosophila* larvae contain a versatile xenobiotic metabolism reflecting the activities found in extrahepatic mammalian tissues. In the near future the continuation of the validation studies will allow comparison of the different somatic assays in *Drosophila* with one another and with the reaction of mammals and mammalian cells to the same validation compounds. It can be presumed that assays with somatic cells of *Drosophila* will soon find an important place within the battery of submammalian mutagenicity assay systems.

ACKNOWLEDGMENTS

We thank all our colleagues who allowed us to cite unpublished results in Table 1. We thank H. Frei for permission to reproduce the data shown in Figure 2 and M. A. Yoo, H. Ryo, T. Todo, and S. Kondo for the data shown in Figure 3. We thank the Environmental Mutagen, Carcinogen, and Teratogen Information Center, Oak Ridge National Laboratory, Oak Ridge, Tennessee, for supplying the carcinogenesis information in Table 1. For discussion and critical reading of the manuscript, we thank U. Graf, H. Frei, and K. Fujikawa.

The experimental work at the Institute of Toxicology was supported by grant 3.670-0.80 of the Swiss National Science Foundation and the Julius-Klaus-Stiftung für Genetik und Sozialanthropologie, Zurich, Switzerland.

9. References

1. C. S. Aaron, H. E. Nardin, and W. R. Lee, EMS effect on consumption behavior in *Drosophila, Drosophila Inform. Serv. 52,* 166 (1977).
2. C. S. Aaron, W. R. Lee, P. M. Seamster, and F. Janca, A non-aqueous gas exposure technique for microliter quantities of chemical mutagens, *Drosophila Inform. Serv. 52,* 174–175 (1977).
3. C. S. Aaron, A. A. van Zeeland, G. R. Mohn, A. T. Natarajan, A. G. A. C. Knaap, A. D. Tates, and B. W. Glickman, Molecular dosimetry of the chemical mutagen ethyl methanesulfonate. Quantitative comparison of the mutation induction in *Escherichia coli,* V79 Chinese hamster cells and L5178Y mouse lymphoma cells and some cytological results *in vivo* and *in vitro, Mutat. Res. 69,* 201–216 (1980).
4. P. Alba, M. D. Ferres, N. Xamena, A. Creus, and R. Marcos, Genotoxicity of acridine orange and acriflavine in *Drosophila melanogaster, Mutat. Res. 121,* 199–203 (1983).

5. T. Alderson, Induction of genetically recombinant chromosomes in the absence of induced mutation, *Nature 215*, 1281–1283 (1967).
6. M. Ashburner and T. R. F. Wright (eds.), *The Genetics and Biology of Drosophila*, Vol. 2a, Academic Press, New York (1978).
7. M. Ashburner and T. R. F. Wright (eds.), *The Genetics and Biology of Drosophila*, Vol. 2c, Academic Press, New York (1978).
8. C. Auerbach, The problem of chromosome rearrangements in somatic cells of *Drosophila melanogaster, Proc. R. Soc. Ed. B 62*, 120–127 (1945).
9. A. E. Aust, N. R. Drinkwater, V. M. Maher, and J. J. McCormick, The kinds of mutations induced by carcinogens in human cells differ from those in bacteria (abstract), *Environ. Mutagen. 5*, 458 (1983).
10. T. Ayaki, K. Ohshima, Y. Okumura, I. Yoshikawa, and T. Shiomi, The relationship between lethal mutation yield and intake of ethylnitrosourea (ENU) in *Drosophila melanogaster, Environ. Mutagen. 6*, 483–488 (1984).
11. A. J. Baars, Biotransformation of xenobiotics in *Drosophila melanogaster* and its relevance for mutagenicity testing, *Drug Metab. Rev. II*(2), 191–121 (1980).
12. B. S. Baker, J. B. Boyd, A. T. C. Carpenter, M. M. Green, T. D. Nguyen, and P. D. Smith, Genetic control of meiotic recombination and somatic DNA metabolism in *Drosophila melanogaster, Proc. Natl. Acad. Sci.* (*USA*) *73*, 4140–4144 (1976).
13. B. S. Baker, A. T. C. Carpenter, and P. Ripoll, The utilization during mitotic cell division of loci controlling meiotic recombination and disjunction in *Drosophila melanogaster, Genetics 90*, 531–578 (1978).
14. H. J. Becker, Ueber Röntgenmosaikflecken und Defektmutationen am Auge von *Drosophila* und die Entwicklungsphysiologie des Auges, *Z. Vererbungsl. 88*, 333–373 (1957).
15. H. J. Becker, Genetic and variegation mosaics in the eye of *Drosophila*, in: *Current Topics in Developmental Biology* (A. A. Moscona and A. Monroy, eds.), Vol. 1, pp. 155–171, Academic Press, New York (1966).
16. H. J. Becker, The influence of heterochromatin, inversion-heterozygosity and somatic pairing on X-ray induced mitotic recombination in *Drosophila melanogaster, Mol. Gen. Genet. 105*, 203–218 (1969).
17. H. J. Becker, X-ray- and TEM-induced mitotic recombination in *Drosophila melanogaster*: Unequal and sister-strand recombination, *Mol. Gen. Genet. 138*, 11–24 (1975).
18. H. J. Becker, Mitotic recombination, in: *Genetics and Biology of Drosophila* (M. Ashburner and E. Novitski, eds.), Vol. 1c, pp. 1020–1087, Academic Press, New York (1976).
19. C. Benyajati, A. R. Place, and W. Sofer, Formaldehyde mutagenesis in *Drosophila*. Molecular analysis of ADH-negative mutants, *Mutat. Res. 111*, 1–7 (1983).
20. S. Benzer, Genetic dissection of behavior, *Sci. Am. 299*, 24–37 (1973).
21. F. Bieri, P. Bently, F. Waechter, and W. Stäubli, Use of primary cultures of adult rat hepatocytes to investigate mechanisms of action of nafenopin, a hepatocarcinogenic peroxisome proliferator, *Carcinogenesis 5*, 1033–1039 (1984).
22. P. Binz, C. Büchi, I. Cordt, U. Frangi, H. Fritz-Niggli, K. Schäppi, and P. Schweizer, RBE of pions for the induction of non lethal, clonally propagated somatic mutations and recombinations in *Drosophila, Swiss Inst. Nuclear Res., SIN Villingen, Med. Newsl. 1983*(5), 46–47 (1983).
23. W. G. H. Blijleven, Metabolic activation of DEN and procarbazine in germ-free *Drosophila, Mutat. Res. 63*, 413–415 (1979).
24. W. G. H. Blijleven and E. Vogel, The mutational spectrum of procarbazine in *Drosophila melanogaster, Mutat. Res. 45*, 47–59 (1977).
25. Z. Bodi, K. Antal, and J. Szabad, Nonmutagenic activity of the nitroprusside, *Acta Biol. Acad. Sci. Hung. 32*, 61–68 (1981).

26. J. B. Boyd, P. V. Harris, J. M. Presley, and M. Narachi, *Drosophila melanogaster*: A model eukaryote for the study of DNA repair, in: *Cellular Response to DNA Damage* (E. C. Friedberg and B. A. Bridges, eds.), pp. 107–123, Liss, New York (1983).
27. R. Braun, E. Hütter, and J. Schöneich, Genetic endpoints in the mammalian spot test—Chromosome or gene mutations, in *Abstracts XIV Annual Meeting of the European Environmental Mutagen Society, 11–14 September 1984, Moscow* (P. N. Burgasov, ed.), p. 74 (1984).
28. S. W. Brown and W. Welshons, Maternal aging and crossing over of attached X chromosomes, *Proc. Natl. Acad. Sci.* (*USA*) *41*, 209–245 (1955).
29. D. Brusick and H. Andrews, Comparison of the genetic activity of dimethylnitrosamine, ethyl methanesulfonate, 2-acetylaminofluorene and ICR-170 in *Saccharomyces cerevisiae* strains D3, D4 and D5 using *in vitro* assays with and without metabolic activation, *Mutat. Res. 26*, 491–500 (1974).
30. P. J. Bryant, Cell lineage relationships in the imaginal wing disc of *Drosophila melanogaster, Dev. Biol. 22*, 389–411 (1970).
31. B. Burnet, D. Sewell, and M. Bos, Genetic analysis of larval feeding behaviour in *Drosophila melanogaster, Genet. Res.* (*Camb.*) *30*, 149–161 (1977).
32. J. Cairns, The origin of human cancers, *Nature 289*, 353–357 (1981).
33. U. Candrian, J. Lüthy, U. Graf, and Ch. Schlatter, Mutagenic activity of the pyrrolizidine alkaloids seneciphylline and senkirkine in *Drosophila* and their transfer into rat milk, *Food Chem. Toxicol. 22*, 223–225 (1984).
34. P. S. Carlson, A genetic analysis of the *rudimentary* locus of *Drosophila melanogaster, Genet. Res. 17*, 53–81 (1971).
35. J. E. Cassida, Insect microsomes and insecticide chemical oxidations, in: *Microsomes and Drug Oxidations* (J. R. Gillette, A. H. Conney, G. J. Cosmides, B. W. Eastbrook, J. R. Fouts, and G. J. Mannerling, eds.), pp. 517–531, Academic Press, New York (1969).
36. A. Chovnick, M. McCarron, A. Hilliker, J. O'Donnell, W. Gelbart, and S. Clark, Gene organization in *Drosophila, Cold Spring Harbor Symp. Quant. Biol. 42*, 1011–1021 (1977).
37. J. Clements, M. Phillips, and N. Todd, Mutagenicity of adriamycin in *Drosophila, Mutat. Res. 135,* 175–179 (1984).
38. B. Cotton-Menzl, F. E. Würgler, and U. Graf, MMS-sensitivity of *Drosophila* larvae, *Arch. Genet. (Zürich) 49/50*, 70–86 (1976, 1977).
39. A. Creus, N. Xamena, and R. Marcos, Positive response of diethylstilbestrol in the sex-linked recessive lethal assay in *Drosophila* after larval feeding, *Mutat. Res. 122*, 309–313 (1983).
40. C. C. J. Culvenor, Estimated intakes of pyrrolizidine alkaloids by humans. A comparison with dose rates causing tumors in rats, *J. Toxicol. Environ. Health 11*, 625–635 (1983).
41. A. R. Datta, B. W. Randolph, and J. L. Rosner, Detection of chemicals that stimulate *Tn9* transposition in *Escherichia coli* K12, *Mol. Gen. Genet. 189*, 245–250 (1983).
42. N. A. Demopoulos, N. D. Stamatis, and G. Yannopoulos, Induction of somatic and male crossing-over by bleomycin in *Drosophila melanogaster, Mutat. Res. 78*, 347–351 (1980).
43. F. J. de Serres, G. R. Hoffmann, J. von Borstel, R. Jagannath, N. Loprieno, J. Parry, D. Sharp, I. A. Zakharov, and F. Zimmermann, Summary report on the performance of yeast assays, in: *Evaluation of Short-Term Tests for Carcinogens, Report of the International Collaborative Program* (F. J. de Serres and J. Ashby, eds.), pp. 68–76, Elsevier/North-Holland, New York (1981).

44. C. N. Edwards and R. D. Combes, An evaluation of the *Drosophila* zeste somatic eye mutation test for the detection of genotoxic azo dyes and an aromatic amine, *Mutat. Res. 84,* 221–226 (1981).
45. L. Ehrenberg, E. Moustacchi, and S. Osterman-Golkar, with an Appendix (II) by G. Ekman, Dosimetry of genotoxic agents and dose–response relationships of their effects, *Mutat. Res. 123,* 121–182 (1983).
46. M. J. Fahmy and O. G. Fahmy, Altered control of gene activity in the soma by carcinogens, *Mutat. Res. 72,* 165–172 (1980).
47. M. J. Fahmy and O. G. Fahmy, Intervening DNA insertions and the alteration of gene expression by carcinogens, *Cancer Res. 40,* 3374–3382 (1980).
48. M. J. Fahmy and O. G. Fahmy, Genetic activities of 4-chloromethylbiphenyl, the 4-hydroxy derivative and benzyl chloride in the soma and germ line of *Drosophila melanogaster, Mutat. Res. 100,* 339–344 (1982).
49. M. J. Fahmy and O. G. Fahmy, Genetic activity of the carcinogen *N*-nitrosodiethanolamine (NDELA) on unstable mutant alleles of the *white* locus in *Drosophila melanogaster, Teratogen. Carcinogen. Mutagen. 4,* 437–447 (1984).
50. R. Fahrig, Genetic mode of action of cocarcinogens and tumor promotors in yeast and mice, *Mol. Gen. Genet. 194,* 7–14 (1984).
51. V. Finnerty, Gene conversion in *Drosophila,* in: *The Genetics and Biology of Drosophila* (M. Ashburner and E. Novitski, eds.), Vol. 1a, pp. 331–349, Academic Press, New York (1976).
52. C. Forbes, Sex-linked lethal mutations induced in *Drosophila melanogaster* by 7,12-dimethylbenz(*a*)anthracene, *Mutat. Res. 79,* 231–237 (1980).
53. R. E. Förster, *In vitro* Untersuchungen über den Metabolismus von Aflatoxin B1 und Aldrin in Hodengewebe von genetisch verschiedenen Stämmen von *Drosophila melanogaster,* Dissertation ETH-Zürich no. 7081, Zurich (1982).
54. R. E. Förster and F. E. Würgler, *In vitro* studies on the metabolism of aflatoxin B1 and aldrin in testes of genetically different strains of *Drosophila melanogaster, Arch. Toxicol. 56,* 12–17 (1984).
55. H. Frei, J. Brauchli, J. Lüthy, and U. Zweifel, Genotoxicity of *Symphytum* pyrrolizidine alkaloids in somatic cells of *Drosophila* (abstract), *Experientia 40,* 645 (1984).
56. H. Frei, F. E. Würgler, H. Juon, C. Hall, and U. Graf, Aristolochic acid is mutagenic and recombinogenic in *Drosophila* genotoxicity tests, *Arch. Toxicol. 56,* 158–166 (1985).
57. D. K. Fristrom and W. L. Rickoll. The morphogenesis of imaginal discs of *Drosophila,* in: *Insect Ultrastructure* (R. C. King and H. Akai, eds.), Vol. 1, pp. 247–277, Plenum Press, New York (1982).
58. K. Fujikawa, Induction of somatic eye-color mutations in *Drosophila melanogaster* by trytptophan pyrolysis products and furylfuramide, in: *Abstracts, Third International Conference on Environmental Mutagens,* p. 45 (1981).
59. K. Fujikawa, Mutagenicity of furylfuramide in somatic and germ cells of *Drosophila melanogaster, Jpn. J. Genet. 57,* 569–574 (1982).
60. K. Fujikawa and S. Kondo, Somatic mutagenesis by ethyl methanesulfonate in repair-deficient strains of *Drosophila melanogaster,* in: *Abstracts, Third International Conference on Environmental Mutagenesis,* p. 31 (1981).
61. K. Fujikawa and S, Kondo, in preparation (1984).
62. K. Fujikawa, H. Ryo, and S. Kondo, Somatic mutagenesis induced in *Drosophila melanogaster* in relation to DNA repair, *Radiat. Biol. Res. Commun. 17,* 64 (1982).
63. K. Fujikawa, E. Inagaki, M. Uchibori, and S. Kondo, Comparative induction of somatic eye-color mutations and sex-linked recessive lethals in *Drosophila melanogaster* by tryptophan pyrolysates, *Mutat. Res. 122,* 315–320 (1983).

64. K. Fujikawa, H. Ryo, and S. Kondo, The *Drosophila* reversion assay using the unstable white–zeste somatic eye color system, in: *Progress in Mutation Research, Collaborative Study of Short-Term Tests for Carcinogens* (J. Ashby, F. J. de Serres, M. Draper, M. Ishidate Jr., B. Margolin, B. Matter, and M. D. Shelby, eds.) Vol. 5, pp. 319–324, Elsevier/North-Holland, Amsterdam (1985).
65. K. Fujikawa, H. Ryo, and S. Kondo, in preparation.
66. A. Garcia-Bellido, Some parameters of mitotic recombination in *Drosophila melanogaster, Mol. Gen. Genet. 115*, 54–72 (1972).
67. A. Garcia-Bellido and J. Dapena, Induction, detection and characterization of cell differentiation mutants in *Drosophila, Mol. Gen. Genet. 128*, 117–130 (1974).
68. A. Garcia-Bellido and J. R. Merriam, Parameters of the wing imaginal disc development of *Drosophila melanogaster, Dev. Biol. 24*, 61–87 (1971).
69. A. Garcia-Bellido and J. R. Merriam, Clonal parameters of tergite development in *Drosophila, Dev. Biol. 26*, 265–276 (1971).
70. A. Garcia-Bellido and J. R. Merriam, Genetic analyses of cell heredity in imaginal discs of *Drosophila melanogaster, Proc. Natl. Acad. Sci. USA 68*, 2222–2226 (1971).
71. M. Gatti, S. Pimpinelli, A. de Marco, and C. Tanzarella, Chemical induction of chromosome aberrations in somatic cells of *Drosophila melanogaster, Mutat. Res. 33*, 201–212 (1975).
72. J. R. Girton and P. J. Bryant, The use of cell lethal mutations in the study of *Drosophila* development, *Dev. Biol. 77*, 233–243 (1980).
73. E. Gocke, D. Wild, K. Eckhardt, and M.-T. King, Mutagenicity studies with the mouse spot test, *Mutat. Res. 117*, 201–212 (1983).
74. B. B. Gollapudi, R. J. Bruce, and A. K. Sinha, The role of feeding rejection in *Drosophila* mutation assays (abstract), *Environ. Mutagen. 6*, 439 (1984).
75. U. Graf and F. E. Würgler, DNA repair dependent mutagenesis in *Drosophila melanogaster*, in: *Advances in Genetics, Development, and Evolution of Drosophila* (S. Lakovaara, ed.), pp. 85–99, Plenum Press, New York (1982).
76. U. Graf and F. E. Würgler, Excision repair defect sensitizes a novel *Drosophila* mutagenicity test (abstract), *Experientia 39*, 685 (1983).
77. U. Graf, E. Vogel, U. P. Biber, and F. E. Würgler, Genetic control of mutagen sensitivity in *Drosophila*. A new allele at the *mei-9* locus on the X-chromosome, *Mutat. Res. 59*, 129–133 (1979).
78. U. Graf, H. Juon, A. J. Katz, H. J. Frei, and F. E. Würgler, A pilot study on a new *Drosophila* spot test, *Mutat. Res. 120*, 233–239 (1983).
79. U. Graf, C. Hall, and F. E. Würgler, Mutagenicity testing with excision repair-deficient somatic cells in *Drosophila* (abstract), *Mutat. Res. 130,* 198 (1984).
80. U. Graf, F. E. Würgler, A. J. Katz, H. Frei, H. Juon, C. B. Hall, and P. G. Kale, Somatic mutation and recombination test in *Drosophila melanogaster, Environ. Mutagen. 6*, 153–188 (1984).
81. M. M. Green, Back mutation in *Drosophila melanogaster*. II. Data on additional *yellow* and *white* mutants, *Genetics 47,* 483–488 (1962).
82. H. Grüneberg, The case for somatic crossing over in the mouse, *Genet. Res.* (*Camb.*) *7*, 58–75 (1966).
83. M. Guerra, J. H. Postlethwait, and H. A. Schneiderman, The development of the imaginal abdomen of *Drosophila melanogaster, Dev. Biol. 32*, 361–372 (1973).
84. J. Haendle, Röntgeninduzierte mitotische Rekombination bei *Drosophila melanogaster, Mol. Gen. Genet. 113*, 114–131 (1971).
85. I. Hällström and R. Grafström, The metabolism of drugs and carcinogens in isolated subcellular fractions of *Drosophila melanogaster*. II. Enzyme induction and metabolism of benzo(*a*)pyrene, *Chem. Biol. Interact. 34,* 145–159 (1981).

86. I. Hällström, A. Sundvall, U. Rannug, R. Grafström, and C. Ramel, The metabolism of drugs and carcinogens in isolated subcellular fractions of *Drosophila melanogaster*. I. Activation of vinyl chloride, 2-aminoanthracene and benzo(*a*)pyrene as measured by mutagenic effects in *Salmonella typhimurium, Chem. Biol. Interact. 34*, 129–143 (1981).
87. Y. Hamnerius, H. Olofsson, A. Rasmuson, and B. Rasmuson, A negative test for mutagenic action of microwave radiation in *Drosophila melanogaster, Mutat. Res. 68*, 217–223 (1979).
88. J. L. Haynie and P. J. Bryant, Intercalary regeneration in imaginal wing disc of *Drosophila melanogaster, Nature 259*, 659–662 (1976).
89. J. L. Haynie and P. J. Bryant, The effects of X-rays on the proliferation dynamics in the imaginal wing disc of *Drosophila melanogaster, Wilhelm Roux' Arch. Entwicklungsmech. Org. 183*, 85–100 (1977).
90. W. M. Hexter, Nonreciprocal events at the *garnet* locus in *Drosophila melanogaster, Proc. Natl. Acad. Sci. USA 50*, 372–379 (1963).
91. C. N. Holzer, Untersuchungen über Metabolismus und Mutagenität von Nitrosaminen und Aflatoxin B1 in *Drosophila melanogaster*, Dissertation ETH-Zürich, no. 6893, Zurich (1981).
92. S. Hosaka, T. Matsushima, I. Hirono, and T. Sugimura, Carcinogenic activity of 3-amino-1-methyl-5*H*-pyrido-(4,3-b)indole (Trp-P-2), a pyrolysis product of tryptophan, *Cancer Lett. 13*, 23–28 (1981).
93. K. M. Huttner and F. H. Ruddle, Study of mitomycin C-induced chromosomal exchange, *Chromosoma 56*, 1–13 (1976).
94. IARC, *Monographs on the Evaluation of the Carcinogenic Risk of Chemicals to Humans* (IARC Monographs Supplement 1), International Agency for Research on Cancer, Lyon (1979).
95. T. Ishikawa, S. Takayama, T. Kitagawa, T. Kawachi, M. Kinebuchi, N. Matsukura, E. Uchida, and T. Sugimura, *In vivo* experiments on tryptophan pyrolysis products, in: *Naturally Occurring Carcinogens–Mutagens and Modulators of Carcinogenesis*, (E. C. Miller, J. C. Miller, I. Hirono, T. Sugimura, and S. Takayama, eds.), pp. 159–167, University Park Press, Baltimore (1979).
96. J. W. Jack and B. H. Judd, Allelic pairing and gene regulation: A model for the *zeste–white* interaction in *Drosophila melanogaster, Proc. Natl. Acad. Sci. USA 76*, 1368–1372 (1979).
97. E. Käfer, B. R. Scott, G. L. Dorn, and R. Stafford, *Aspergillus nidulans*: Systems and results of tests for chemical induction of mitotic segregation and mutation. I. Diploid and duplication assay systems. A Report of the U. S. EPA Gene-Tox Program, *Mutat. Res. 98*, 1–48 (1982).
98. A. Kahn and K. Sick, Gene conversion in the *shibire*(*ts*) locus of *Drosophila melanogaster, Hereditas 97*, 59–63 (1982).
99. P. G. Kale and J. W. Baum, Sensitivity of *Drosophila melanogaster* to low concentrations of gaseous mutagens. II. Chronic exposures, *Mutat. Res. 68*, 59–68 (1979).
100. P. G. Kale and J. W. Baum, Genetic effects of 1,2-dibromo-3-chloropropane (DBCP) in *Drosophila, Environ. Mutagen. 4*, 681–687 (1982).
101. M. A. Kastenbaum and K. O. Bowman, Tables for determining the statistical significance of mutation frequencies, *Mutat. Res. 9*, 527–549 (1970).
102. A. J. Katz, Genotoxicity of isopropyl methanesulfonate (iPMS) in somatic cells of *Drosophila* (abstract), *Genetics 107*, s55 (1984).
103. A. J. Katz and F. E. Würgler, Distinguishing the mutagenic and recombinogenic effects of mutagens in somatic cells of *Drosophila* larvae (abstract), *Mutat. Res. 130*, 184 (1984).

104. A. Kinsella and U. Radman, Tumor promotor induces sister chromatid exchanges: Relevance to mechanisms of carcinogenesis, *Proc. Natl. Acad. Sci. USA 75,* 6149–6153 (1978).
105. M. J. H. Kortselius, Induction of sex-linked recessive lethals and autosomal translocations by beta-propiolactone in *Drosophila*: Influence of the route of administration on mutagenic activity, *Mutat. Res. 66,* 55–63 (1979).
106. P. G. N. Kramers, Studies on the induction of sex-linked recessive lethal mutations in *Drosophila melanogaster* by nitroheterocyclic compounds, *Mutat. Res. 101,* 209–236 (1982).
107. P. G. N. Kramers and A. G. L. Burm, Mutagenicity studies with halothane in *Drosophila melanogaster, Anesthesiology 50,* 510–513 (1979).
108. P. G. N. Kramers and A. G. A. C. Knaap, Absence of a mutagenic effect after feeding dichlorvos to larvae of *Drosophila melanogaster, Mutat. Res. 57,* 103–105 (1978).
109. P. G. N. Kramers, A. P. Schalet, E. Paradi, and L. Huiser-Hoogteyling, High proportion of multi-locus deletions among hycanthone-induced X-linked recessive lethals in *Drosophila melanogaster, Mutat. Res. 107,* 187–201 (1983).
110. B. A. Kunz, Genetic effects of deoxyribonucleotide pool imbalances, *Environ. Mutagen. 4,* 695–725 (1982).
111. W. R. Lee, Chemical mutagenesis, in: *The Genetics and Biology of Drosophila* (M. Ashburner and E. Novitski, eds.), Vol. 1c, pp. 1299–1341, Academic Press, New York (1976).
112. W. R. Lee, S. Abrahamson, R. Valencia, E. S. von Halle, F. E. Würgler, and S. Zimmering, The sex-linked recessive lethal test for mutagenesis in *Drosophila melanogaster,* A report of the U. S. Environmental Protection Agency Gene-Tox Program, *Mutat. Res. 123,* 183–279 (1983).
113. R. Levis, M. Collins, and G. M. Rubin, FB elements are the common basis for the instability of the *w(DZL)* and *w(c)* Drosophila mutations, *Cell 30,* 551–565 (1982).
114. E. Lifschytz and R. Falk, Fine structure analysis of a chromosome segment in *Drosophila melanogaster*: Analysis of ethyl methanesulfonate-induced lethals, *Mutat. Res. 8,* 147–155 (1969).
115. D. L. Lindsley and E. H. Grell, *Genetic Variations of Drosophila melanogaster,* Carnegie Institution of Washington (1968).
116. J. Lopez de Sepulveda, R. Marcos, N. Xamena, and A. Creus, Mutagenicity of ethidium bromide in the sex-linked recessive lethal assay in *Drosophila melanogaster, Mutat. Res. 91,* 337–340 (1981).
117. D. MacDonald and K. Telford, *Drosophila* sex-linked recessive lethal tests with 4-chloromethylbiphenyl and benzyl chloride, *Mutat. Res. 100,* 335–338 (1982).
118. M. M. Madhavan and K. Madhavan, Morphogenesis of the epidermis of adult abdomen of *Drosophila, J. Embryol. Exp. Morphol. 60,* 1–31 (1980).
119. M. M. Madhavan and H. A. Schneiderman, Histological analysis of the dynamics of growth of imaginal discs and histoblast nests during the larval development of *Drosophila melanogaster, Wilhelm Roux' Arch. Entwicklungsmech. Org. 183,* 269–302 (1977).
120. J. Magnusson, I. Hällström, and C. Ramel, Studies on metabolic activation of vinyl chloride in *Drosophila melanogaster* after pretreatment with phenobarbital and polychlorinated biphenyls, *Chem. Biol. Interact. 24,* 287–298 (1979).
121. J. Mason, E. Strobel, and M. M. Green, *mu-2*: A mutator in *Drosophila* that potentiates the induction of terminal deficiencies, *Proc. Natl. Acad. Sci. USA 81,* 6090–6094 (1984).
122. U. Mengs, On the histopathogenesis of rat forestomach carcinoma caused by aristolochic acid, *Arch. Toxicol. 52,* 209–220 (1983).

123. U. Mengs, W. Lang, and J.-A. Poch, The carcinogenic action of aristolochic acid in rats, *Arch. Toxicol. 51*, 107–119 (1982).
124. J. R. Merriam and W. E. Fyffe, The kinetics of X-ray induced somatic crossing-over in *Drosophila melanogaster* and the effects of dose fractionation, *Mutat. Res. 14*, 309–314 (1972).
125. J. R. Merriam and A. Garcia-Bellido, A model for somatic pairing derived from somatic crossing over with the third chromosome rearrangements in *Drosophila melanogaster*, *Mol. Gen. Genet. 115*, 302–313 (1972).
126. J. R. Merriam, R. Nöthiger, and A. Garcia-Bellido, Are dicentric anaphase bridges formed by somatic recombination in X chromosome inversion heterozygotes of *Drosophila melanogaster*?, *Mol. Gen. Genet. 86*, 294–301 (1972).
127. I. de G. Mitchell, J. P. Gilbert, A. J. Brice, and D. J. White, Somatic eye mutation in *Drosophila melanogaster* as a short-term test for mutagens and carcinogens, *Carcinogeneis 2*, 783–786 (1981).
128. J. Modolell, W. Bender and M. Meselson, *Drosophila melanogaster* mutations suppressible by the suppressor of *Hairy-wing* are insertions of a 7.3 kilobase mobile element, *Proc. Natl. Acad. Sci. USA 80*, 1678–1682 (1983).
129. P. Mollet and J. Szabad, Induction and detection of mosaicism by MMS in the eye disc and female germ line of *Drosophila*, *Mutat. Res. 51*, 293–296 (1978).
130. P. Mollet and W. Weilenmann, Characteristics of a new mutagenicity test—Induction of somatic recombination and mutation in *Drosophila* by different chemicals (abstract), *Mutat. Res. 38*, 131–132 (1976).
131. P. Mollet and F. E. Würgler, Detection of somatic recombination and mutation in *Drosophila*. A method for testing genetic activity of chemical compounds, *Mutat. Res. 25*, 421–424 (1974).
132. G. Morata and P. Ripoll, *Minutes*: Mutants of *Drosophila* autonomously affecting cell division rate, *Dev. Biol. 42*, 211–221 (1975).
133. M. Nagao, Mutagenic substances produced by pyrolysis, in: *Seitai to Yugai-Hatugan Bussitu* (H. Hayatsu, ed.) (in Japanese), pp. 128–151, Kodansha Scientific, Tokyo (1982).
134. S. Nesnow, M. Argus, H. Bergman, K. Chu, C. Frith, T. Helmes, R. McGaughey, V. Ray, T. J. Slaga, R. Tennant, and E. Weisburger, Chemical carcinogens: A review and analysis of the literature of selected chemicals and the establishment of the Gene-Tox Carcinogen data base, in preparation.
135. J. Neugebauer, *Arzneimittel-Kompendium der Schweiz 1981*, Verlag Documed AG, Basel (1982).
136. R. Nöthiger, Sucrose density separation: A method for collecting large numbers of *Drosophila* larvae, *Drosophila Inform. Serv.* 45, 177 (1970).
137. F. K. Zimmermann, R. C. von Borstel, E. S. von Halle, J. M. Perry, D. Siebert, G. Zetterberg, R. Barale, and N. Loprieno, Testing of chemicals for genetic activity with *Saccharomyces cerevisiae*: A report of the U. S. Environmental Protection Agency Gene-Tox Program, *Mutat. Res. 133*, 199–244 (1984).
138. P.-O. Nylander, H. Olofsson, and B. Rasmuson, The use of *Drosophila melanogaster* as a test system for indirect mutagens, *Mutat. Res. 64*, 122–123 (1979).
139. P.-O. Nylander, H. Olofsson, B. Rasmuson, and S. Svahlin, The use of *Drosophila melanogaster*. I. Effects of benzene and 1,2-dichloroethane, *Mutat. Res. 57*, 163–167 (1978).
140. G. Obe and Hj. Ristow, Mutagenic, cancerogenic and teratogenic effects of alcohol, *Mutat. Res. 65*, 229–259 (1979).
141. A. Oenfelt and C. Ramel, Some aspects on the organization of microfilaments and microtubules in relation to nondisjunction, *Environ. Health Perspect. 31*, 45–52 (1979).

142. H. Ohgaki, N. Matsukura, K. Morino, T. Kawachi, T. Sugimura, and S. Takayama, Carcinogenicity in mice of mutagenic compounds from glutamic acid and soya bean globulin pyrolysates, *Carcinogenesis 5*, 815–819 (1984).
143. H. Ohgaki, K. Kusama, N. Matsukura, K. Morino, H. Hasegawa, S. Sato, S. Takayama, and T. Sugimura, Carcinogenicity in mice of a mutagenic compound, 2-amino-3-methylimidazo(4,5-f)quinoline, from broiled sardine, cooked beef and beef extract, *Carcinogenesis 5*, 921–924 (1984).
144. S. Ohnishi, Relationship between larval feeding behavior and viability in *Drosophila melanogaster* and *D. simulans, Behav. Genet. 9*, 129–134 (1979).
145. S. Ohno, Aneuploidy as a possible means employed by malignant cells to express recessive phenotypes, in: *Chromosomes and Cancer* (J. German, ed.), pp. 77–94, Wiley, New York (1974).
146. O. A. Olsen and M. M. Green, The mutagenic effects of diepoxybutane in wild-type and mutagen-sensitive mutants of *Drosophila melanogaster, Mutat. Res. 92*, 107–115 (1982).
147. E. Paradi, E. W. Vogel, and E. Szilagyi, Effect of storage and dose on MMS-induced deletions. Complementation analysis of X-chromosomal recessive lethals in the *zeste-white* and *maroon-like* regions of *Drosophila melanogaster, Mutat. Res. 111*, 145–159 (1983).
148. J. T. Patterson, The production of mutations in somatic cells of *Drosophila melanogaster* by means of X-rays, *J. Exp. Zool. 53*, 327–372 (1929).
149. B. Peyer and E. Hadorn, Zum Manifestationsmuster der Mutante "multiple wing hairs" (*mwh*) von *Drosophila melanogaster, Arch. Julius Klaus-Stift. Vererbungsforsch. 40*, 19–26 (1965).
150. I. Porsch-Hällström, Cytochrome P-450 in *Drosophila melanogaster*, Activity, Genetic Variation and Regulation, Thesis, University of Stockholm, Graphic Systems, Göteborg (1984).
151. J. H. Postlethwait, Clonal analysis of *Drosophila* cuticular patterns, in: *The Genetics and Biology of Drosophila* (M. Ashburner and T. R. F. Wright, eds.), Vol. 2c, pp. 359–441, Academic Press, New York (1978).
152. M. Radman and A. R. Kinsella, Chromosomal events in carcinogenic initiation and promotion: Implications for carcinogenicity testing and cancer prevention strategies, in: *Molecular and Cellular Aspects of Carcinogen Screening Tests* (R. Montesano, H. Bartsch, and L. Tomatis, eds.), pp. 75–90, International Agency for Research on Cancer, Lyon (1980).
153. C. Ramel and J. Magnusson, Chemical induction of nondisjunction in *Drosophila, Environ. Health Perspect. 31*, 59–66 (1979).
154. R. Ransom (ed.), *A Handbook of Drosophila Development*, Elsevier Biomedical Press, Amsterdam (1982).
155. R. Ransom, Techniques, in: *A Handbook of Drosophila Development* (R. Ransom, ed.), pp. 1–29, Elsevier Biomedical Press, Amsterdam (1982).
156. A. Rasmuson, Effects of DNA-repair-deficient mutants on somatic and germ line mutagenesis in the UZ system in *Drosophila melanogaster, Mutat. Res. 141*, 29–33 (1984).
157. A. Rasmuson, Comparative studies of the induction of somatic eye-color mutations in an unstable strain of *Drosophila melanogaster* by MMS and X-rays at different developmental stages, *Mutat. Res. 148*, 65–70 (1985).
158. B. Rasmuson and M. M. Green, Genetic instability in *Drosophila melanogaster*: A mutable tandem duplication, *Mol. Gen. Genet. 133*, 249–260 (1974).
159. B. Rasmuson and H. Svahlin, Mutagenicity tests of 2,4-dichlorophenoxyacetic acid

and 2,4,5-trichlorophenoxyacetic acid in genetically stable and unstable strains of *Drosophila melanogaster, Ecol. Bull. (Stockholm) 27,* 190–192 (1978).
160. B. Rasmuson, H. Svahlin, A. Rasmuson, I. Montell, and H. Olofsson, The use of a mutationally unstable X-chromosome in *Drosophila melanogaster* for mutagenicity testing, *Mutat. Res. 54,* 33–38 (1978).
161. B. Rasmuson, B. M. Westerberg, A. Rasmuson, B. A. Gvozdev, E. S. Belyava, and Y. V. Ilyin, Transpositions, mutable genes, and the dispersed gene family Dm255 in *Drosophila melanogaster, Cold Spring Harbor Symp Quant. Biol. 45,* 545–551 (1981).
162. B. Rasmuson, A. Rasmuson, and J. Nygren, Eye pigmentation changes in *Drosophila melanogaster* a sensitive test for mutagenicity, in: *Handbook of Mutagenicity Test Procedures* (B. Kilbey, M. Legator, W. Nichols, and C. Ramel, eds.), pp. 603–613, Elsevier Science Publishers, Amsterdam (1984).
163. J. K. Reddy, D. L. Azarnoff, and C. E. Hignite, Hypolipidaemic hepatic peroxisome proliferators form a novel class of chemical carcinogens, *Nature 283,* 397–398 (1980).
164. J. K. Reddy, N. D. Lalwani, M. K. Reddy, and S. A. Qureshi, Excessive accumulation of autofluorescent lipofuscin in the liver during hepatocarcinogenesis by methyl clofenapate and other hypolipidemic peroxysome proliferators, *Cancer Res. 42,* 259–266 (1982).
165. P. Ripoll, Behavior of somatic cells homozygous for zygotic lethals in *Drosophila melanogaster, Genetics 86,* 357–376 (1977).
166. P. Ripoll, Effect of terminal aneuploidy on epidermal cell viability in *Drosophila melanogaster, Genetics 94,* 135–152 (1980).
167. P. Ripoll and A. Garcia-Bellido, Mitotic recombination in the heterochromatin of the sex chromosomes of *Drosophila melanogaster, Genetics 90,* 93–104 (1978).
168. B. Robert de Saint Vincent and G. M. Wahl, Homologous recombination in mammalian cells mediates formation of a functional gene from two overlapping gene fragments, *Proc. Natl. Acad. Sci. USA 80,* 2002–2006 (1983).
169. P. A. Roberts, Are stable terminal deletions recoverable in *Drosophila*? (abstract), *Genetics 74,* s231 (1973).
170. P. A. Roberts, A cytogenetic analysis of X-ray induced "visible" mutations at the *yellow* locus of *Drosophila melanogaster, Mutat. Res. 22,* 139–144 (1974).
171. C. Roseland and H. A. Schneiderman, Regulation and metamorphosis of the abdominal histoblasts of *Drosophila melanogaster, Wilhelm Roux' Arch. Entwicklungsmech. Org. 186,* 235–265 (1979).
172. L. B. Russell, P. B. Selby, E. von Halle, W. Sheridan, and L. Valcovic, Use of the mouse spot test in chemical mutagenesis: Interpretation of past data and recommendations for future work, *Mutat Res. 86,* 355–379 (1981).
173. H. Ryo, S. Kondo, and B. Rasmuson, Enhanced susceptibility of a transposable-element-bearing strain of *Drosophila melanogaster* to somatic eye-color mutations by ethyl nitrosourea, methyl nitrosourea, and X-rays, *Mutat. Res. 122,* 123–128 (1983).
174. L. Sanchez and R. Nöthiger, Sex determination and dosage compensation in *Drosophila melanogaster*: Production of male clones in XX females, *EMBO J. 2,* 485–491 (1983).
175. K. Sankaranarayanan and F. H. Sobels, Radiation genetics, in: *The Genetics and Biology of Drosophila* (M. Ashburner and E. Novitski, eds.), Vol. 1c, pp. 1090–1250, Academic Press, New York (1976).
176. N. van Schaik, A. Grant, I. Rubenchik, and U. Graf, Use of *Drosophila* test systems for genotoxicity testing of herbal teas, *Immunobiol. Hematol. Res. 3,* 207–210 (1984).
177. A. Schalet and G. Lefevre Jr., The location of "ordinary" sex-linked genes in section 20 of the polytene X chromosome of *Drosophila melanogaster, Chromosoma 44,* 183–202 (1973).

178. P. Schweizer, Wirkung von Röntgenstrahlen auf die Entwicklung der männlichen Genitalprimordien von *Drosophila melanogaster* und Untersuchung von Erholungsvorgängen durch Zellklon-Analyse, *Biophysik 8,* 153–188 (1972).
179. G. A. Sega and W. R. Lee, A vacuum injection technique for obtaining uniform dosages in *Drosophila melanogaster, Drosophila Inform. Serv. 45,* 179 (1970).
180. P. B. Selby and W. H. Olson, Methods and criteria for deciding whether specific-locus mutation-rate data in mice indicate a positive, negative, or inconclusive result, *Mutat. Res. 83,* 403–418 (1981).
181. P. T. Shukla and C. Auerbach, Genetic tests for the detection of chemically induced small deletions in *Drosophila* chromosomes, *Mutat. Res. 72,* 231–243 (1980).
182. V. F. Simmon, *In vitro* assay for recombinogenic activity of chemical carcinogens and related compounds with *Saccharomyces cerevisiae* D3, *J. Natl. Cancer Inst. 62.* 901–909 (1979).
183. P. Simpson, Parameters of cell competition in the compartments of the wing disc of *Drosophila, Dev. Biol. 69,* 182–193 (1979).
184. P. Simpson, Growth and cell competition in *Drosophila, J. Embryol. Exp. Morphol. 65*(Suppl.), 77–88 (1981).
185. J. Small and G. Scangos, Recombination during gene transfer into mouse cells can restore the function of deleted genes, *Science 219,* 174–176 (1983).
186. J. A. Snyder and J. R. McIntosh, Biochemistry and physiology of microtubules, *Annu. Rev. Biochem. 45,* 699–720 (1976).
187. T. E. Spreij, Cell death during development of the imaginal discs of *Calliphora erythrocephala, Neth. J. Zool. 21,* 221–264 (1971).
188. T. Sugimura, Mutagens, carcinogens, and tumor promotors in our daily food, *Cancer 49,* 1970–1984 (1982).
189. A. Surjan, Zs. Kocsis, M. Börzsönyi, M. Csik, A. Pinter, and G. Török, Genotoxic activity of some N-nitroso compounds in a *Drosophila* mosaic test (abstract), in: *Abstracts XIV Annual Meeting of the European Environmental Mutagen Society, 11–14 September 1984, Moscow* (P. N. Burgasov, ed.) p. 510 (1984).
190. A. Surjan, Z. Kocsis, M. Csik, A. Pinter, G. Török, M. Görzsönyi, and J. Szabad, Analysis of the genotoxic activity of four *N*-nitroso-compounds by the *Drosophila* mosaic test, *Mutat. Res. 144,* 177–181 (1985).
191. D. T. Suzuki, L. K. Piternick, S. Hayashi, M. Tarasoff, D. Baillie, and U. Erasmus, Temperature-sensitive mutations in *Drosophila melanogaster*. I. Relative frequency among X-ray and chemically-induced sex-linked recessive lethals and semi-lethals, *Proc. Natl. Acad. Sci. USA 57,* 907–912 (1967).
192. D. T. Suzuki, T. C. Kaufmann, D. Falk, and the UBC *Drosophila* Research Group, Conditionally expressed mutations in *Drosophila melanogaster*, in: *The Genetics and Biology of Drosophila* (M. Ashburner and E. Novitski, eds.), Vol. 1a, pp. 208–264, Academic Press, New York (1976).
193. J. Szabad and B. Bennettova, Analysis of the genotoxic activity of five compounds affecting insect fertility, *Mutat. Res.* (submitted) (1985).
194. J. Szabad and P. J. Bryant, The mode of action of "discless" mutations in *Drosophila melanaogaster, Dev. Biol. 93*, 240–256 (1982).
195. J. Szabad, T. Schüpbach, and E. Wieschaus, Cell lineage and development in the larval epidermis of *Drosophila melanogaster, Dev. Biol. 73,* 256–271 (1979).
196. J. Szabad, I. Soos, G. Polgar, and G. Hejja, Testing the mutagenicity of malondialdehyde and formaldehyde by the *Drosophila* mosaic and the sex-linked recessive lethal tests, *Mutat. Res. 113,* 117–133 (1983).
197. E. Therman and E. M. Kuhn, Cytological demonstration of mitotic crossing-over in man, *Cytogenet Cell Genet. 17*, 245–267 (1976).

198. J. N. Thompson, M. Ashburner, and R. C. Woodruff, Presumptive control mutation for alcohol dehydrogenase in *Drosophila melanogaster, Nature 270*, 363 (1977).

199. N. Todd, J. Clements, P. Zoeller, and M. Phillips, Absence of a mutagenic effect after feeding 4 anti-cancer drugs to *Drosophila melanogaster, Mutat. Res. 120,* 12–125 (1983).

200. E. Ulrich, Cell lineage, Determination und Regulation in der weiblichen Genitalscheibe von *Drosophila melanogaster, Wilhelm Roux' Arch. Entwicklungsmech. Org. 167,* 64–82 (1971).

201. H. Ursprung, *In vivo* culture of *Drosophila* imaginal discs, in: *Methods in Developmental Biology* (F. H. Wilt and M. K. Wessells, eds.), pp. 485–492, Crowell, New York (1967).

202. H. Ursprung, The fine structure of imaginal discs, in: *Results and Problems in Cell Differentiation* (H. Ursprung and R. Nöthiger, eds.), Vol. 5, pp. 93–107, Springer, Berlin (1972).

203. R. Valencia, S. Abrahamson, W. R. Lee, E. S. Von Halle, R. C. Woodruff, F. E. Würgler, and S. Zimmering, Chromosome mutation tests for mutagenesis in *Drosophila melanogaster.* A report of the U. S. Environmental Protection Agency Gene-Tox Program, *Mutat. Res. 134,* 61–88 (1984).

204. M. Vasudeva and R. K. Vashishat, Mutagenic and recombinogenic activity of hydrazine sulphate in *Saccharomyces cerevisiae, Mutat. Res. 155,* 113–115 (1985).

205. F. G. Verburgt and E. Vogel, Vinyl chloride mutagenesis in *Drosophila melanogaster, Mutat. Res. 48,* 327–336 (1977).

206. B. K. Vig, Genetic toxicology of mitomycin C, adriamycins, daunomycin, and adriamycin, *Mutat. Res. 49,* 189–238 (1977).

207. E. Vogel, Mutagenicity of carcinogens in *Drosophila* as a function of genotype-controlled metabolism, in: *In Vitro Metabolic Activation in Mutagenicity Testing* (F. J. de Serres, J. R. Fouts, J. R. Bend, and R. M. Philpot, eds.), pp. 63–79, Elsevier/North-Holland Biomedical Press, Amsterdam (1976).

208. E. W. Vogel, A comparison of genotoxic activity in somatic tissue and in germ cells of *Drosophila melanogaster,* in: *Mutation, Cancer, and Malformation* (E. H. Y. Chu and W. M. Generoso, eds.), pp. 233–255, Plenum Press, New York (1984).

209. E. W. Vogel, Activity of 10 compounds in the *Drosophila* somatic recombination and mutation assay (SRM) using the *white/white-coral* system, in: *Progress in Mutation Research, Collaborative Study of Short-Term Tests for Carcinogens* (J. Ashby, F. J. de Serres, M. Draper, M. Ishidate Jr., B. Margolin, B. Matter, and M. D. Shelby, eds.) Vol. 5, pp. 313–317, Elsevier/North-Holland, Amsterdam (1985).

210. E. Vogel and C. Ramel, Mutagenesis assays with *Drosophila,* in: *Long-Term and Short-Term Screening Assays for Carcinogens: A Critical Appraisal* (R. Montesano, H. Bartsch, and L. Tomatis, eds.), International Agency for Research on Cancer Monograph, Supplement 2, pp. 157–183, International Agency for Research on Cancer, Lyon (1980).

211. E. Vogel, W. G. H. Blijleven, P. M. Klapwijk, and J. A. Zijlstra, Some current perspectives of the application of *Drosophila* in the evaluation of carcinogens, in: *The Predictive Value of Short-Term Screening Tests in Carcinogenicity* (G. M. Williams, R. Kroes, H. W. Waaijers, and K. W. van de Poll, eds.), pp. 125–147, Elsevier, Amsterdam (1980).

212. E. Vogel, A. Schalet, W. R. Lee, and F. Würgler, Mutagenicity of selected chemicals in *Drosophila,* in: *Comparative Chemical Mutagenesis* (F. J. de Serres and M. D. Shelby, eds.), pp. 175–256, Plenum Press, New York (1981).

213. E. Vogel, W. G. H. Blijleven, M. J. H. Kortselius, and J. A. Zijlstra, A search for

some common characteristics of the effects of chemical mutagens in *Drosophila, Mutat. Res. 92,* 69–87 (1982).

214. E. Vogel, J. A. Zijlstra, and W. G. H. Blijleven, Mutagenic activity of selected aromatic amines and polycyclic hydrocarbons in *Drosophila melanogaster, Mutat. Res. 107,* 53–77 (1983).

215. E. W. Vogel, H. J. Frei, K. Fujikawa, U. Graf, S. Kondo, H. Ryo, and F. E. Würgler, Summary report on the performance of *Drosophila* mitotic recombination and somatic mutation assays, in: *Progress in Mutation Research, Collaborative Study of Short-Term Tests for Carcinogens* (J. Ashby, F. J. de Serres, M. Draper, M. Ishidate, Jr., B. Margolin, B. Matter, and M. D. Shelby, eds.) Vol. 5, pp. 47–57, Elsevier/North-Holland, Amsterdam (1985).

216. F. Vogel and A. G. Motulsky, *Human Genetics*, Springer, Heidelberg (1979).

217. R. C. von Borstel, Differential response of meiotic stages of *Habrobracon* eggs to nitrogen mustard, *Genetics 40,* 107–116 (1950).

218. L. C. Waters, C. E. Nix, K. M. Solden, and J. L. Epler, Effects of genotype and age on mixed-function oxidase activities in adult *Drosphila melanogaster, Mutat. Res. 139,* 51–55 (1984).

219. E. C. Weaver, Somatic crossing over and its genetic control in *Drosophila, Genetics 45,* 345–357 (1960).

220. M. Whittinghill and B. M. Lewis, Clustered crossovers from male *Drosophila* raised on formaldehyde media, *Genetics 46,* 459–462 (1961).

221. L. Wilson, K. Anderson, L. Grisham, and D. Chin, Biochemical mechanisms of action of microtubule inhibitors, in: *Microtubules and Microtubule Inhibitors* (M. Borgers and M. de Brabander, eds.), pp. 103–113, North-Holland, Amsterdam (1975).

222. R. C. Woodruff, R. Valencia, R. F. Lyman, B. A. Earle, and J. T. Boyce, The mutagenic effect of platinum compounds in *Drosophila melanogaster, Environ. Mutagen. 2,* 133–138 (1980).

223. T. R. F. Wright, R. B. Hodgetts, and A. F. Sherald, The genetics of dopa decarboxylase in *Drosophila melanogaster.* I. Isolation and characterization of deficiencies that delete the dopa-decarboxylase-dosage-sensitive region and the alpha-methyl-dopa-hypersensitive locus, *Genetics 84,* 267–285 (1976).

224. F. E. Würgler, U. Graf, H. Frei, and H. Juon, Genotoxic activity of the anti-cancer drug methotrexate in somatic cells of *Drosophila melanogaster, Mutat. Res. 122,* 321–328 (1983).

225. F. E. Würgler, U. Graf, H. Frei, and H. Juon, Detection of point mutations, chromosome aberrations, and somatic recombination in *Drosophila melanogaster*, in: *Critical Evaluation of Mutagenicity Tests* (R. Bass, V. Glocklin, P. Grosdanoff, D. Henschler, B. Kilbey, D. Müller, and D. Neubert, eds.), BGA Schriften, pp. 185–223, MMV Medizin, Munich (1984).

226. F. E. Würgler, U. Graf, and H. Frei, Somatic mutation and recombination in wings of *Drosophila melanogaster*, in: *Progress in Mutation Research, Collaborative Study of Short-Term Tests for Carcinogens* (J. Ashby, F. J. de Serres, M. Draper, M. Ishidate, Jr., B. Margolin, B. Matter, and M. D. Shelby, eds.) Vol. 5, pp. 325–340, Elsevier/North-Holland, Amsterdam (1985).

227. N. Xamena, A. Creus, and R. Marcos, Mutagenic activity of some intercalating compounds in the *Drosophila zeste* somatic eye mutation test, *Mutat. Res. 138,* 169–173 (1984).

228. M. A. Yoo, H. Ryo, T. Todo, and S. Kondo, Mutagenic potency of amino acid pyrolysates in the *Drosophila* wing spot test and its correlation to carcinogenic potency, *Jpn. J. Cancer Res.* (Gann) *76,* 468–473 (1985).

229. Z. Zachar and P. M. Bingham, Regulation of *white* locus expression: The structure of mutant alleles at the *white* locus of *Drosophila melanogaster, Cell 30*, 529–541 (1982).
230. J. A. Zijlstra and E. W. Vogel, Mutagenicity of 7,12-dimethylbenz(*a*)anthracene and some aromatic mutagens in *Drosophila melanogaster, Mutat. Res. 125*, 243–261 (1984).
231. J. A. Zijlstra, E. Vogel, and D. D. Breimer, Occurrence and inducibility of cytochrome P-450 and mixed-function oxidase activities in microsomes from *Drosophila* larvae, *Mutat. Res. 64,* 151–152 (1979).
232. J. A. Zijlstra, E. W. Vogel, and D. D. Breimer, Strain-differences and inducibility of microsomal oxidative enzymes in *Drosophila melanogaster* flies, *Chem. Biol. Interact. 48,* 317–338 (1984).
233. S. Zimmering, The *mei-9*(*a*) test for chromosome loss in Drosophila: A review of assays of 21 chemicals for chromosome breakage, *Environ. Mutagen. 5,* 907–921 (1983).
234. S. Zimmering and K. L. Kammermeyer, On the nature of partial losses of the Y chromosome from treatment of *ring-X*/B^S $Y y^+$ males with diethylnitrosamine (DEN) or procarbazine and mating with repair-deficient *st mus302* females of *Drosophila, Mutat. Res. 104,* 121–123 (1982).
235. F. K. Zimmermann, R. Kern, and H. Rasenberger, A yeast strain for simultaneous detection of induced mitotic crossing over, mitotic gene conversion and reverse mutation, *Mutat. Res. 28,* 381–388 (1975).

CHAPTER 2

Structural and Metabolic Parameters Governing the Mutagenicity of Polycyclic Aromatic Hydrocarbons

Hansruedi Glatt and Franz Oesch

1. Introduction

1.1. Scope of This Review

Polycyclic aromatic hydrocarbons (PAHs) are common pyrolysis products of organic matter and are therefore ubiquitous in the human environment. That some PAHs are potent carcinogens in animal experiments has been known for many decades. In contrast, extensive mutagenicity research on PAHs has been performed only in the last 12 years, since the development of sensitive tests using bacteria or cultured mammalian cells in combination with metabolic activating systems. The synthesis of many potential metabolites of PAHs, the purification of various drug-metabolizing enzymes, and the development of efficient analytical methods for the separation, quantification, and identification of metabolites and DNA adducts are further important developments which have led to a more detailed understanding of the ultimate mutagens derived from PAHs and their metabolic control.

Hansruedi Glatt and Franz Oesch • Institute of Toxicology, University of Mainz, D-6500 Mainz, West Germany.

Most carcinogens are also mutagens, particularly PAHs.[84] While the carcinogenicity of a large number of PAHs and their metabolites has been established *in vivo*, the mutagenic potential of most of these compounds has been determined only *in vitro*. Therefore, in this review we will refer to carcinogenicity results to indicate that the compound may cause (somatic) mutations *in vivo*.

1.2. Terms, Short Names, and Abbreviations

1.2.1. Metabolic Activation

As we shall see later, PAHs generally have to be metabolized in order to exert mutagenic effects. The conversion from a promutagen to an ultimate mutagen is called *activation*. The formation of ultimate mutagens may involve several metabolic steps. In this case, the term may be used for the overall process or only for the last step directly resulting in the active metabolite. We propose to name the preceding steps *preactivation*. Frequently there exist metabolic steps that compete with the activation (at any level) for promutagens, converting them to less harmful or inactive products. This *sequestration* leads to a decreased rate of formation of active metabolite, while *inactivation* denotes metabolic removal of the formed active intermediate before it exerts harmful effects. Since ultimate mutagens usually are reactive, inactivation is a very common phenomenon even in the absence of enzymes.

1.2.2. Molecular Regions of PAHs

PAHs formally consist of anellated benzene rings and, in some cases, other hydrocarbon moieties. In contrast to benzene, not all carbon positions are equivalent and ease of chemical or metabolic attack can vary greatly. Furthermore, functional groups introduced in PAHs can also have greatly different reactivities, depending on their position. The terms K, L, bay, prebay, fjord, and prefjord region denote structural elements of PAHs (Figure 1) and have been useful for characterizing similarities in chemical and metabolic behavior of carbon atoms in different PAHs.

1.2.3. Abbreviations and Short Names for Some Complex PAH Metabolites

Since PAH metabolites may have complex structures, especially with respect to their steric arrangements, it is often convenient to use

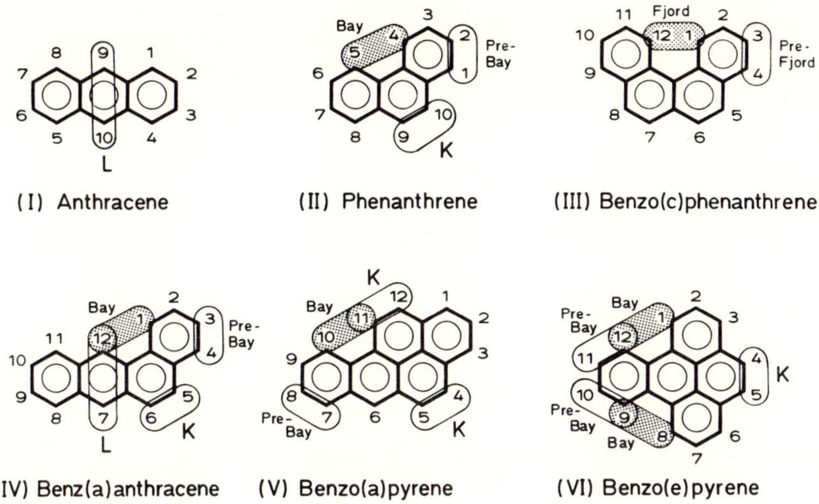

FIGURE 1. Molecular regions of PAHs. The terms L, K, bay, and fjord region have been introduced into the literature by giving prototype positions (consisting of two carbon centers) in prototype molecules[9,73,109] (upper row). In molecules that formally contain the aromatic systems of the prototype molecules, they denote the positions corresponding to the prototype positions, if *both* carbon centers are not anellated with other aromatic structures. Examples are shown in the lower row. Slight variations in the use of the terms occur, since, for example, the term bay region sometimes also includes fjord regions. In the older literature, the term K region primarily involved electron-richness rather than geometric properties. The terms prebay and prefjord region are introduced here as regions which presuppose bay and fjord regions and correspond to the 1,2 positions of phenanthrene and the 3,4 positions of benzo(*c*)phenanthrene, respectively.

short names. Characteristic examples together with the full names and the structures of the compounds are given in Figure 2. They also will allow unambiguous interpretation of the other analogously constructed short names used here. The abbreviations BP, BA, and DMBA will be used not only for the parent hydrocarbons benzo(*a*)pyrene (structure V in Fig. 1), benz(*a*)anthracene (IV), and 7,12-dimethylbenz(*a*)anthracene, but also in composed names for their derivatives.

1.3. Overview of the Metabolism of PAHs in Mammals

In the simplest case, a functional group is introduced into the PAH and then used for conjugation with a hydrophilic endogenous com-

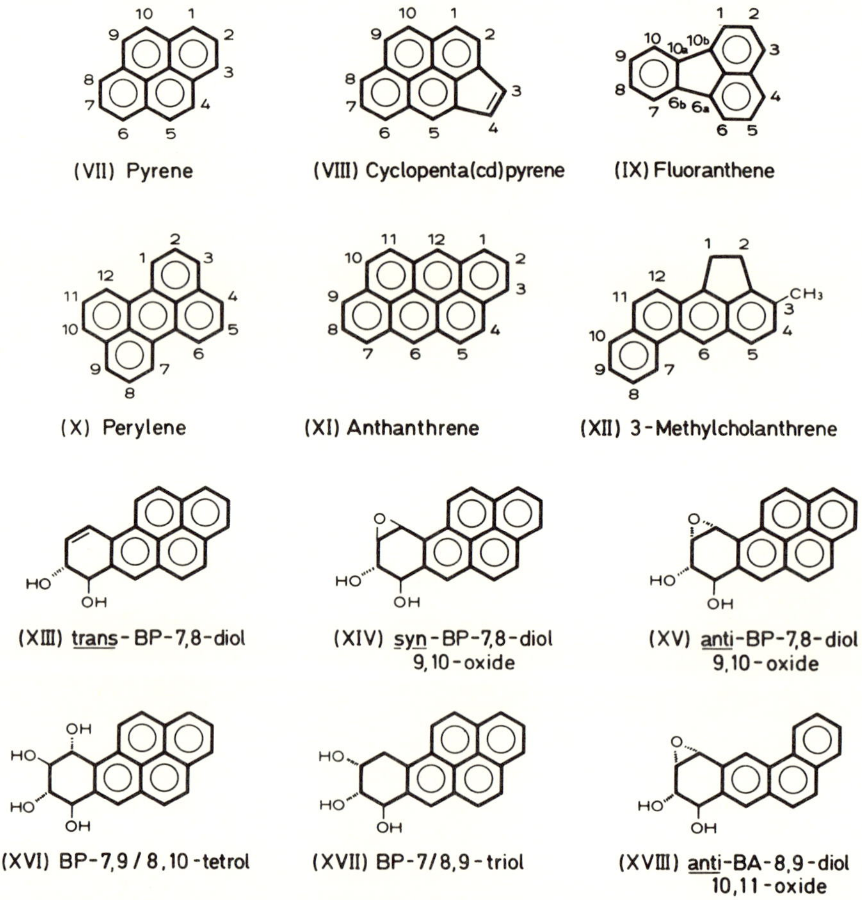

FIGURE 2. Structures and short names of PAHs and PAH metabolites discussed. Full names for the compounds (XIII)–(XVIII) are *trans*-7,8-dihydro-7,8-dihydroxybenzo(*a*)pyrene, *r*-7,*t*-8-dihydroxy-*c*-9,10-oxy-7,8,9,10-tetrahydrobenzo(*a*)pyrene, *r*-7,*t*-8-dihydroxy-*t*-9,10-oxy-7,8,9,10-tetrahydrobenzo(*a*)pyrene, *r*-7,*t*-8,*c*-9,*t*-10-tetrahydroxy-7,8,9,10-tetrahydrobenzo(*a*)pyrene, *r*-7,*t*-8,*t*-9-trihydroxy-7,8,9,10,10-pentahydrobenzo(*a*)pyrene, and *r*-8,*t*-9-dihydroxy-*t*-10,11-oxy-8,9,10,11-tetrahydrobenz(*a*)anthracene. Analogous abbreviations are used for corresponding derivatives from other PAHs.

pound. The conjugate resulting from this two-step metabolism is then excreted from the organism. In reality the situation is more complicated:

1. A functional group can be introduced at different sites on the molecule, and therefore several different primary metabolites are usually formed. For example, the primary isolatable metabolites of BP include 1-hydroxy-BP (possibly formed via BP 1,2-oxide); 3-hydroxy-BP (formed at least partially via BP 2,3-oxide[151]); BP 4,5-oxide (the stereospecific formation of the (+)-(4*S*,5*R*) enantiomer was observed with purified rat liver cytochrome P-450c,[4] the only situation where the stereochemistry of this metabolite has been studied); BP-1,6-quinone, BP-3,6-quinone, and BP-6,12-quinone (formed via 6-hydroxy-BP[78]); 6-hydroxymethyl-BP, 7-hydroxy-BP, (+)- and (−)-BP-7,8-diol [formed via (−)- and (+)-BP 7,8-oxide]; 9-hydroxy-BP, (+)- and (−)-BP-9,10-diol [formed via (−)- and (+)-BP 9,10-oxide]; and (+)- and/or (−)-BP-11,12-diol (probably formed via BP 11,12-oxide).[35,52,55,75,115,122,129,152]
2. The primary functional group is not always immediately used for conjugation, but may be modified or, in rare cases, removed, yielding again the parent hydrocarbon. Alternatively, additional functional groups may be introduced. By permutation of the sites of metabolic attack, numerous metabolites with several functional groups can occur. The number of possible multifunctional metabolites is further increased because functionalization often introduces chiral centers and different stereoisomeric metabolites may be realized. Extensive functionalization is pronounced under nonphysiological *in vitro* conditions, particularly if the compound is incubated for long periods or with large amounts of microsomal enzyme preparation in the presence of cofactors for oxidative metabolism (i.e., NADPH), but without cofactors for conjugation.[55]
3. While conjugates are not usually substrates for enzymes that introduce functional groups, metabolism of the hydrophilic moiety may occur or the conjugate may even be cleaved.

A list of biotransformations of PAHs at these various levels, including the enzymes involved and the cofactors required, is presented in Table 1. Although the number of principal reaction types is small, a virtually infinite number of products can be constructed by permutation of the possible kind, site, and stereochemistry of the reactions. Whereas only a few metabolites may be quantititatively important, numerous others can be found in small quantities.

TABLE 1. Metabolic Steps in the Biotransformation of PAHs[a]

Reaction	Enzymes involved
1. Introduction of functional groups a. *Epoxidation of aromatic and olefinic double bonds to arene oxides and alkene oxides, respectively* b. Direct ring hydroxylation c. *Hydroxylation of aliphatic moieties*[26]	Various cytochrome P-450 isozymes, requiring NADPH, O_2, and cytochrome P-450 reductase for monooxygenase activity; for some reactions, co-oxidation during prostaglandin synthesis is an alternative mechanism[82,118]
d. Introduction of hydroxymethyl[35,119]	Little known enzyme(s) (hydroxymethylsynthetase), *S*-adenosylmethionine can serve as carbon donor[35]
2. Removal of functional groups Reduction of the oxirane ring in arene oxides to an aromatic double bond[120]	Cytochrome P-450 with NADPH as cofactor, anaerobic conditions are favorable
3. Modification of functional groups a. Epoxides	
Isomerization to phenols and ketones	Nonenzymatic reactions only are known
Hydrolysis to diols[94]	Most arene oxides are hydrolyzed at significant rates only if catalyzed by microsomal epoxide hydrolase or (less efficiently) by cytosolic epoxide hydrolase[96]; in contrast, virtually only nonenzymatic hydrolysis is known for diol-epoxides
Reduction of diol-epoxides to pentahydrotriols (e.g., XV to XVII) by NADPH and NADH[150]	Nonenzymatic reactions only are known
b. Diols to catechols	Dihydrodiol dehydrogenase with NAD or NADP as cofactor[101]
c. Phenols Oxidation to oxy radicals[78,88]	Nonenzymatic and enzymatic mechanisms
d. Diphenols (catechols, hydroquinones) *O*-Methylation[76]	Catechol-*O*-methyltransferase with *S*-adenosylmethionine as cofactor
Oxidation to quinones, direct and via semiquinones[20,77] e. Quinones *1e⁻ and 2e⁻ reductions* to semiquinones and diphenols[20,77,101]	Various incompletely eludicated mechanisms, including autoxidations, $2e^-$ reduction of *o*-quinones by dihydrodiol dehydrogenase, $2e^-$ reductions by DT-diaphorase (vitamin K reductase), $1e^-$ reductions by cytochrome P-450, and $1e^-$ reductions by cytochrome P-450 reductase

TABLE 1. (*continued*)

Reaction	Enzymes involved
f. Aliphatic hydroxyl groups	
Oxidation to formyl and carbonyl groups (26)	Probably enzymatic
4. Covalent binding of reactive metabolites to macromolecules	Nonenzymatic
5. Conjugation with hydrophilic endogenous compounds	
a. Epoxides	
With glutathione[10,30,127]	Sometimes nonenzymatic; usually catalyzed by various glutathione transferases
b. Diols	
With glucuronic acid[15,89]	Various UDP-glucuronosyl transferases, using UDP-glucuronic acid as cofactor
With sulfate[90]	Sulfotransferase(s) using PAPS or a sulfo conjugate as sulfate donor
c. Quinones	
With glutathione[86]	Glutathione transferase B and possibly other isozymes; also nonenzymatic
With sulfate[90]	Sulfotransferases, using PAPS or a sulfo conjugate as sulfate donor
d. Phenols, diphenols	
With sulfate[90]	Sulfotransferase(s); using PAPS or a sulfo conjugate as sulfate donor
With glucuronic acid[15,89,103]	Various UDP-glucuronosyl transferases, using UDP-glucuronic acid as cofactor
e. Hydroxymethyl groups	
With sulfate[128]	Sulfotransferase(s); using PAPS or a sulfoconjugate as sulfate donor
With phosphate[111]	Nonenzymatic, with ADP and ATP as phosphate donors
6. Modification of conjugates	
a. Hydrolysis of some glucuronic acid and sulfate conjugates	β-Glucuronidases and sulfatases (present particularly in lysosomes and intestinal bacteria)
b. Glutathione conjugates: splicing of glutamic acid and glycine, acetylation of the amino group of cysteine	Enzymatic

[a] Italics denote very common, well-known metabolic routes. References are given for less known reactions preferentially.

2. Activation Pathways of PAHs

2.1. Requirement of Activation for Mutagenicity to Occur

Results of bacterial mutagenicity studies on more than 100 PAHs, exclusively consisting of hydrogen and carbon atoms, have been reported.[3,17,25,29,43,51,70–72,84,157,158] In the presence of mammalian enzyme preparations, most were mutagenic; some (e.g. BP, DMBA) showed potent effects with a variety of bacteria, genetic endpoints, and metabolic conditions; others [e.g., pyrene (VII), anthracene (I), phenanthrene (II)] showed mutagenicity only under special conditions.[17,100,158] Only a few PAHs, which all belong to the bicyclics and tricyclics (naphthalene, fluorene, and some methylated congeners of phenanthrene, fluorene, and anthracene), were negative in all studies performed.[70–72,84]

Most PAHs that were mutagenic in the presence of mammalian enzymes were also tested in their absence and produced negative results. To our knowledge, only five PAHs, BP,[54] 1,2,3,4-tetrahydro-DMBA (cf. IV),[62] 1,2-dihydro-DMBA,[62] 6,7,8,9,10,12b-hexahydro-3-methylcholanthrene (cf. XII),[3] and 9,10-dimethylanthracene (cf. I),[71] showed mutagenicity in bacteria without use of an activating system. Direct mutagenicity of BP was seen in *Escherichia coli* using a continuous culture technique in suspension, while conventional assays with various strains of *E. coli* and *Salmonella typhimurium* regularly yielded negative results. The other four compounds were mutagenic with *S. typhimurium* in conventional plate-incorporation assays. However, even if these compounds are truely ultimate mutagens, i.e., were not activated by solvents, media, light, or bacterial enzymes, an overwhelming fraction of the mutagenic effects observed with PAHs is not due to the parent compounds but to metabolites or other activation products.

Table 2 summarizes results of mutagenicity studies in *S. typhimurium* on known and potential BP metabolites. It shows that, albeit with drastic differences in potency, many BP metabolites are mutagenic, e.g., several simple epoxides, the diastereomeric BP-7,8-diol 9,10-oxides (XIV, XV) and their enantiomers, several phenols, and some unknown metabolites generated via phenols. For comparison, Table 2 includes results on mammalian (V79) cells, which do not metabolize BP. Qualitatively the results are similar to those obtained in bacteria. However, differences in the relative potencies occur. In V79 cells *anti*-BP-7,8-diol 9,10-oxide (XV) is the strongest mutagen, while in bacteria the *syn*-diastereomer (XIV) is more active. Also, BP 4,5-oxide, which is a very potent mutagen in bacteria, is less active in V79 cells. Therefore, it is concluded that metabolic pathways leading to mutagenic products in bacteria also may

TABLE 2. Mutagenicity of Metabolites and Related Derivatives of BP in Bacterial and Mammalian Cells[a]

	Mutagenicity							
	Salmonella typhimurium				V79 cells			
Compound	Direct	P-450	S9	Cells	Direct	S9	Cells	Remarks
BP	0[84]	↑ ↑[93,135,139]	↑ ↑[84]	↑ ↑[44]	0[56]	↑[23,81]	↑[56,58]	EH ↓ ↑ =[b], DH ↓ =[b]
(±)-BP 4,5-oxide[c]	+ +[40,84,133]	—	↓ ↓[k]	—	+[50,56,91,133]	—	—	(−) > (+),[d] TA1537/8,[e] EH ↓[f]
(±)-BP 4,5-aziridine	+ + +[50]	—	—	—	+ + +[50]	—	—	EH =[f]
(±)-BP 7,8-oxide[c]	W[133,135]	W[135]	—	—	0[134]	—	—	P-450 + EH ↑[g]
(±)-BP 9,10-oxide[c]	W[133]	—	—	—	W[134]	—	—	—
(±)-BP 11,12-oxide	W[133]	—	—	—	+[133]	—	—	—
1-Hydroxy-BP[c]	W[38,133]	↑[135]	↑ ↑[104]	—	W[56,133]	—	0[56]	EH =,[h] EHI =,[i] GSH =,[i] UDPGA ↓,[i] S ↓[i]
2-Hydroxy-BP	0[133]	↑[135]	↑,[104] ↑ ↑[23]	—	W[133]	↑[23]	—	EH =[h]
3-Hydroxy-BP[c]	W,[38,84,133] +[44]	↑[135]	↑ ↑[44,79,84,104]	(↑)[44]	0,[133] W[56]	—	0[56]	EH =,[h] EHI = ↓,[i] S ↓,[i] UDPGA ↓,[i] GSH ↓[i]
4-Hydroxy-BP	0,[133] W[38]	0[135]	—	—	0[134]	—	—	—
5-Hydroxy-BP	0[133]	0[135]	—	—	0[134]	—	—	—
6-Hydroxy-BP[c]	W,[38] +[133]	↑[135]	—	—	0,[56] W[133]	—	0[56]	EH =[h]
7-Hydroxy-BP[c]	0[133]	0[135]	0[104]	—	0[56,134]	—	0[56]	—
8-Hydroxy-BP	0[133]	0[135]	—	—	0[56,134]	—	0[56]	—
9-Hydroxy-BP[c]	0,[133] W[38,44]	↑[135]	↑,[104] ↑ ↑[44,79]	(↑)[44]	0[56,134]	—	0[56]	EH =,[h,i] UDPGA ↓,[i] GSH ↓,[i] S =[i]
10-Hydroxy-BP	0[133]	0[135]	—	—	0[134]	—	—	—
11-Hydroxy-BP	0[133]	0[135]	—	—	0[133]	—	—	—
12-Hydroxy-BP	+[133]	↑[135]	↑[104]	—	0[133]	—	—	EH =[h]
6-Hydroxymethyl-BP[c]	0[36,84]	—	↑ ↑[36,84]	—	0[56]	—	0[56]	S ↑[i]
6-Methyl-BP[c]	0[36]	—	↑ ↑[36]	—	—	—	—	—
6-Acetoxymethyl-BP	+ +[36]	—	↓[36]	—	—	—	—	—
6-Methoxymethyl-BP	0[36]	—	↑[36]	—	—	—	—	—
6-Chloromethyl-BP	+ +[36]	—	—	—	—	—	—	—
BP-1,6-quinone[c]	0[38,133]	—	↑[153]	—	0[133]	—	—	TA104[e]
BP-3,6-quinone[c]	0[38,133]	—	↑[153]	—	0[133]	—	—	TA104,[e] RED ↑,[j] SOD ↓,[j] CAT ↓[j]

(continued)

TABLE 2. *(continued)*

	Mutagenicity							
	Salmonella typhimurium				V79 cells			
Compound	Direct	P-450	S9	Cells	Direct	S9	Cells	Remarks
BP-6,12-quinone[c]	0[38,133]	—	(↑)[153]	—	0[133]	—	—	TA104[e]
BP-4,5-quinone	0[133]	—	—	—	0[133]	—	—	—
BP-7,8-quinone	0[139]	0[139]	0[139]	—	—	—	—	—
BP-9,10-quinone	0[j]	—	↑[k]	—	—	—	—	—
BP-11,12-quinone	0[133]	—	—	—	0[133]	—	—	—
(±)-*cis*-BP-4,5-diol	0[133]	—	—	—	0[56,133]	—	0[56]	—
(±)-*trans*-BP-4,5-diol[c]	0[133]	0[135]	—	—	0[56,133]	—	0[56]	—
(±)-*cis*-BP-7,8-diol	0[133,139]	↑↑↑[139]	—	—	—	—	—	—
(±)-*trans*-BP-7,8-diol[c]	0,[133] +[87]	↑↑↑[135,139]	↑↑↑[44]	↑↑↑[44]	0,[56,58] W[133]	—	↑↑[56,58]	EH =,[i] GSH↓,[i] UDPGA↓,[i] S↑[i]
(±)-*cis*-BP-9,10-diol	0[133]	—	—	—	0[133]	—	—	—
(±)-*trans*-BP-9,10-diol[c]	0[44,133]	(↑)[135]	↑[44]	(↑)[44]	0[56]	—	0[56]	EH↓,[i] GSH↓,[i] UDPGA↓,[i] S↑[i]
(±)-*cis*-BP-11,12-diol	0[133]	—	—	—	0[133]	—	—	—
(±)-*trans*-BP-11,12-diol[c]	0[133]	(↑)[135]	—	—	0[133]	—	—	—
(±)-*anti*-BP-7,8-diol 9,10-oxide[c]	+++[81,136]	—	—	—	+++[50,56,81,83,91,136]	—	—	(+)>(−),[d] EH =,[f] TA100[e]
(±)-*syn*-BP-7,8-diol 9,10-oxide[c]	+++[81,123]	—	—	—	+,[51] ++,[83,136] +++[81,91]	—	—	(−)≥(+),[d] EH =,[f] TA100[e]
(±)-*anti*-BP-9,10-diol 7,8-oxide[c]	++[81,123]	—	—	—	0,[81] W,[123] +[83]	—	—	TA98[e]
(±)-*syn*-BP-9,10-diol 7,8-oxide[c]	+[75,114]	—	—	—	0,[81,123,133] +[83]	—	—	TA98[e]
(±)-BP-*r*-7,*t*-10-diol *c*-8,9-oxide	W[123]	—	—	—	0[123]	—	—	—
(±)-BP-7,10/8,9-tetrol[c]	0[87]	—	0[87]	—	0[58]	—	0[58]	—
(±)-BP-7/8,9,10-tetrol[c]	0[87]	—	0[87]	—	0[58]	—	0[58]	—
(±)-BP-7,9/8,10-tetrol[c]	0[87]	—	0[87]	—	0[58]	—	0[58]	—
(±)-BP-7,9,10/8-tetrol[c]	0[87]	—	0[87]	—	0[58]	—	0[58]	—
(±)-BP-7/8,9-triol[c]	0[87]	—	(↑)[87]	—	0[58]	—	0[58]	—
(±)-BP-7,9/8-triol[c]	0[87]	—	(↑)[87]	—	0[58]	—	0[58]	—
4,5-Dihydro-BP	0[139]	(↑)[139]	↑[139]	—	—	—	—	—

7,8,9,10-Tetrahydro-BP	0[139]	(↑)[139]	—	—	—	—	—	—
(±)-7,8,9,10-Tetrahydro-BP 4,5-oxide	+[139]	—	—	—	—	—	—	—
9,10-Dihydro-BP	W[139]	↑↑[139]	—	—	—	—	—	—
(±)-9,10-Dihydro-BP 7,8-oxide	++[123,136]	—	—	—	+[92,136]	—	—	EH↓[f]
(±)-9,10-Dihydro-BP 7,8-diol	0[139]	(↑)[139]	—	—	—	—	—	—
7,8-Dihydro-BP	0,[84] W[139]	↑↑↑[139]	↑[84]	—	—	—	—	EH↓[h]
(±)-7,8-Dihydro-BP 9,10-oxide	+++[136]	—	—	—	++,[136] +[92]	—	—	EH↓[f]

[a] Mutagenicity in *S. typhimurium* was studied using reversion to histidine independence as marker. Different batteries of strains were used in different studies, but including TA98 and TA100 with most compounds. In the experiments on V79 cells, aquisition of resistance to 6-thioguanine, 8-azaguanine, or ouabain served as markers for mutagenicity. Tests were performed (1) in the absence of exogenous metabolizing systems (column headed "Direct"), (2) in the presence of highly purified cytochrome P-450 systems ("P-450"), (3) in the presence of S9 or related subcellular systems, such as liver microsomes or cell homogenate supplemented with NADPH ("S9"), and (4) in the presence of intact, metabolically active cells, hepatocytes in the bacterial studies, fetal Syrian hamster cells in the studies with V79 cells ("Cells"). The relative mutagenic potency of different BP derivatives depends on many factors, including incubation period, buffer, and bacterial strain (see footnote *e*). Despite this problem and the additional complication that the data are compiled from different studies using different methods, a rough quantitation was attempted: 0; W, +, ++, +++: No effect; weak, moderate, strong, very strong mutagens. (↑), ↑, ↑↑, ↑↑↑: The metabolizing system activated the compound and the resulting mutagenic potencies were weak, moderate, strong, and very strong, respectively. ↓: The metabolizing system reduced the mutagenicity.

[b] Depending on the experimental conditions (e.g., metabolizing system, BP concentration, bacterial strain) the same metabolic modulation (e.g., addition of purified microsomal epoxide hydrolase or dihydrodiol dehydrogenase[12,41,97]) decreased (↓), increased (↑), or had no effect (=) on the mutagenicity, indicating the predominance of different activation pathways.

[c] Known metabolite of BP.[35,55,78,115,129,151,152] However, in the case of chiral molecules, one enantiomer may be formed predominantly or exclusively.

[d] The (+) enantiomer was more mutagenic than the (−) enantiomer in various *S. typhimurium* strains as well as in V79 cells [(+) > (−)] or *vice versa* [(−) > (+)].[24,87,138]

[e] Among the known BP metabolites, the BP-7,8-diol 9,10-oxides are clearly the strongest mutagens in TA100 and TA98. However, in TA1537 and TA1538, BP 4,5-oxide shows a similar or stronger mutagenic potency. The BP-9,10-diol 7,8-oxides produce substantially stronger effects in TA98 than in TA100. The mutagenicity of BP-1,6-quinone, BP-3,6-quinone, and BP-6,12-quinone, in the presence of liver enzyme preparations, was detected using *S. typhimurium* TA104, a strain responsive to reactive oxygen species.

[f] Addition of purified epoxide hydrolase efficiently inactivated (EH↓) or had no appreciable effect (EH=) on the direct bacterial mutagenicity.[50,135,136]

[g] If both a purified cytochrome P-450 system and an epoxide hydrolase were used, a strong activation resulted.[135]

[h] Addition of purified epoxide hydrolase to the purified cytochrome P-450 system did not appreciably affect the mutagenicity.[135,136]

[i] Addition of purified microsomal epoxide hydrolase (EH), an epoxide hydrolase inhibitor (EHI), glutathione (GSH), UDP-glucuronic acid (UDPGA), and a system providing active sulfate (S) decreased (↓), increased (↑), or had no effect (=) on the S9- or homogenate-mediated bacterial mutagenicity.[36,39,44,79,104] In the case of 3-hydroxy-BP, inhibition of epoxide hydrolase had no effect with most metabolizing systems, but decreased the mutagenicity with one particular metabolizing system (EHI = ↓).[104]

[j] BP-3,6-quinone was activated to a mutagen by purified cytochrome P-450 reductase. This mutagenicity was inhibited almost completely by catalase and superoxide dismutase.[153]

[k] H. R. Glatt and F. Oesch, unpublished results.

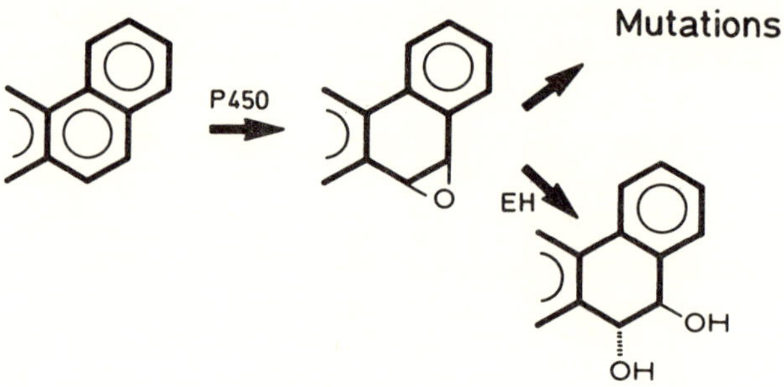

FIGURE 3. Activation of PAHs to monofunctional epoxides. The activation is usually conducted by cytochrome P-450-dependent monooxygenases (P450). The mutagen can be inactivated by epoxide hydrolases (EH) and by less specific enzymatic and nonenzymatic mechanisms (not shown).

be relevant for mammalian mutagenicity, but that the contribution of different activation pathways may be different.

In the following sections we discuss the known pathways of PAHs that can lead to mutagenic metabolites.

2.2. Activation to Monofunctional Epoxides

The most common first metabolic step on aromatic structures is monooxygenation to arene oxides (Figure 3). All arene oxides tested showed mutagenic activities, but large differences in potency occurred (Table 3). Most investigated arene oxides were K-region epoxides.

TABLE 3. Direct Mutagenicity of K-Region Epoxides in *Salmonella typhimurium*

Compound	$\Delta E_{deloc}/\beta$[a]	his$^+$ revertants per nmole[b]			Ref.[c]
		TA100	TA98	TA1537	
BP 4,5-oxide	0.623	375	563	174	50
Dibenz(*a,h*)anthracene 5,6-oxide	0.613	73	3	9	50
Benzo(*e*)pyrene 4,5-oxide	0.586	50	40	50	47
BA 5,6-oxide	0.576	123	55	12	50
7-Methyl-BA 5,6-oxide	0.576	253	35	3	50
DMBA 5,6-oxide	0.576	130	4	2	47
3-Methylcholanthrene 11,12-oxide	0.576	0.5	0.8	<0.05	50
Phenanthrene 9,10-oxide	0.563	1.1	0.2	0.4	17

[a] If different for the two possible carbonium ions, the higher value is given.
[b] Calculated from the linear part of the dose–response curve.
[c] All epoxides were tested under the same experimental conditions.

Exceptions are some non-K-region epoxides of BP (Table 2) and BA.[1] They were much less active than their K-region counterparts. Whether this is also the case with other PAHs is unknown, but it would be in accordance with chemical properties of these molecules. While non-K-region arene oxides strongly tend to regain aromaticity by isomerization to phenols, K-region epoxides are rather stable and prefer to undergo addition reactions with nucleophiles.

PAHs that contain olefinic moieties may be epoxidized to alkene oxides. In particular, potent electrophiles and mutagens may arise if the epoxidized olefinic bond includes a benzylic position. Examples are the 3,4-oxide of cyclopenta(*c,d*)pyrene (VIII)[51], the 1,2-oxide of benz-(*l*)aceanthrylene[157] and the 9,10-oxide of 7,8-dihydro-BP (Table 2).

Mammals possess enzymes that can efficiently catalyze the addition of water (epoxide hydrolases) or glutathione (glutathione transferases) to epoxides, thereby protecting more critical nucleophiles such as DNA. The glutathione transferases can also conjugate many other electrophiles. In contrast, epoxides are the only known substrates for epoxide hydrolases. Therefore, in mutagenicity tests where the active metabolites are generated by a complex metabolizing system, addition of epoxide hydrolases can serve as a specific (but not general) probe for the involvement of epoxides.

Epoxide hydrolases and glutathione transferases appear to be ubiquitous in mammalian tissues.[48,69,98] Even V79 cells, which do not have any cytochrome P-450 and are unable to metabolize parent PAHs,[57,132] exhibit substantial activities of these enzymes.[114] In contrast, no measurable levels of epoxide hydrolase activities and only low glutathione transferase activities (to date only measured with substrates other than epoxides) were observed in *S. typhimurium*.[85,121] At least in part, these differences may explain the lowered mutagenic activity of monofunctional arene oxides in V79 cells as opposed to their potent mutagenicities in bacteria. As we shall see in the next section, all epoxides tested carrying hydroxyl groups in the vicinity of the epoxide group are poor substrates of epoxide hydrolases and often are potent mutagens both in bacterial and mammalian cells.

Most epoxides derived from PAHs are chiral molecules. The enantiomers may differ in mutagenicity and metabolic properties, as shown, for example, with BP 4,5-oxide, where in all investigated *S. typhimurium* strains as well as in V79 cells the (−) enantiomer was more mutagenic than the (+) enantiomer.[24] Interestingly, in V79 cells a mixture of (+)- and (−)-BP 4,5-oxide was more active than even the pure (−) enantiomer. The mechanism is unknown, but may involve interference at metabolic inactivation.

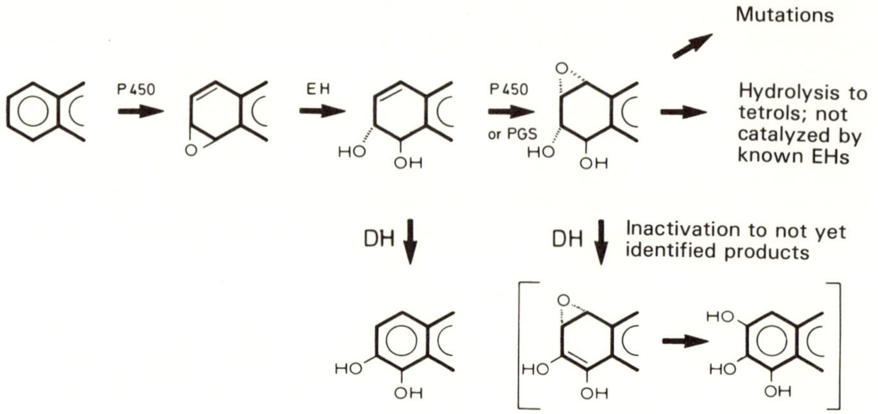

FIGURE 4. Activation of PAHs to vicinal diol-epoxides. The first step is a monooxygenation, usually conducted by cytochrome P-450 (P450). The second step, hydrolysis to the *trans*-diol, requires epoxide hydrolase (EH). The final activation step to the diol-epoxide can be performed by cytochrome P-450 or, alternatively, prostaglandin synthase (PGS). Among various metabolic pathways that sequester precursors of the diol-epoxide or inactivate it, those involving dihydrodiol dehydrogenase (DH) appear most specific for diol-epoxides compared to other PAH-derived mutagens (albeit not necessarily most efficient). It should be noted that the monofunctional epoxides, *trans*-diols, and the diol-epoxides in this activation pathway are chiral molecules and that two diastereomers of the diol-epoxides can occur, that with the epoxy group *cis* to the benzylic hydroxy group (*syn*-diol-epoxide, e.g., XIV in Figure 2) and that with the epoxy group *trans* to the benzylic hydroxy group (*anti*-diol-epoxide, e.g., XV and XVIII in Figure 2). The different enantiomers and diastereomers may differ substantially in their metabolism and in their intrinsic mutagenic activities.

2.3. Activation to Vicinal Diol-Epoxides

Many metabolites or synthetic derivatives of PAHs already containing functional groups may be activated by liver homogenate fractions or purified cytochrome P-450 systems to mutagens. Examples are shown in Table 2. Various authors have proposed phenol-epoxides, quinone-epoxides, diol-epoxides, etc., as active metabolites.[60,64,68,106,117] However, in most cases these hypotheses have not been experimentally confirmed. Vicinal diol-epoxides, however, are metabolites that can be formed on terminal benzo rings of PAHs. This activation pathway, shown in Figure 4, involves the following noteworthy aspects.

2.3.1. Formation of an Olefinic Double Bond

Epoxidation on a terminal benzo ring leads to olefinization of the adjacent aromatic double bond. The resulting non-K-region arene oxide may spontaneously rearrange to a phenol and thereby reestablish aromaticity in the ring. Alternatively, the nonaromatic system may be fixed through the enzymatic hydrolysis of the epoxide to the much more stable diol function.

2.3.2. Epoxidation of the Olefinic Double Bond

Whether an olefinic bond on a benzo ring is metabolically epoxidized depends on the substituents on the neighboring carbon atoms, the aromatic system, and the metabolic system. The 7,8-saturated BP derivatives 7,8-dihydro-BP, *trans-BP-7,8-diol (XIII),* and *cis*-BP-7,8-diol, but not BP 7,8-oxide and BP-7,8-quinone, were activated by a purified cytochrome P-450 system to potent bacterial mutagens.[(139)] *Trans*-BP-9,10-diol was metabolized by various enzyme preparations mainly at sites other than the olefinic 7,8 bond,[(123)] a phenomenon attributed to the fact that the hydroxy groups in the sterically hindered 9,10 (bay) region prefer the quasidiaxial conformation, while other diols, including *trans*-BP-7,8-diol, prefer the quasidiequatorial conformation. In agreement with this observation is that *trans*-benzo(*e*)pyrene-9,10-diol (cf. VI) was poorly metabolized at the adjacent double bond and was a poor promutagen with enzymes that readily activated 9,10-dihydrobenzo(*e*)pyrene.[(142)] However, poor metabolism to vicinal diol-epoxides of quasidiaxial diols is not a general phenomenon. For example, liver microsomes from the Syrian hamster, in contrast to microsomes from many other species, including rat, mouse, and man, preferentially metabolized *trans*-benzo(*e*)pyrene-9,10-diol at the 11,12 position.[(124)] Moreover, the quasidiaxial bay-region diols *trans*-dibenz(*a,h*)anthracene-1,2-diol and *trans*-chrysene-3,4-diol were predominantly epoxidized at the adjacent double bond even by rat microsomes.[(27)]

The epoxide group can be introduced from above or below with respect to the plane of the aromatic system, resulting in different (diastereometric) products: diol-epoxides, such as XIV, with the epoxy function *cis* to the benzylic hydroxy group are termed *syn*-diol-epoxides (or diol epoxide-1 in Jerina's nomenclature[(64)] and diol epoxide II in Gelboin's nomenclature[(56,150)]), while *anti*-diol-epoxides (or diol epoxide-2 and diol epoxide I, respectively), such as XV and XVIII, have the two functions in *trans* position. Which of these diol-epoxides is formed depends on the diol and on the metabolizing enzyme. As an example (+)-*trans*-BP-7,8,-diol is preferentially converted by cytochrome P-450 to the *syn*-diol-epoxide, whereas (−)-*trans*-BP-7,8-diol is preferentially metabolized to the *anti*-diol-epoxide.[(122)] If (±)-*trans*-BP-7,8-diol is metabolized through co-oxidation during prostaglandin

TABLE 4. Direct Mutagenicity of Diol-Epoxides and Tetrahydro-Epoxides in *Salmonella typhimurium*

Diol-epoxide[a]	$\Delta E_{deloc}/\beta$	his+ revertants per nmole[b]						Ref.
		syn-Diol-epoxide		*anti*-Diol-epoxide		Tetrahydro-epoxide[c]		
		TA100	TA98	TA100	TA98	TA100	TA98	
(±)-Anthracene-1,2-diol 3,4-oxide	0.545	—	—	—	—	—	2	71[f]
(±)-9,10-Dimethylanthracene-1,2-diol 3,4-oxide	0.545	—	—	—	—	—	15	71[f]
(±)-Phenanthrene-1,2-diol 3,4*-oxide	0.658	*114*	*12*	*270*	*12*	*1875*	*690*	141
(±)-Phenanthrene-3,4*-diol 1,2-oxide	0.545	—	—	—	—	*230*	*52*	141
(±)-Chrysene-1,2-diol 3,4*-oxide	0.639	*258*	*32*	*1768*	*252*	*6650*	*1150*	141, 148
(+)-(4*R*) enantiomer	—	*137*	*33*	*3193*	*72*	*6000*	*1200*	148
(−)-(4*S*) enantiomer	—	*368*	*32*	*590*	*377*	*6000*	*1200*	148
(±)-Chrysene-3,4*-diol 1,2-oxide	0.526	—	—	—	—	*1250*	*64*	141
(±)-Benzo(*c*)phenanthrene-3,4-diol 1,2**-oxide	0.600	*2700*	*1000*	*5200*	*800*	—	—	144
(±)-Triphenylene-1*,2-diol 3,4*-oxide	0.664	*80*	*25*	*150*	*25*	*5000*	*3750*	143[f]
(±)-BA-3,4-diol 1*,2-oxide	0.766	*1650*	—	*850*	—	*2300*	—	137
(±)-BA-1*,2-diol 3,4-oxide	0.628	—	—	—	—	*460*	—	137
	—	—	—	*2§*	*3§*	—	—	[e]
(±)-BA-8,9-diol 10,11-oxide	0.572	11	2.7	38	85	—	—	81
	—	*17*	—	*50*	—	—	—	137
	—	—	—	*152§*	*140§*	—	—	45, [e]
(±)-BA-10,11-diol 8,9-oxide	0.526	*6*	—	*50*	—	—	—	137
	—	—	—	*110§*	*900§*	—	—	[e]

(±)-Dibenz(*a,c*)anthracene-10,11-diol 12,13-oxide	0.545	—	—	*80§*	*150§*	—	—	[e]
(±)-BP-7,8-diol 9,10*-oxide	0.794	2740	376	987	74	—	—	81
	—	2500	1500	1600	350	2000	1800	136[f]
	—	*9700*	*3460*	*4070*	*1330*	—	*10000*	123, 136[f]
(+)-(10*R*) enantiomer	—	*5100*	*3900*	*6000*	*2500*	—	—	138
(−)-(10*S*) enantiomer	—	*9500*	*5200*	*950*	*1800*	—	—	138
(±)-BP-9,10*-diol 7,8-oxide	0.488	7	13	20	152	—	—	81
	—	*38*	*51*	*110*	*380*	*260*	*320*	123
	—	—	—	*210§*	*850§*	—	—	[e]
(±)-BP-*r*-7,*t*-10*-diol *c*-8,9-oxide	[d]	*40*	2	—	—	—	—	123
(±)-Benzo(*e*)pyrene-9*,10-diol 11,12*-oxide	0.714	*300*	*200*	*480*	*200*	*5000*	*6000*	143[f]
(±)-Dibenzo(*a,h*)pyrene-1,2-diol 3,4*-oxide	0.845	—	—	*4900*	*580*	—	—	147
(±)-Dibenzo(*a,h*)pyrene-3,4*-diol 1,2-oxide	0.500	—	—	—	—	*70*	*60*	147[f]
(±)-Dibenzo(*a,i*)pyrene-3,4-diol 1*,2-oxide	0.866	—	—	*7700*	*1940*	—	—	147
(±)-Dibenzo(*a,i*)pyrene-1*,2-diol 3,4-oxide	0.500	—	—	—	—	*100*	*70*	147[f]
(±)-Fluoranthene-1,10b-diol 2,3-oxide	0.488	[g]	—	[g]	—	—	—	110
(±)-Fluoranthene-2,3-diol 1,10b-oxide	[d]	[h]	—	[h]	—	—	—	110

[a] Bay-region positions are labeled with one asterisk, fjord-region positions with two asterisks. The structure and numbering of several PAHs are presented in Figs. 1 and 2.
[b] Different assay methods were used by different groups: standard plate incorporation assay without preincubation (standard letters), assays with a 5-min preincubation (italics), or 20-min preincubation (italics, §) of bacteria, test compound, and buffer before addition of the top agar and plating. Usually, preincubation markedly increased the number of induced mutants, as seen, for example, in the table for (±)-*anti*-BA-8,9-diol 10,11-oxide.
[c] Hydroxyl groups replaced by hydrogen.
[d] Not applicable.
[e] H. R. Glatt and F. Oesch (unpublished results). The method described in ref. 45 with the buffer of pH 7.4 was used.
[f] Calculated from figures presented in the quoted papers.
[g] The *anti* isomer did not mutate *S. typhimurium* TM677 to 8-azaguanine resistance; the *syn* isomer was weakly active.(110)
[h] Both the *syn* and *anti* isomers efficiently mutated *S. typhimurium* TM677 to 8-azaguanine resistance.(110)

synthesis (an alternative to the activation by cytochrome P-450), exclusively the *anti*-diol-epoxide appears to be formed.[118]

2.3.3. Roles of the Hydroxyl Groups for Reactivity and Mutagenicity of Diol-Epoxides

In *syn*-diol-epoxides (but not in *anti*-diol-epoxides), the benzylic hydroxyl group can form a hydrogen bond to the oxiranyl group and thereby facilitate the opening of the epoxide ring.[59] This may explain why *syn* isomers are more reactive and shorter lived in an aqueous environment than the corresponding *anti* isomers. Interestingly, in *S. typhimurium*, *syn*-BP-7,8-diol 9,10-oxide was a stronger mutagen than its *anti*-diastereomer, while in V79 cells the opposite was observed.[56,81,83,91,136] However, despite modulating effects, the hydroxyl groups are not a prime requirement for mutagenicity, since their substitution by hydrogen usually does not diminish the mutagenic effect (Table 4). In some cases [e.g. the bay-region alkene oxides of phenanthrene (II), chrysene, triphenylene, benzo(*e*)pyrene (VI)] the diol group even drastically reduces the mutagenicity (Table 4). This effect may have a steric basis (e.g., during the interaction with DNA), since the half-lives in water usually were of the same order of magnitude as those of the corresponding strongly mutagenic tetrahydro-epoxides.[73] Changes in mutagenicity by the diol group appear especially pronounced if the diol group is localized in a bay region preferring a quasidiaxial conformation due to crowding. This is the case with the diol-epoxides of benzo(*e*)pyrene (VI) and triphenylene, where not only the epoxy, but also the diol functions, are localized in bay regions.

2.3.4. Significance of the Diol Function for the Inactivation of Diol-Epoxides

Diol-epoxides, if metabolized at all, are poor substrates for the known epoxide hydrolases (Figure 5, Table 5). The reason they are poor substrates is not their short life spans, because similarly short-lived tetrahydro-epoxides are inactivated by this enzyme (Table 5). Hence, the known epoxide hydrolases play opposing roles in the activation of PAHs to monofunctional epoxides and diol-epoxides: epoxide hydrolase is not involved in the formation of monofunctional epoxides, but can detoxify them. By contrast, epoxide hydrolase is required for the formation of diol-epoxides, but does not inactivate them.

Hydroxyl groups of diol-epoxides may offer a site of attack for

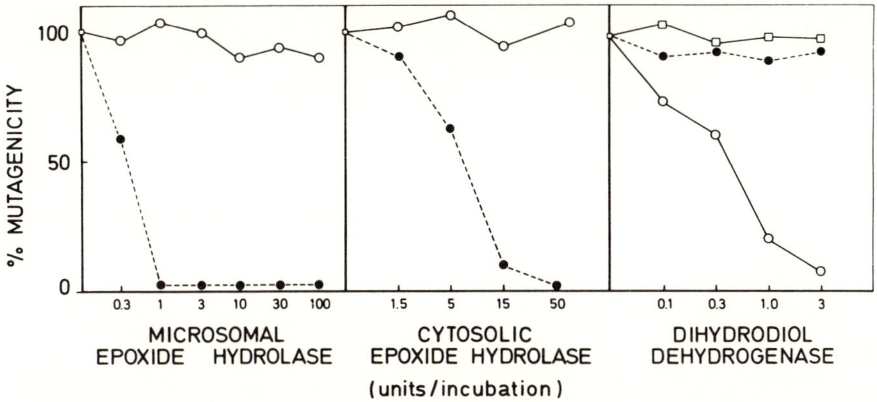

FIGURE 5. Effect of purified enzymes on the mutagenicity of BA 5,6-oxide (●) and BA-8,9-diol 10,11-oxide (○, in the presence of NADP; □, in the absence of NADP). Mutagenicity was determined in *Salmonella typhimurium* TA100. Microsomal epoxide hydrolase and dihydrodiol dehydrogenase were purified from rat liver to apparent homogeneity. Cytosolic epoxide hydrolase was purified from rabbit liver. For details see ref. 45.

various enzymes, including sulfotransferases, UDP-glucuronosyltransferases, and dihydrodiol dehydrogenases. In contrast to the first two groups of enzymes, dihydrodiol dehydrogenase is relatively specific for diol (and keto) groups. Therefore, this enzyme can be used as a diagnostic probe in tests mediated by subcellular metabolizing systems. A dihydrodiol dehydrogenase purified in our laboratory was able to sequester some diols[101] and to inactivate some diol-epoxides (Figure 5).

2.3.5. Role of the Aromatic System for Reactivity and Mutagenicity of Diol-Epoxides and Other PAH Derivatives

Diol-epoxides vary enormously in mutagenic potency (Table 4). Jerina *et al.*[64] recognized that diol-epoxides with the epoxide group in a bay region are particularly reactive and mutagenic. Perturbational molecular orbital calculations according to Dewar[33] show that benzylic carbonium ions generated through the opening of bay-region alkene oxides are stabilized by a high gain in energy (ΔE_{deloc}) obtained by extension of the aromatic system by one carbon center (Figure 6). By comparing different diol-epoxides (or tetrahydro-epoxides) from different PAHs, Lehr *et al.*[73] found a close positive correlation of the

TABLE 5. Effects of Highly Purified Microsomal Epoxide Hydrolase in Bacterial Mutagenicity Tests on Derivatives of PAHs

Compound	Activating system[a]	Percent changes in mutagenicity upon addition of epoxide hydrolase[b]	Ref.
K-region epoxides and aziridines[c]			
(±)-BA 5,6-oxide	—	−100 (1–6 units)	45, 137, Fig. 5
(±)-BA 5,6-aziridine	—	n.s. (30 units)	50
(±)-BP 4,5-oxide	—	−95 (0.9 unit)	135
(±)-BP 4,5-oxide	—	−98 (10 units)	50
(±)-BP 4,5-aziridine	—	n.s. (10 units)	50
Vicinal diol-epoxides[h]			
(±)-*syn*-Phenanthrene-1,2-diol 3,4-oxide	—	n.s. (5 units)	141
(±)-*anti*-Phenanthrene-1,2-diol 3,4-oxide	—	n.s. (5 units)	141
(±)-*syn*-Chrysene-1,2-diol 3,4-oxide	—	n.s. (5 units)	141
(±)-*anti*-Chrysene-1,2-diol 3,4-oxide	—	n.s. (5 units)	141
(±)-*syn*-Benzo(*c*)phenanthrene 3,4-diol 1,2-oxide	—	n.s. (5 units)	144
(±)-*anti*-Benzo(*c*)phenanthrene 3,4-diol 1,2-oxide	—	n.s. (5 units)	144
(±)-*syn*-BA-3,4-diol 1,2-oxide	—	n.s. (20 units)	137
(±)-*anti*-BA-3,4-diol 1,2-oxide	—	n.s. (20 units)	137
(±)-*syn*-BA-8,9-diol 10,11-oxide	—	n.s. (20 units)	137
(±)-*anti*-BA-8,9-diol 10,11-oxide	—	n.s. (20–100 units)	45, 137, Fig. 5
(±)-*syn*-BA-10,11-diol 8,9-oxide	—	n.s. (20 units)	137
(±)-*anti*-BA-10,11-diol 8,9-oxide	—	n.s. (20 units)	137
(±)-*syn*-BP-7,8-diol 9,10-oxide	—	n.s. (20 units)	135, 136
(±)-*anti*-BP-7,8-diol 9,10-oxide	—	n.s. (20 units)	136
(±)-*anti*-Dibenzo(*a,h*)pyrene-1,2-diol 3,4-oxide	—	n.s. (5 units)	147
(±)-*anti*-Dibenzo(*a,i*)pyrene-3,4-diol 1,2-oxide	—	n.s. (5 units)	147
Tetrahydro-epoxides			
(±)-1,2-Dihydrophenanthrene 3,4-oxide	—	−63 (5 units)	141
(±)-3,4-Dihydrophenanthrene 1,2-oxide	—	−97 (5 units)	141
(+)-1,2-Dihydrochrysene 3,4-oxide	—	−95 (10 units)	148

(−)-1,2-Dihydrochrysene 3,4-oxide	—	−60 (10 units)	148
(±)-3,4-Dihydrochrysene 1,2-oxide	—	−93 (5 units)	141
(±)-1,2-Dihydro-BA 3,4-oxide	—	−100 (6 units)	137
(±)-3,4-Dihydro-BA 1,2-oxide	—	−50 (20 units)	137
(±)-7,8-Dihydro-BP 9,10-oxide	—	−97 (20 units)	136
(±)-9,10-Dihydro-BP 7,8-oxide	—	−99 (20 units)	136
Miscellaneous other epoxides			
1-Pyrenyl oxirane	—	−99.5 (5 units)	146
4-Pyrenyl oxirane	—	−92 (5 units)	146
(±)-10,11-Dihydrocyclohepta(*c,d*)pyrene 8,9-oxide	—	−94 (5 units)	146
PAHs, in the presence of activating systems			
Phenanthrene	S9	n.s. (64 units)	100
Cyclopenta(*c,d*)pyrene	P-450	−90 (10 units)	146
BA	Mi	−95[d] (32 units)	97
BA	S9	−17[d] (50 units)	43
1-Methyl-BA	S9	−48 (50 units)	43
2-Methyl-BA	S9	−94 (50 units)	43
3-Methyl-BA	S9	−29 (50 units)	43
4-Methyl-BA	S9	−73 (50 units)	43
5-Methyl-BA	S9	+116 (50 units)	43
6-Methyl-BA	S9	+209 (50 units)	43
7-Methyl-BA	S9	−36 (50 units)	43
8-Methyl-BA	S9	−22 (50 units)	43
9-Methyl-BA	S9	−50 (50 units)	43
10-Methyl-BA	S9	−47 (50 units)	43
11-Methyl-BA	S9	−79 (50 units)	43
12-Methyl-BA	S9	−16 (50 units)	43
7,12-Dimethyl-BA	S9	n.s. (50 units)	43
BP	P-450	n.s. (10 units) to 50 (1.2–20 units)[d]	135, 146
BP	Mi	−98 (32 units) to +23 (32 units)[e]	12, 97
BP	Ho	n.s. (70 units)	44
9,10-Dihydro-BP	P-450	−85 (10 units)	139

(continued)

TABLE 5. (*continued*)

Compound	Activating system[a]	Percent changes in mutagenicity upon addition of epoxide hydrolase[b]	Ref.
PAH metabolites, in the presence of activating systems			
1-Hydroxy-BP	P-450	n.s. (5 units)	135
2-Hydroxy-BP	P-450	n.s. (5 units)	135
3-Hydroxy-BP	P-450, Ho	n.s. (5–70 units)	44, 135
6-Hydroxy-BP	P-450	n.s. (5 units)	135
9-Hydroxy-BP	P-450, Ho	n.s. (5–70 units)	44, 135
12-Hydroxy-BP	P-450	n.s. (5 units)	135
(±)-BP 7,8-oxide	P-450	+[f]	135
(±)-BP-7,8-diol	Ho	n.s. (70 units)	44
(±)-BP-9,10-diol	Ho	−36 (70 units)	44
(±)-Dibenz(*a,h*)anthracene-5,6-diol	P-450	n.s. (—[g])	140

[a] Where used, the activating system consisted of NADPH and either a highly purified cytochrome P-450 system (P-450), liver microsomes (Mi), liver postmitochondrial supernatant fraction (S9), or total homogenate of hepatocytes (Ho) from rats or mice.

[b] Values are the maximal percent increases or decreases in mutagenicity observed and, in parentheses, the number of units required to cause these effects (1 unit hydrolyzes 1 nmole styrene oxide per min). Stronger effects by larger amounts of enzyme may be possible in some cases, since saturation levels different from complete inactivation were demonstrated only in the case of BP during activation with microsomes or purified cytochrome P-450. Where no significant or appreciable effects (n.s.) were obtained according to the investigators' opinion, values in parentheses show the maximal amount of enzyme used. Since different mutagens were tested at different concentrations and often under variable incubation conditions, quantitative comparisons should be done with caution.

[c] For various other K-region oxides, efficient hydrolysis by purified microsomal epoxide hydrolase has been demonstrated with biochemical methods.[11]

[d] Depending on the bacterial strain used.

[e] Depending on the bacterial strain, the concentration of test compound, and the source of microsomes (untreated, phenobarbital-treated, or 3-methylcholanthrene-treated mice).

[f] Activation only occurred if both cytochrome P-450 and epoxide hydrolase were present.

[g] Data not given by the investigators.

[h] However, using spectrophotometric methods, it could be demonstrated that (±)*anti*-BA-1,2-diol 3,4-oxide as well as (±)*syn*-BA-diol 3,4-oxide are hydrolyzed by purified microsomal epoxide hydrolase.[159] While the maximal rates of conversion were comparable to those of other moderately good substrates, the K_m values were high. The situation with these diol-epoxides may be exceptional as one enantiomer of the corresponding tetrahydro-epoxide, (−)tetrahydro-BA 3,4-oxide, is the best substrate known for microsomal epoxide hydrolase.[159]

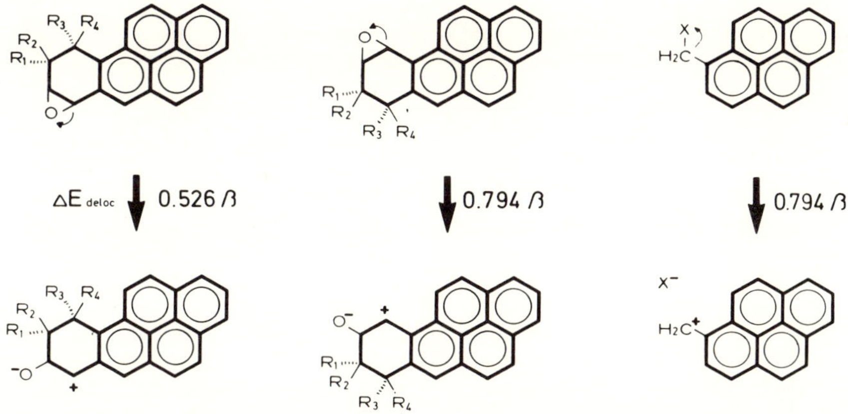

FIGURE 6. Reactivity of the same functional group depending on its location: predictions from perturbational molecular orbital calculations. Reactions of electrophilic PAH derivatives, such as epoxides (upper row, left two structures) or benzylic sulfate esters (upper row, rightmost structure, X = OSO_3^-) with cellular targets may involve carbonium ion intermediates (lower row). These carbonium ions are stabilized by the aromatic system. The energy ΔE_{deloc} that is gained when the π system is expanded by the additional carbon center can serve as a measure of this stabilization and can be estimated by perturbational molecular orbital calculations according to Dewar.[33] As illustrated in the figure, this calculation takes into account the kind and orientation of the aromatic system (cf. first and second columns), but does not take into account the reaction that leads to the carbonium ion (cf. second and third columns) or any substituents not conjugated with the carbonium ions (e.g., $R_1 = R_4 =$ H, $R_2 = R_3 =$ OH: *anti*-diol-epoxide; $R_1 = R_4 =$ OH, $R_2 = R_3 =$ H: *syn*-diol-epoxide; $R_1 = R_2 = R_3 = R_4$: tetrahydro-epoxide). Hence, the calculation is suitable for comparison of identical (or sufficiently similar) reactive groups located at different positions of the same PAH or different PAHs.

ΔE_{deloc} value with reactivity (uncatalyzed hydrolysis). The ΔE_{deloc} value also correlated with mutagenic potency, but this correlation was weaker than that with reactivity. It is most likely that other, probably steric factors, in addition to intrinsic reactivities, play important roles in reactions with DNA.

Since carbonium ion intermediates can occur not only with diol-epoxides but also with other mutagenic metabolites of PAHs, such as K-region epoxides (Section 2.2) and benzylic esters (Section 2.4), it is reasonable to check whether structure–mutagenicity relationships can be established on the basis of their ΔE_{deloc} values. To date, the number of tested benzylic sulfate esters is too small for a correlation study. In Table 3, which shows the mutagenicity of various K-region epoxides, we have added the ΔE_{deloc} values. Although a weak correlation with

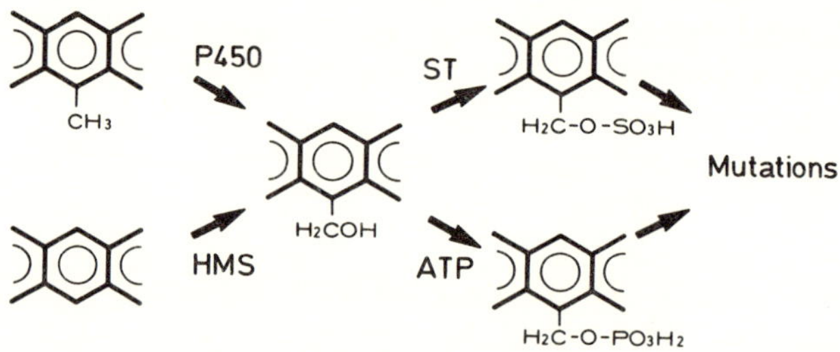

FIGURE 7. Activation of PAHs to benzylic esters. Methyl groups are frequently hydroxylated by cytochrome P-450 (P450). Sulfotransferases (ST) can then form mutagenic sulfate esters.[36,128] Incubation of hydroxymethyl-PAHs with ATP or ADP also can result in nonenzymatic formation of DNA-binding products, probably phosphate esters.[111] Although presumably not of practical significance, formation of reactive benzylic esters may also occur in the case of PAHs not possessing methyl groups, since insertion of a hydroxymethyl substituent by enzymes termed hydroxymethylsynthetases (HMS) has been reported.[35,119] In various mutagenicity studies, bromomethyl-PAHs, chloromethyl-PAHs, and acetate esters of hydroxymethyl-PAHs were used as models for metabolically postulated sulfate esters when these were not synthetically available.

mutagenicity may exist, some K-region epoxides with identical ΔE_{deloc} values differ markedly in mutagenic potency.

2.4. Activation to Benzylic Esters

Overall, BA (IV) is a very weak carcinogen. In contrast, DMBA is one of the strongest carcinogenic PAHs. Also, while anthracene (I) is considered inactive, its 9,10-dimethylated congener is carcinogenic. Therefore, activation at the side chain was suspected two decades ago, the proposal being that activation resulted from the hydroxylation of the methyl group and subsequent conjugation with sulfate (Figure 7). For a long time, such sulfate esters were not available for testing and instead model compounds with presumably similar chemical reactivity, choromethyl- and bromomethyl-PAHs were synthesized and tested. Although these compounds usually were less carcinogenic than the corresponding methyl-PAHs, they exerted potent direct mutagenic effects in bacterial and mammalian cells.[36,84,92] Recently, Watabe *et al.*[128] synthesized various benzylic sulfate esters and demonstrated that they are mutagenic. These researchers also showed that 7-hydroxymethyl-12-methyl-BA can be activated in bacterial mutagenicity tests by cytosolic sulfotransferases. Using the same metabolizing system, Flesher *et al.*[36] observed mutagenicity of 6-hydroxymethyl-BP. Other conjugates may also be of significance: benzylic acetate esters bind to DNA and are mutagenic in bacteria,[36] and in the presence of ATP and ADP, hydroxymethyl-PAHs nonenzymatically form intermediates, probably phosphate esters, which react with DNA.[111]

Formation of benzylic esters may not be completely limited to PAHs with a side chain. With BP enzymatic insertion of a hydroxymethyl group has been observed.[35,119] However, the extent of this reaction appears rather small and hydroxymethylation has not yet been demonstrated with other PAHs.

2.5. Activation to Free Radicals

Not only can free radicals react directly with critical target macromolecules, but alternatively they can generate secondary free radicals by one-electron transfer, e.g., to molecular oxygen, yielding the superoxide free radical. With the limitation that the bulk of the experimental evidence stems from studies with subcellular metabolizing systems, at least three types of primary free radicals seem to occur in the metabolism of PAHs (Figure 8):

1. Based on correlations between the tumorigenicity of various PAHs in mammary tissue and their ionization potential, it has been postulated that one-electron oxidation products directly derived from PAHs [such as benzo(*a*)pyrenyl, XXII] may play a decisive role in the development of these tumors.[21] Enzymes that can form these radicals, such as prostaglandin synthase and other peroxidases, indeed are present at high concentrations in mammary tissue.[21]

2. In the presence of subcellular liver preparations, various phenols are converted to free radicals.[78,88] In the case of 6-hydroxy-BP (XIX), this conversion occurs even in the absence of enzymes. The resulting free radical was isolated and identified as 6-oxy-BP (XXIII).[78]

3. Possibly the most interesting types of free radicals observed in the metabolism of PAHs are semiquinones (e.g., XXIV). They are one-electron oxidation products of hydroquinones (e.g., XX) and one-electron reduction products of quinones (e.g., XXI). As is apparent in Figure 8, semiquinones can be continuously generated in a redox cycle with these two species by one-electron transfers. Although the mechanisms of this cycle are not fully elucidated, it appears that the oxidations of the PAH derivatives usually occur spontaneously with molecular oxygen as the oxidant, while the reductions generally are enzymatic with NADPH as the reductant. Hence, additional adverse effects of the quinone/semiquinone/hydroquinone cycle are the consumption of NADPH and oxygen and the formation of various reactive oxygen species, such as superoxide, hydrogen peroxide, singlet oxygen, and

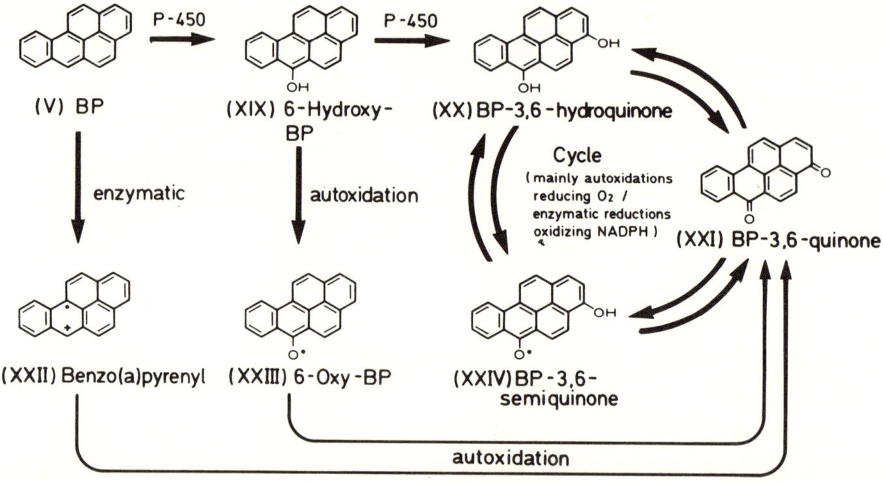

FIGURE 8. Activation of PAHs to free radicals. There is evidence for the occurrence of three types of free radical metabolites of PAHs (lower row), which are formally derived from parent PAHs, phenols, or hydroquinones by one-electron oxidation. Benzo(*a*)pyrenyl (a metabolite of BP postulated on indirect grounds) and 6-oxy-BP (a metabolite which has been isolated) readily autoxidize to the 1,6-, 3,6-, and 6,12-quinones of BP (only one is shown). The quinones may be reduced by various enzymes to hydroquinones under conversion of NADPH to $NADP^+$. Certain hydroquinones, including the 1,6-, 3,6-, and 6,12-hydroquinones of BP, readily autoxidize to the quinones. The reduction of the quinones and/or the oxidation of the hydroquinones may involve one-electron transfer, resulting in intermediate semiquinone radicals. Concomitantly, free oxygen radicals may be produced. Hence, harmful effects of the redox cycle are (1) the consumption of valuable reduction equivalents of NADPH and of molecular oxygen, (2) the intermediate occurrence of reactive semiquinone radicals, and (3) the production of reactive oxygen species. P-450: cytochrome P-450.

hydroxyl. Unless detoxified by enzymes such as catalases, superoxide dismutases, and peroxidases, these species can also react with DNA. Indeed, increased rates of DNA modifications that are characteristic for reactive oxygen species have been found in cultured human and animal cells treated with BP.[22,61]

Free radical metabolites of PAHs are not available synthetically. For assessement of their mutagenicity, therefore, they have to be generated *in situ*. So far very few mutagenic effects of PAH derivatives have been attributed to free radicals. The clearest case may be 6-hydroxy-BP (XIX), which exerts marked direct mutagenic effects on *S. typhimurium* (Table 2) and, in contrast to less mutagenic BP phenols, readily autoxidizes to an oxy radical (XXIII).[78,88] In contrast, BP tested

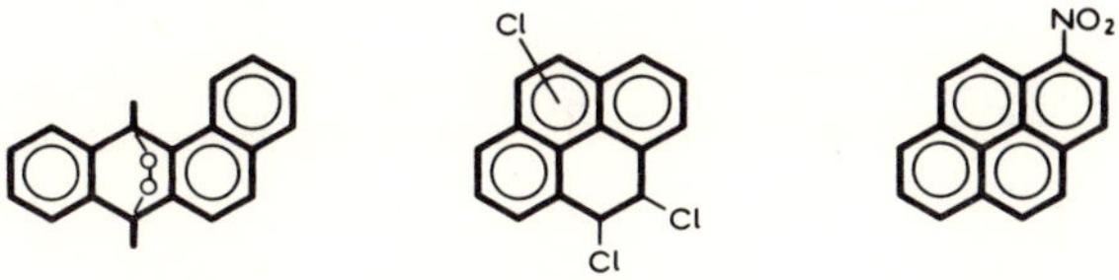

(XXV) DMBA-7,12-epidioxide | (XXVI) 4,5,X-Trichloro-4,5-dihydropyrene | (XXVII) 1-Nitropyrene

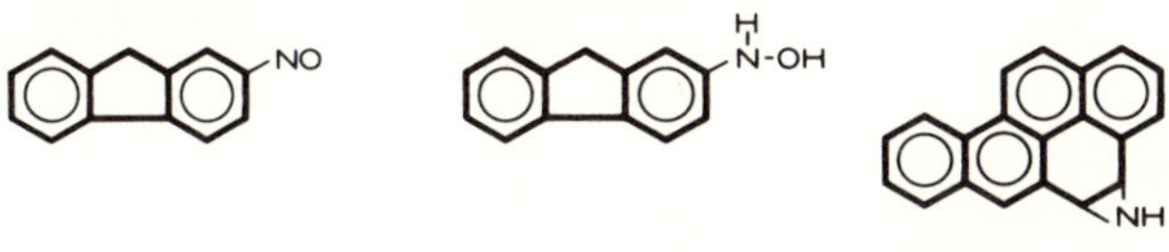

(XXVIII) 2-Nitrosofluorene | (XXIX) N-Hydroxy-2-aminofluorene | (XXX) BP 4,5-aziridine

FIGURE 9. Examples of potent mutagenic PAH derivatives other than the known metabolites of PAHs. Some of them (upper row) may readily be formed from PAHs in the human environment.

in the presence of ram seminal vesicles and arachidonic acid was not mutagenic to *S. typhimurium*.[53,82] In our laboratory, we have investigated many PAH quinones for direct and S9-mediated (postmitochondrial supernatant from liver homogenate, supplemented with an NADPH-generating system) mutagenicity with *S. typhimurium* TA100, TA98, and TA1537 and observed that most were inactive or very poor mutagens (H. R. Glatt, A. Seidel, K. L. Platt, and F. Oesch, unpublished results). Hence, in bacterial tests with exogenous metabolizing systems, activation of PAHs to free radicals may not be important. However, these experiments do not rule out that in other systems free radicals, especially if formed directly in the target cell, could be of significance.

2.6. Other Activation Pathways

In addition to the above-mentioned metabolites, a variety of PAH derivatives exert direct mutagenic effects. Examples are shown in Figure 9.

Exposure of DMBA to light in the presence of oxygen yields DMBA 7,12-epidioxide (XXV). This compound is probably responsible for the photomediated mutagenicity of DMBA to *S. typhimurium*.[126] Also, 9,10-dimethylanthracene 9,10-epidioxide, a direct mutagen, may be the actual cause for the weak mutagenic effects observed with 9,10-dimethylanthracene (cf. I) on *S. typhimurium* TA98 in the absence of mammalian enzymes.[71]

Treatment of pyrene (VII) or BP (V) with chlorine yields a variety of substitution and addition products.[28] Some of the latter (e.g., XXVI) are potent direct mutagens to *S. typhimurium*. Exposure of PAHs to air containing nitrogen dioxide can lead to the formation of nitro arenes.[108] Some of them (e.g., XXVII) are the strongest known mutagens to *S. typhimurium*.[112] Although various nitro arenes are directly mutagenic in this system, they probably are not the ultimate mutagens, but undergo the final activation through the action of bacterial enzymes.[112] Also, various nitroso- (e.g., XXVIII) and hydroxylamino- (e.g., XXIX) substituted aromatic compounds are potent direct bacterial mutagens[42,84] and probably require activation by bacterial metabolism (possibly to the same ultimate mutagens as nitro arenes).

Aziridines are nitrogen analogues of epoxides. Recently we observed that K-region aziridines of PAHs are exceptionally potent mutagens in bacterial and mammalian cells.[50] BP 4,5-aziridine (XXX) was several times more mutagenic in mammalian V79 cells than *anti*-BP-7,8-diol 9,10-oxide (XV), which is one of the most active compounds previously tested in this system.[56,81,83,91,136]

Obviously, mutagenicity is not a rare property of PAH derivatives. This propensity to mutagenicity and the existence of PAH derivatives that are more mutagenic than even the most mutagenic known PAH metabolites should warn us not to brashly attribute the mutagenicity of a PAH to a metabolite for which proximate or ultimate mutagenicity can be demonstrated.

3. Contribution of Different Activation Pathways to the Mutagenicity of PAHs

A first approach to assessing the relative importance of different activation pathways is to examine the known metabolites individually for mutagenicity and to compare their activities, relative to the amounts at which they are formed, with the activity of the parent compound. Obviously, the method presupposes substantial information on metabolism and availability of standards for testing. Moreover, even in the most defined metabolic system only a crude quantitation is possible, because the effectiveness of a mutagen may depend on pharmacokinetic parameters. For example, different effects may be observed if the mutagen is added as a single large dose or is continuously produced from the metabolic precursor. Hence, the importance of unknown metabolites can never be excluded. A final consideration is that when several metabolites exert mutagenic activities, as is often the case with PAHs, it is difficult to estimate their contribution. Therefore, additional methods have been employed, including (1) analysis of the DNA adducts, (2) modification of the test compound to block individual metabolic pathways, and (3) modulation of the metabolizing system. For the first two methods we limit our discussion essentially to BP as the most thoroughly investigated PAH. We give wider attention to the third method, since it is also applicable to PAHs where few or no metabolic standards are available.

3.1. DNA Adducts

3.1.1. Patterns of DNA Adducts

Adducts of PAHs may be studied directly in the undigested DNA by spectroscopic methods or, as is more common, after hydrolysis to nucleosides and chromatographic separation. Using the latter method, Pelkonen *et al.*[106] separated nine different adducts (or groups of adducts) from incubations of DNA and BP with hepatic microsomal systems and tentatively identified them as adducts of different BP metabolites. Owens *et al.*[105] observed similar numbers of different adducts when they performed microsomal incubations using 3-hydroxy-BP instead of BP. Since other BP metabolites are able to bind to DNA directly or in the presence of microsomal systems, microsome-mediated binding to DNA can be very complex and highly variable, depending on both the substrate concentration and the source and amount of microsomal preparations used.

Cells in culture,[5,116,117,125] tissue explants,[6,7,32,63] perfused organs,[66] and tissues from animals[5,8,16] exposed to BP usually display less complex adduct patterns than do subcellular systems. The predominant adducts cochromatographed in most but not all cases with adducts of *anti*-BP-7,8-diol 9,10-oxide. Where minor adducts were observed, they usually cochromatographed with adducts from *syn*-BP-7,8-diol 9,10-oxide, BP 4,5-oxide, or a derivative of 9-hydroxy-BP. From the chrysene like chromophore of 9-hydroxy-BP-derived adducts it is speculated that 9-hydroxy-BP 4,5-oxide[68] or a quinone-methide, a postulated rearrangement product of the phenol-epoxide,[60] might represent the reactive species. In rare cases, such as in baby hamster kidney cells,[116] the contribution of *syn*-BP-7,8-diol 9,10-oxide to the DNA binding was comparable to that of the *anti* isomer. Another exception to the rule that *anti*-BP-7,8-diol 9,10-oxide is mainly responsible for BP-derived DNA adducts in cells and tissues has been reported by Boroujerdi *et al.*,[16] who could not find any of these DNA adducts in the liver from rats after intravenous application of BP. In lungs from the same animals BP-7,8-diol 9,10-oxide–DNA adducts were formed, but were less frequent than other adducts.

When BP was incubated with *S. typhimurium* in the presence of liver microsomal enzymes from Aroclor 1254- or 3-methylcholanthrene-treated animals, the predominant adducts to the bacterial DNA cochromatographed with BP-7,8-diol 9,10-oxide adducts,[34,113] and adducts that may be derived from bioactivated 9-hydroxy-BP were also found.[34] In contrast, enzyme preparations from phenobarbital-treated animals produced a different profile of adducts.[113]

Similarly to BP, DMBA forms more complex and variable patterns of DNA adducts in subcellular systems than in cellular systems.[13] The major adducts found in cellular systems were derived from bay-region diol-epoxides. In addition to the simple diol-epoxide, 7-hydroxymethyl-12-methyl-BA-3,4-diol 1,2-oxide(s), a metabolite requiring three monooxygenase activation steps was identified as a major DNA-binding species *in vivo*.[65]

BA differs from DMBA not only in that a much smaller extent of binding to DNA occurs *in vivo*, but also in that a non-bay-region diol-epoxide, *anti*-BA-8,9-diol 10,11-oxide (XVIII), often represents the major DNA-bound product.[31,80]

3.1.2. Biological Significance of DNA Adducts

In principle, DNA adducts can be studied in any cell population, including target cells; one *caveat*, however, is that little is known of the relative importance of specific DNA adducts to mutagenicity. It is conceivable that some adducts may not lead to mutations at all. This implies that the contribution of a metabolite to the DNA binding may substantially differ from its contribution to mutagenic effects. The estimation of the effects of a metabolite in a potential target cell where only adducts can be studied can be improved by comparison with a system where both adducts and mutagenic effects are investigated. Using this calibration method (frequently termed *molecular dosimetry*), Fahl *et al.*[34] found that DNA adducts caused by *syn*-BP-7,8-diol 9,10-oxide (XIV) and bioactivated 9-OH-BP resulted in about four times as many revertants in *S. typhimurium* TA98 as did adducts produced by *anti*-BP-7,8-diol 9,10-oxide (XV). In the strain TA100, adducts produced from *syn*-BP-7,8-diol 9,10-oxide and bioactivated 9-hydroxy-BP were 1.5 and 2.3 times more efficient than adducts derived from *anti*-BP-

7,8-diol 9,10-oxide. Since only reversions (requiring a base substitution in TA100 and a frameshift mutation in TA98) were studied, and the strains used represent particular repair mutants (*uvrB*; pKM101), the relative efficiencies of these different adducts in *S. typhimurium* may not be representative of other types of mutations.

3.2. Modified Test Compounds to Elucidate the Activation Process

Saturation of a double bond prevents metabolic epoxidation of this site. Since both the 4,5-epoxide and vicinal diol-epoxides on the 7,8,9,10-benzo ring of BP are potent direct mutagens to *S. typhimurium*, 4,5-dihydro-BP and 7,8,9,10-tetrahydro-BP were synthesized and tested for mutagenicity in order to distinguish activation at these two sites of BP.[139] At first glance, it is surprising that both these hydrogenated congeners of BP are very poor promutagens. However, in Section 2.3.5 we saw that the aromatic system is critical for reactivity and mutagenicity of epoxides. Indeed, 7,8,9,10-tetrahydro-BP 4,5-oxide (which has the same aromatic system as the weakly mutagenic K-region epoxide of pyrene) had only very weak mutagenic activity.[139] Also, the 7,8-diol 9,10-oxides of 4,5-dihydro-BP, if formed at all, would be expected to possess moderate reactivity, since the aromatic system corresponds to that of the bay-region diol-epoxide of chrysene.

Without disruption of the aromaticity of the molecule, derivatization may be achieved by substitution of hydrogen atoms by fluorine. The 7-, 8-, 9-, and 10-fluoro derivatives of BP have been studied in S9- and microsome-mediated mutagenicity tests with *S. typhimurium* TA98 and TA1538.[37] 9-Fluoro-BP and 11-fluoro-BP showed mutagenic potencies similar to those of BP, whereas 10-fluoro-BP was weaker and 8-fluoro-BP was much more active. These results are in marked contrast to experiments on tumor initiation, where only BP, but not the four fluoro derivatives, was active.[19] In metabolic studies with microsomes,[19] 7,8-diols were not detected as metabolites of 7- and 8-fluoro-BP, implying blockage of the bay-region diol-epoxide pathway. However, 7,8-diols were found as metabolites of 9- and 10-fluoro-BP. Possibly final activation through oxidation of the 9,10 double bond (with a fluorine substituent) does not occur. Hence, with 7- and 8-fluoro-BP, and probably also with 9- and 10-fluoro-BP, bay-region diol-epoxides may be unimportant for the bacterial mutagenicity. Interestingly, large quantities of 4,5-diol were observed after incubation of 8-fluoro-BP with microsomes, strongly suggesting that the 4,5-oxide was a quantitatively important metabolite and responsible for the strong mutagenicity.

10-Aza-BP is an environmental carcinogen which, by virtue of its structure, cannot be metabolized to a vicinal diol-epoxide. It is a potent bacterial mutagen when activated by microsomal preparations.[102] The K-region oxide was isolated from such incubations when epoxide hydrolase was inhibited and showed strong direct mutagenicity in *S. typhimurium* TA98 and TA100.[102]

3.3. Use of Diagnostic Enzymes and Enzyme Inhibitors

As mentioned in previous sections, microsomes contain an enzyme, epoxide hydrolase, that can hydrolyze many monofunctional epoxides to diols but is ineffective toward diol-epoxides (Section 2.3.4). Changes in the activity of this enzyme can be achieved by enzyme induction, inhibition, or activation. Along with the use of subcellular metabolizing systems, purified enzyme may be added. Changes in mutagenicity when epoxide hydrolase activity is altered indicate that the activation pathway includes formation of an epoxide. A decrease in mutagenicity with increasing epoxide hydrolase activity implies that the ultimate mutagen is an epoxide or a metabolite formed via an epoxide but not via a diol. Inversely, an increase in mutagenicity with increasing epoxide hydrolase levels indicates that the ultimate mutagen is a diol or a diol-derived metabolite. A second enzyme, dihydrodiol dehydrogenase, can also be used to provide evidence for the involvement of diols in the activation (Section 2.3.4). However, since some diols are poor substrates for dihydrodiol dehydrogenase, the lack of an effect is of little diagnostic value.

3.3.1. Experiments with BP

When BP was incubated with liver microsomes derived from untreated mice, inhibition of epoxide hydrolase potentiated the bacterial mutagenicity, while addition of purified epoxide hydrolase decreased it.[95,97] Using *S. typhimurium* TA1537 as the target organism, it was found that the mutagenicity of BP was reduced by more than 95% by the addition of purified epoxide hydrolase.[12,97] With the strain TA98, only 75% of the mutagenic effect could be abolished.[12] One may conclude that at least two metabolites measurably contributed to the mutagenicity of BP in this system: the major mutagen(s) readily caused both *S. typhimurium* strains TA1537 and TA98 to revert and was epoxide hydrolase-sensitive; the minor mutagen(s) was inefficient toward TA1537 and was epoxide hydrolase-resistant. There is good evidence that the major mutagen was BP 4,5-oxide, because this mutagen is very active both with TA1537 and TA98 and is epoxide hydrolase-sensitive. Also, incubation of BP with the microsomal system in the absence of bacteria yielded an isolatable mutagen with the chromatographic properties of BP 4,5-oxide.[95] Moreover, when epoxide hydrolase was inhibited in the bacteria-less preparations, three times as much of this mutagenic metabolite was found. This corresponds precisely to the three-fold potentiation of mutagenicity observed after epoxide hydrolase was inhibited in the preparations containing bacteria.[95]

When microsomes from mice treated with the cytochrome P-450-inducer phenobarbital were used, the effects of purified epoxide hydrolase and epoxide hydrolase inhibition were similar to those with control microsomes, except that a small epoxide hydrolase-resistant mutagenic effect with TA1537 was clearly observed.[12]

When microsomes from mice treated with 3-methylcholanthrene, a compound that induces other cytochrome P-450s, were used the effect of BP concentration became important.[12] At high BP concentrations, the addition of epoxide hydrolase decreased the mutagenicity, as with control microsomes. However, there were marked differences in the relative sensitivities of the strains TA98 and TA1537, demonstrating that the ultimate mutagens differ from those produced by control microsomes. At low concentrations of BP, addition of purified epoxide hydrolase increased the mutagenicity, while the inhibition of epoxide hydrolase reduced it. This implied that diol-derived mutagens, probably diol-epoxides, were involved. Because two monooxygenase steps (from BP to monofunctional arene oxides and from diols to diol-epoxides) are required for the formation of these mutagens, an excess of BP may competitively inhibit the second oxidation step. Consistent with the involvement of diol-epoxides in these experiments is the observation that the addition of purified dihydrodiol dehydrogenase reduced the mutagenicity.[41] On the other hand, dihydrodiol dehydrogenase had no effect when BP was activated by control microsomes or when diol formation was prevented by an epoxide hydrolase inhibitor.

When liver microsomes from untreated rats were used for activation, BP was only slightly mutagenic to *S. typhimurium* TA1537,[18,99] but showed marked effects with other strains. This weak mutagenicity to TA1537 compared to experiments where mouse microsomes were used can be explained by the higher endogenous levels of epoxide hydrolase in rat than in mouse liver microsomes.[99]

In rat liver, microsomal epoxide hydrolase can be induced by *trans*-stilbene oxide. This compound has a second remarkable effect in that it shifts the metabolism of BP from the terminal benzo ring to the K-region.[18] This shift is opposite to that observed after induction by 3-

methylcholanthrene. The shift in BP metabolism by *trans*-stilbene oxide treatment of rats resulted in extremely weak mutagenicity of BP not only with *S. typhimurium* TA1537 (the strain with preferential detection of K-region oxides), but also with TA100 and TA98 (strains responsive to various mutagenic BP metabolites, including diol-epoxides).[18] The decrease in mutagenicity was dependent on the concomitant induction of epoxide hydrolase. When epoxide hydrolase was inhibited, microsomes from *trans*-stilbene oxide-treated rats activated BP to a similar or stronger extent (depending on the target strain) than did control microsomes.

An important conclusion from these results is that the metabolites responsible for the mutagenicity of BP in the Ames test can be very different, depending on the metabolizing system, the substrate concentration, and the bacterial strain used.

3.3.2. Experiments with BA and Methylated Congeners

When BA (IV), its 12 isomeric monomethylated congeners, and DMBA were activated by liver S9 from Aroclor 1254-treated mice, the addition of purified epoxide hydrolase reduced the mutagenicity of most of these compounds (Table 5). The only exceptions were DMBA, whose mutagenicity was not measurably affected, and the compounds methylated at the K-region, 5-methyl-BA and 6-methyl-BA, whose mutagenicity was markedly potentiated by added epoxide hydrolase. This striking behavior of the K-region methyl-BAs, along with the potent (Table 3) epoxide hydrolase-sensitive (Figure 5) mutagenicity of BA 5,6-oxide, strongly suggests (1) that K-region epoxides played an important role in the mutagenicity of BA and those monomethyl-BAs that carry the methyl group at a non-K-region position and (2) that, at least for the K-region methyl-BAs, diol-derived mutagens were also important. Moreover, in these investigations, dihydrodiol dehydrogenase reduced the mutagenicity of 7-methyl-BA, 12-methyl-BA, and DMBA, indicating that diol-derived metabolites contributed to the mutagenicity.[43]

Besides providing specific insight into the activation of these compounds, these findings demonstrate that compounds that are closely related structurally may be differentially activated and inactivated.

3.3.3. Experiments with Miscellaneous PAHs

Two exemplary experiments where modulation of epoxide hydrolase activity provided interesting new insight into PAH mutagenicity were performed with cyclopenta(*c,d*)pyrene (VIII) and phenanthrene (II). Cyclopenta(*c,d*)pyrene, a moderately potent carcinogen, is a potent bacterial mutagen requiring activation.(51) Mutagenicity of this compound can be completely abolished by added epoxide hydrolase.(146) Consistent with this observation is the postulate that cyclopenta(*c,d*)pyrene 3,4-oxide is the active species involved.(51) Phenanthrene (II), generally considered a harmless compound, was reported negative in the Ames test.(72,84) Nevertheless, we demonstrated mutagenicity of phenanthrene by blocking its inactivation by microsomal epoxide hydrolase.(17) A second, epoxide hydrolase-resistant mutagen was produced from phenanthrene when very high concentrations of S9 were used.(100)

4. Metabolic Control of Mutagenic Intermediates of PAHs

One point in studying an activation process is to identify the active and inactive metabolites of the chemical. It should be understood that such metabolites do not arise independently of one another, but their relative abundance is under the direct influence of such biological factors as the relative activities of xenobiotic-metabolizing enzymes, the availability of cofactors, and the compartmentalization of precursors. Therefore, experiments in which the effects of modulations of these primary biological variables on the mutagenicity are investigated provide important information exceeding (and partially bypassing) the mere identification of the mutagenic intermediates. Elucidation of metabolic control can be of great value in understanding differences in susceptibility among species, tissues, or individuals toward mutagens and carcinogens.

4.1. Mutagenicity Experiments with Subcellular Metabolizing Systems

Cytochrome P-450-dependent monooxygenases are of prime importance for activation of PAHs, as demonstrated by the dependence of activation on the presence of NADPH,(44,95) reduction of mutagenicity by antibodies and other inhibitors of cytochrome P-450,(67) and activation with purified cytochrome P-450 systems.(93,135) Cytochrome P-450 isozymes differ greatly in their ability to activate PAHs.(93,135) One reason is that different isozymes preferentially attack PAHs at different sites of the molecule.(52,74,131) While the pattern (but not the concentration) of cytochrome P-450 isozymes present is of critical importance for the primary metabolism under substrate-limited conditions, the enzyme concentration may become critical for secondary oxidation steps. In the latter case cytochrome P-450-dependent metabolism is in competition with alternative pathways, such as conjugation.

Some biotransformations performed by cytochrome P-450 may alternatively be conducted through co-oxidation during prostaglandin synthesis.[82,118] It appears, however, that the specificity of prostaglandin synthase-dependent activation of PAH derivatives to mutagens is narrower both with respect to substrates[53] and products[118] than those performed by the cytochrome P-450s.

Epoxide hydrolases and dihydrodiol dehydrogenases are enzymes that metabolize specific chemical classes of PAH metabolites. Their role in the activation and inactivation of PAHs has been detailed in Sections 2.3.4 and 3.3. In contrast, glutathione transferases represent a family of enzymes that can metabolize many different electrophiles and therefore can inactivate widely different mutagens. Examples of PAH derivatives whose mutagenicity was reduced by the addition of glutathione are bioactivated BP, *trans*-BP-7,8-diol, *trans*-BP-9,10-diol, 3-hydroxy-BP,[39,44,104] and the ultimate mutagens BA 5,6-oxide and *anti*-BA-8,9-diol 10,11-oxide.[46]

In order to obtain a first approximation of the relative roles of the different enzymes, we calculated the quantity of liver tissue that contains the amount of enzyme required for a 50% reduction in mutagenicity of a K-region epoxide, BA 5,6-oxide, and a vicinal diol-epoxide, *anti*-BA-8,9-diol 10,11-oxide (Table 6). Three aspects of this comparison are particularly noteworthy:

1. Both compounds were more efficiently inactivated by glutathione transferases than by the other effective enzymes, epoxide hydrolases (toward BA 5,6-oxide) and dihydrodiol dehydrogenase (toward *anti*-BA-8,9-diol 10,11-oxide). Therefore, it is quite likely that glutathione transferases play an important role in the *in vivo* inactivation of epoxides derived from PAHs. An accurate quantitative estimation of the contribution of different enzymes is impossible because differences in cofactor concentration, pH optima, and other environmental factors may lead to substantial differences between enzyme activities *in vivo* and *in vitro.* For example, microsomal epoxide hydrolase was tested as the solubilized purified enzyme, whereas *in vivo* it is situated in the endoplasmic reticulum and other membranes. The localization in membranes may be of advantage in comparison with cytosolic enzymes, since it increases the opportunities for reaction with epoxides, which are themselves generated in these membranes and tend to stay there because of their

TABLE 6. Quantitative Comparison of the Inactivation of BA 5,6-Oxide and BA-8,9-diol 10,11-Oxide by Different Highly Purified Enzymes[a]

		Amounts of enzyme required for a 50% reduction in mutagenicity			
		BA 5,6-oxide		BA-8,9-diol 10,11-oxide	
Enzyme	Concentration in liver, μg/mg tissue	Micrograms	Milligram liver equivalents	Micrograms	Milligram liver equivalents
Glutathione transferase A	0.5	0.11	0.2	110	200
Glutathione transferase B	2.2	0.5	0.2	130	60
Glutathione transferase C	1.1	0.02	0.017	30	20
Glutathione transferase X	0.25	0.003	0.011	6	20
Microsomal epoxide hydrolase	0.5	0.7	1.3	$\geqslant$170[b]	$\geqslant$300[b]
Cytosolic epoxide hydrolase	0.16	4	30	$\geqslant$30[b]	$\geqslant$200[b]
Dihydrodiol dehydrogenase	0.45	$\geqslant$3[b]	$\geqslant$1000[b]	70	170

[a] Data from ref. 46. Glutathione transferases, microsomal epoxide hydrolase, and dihydrodiol dehydrogenase were purified from male Sprague-Dawley rats; cytosolic epoxide hydrolase was purified from rabbits. Concentrations in liver refer to untreated adult males of the species from which the enzyme was purified.

[b] The enzymes, tested up to the indicated amounts, did not show any significant effect.

relative lipophilicity.[39] In the case of dihydrodiol dehydrogenase, one should bear in mind that not only may it inactivate diol-epoxides, but it may also sequester precursor diols, an effect that has not been taken into account in the experimental model used.

2. The different glutathione transferases investigated differed greatly in their abilities to inactivate the two epoxides studied. In this connection it is noteworthy that the relative efficiency of the different isozymes was similar (X > C > A ⩾ B) with both substrates despite the structural differences in the epoxides and the large differences (about 1000-fold) in the amount of enzyme required for the inactivation of the two epoxides.

3. BA 5,6-oxide, a potent bacterial mutagen which contributes little to the effects of BA in mammalian systems, was inactivated very efficiently by various mechanisms. In contrast, *anti*-BA-8,9-diol 10,11-oxide, which is a weaker mutagen in bacteria than is BA 5,6-oxide, but often represents the major DNA-binding species derived from BA in animals and in cultured cells,[31,80] was inactivated with only moderate efficiency by all investigated enzymes (Table 6).

While the studies on the roles of glutathione transferases also included experiments with purified enzymes, the role of other conjugating enzymes has been investigated only indirectly by studying the effects of addition or omission of cofactors. *S*-Adenosylmethionine decreased the S9-mediated mutagenicity of BP by an unknown, probably enzymatic mechanism.[76] Cofactors for sulfo- and glucuronosyl transferases reduced the mutagenicity of bioactivated BP[44] and of some of its metabolites (Table 2). In other experiments potentiation of the mutagenic effects of PAHs was observed. From some hydroxymethyl-PAHs mutagenic benzylic esters are formed (Section 2.4). In other cases, the potentiation of mutagenicity may be an indirect effect, since many phenols (which often are of low immediate significance for mutagenicity) are potent inhibitors of cytochrome P-450-dependent monooxygenases and therefore their removal by conjugation can stimulate the metabolism of the parent PAH.[14]

Nonenzymatic factors can also play a role. BA 5,6-oxide[46] and BP 4,5-oxide[127] can be conjugated with glutathione in the absence of enzyme and the rate of this reaction strongly depends on pH.[127] NADPH can nonenzymatically reduce diol-epoxides.[150] The stability of diol-epoxides strongly depends on the pH and even a mildly acidic milieu can drastically reduce their life spans.[130] Moreover, some acids, such as the dihydrogen phosphate ion ($H_2PO_4^-$), acetic acid, and phenol, catalyze the hydrolysis of diol-epoxides by direct transfer of a proton.[130] Planar aromatic molecules that allow noncovalent binding of PAH derivatives and possess an acidic or other reactive group, such as riboflavin 5′-phosphate and ellagic acid, efficiently inactivate some diol-epoxides by hydrolysis and conjugation, respectively.[145,149]

Another factor to be considered is the compartmentalization of drug metabolism. There is evidence that localization of an enzyme in the cytosol may be unfavorable for metabolism of PAH derivatives because the primary metabolism of PAHs usually occurs in membranes and the metabolites are so lipophilic that they tend to stay there. In an experiment where bacteria and microsomes were in direct contact, inactivation of mutagenic metabolites formed from BP and *trans*-BP-7,8-diol was much less efficient than in a situation where bacteria and microsomes were clearly separated and the mutagenic metabolite had to diffuse through a soluble compartment containing inactivating enzymes.[39]

4.2. Mutagenicity Experiments Using Intact Cells as the Metabolizing System

Since many enzymes are involved in the metabolism of PAHs, it may be expected that the patterns of metabolites and the steady-state concentrations of mutagenic intermediates produced by subcellular metabolizing systems will differ widely from those obtained *in vivo* or in intact isolated cells. Important enzymes may be localized in cellular components not included in the subcellular metabolizing systems, or cofactors for some enzymes may be too dilute for an appreciable activity. Even if cofactors for known enzymes are added, the requirement of special cofactors for unknown enzymes cannot be ruled out. The disruption of cellular compartmentalization, as discussed above, is another source for changes in the metabolism. For these reasons, we investigated the bacterial mutagenicity of PAHs using intact hepatocytes as the metabolizing system and compared the results with those from conventional tests. In the first study,[44] which was performed on BP and four of its major promutagenic metabolites, we observed that all five compounds produced strong mutagenic effects if they were activated by S9 or NADPH-fortified hepatocyte homogenate. However, with intact hepatocytes, only the potent carcinogens BP and *trans*-BP-7,8-diol demonstrated strong mutagenic effects, while the noncarcinogens (or very weak carcinogens) 3-hydroxy-BP, 9-hydroxy-BP, and *trans*-BP-9,10-diol were only very weakly mutagenic. Experiments where the subcellular metabolizing system was made similar to the cellular system by addition of cofactors for conjugating enzymes indicated that differential inactivation accounted for a substantial part of the differ-

TABLE 7. Mutagenicity of BA, the 12 Isomeric Monomethyl-BAs, and DMBA Using Various Metabolizing Systems

		Mutagenicity with *Salmonella typhimurium* TA100							
		Rat liver S9[b]		Mouse liver S9[c]		Hepatocyte homogenate + NADPH[d]		Intact hepatocytes[d]	
Compound	Carcinogenicity[a]	Potency[e]	Max effect[f]	Potency[e]	Max effect[f]	Potency[e]	Max effect[f]	Potency[e]	Max effect[f]
DMBA	++++	5.6	437	5.8	460	35.6	1130	13.3	1100
7-Methyl-BA	+++	15.4	1275	11.8	1000	10.8	1810	3.0	960
12-Methyl-BA	++	5.7	460	24.0	2000	8.3	820	3.0	380
6-Methyl-BA	++	7.7	318	8.0	660	11.4	1560	1.2	210
8-Methyl-BA	++	4.0	330	6.2	530	16.7	1150	1.4	120
5-Methyl-BA	+	16.9	699	14.4	1200	12.5	2450	3.7	500
11-Methyl-BA	+	2.3	464	6.7	550	59.1	3080	1.5	320
9-Methyl-BA	+	11.4	348	11.8	1000	8.9	890	1.4	150
BA	+	11.3	994	11.3	1000	50.9	3610	3.6	490
10-Methyl-BA	+	3.0	122	5.1	440	28.8	1970	2.0	210
4-Methyl-BA	+	30.2	2492	16.7	1400	10.5	1230	1.7	160
1-Methyl-BA	±	6.0	500	15.6	1300	23.8	2480	0.9	240
2-Methyl-BA	±	7.6	313	32.2	3100	10.9	450	0.3	58
3-Methyl-BA	±	9.4	194	3.0	250	6.8	430	0.2	35

[a] The compounds are arranged in the order of their carcinogenic potencies. DMBA is one of the most carcinogenic polycyclic aromtic hydrocarbons; 7-methyl-BA is strongly carcinogenic; 12-, 6-, and 8-methyl-BA are "moderately potent carcinogens." Most of the other compounds were weakly carcinogenic in several studies ("clear but weak carcinogens"). 1-, 2-, and 3-MBA produced a total of three, one, and one tumors in the combined carcinogenicity studies ("questionable carcinogens"). For references see ref. 49.
[b] Data from ref. 29.
[c] Data from ref. 43.
[d] Data from ref. 49.
[e] Number of revertants per nmole test compound under optimal conditions.
[f] Maximal increase in the number of revertants above background. The numbers of spontaneous colonies (70–130) were subtracted.

ences in the results of hepatocyte- and standard S9-mediated experiments. Since the number of compounds tested in this study was small, the good correlation of hepatocyte-mediated bacterial mutagenicity with carcinogenicity in animals could be fortuitous. Therefore, we conducted a second study using a larger series of compounds, BA, the 12 isomeric methyl-BAs, and DMBA (Table 7).[49] These compounds differ widely in their carcinogenicity despite their physicochemical and structural similarities. Moreover, several comparative carcinogenicity tests have been performed with the compounds with reasonable consistency of the results, as indicated by values of Spearman's rank correlation coefficient r_S of 0.40–0.89 in pairs of test.[49] In contrast, the results of two S9-mediated and one hepatocyte homogenate-mediated bacterial mutagenicity studies did not show any significant correlation with any carcinogenicity study (r_S = −0.20 to +0.23). Bacterial mutagenicity mediated by intact hepatocytes poorly correlated with mutagenicity mediated by subcellular preparations (r_S = 0.10–0.17), but correlated well with animal carcinogenicity (r_S = 0.48–0.79, i.e., as good as the correlations among the carcinogenicity studies themselves). These results imply (1) that the poor correlation between animal carcinogenicity and mutagenicity in the standard Ames test of these PAHs is predominantly due to differences in the metabolism in the two systems rather than to differences in the biological endpoint, i.e., bacterial mutagenicity versus mammalian carcinogenicity, and (2) that differences in metabolism under various conditions with cellular integrity (i.e., the different target tissues in the carcinogenicity experiments and the hepatocytes in the mutagenicity experiments) were substantially smaller than were the differences between metabolism in cellular and subcellular preparations from the same tissue. Moreover, because the mutagenic effects generally were stronger with subcellular rather than with cellular metabolizing systems, it is likely that the differences in metabolism by these two systems primarily occur at the levels of inactivation and sequestration rather than at the level of activation.

5. Summary and Conclusions

Few, if any, mutagenic effects are caused by PAHs themselves. However, data from experiments on bacteria and mammalian cells in culture suggest that nearly all PAHs can be converted by mammalian enzymes to mutagens. The mutagenic potency can vary widely among different PAHs and for the same PAH under different metabolic conditions. Some PAHs, such as BP, can be metabolized to a large number of different ultimate mutagens. In bacterial tests mediated by subcellular mammalian enzyme systems (e.g., Ames test) K-region epoxides, simple alkene oxides, and vicinal diol-epoxides account for a large proportion of the mutagenic effects of various PAHs, as demonstrated by modulation of epoxide hydrolase activity. However, since certain PAHs whose structure does not allow formation of the mentioned epoxides, e.g., perylene (X), can exert potent mutagenic effects,[107,158] other metabolites must also be important. In a few cases, mutagenicity of metabolically generated free radicals as well as benzylic esters and photochemically generated epidioxides has been demonstrated.

Various carcinogenicity tests on PAH metabolites have been performed, but practically no *in vivo* mammalian mutagenicity tests have been conducted. The mode of activation of PAHs *in vivo* has been investigated mainly by analysis of DNA adducts. It appears that bay-region diol-epoxides represent the major active species of the investigated PAHs that are potent carcinogens. However, other DNA adducts have been found. In certain cases they could be associated with known metabolites (e.g., K-region epoxides, non-bay-region diol-epoxides), but in other cases they could not. It is apparent that metabolites other than bay-region diol-epoxides contribute to *in vivo* effects of PAHs, because various carcinogenic polycyclic aromatic compounds (usually of moderate potency), such as 10-aza-BP, anthanthrene (XI), 9,10-dimethylanthracene (cf. I), and cyclopenta(*c,d*)pyrene (VIII), cannot be metabolized to bay-region diol-epoxides.

Although a variety of metabolites contribute to both animal carcinogenicity and S9-mediated bacterial mutagenicity of PAHs, the contribution of a metabolite may be very different in the two systems. Consistent with this statement is the very poor correlation between carcinogenic potency and mutagenic potency in the Ames test of PAHs. Bacterial mutagenicity mediated by intact mammalian cells displays a better correlation with animal carcinogenicity than does S9-mediated mutagenicity. This strongly suggests that the poor correlation (in the case of PAHs) between carcinogenicity and mutagenicity in the standard Ames test is predominantly due to differences in metabolism of the test compounds in the two systems rather than to the differences in the biological endpoint, i.e., bacterial mutagenicity versus mammalian carcinogenicity.

One important difference between cellular and subcellular metabolizing systems is that in the latter many inactivation mechanisms are not operative and hence there is no discrimination between active intermediates that are readily inactivated in tissues and those that are poorly inactivated.

From the relatively good correlation between cell-mediated bacterial mutagenicity and carcinogenicity of PAHs, one would expect that mutagenicity in mammals *in vivo* should correlate better with these two effects than with S9-mediated bacterial mutagenicity. Despite the correlations, metabolism *in vivo* may vary, depending on species, tissue, enzyme induction state, dose and mode of exposure, and other factors. Understanding the metabolic control of mutagenic intermediates in test systems and in man should be very useful for risk estimations. However, numerous findings in the last decade have shown that this control is much more complex than may have been expected.

6. Addendum: Recent Developments

6.1. Mutagenicity of Quinones

TA104 is a new *his*$^-$ strain of *S. typhimurium*, which is reverted by reactive oxygen species. Tested in the presence of NADPH-fortified postmitochondrial supernantant fraction of liver homogenate, various quinones derived from PAHs, e.g., BP-3,6-quinone, were mutagenic in this strain.[(153)] Activation was also achieved by purified cytochrome P-450 reductase, a one electron reductase. The mutagenic potency of the tested quinones correlated well with the quinone-induced production of superoxide. Moreover, catalase and superoxide dismutase inhibited the cytochrome P-450 reductase-mediated mutagenicity almost completely. Taken together, these results indicate that the observed mutagenic effects of the quinones in *S. typhimurium* TA104 were caused by reactive forms of oxygen. TA97 is another new strain of *S. typhimurium*, which, however, is not sensitive to reactive oxygen. Pyrene-1,6-quinone and pyrene-1,8-quinone elicited potent direct mutagenic effects in this strain.[(158)] It is not known whether the pyrene quinones act as ultimate mutagens. It appears more probable that bacterial metabolites, possibly semiquinones, were responsible for the mutagenicity.

6.2. Mutagenicity of Phenol-diol-epoxides

On electronic grounds it was postulated that the reactivity of epoxides derived from PAHs is enhanced by introduction of a phenolic hydroxy-group at certain positions of the aromatic system.[(60)] Recently, phenol-diol-epoxides were available for investigations for the first time. These compounds, *syn*- and *anti*-9-hydroxychrysene-1,2-diol 3,4-oxide, were more mutagenic in either bacterial or mammalian cells and had shorter half-lives in aqueous medium than *syn*- and *anti*-chrysene-1,2-diol 3,4-oxide.[154] The significance of phenol-diol-epoxides in the mutagenicity and carcinogenicity of chrysene was indicated by two sets of findings: (1) 9-Hydroxychrysene (= 3-hydroxychrysene) and 9-hydroxychrysene-1,2-diol were effectively activated to mutagens by subcellular liver preparations. (2) After local application of chrysene, two major types of DNA adduct could be distinguished in mouse skin, a target tissue of chrysene carcinogenesis.[155] Using various chromatographic methods, these adducts were indistinguishable from the adducts formed with DNA from *anti*-chrysene-1,2-diol 3,4-oxide and *anti*-9-hydroxychrysene-1,2-diol 3,4-oxide, respectively.[156]

These new findings do not fundamentally diverse from previous concepts about the activation of PAHs to mutagens. However, they demonstrate that this activation, in accordance with its inherent biological character, occurs in numerous variations and has many facets.

ACKNOWLEDGMENTS

We thank M. A. Bedell, M. Lafranconi, K. L. Platt, L. W. Robertson, and A. Seidel for critical reading of the manuscript, I. Böhm for her excellent help in preparing this manuscript, and the Deutsche Forschungsgemeinschaft (SFB 302) for financial support.

7. References

1. B. N. Ames, P. Sims, and P. L. Grover, Epoxides of carcinogenic polycyclic hydrocarbons are frameshift mutagens, *Science 176*, 47–49 (1972).
2. B. N. Ames, W. E. Durston, E. Yamasaki, and F. D. Lee, Carcinogens are mutagens: A simple test combining liver homogenates for activation and bacteria for detection, *Proc. Natl. Acad. Sci. USA 70*, 2281–2285 (1973).
3. A. W. Andrews, L. H. Thibault, and W. Lijinski, The relationship between carcinogenicity and mutagenicity of some polynuclear hydrocarbons, *Mutat. Res. 51*, 311–318 (1978).
4. R. N. Armstrong, W. Levin, D. E. Ryan, P. E. Thomas, H. D. Mah, and D. M. Jerina, Stereoselectivity of rat liver cytochrome P-450c on formation of benzo(*a*)pyrene 4,5-oxide, *Biochem. Biophys. Res. Commun. 100*, 1077–1084 (1981).
5. S. W. Ashurst and G. M. Cohen, The formation of benzo(*a*)pyrene–deoxyribonucleoside adducts *in vivo* and *in vitro, Carcinogenesis 3*, 267–273 (1982).
6. H. Autrup, F. C. Wefald, A. M. Jeffrey, H. Tate, R. D. Schwarz, B. F. Trump, and C. C. Harris, Metabolism of benzo(*a*)pyrene by cultured tracheobronchial tissues from mice, rats, hamsters, bovines, and humans, *Int. J. Cancer 25*, 293–300 (1980).
7. H. Autrup, R. C. Grafstrom, M. Brugh, J. F. Lechner, A. Haugen, B. F. Trump, and C. C. Harris, Comparison of benzo(*a*)pyrene metabolism in bronchus, esophagus, colon and duodenum from the same individual, *Cancer Res. 42*, 934–938 (1982).

8. W. Baer-Dubowska, C. Frayssinet, and K. Alexandrov, Formation of covalent deoxyribonucleic acid benzo(*a*)pyrene 4,5-epoxide adduct in mouse and rat skin, *Cancer Lett. 14*, 125–129 (1981).
9. K. D. Bartle and D. W. Jones, The application of proton magnetic resonance spectroscopy to structure identification in polycyclic aromatic molecules, *Adv. Org. Chem. 8*, 317–423 (1972).
10. J. R. Bend, Z. Ben-Zvi, J. Van Anda, P. M. Dansette, and D. M. Jerina, Hepatic and extrahepatic glutathione *S*-transferase activity toward several arene oxides and epoxides in the rat, in: *Carcinogenesis, A Comprehensive Survey* (R. Freudenthal and P. W. Jones, eds.), Vol. 1, pp. 63–75, Raven Press, New York (1976).
11. P. Bentley, H. U. Schmassmann, P. Sims, and F. Oesch, Epoxides derived from various polycyclic hydrocarbons as substrates of homogeneous and microsome-bound epoxide hydratase: A general assay and kinetic properties, *Eur. J. Biochem. 69*, 97–103 (1976).
12. P. Bentley, F. Oesch, and H. R. Glatt, Dual role of epoxide hydratase in both activation and inactivation of benzo(*a*)pyrene, *Arch. Toxicol. 39*, 65–75 (1977).
13. C. A. H. Bigger, J. E. Tomaszewski, and A. Dipple, Variation in route of microsomal activation of 7,12-dimethylbenz(*a*)anthracene with substrate concentration, *Carcinogenesis 1*, 15–20 (1980).
14. K. W. Bock, Increase of liver microsomal benzo(*a*)pyrene monooxygenase activity by subsequent glucuronidation, *Naunyn-Schmiedeberg's Arch. Pharmacol. 304*, 77–79 (1978).
15. K. W. Bock, U. C. von Clausbruch, R. Kaufmann, W. Lilienblum, F. Oesch, H. Pfeil, and K. L. Platt, Functional heterogeneity of UDP-glucuronyltransferase in rat tissues, *Biochem. Pharmacol. 29*, 495–500 (1980).
16. M. Boroujerdi, H.-c. Kung, A. G. E. Wilson, and M. W. Anderson, Metabolism and DNA binding of benzo(*a*)pyrene *in vivo* in the rat, *Cancer Res. 41*, 951–957 (1981).
17. M. Bücker, H. R. Glatt, K. L. Platt, D. Avnir, Y. Ittah, J. Blum, and F. Oesch, Mutagenicity of phenanthrene and phenanthrene K-region derivatives, *Mutat. Res. 66*, 337–348 (1979).
18. M. Bücker, M. Golan, H. U. Schmassmann, H. R. Glatt, P. Stasiecki, and F. Oesch, The epoxide hydratase inducer *trans*-stilbene oxide shifts the metabolic epoxidation of benzo(*a*)pyrene from the bay- to the K-region and reduces its mutagenicity, *Mol. Pharmacol. 16*, 656–666 (1979).
19. D. R. Buhler, F. Unlu, D. R. Thakker, T. J. Slaga, M. S. Newman, W. Levin, A. H. Conney, and D. M. Jerina, Metabolism and tumorigenicity of 7-, 8-, 9- and 10-fluorobenzo(*a*)pyrenes, *Cancer Res. 42*, 4779–4783 (1982).
20. J. Capdevila, R. W. Estrabrook, and R. A. Prough, The existence of a benzo(*a*)pyrene-3,6-quinone reductase in rat liver microsomal fractions, *Biochem. Biophys. Res. Commun. 83*, 1291–1298 (1978).
21. E. Cavalieri and E. Rogan, One-electron oxidation of aromatic hydrocarbons in chemical and biological systems, in: *Polynuclear Aromatic Hydrocarbons: Formation, Metabolism and Measurement* (M. Cooke and A. J. Dennis, eds.), pp. 1–26, Battelle Press, Columbus, Ohio (1983).
22. P. A. Cerutti and J. F. Remsen, Formation and repair of DNA damage induced by oxygen radical species in human cells, in: *Cellular Senescence and Somatic Cell Genetics: DNA Repair Processes* (W. Nichols and D. Murphy, eds.), pp. 147–166, Symposium Specialist, Miami, Florida (1977).
23. R. L. Chang, P. G. Wislocki, J. Kapitulnik, A. W. Wood, W. Levin, H. Yagi, H. D. Mah, D. M. Jerina, and A. H. Conney, Carcinogenicity of 2-hydroxybenzo(*a*)pyrene and 6-hydroxybenzo(*a*)pyrene in newborn mice, *Cancer Res. 39*, 2660–2664 (1979).

24. R. L. Chang, A. W. Wood, W. Levin, H. D. Mah, D. R. Thakker, D. M. Jerina, and A. H. Conney, Differences in mutagenicity and cytotoxicity of (+)- and (−)-benzo(*a*)pyrene 4,5-oxide: A synergistic interaction of enantiomers, *Proc. Natl. Acad. Sci. USA 76*, 4280–4284 (1979).
25. P.-L. Chiu and S. K. Yang, A structure–activity relationship study of monomethylbenzo(*a*)pyrenes by the use of the *Salmonella typhimurium* tester strain TA100 and by analysis of metabolite formation, in: *Polynuclear Aromatic Hydrocarbons: Physical and Biological Chemistry* (M. Cooke, A. J. Dennis, and G. L. Fisher, eds.), pp. 193–200, Battelle Press, Columbus, Ohio (1982).
26. M. W. Chou and S. K. Yang, Combined reversed-phase and normal-phase high-performance liquid chromatography in the purification and identification of 7,12-dimethylbenz(*a*)anthracene metabolites, *J. Chromatogr. 185*, 635–654 (1979).
27. M. W. Chou, P. P. Fu, and S. K. Yang, Metabolic conversion of dibenz(*a,h*)anthracene (±)*trans*-1,2-dihydrodiol and chrysene (±)*trans*-3,4-dihydrodiol to vicinal dihydrodiol epoxides, *Proc. Natl. Acad. Sci. USA 78*, 4270–4273 (1981).
28. A. Colmsjö, A. Rannug, and U. Rannug, Some chloro derivatives of polynuclear aromatic hydrocarbons are potent mutagens in *Salmonella typhimurium, Mutat. Res. 135*, 21–29 (1984).
29. M. M. Coombs, C. Dixon, and A.-M. Kissonerghis, Evaluation of the mutagenicity of compounds of known carcinogenicity, belonging to the benz(*a*)anthracene, chrysene, and cyclopenta(*a*)phenanthrene series, using Ames's test, *Cancer Res. 36*, 4525–4529 (1976).
30. C. S. Cooper, A. Hewer, D. Ribeiro, P. L. Grover, and P. Sims, The enzyme-catalysed conversion of *anti*-benzo(*a*)pyrene-7,8-diol 9,10-oxide into a glutathione conjugate, *Carcinogenesis 1*, 1075–1078 (1980).
31. C. S. Cooper, A. D. MacNicoll, O. Ribeiro, G. P. Gervasi, A. Hewer, C. Walsh, K. Pal, P. L. Grover, and P. Sims, The involvement of a non-'bay-region' diol-epoxide in the metabolic activation of benz(*a*)anthracene in hamster embryo cells, *Cancer Lett. 9*, 53–59 (1980).
32. F. B. Daniel, H. A. J. Schut, D. W. Sandwisch, K. M. Schenck, C. O. Hoffmann, I. R. Patrick, and G. D. Stoner, Interspecies comparisons of benzo(*a*)pyrene metabolism and DNA-adduct formation in cultured human and animal bladder and tracheobronchial tissues, *Cancer Res. 43*, 4723–4729 (1983).
33. M. J. S. Dewar, *The Molecular Orbital Theory of Organic Chemistry*, McGraw-Hill, New York (1969).
34. W. E. Fahl, D. G. Scarpelli, and K. Gill, Relationship between benzo(*a*)pyrene-induced DNA base modification and frequency of reverse mutations in mutant strains of *Salmonella typhimurium, Cancer Res. 41*, 3400–3406 (1981).
35. J. W. Flesher, K. H. Stansbury, and K. L. Sydnor, *S*-Adenosyl-L-methionine is a carbon donor in the conversion of benzo(*a*)pyrene to 6-hydroxymethylbenzo(*a*)pyrene by rat liver S-9, *Cancer Lett. 16*, 91–94 (1982).
36. J. W. Flesher, A. M. Kadry, M. Chien, K. H. Stansbury, C. Gairola, and K. L. Sydnor, Metabolic activation of carcinogenic hydrocarbons in the meso-position (L-region), in: *Polynuclear Aromatic Hydrocarbons: Formation, Metabolism and Measurement* (M. Cooke and A. J. Dennis, eds.), pp. 505–515, Batelle Press, Columbus, Ohio (1983).
37. C. Gairola and J. F. Glass, Mutagenicity of some fluoroderivatives of benzo(*a*)pyrene, *Cancer Lett. 11*, 51–56 (1980).
38. H. R. Glatt and F. Oesch, Phenolic benzo(*a*)pyrene metabolites are mutagens, *Mutat. Res. 36*, 379–384 (1976).

39. H. R. Glatt and F. Oesch, Inactivation of electrophilic metabolites by glutathione *S*-transferases and limitation of the system due to subcellular localization, *Arch. Toxicol. 39*, 87–96 (1977).
40. H. R. Glatt, F. Oesch, A. Frigerio, and S. Garattini, Epoxides metabolically produced from some known carcinogens and from some clinically used drugs: I. Differences in mutagenicity, *Int. J. Cancer 16*, 787–797 (1975).
41. H. R. Glatt, K. Vogel, P. Bentley, and F. Oesch, Reduction of benzo(*a*)pyrene mutagenicity by dihydrodiol dehydrogenase, *Nature 277*, 319–320 (1979).
42. H. R. Glatt, F. Oesch, and H.-G. Neumann, Factors responsible for the metabolic formation and inactivation of bacterial mutagens from *trans*-4-acetylaminostilbene, *Mutat. Res. 73*, 237–250 (1980).
43. H. R. Glatt, K. Vogel, P. Bentley, P. Sims, and F. Oesch, Large differences in metabolic activation and inactivation of chemically closely related compounds: Effects of pure enzymes and enzyme induction on the mutagenicity of the twelve monomethylated benz(*a*)anthracenes, 7,12-dimethylbenz(*a*)anthracene and benz(*a*)anthracene in the Ames test, *Carcinogenesis 2*, 813–821 (1981).
44. H. R. Glatt, R. Billings, K. L. Platt, and F. Oesch, Improvement of the correlation of bacterial mutagenicity with carcinogenicity of benzo(*a*)pyrene and four of its major metabolites by activation with intact liver cells instead of cell homogenate, *Cancer Res. 41*, 270–277 (1981).
45. H. R. Glatt, C. S. Cooper, P. L. Grover, P. Sims, P. Bentley, M. Merdes, F. Waechter, K. Vogel, T. M. Guenthner, and F. Oesch, Inactivation of a diol-epoxide by dihydrodiol dehydrogenase, but not by two epoxide hydrolases, *Science 215*, 1507–1509 (1982).
46. H. R. Glatt, T. Friedberg, P. L. Grover, P. Sims, and F. Oesch, Inactivation of a diol-epoxide and a K-region epoxide with high efficiency by glutathione transferase X, *Cancer Res. 43*, 5713–5717 (1983).
47. H. R. Glatt, R. Jung, and F. Oesch, Bacterial mutagenicity investigation of epoxides: Drugs, drug metabolites, steroids and pesticides, *Mutat. Res. 11*, 99–118 (1983).
48. H. R. Glatt, I. Mertes, T. Wölfel, and F. Oesch, Epoxide hydrolases in laboratory animals and in man, in: *Biochemical Basis of Chemical Carcinogenesis* (H. Greim, R. Jung, M. Kramer, H. Marquardt, and F. Oesch, eds.), pp. 107–121, Raven Press, New York (1984).
49. H. R. Glatt, D. Utesch, and F. Oesch, Hepatocyte but not S9-mediated mutagenicity correlates with the carcinogenicity of methylbenz(*a*)anthracenes, in: *Polynuclear Aromatic Hydrocarbons: Mechanisms, Methods and Metabolism* (M. Cooke and A. J. Dennis, eds.), pp. 475–484, Battelle Press, Columbus, Ohio (1984).
50. H. R. Glatt, G. Ludewig, K. L. Platt, T. Waechter, I. Yona, S. Ben-Shoshan, P. Jerushalmy, J. Blum, and F. Oesch, Arene imines, a new class of exceptionally potent mutagens in bacterial and mammalian cells, *Cancer Res. 45*, 2600–2607 (1985).
51. A. Gold, S. Nesnow, M. Moore, H. Garland, G. Curtis, B. Howard, D. Graham, and E. Eisenstadt, Mutagenesis and morphological transformation of mammalian cells by a non-bay-region polycyclic cyclopenta(*c,d*)pyrene and its 3,4-oxide, *Cancer Res. 40*, 4482–4484 (1980).
52. E. M. Gozukara, F. P. Guengerich, H. Miller, and H. V. Gelboin, Different patterns of benzo(*a*)pyrene metabolism of purified cytochromes P-450 from methylcholanthrene, β-naphthoflavone and phenobarbital treated rats, *Carcinogenesis 3*, 129–133 (1982).
53. J. Guthrie, I. G. C. Robertson, E. Zeiger, J. A. Boyd, and T. E. Eling, Selective activation of some dihydrodiols of several polycyclic aromatic hydrocarbons to mutagenic products by prostaglandin synthetase, *Cancer Res. 42*, 1620–1623 (1982).

54. B. S. Hass, R. B. Webb, and T. B. Gambill, Chemical mutagenesis by benzo(*a*)pyrene in *Escherichia coli* in the absence of any activating agents, *Mutat. Res. 60*, 395–399 (1979).
55. G. M. Holder, H. Yagi, D. M. Jerina, W. Levin, A. Y. H. Lu, and A. H. Conney, Metabolism of benzo(*a*)pyrene: Effect of substrate concentration and 3-methylcholanthrene pretreatment on hepatic metabolism by microsomes from rats and mice, *Arch. Biochem. Biophys. 170*, 557–566 (1975).
56. E. Huberman, L. Sachs, S. K. Yang, and H. V. Gelboin, Identification of mutagenic metabolites of benzo(*a*)pyrene in mammalian cells, *Proc. Natl. Acad. Sci. USA 73*, 607–611 (1976).
57. E. Huberman, H. Yamasaki, and L. Sachs, Independent regulation of two types of aryl hydrocarbon (benzo(*a*)pyrene) hydroxylase in mammalian cells, *Int. J. Cancer 18*, 76–82 (1976).
58. E. Huberman, S. K. Yang, D. W. McCourt, and H. V. Gelboin, Mutagenicity to mammalian cells in culture by (+) and (−) *trans*-7,8-dihydroxy-7,8-dihydrobenzo(*a*)pyrenes and the hydrolysis and reduction products of two stereoisomeric benzo(*a*)pyrene 7,8-diol-9,10-epoxides, *Cancer Lett. 4*, 35–43 (1977).
59. P. B. Hulpert, Carbonium ion as ultimate carcinogens of polycyclic aromatic hydrocarbons, *Nature 256*, 146–148 (1975).
60. P. B. Hulpert and P. L. Grover, Chemical rearrangement of phenol-epoxide metabolites of polycyclic aromatic hydrocarbons to quinone-methides, *Biochem. Biophys. Res. Commun. 117*, 129–134 (1983).
61. M. L. Ide, M. Kaneko, and P. Cerutti, Benzo(*a*)pyrene and ascorbate-copper sulfate induce DNA damage in human cells by indirect action, in: *Protective Agents in Cancer* (D. C. H. McBrian and T. F. Slater, eds.), pp. 125–140, Academic Press, New York (1983).
62. M. N. Inbasekaran, D. T. Witiak, K. Barone, and J. C. Loper, Synthesis and mutagenicity of A-ring reduced analogues of 7,12-dimethylbenz(*a*)anthracene, *J. Med. Chem. 23*, 278–281 (1980).
63. A. M. Jeffrey, I. B. Weinstein, K. W. Jennette, K. Grzeskowiak, and K. Nakanishi, Structures of benzo(*a*)pyrene–nucleic acid adducts formed in human and bovine bronchial explants, *Nature 269*, 348–350 (1977).
64. D. M. Jerina, R. Lehr, M. Schaefer-Ridder, H. Yagi, J. M. Karle, D. R. Thakker, A. W. Wood, A. Y. H. Lu, D. Ryan, S. West, W. Levin, and A. H. Conney, Bay-region epoxides of dihydrodiols: A concept explaining the mutagenic and carcinogenic activity of benzo(*a*)pyrene and benzo(*a*)anthracene, in: *Origins of Human Cancer* (H. H. Hiatt, J. D. Watson, and J. A. Winsten, eds.), Book B, pp. 639–658, Cold Spring Harbor Laboratory, Cold Spring Harbor, New York (1977).
65. N. J. Joyce and F. B. Daniel, 7,12-Dimethylbenz(*a*)anthracene–deoxyribonucleoside adduct formation *in vivo*: Evidence for the formation and binding of a monohydroxymethyl-DMBA metabolite to rat liver DNA, *Carcinogenesis 3*, 297–301 (1982).
66. G. F. Kahl, E. Klaus, C. Legraverand, D. W. Nebert, and O. Pelkonen, Formation of benzo(*a*)pyrene metabolite–nucleoside adducts in isolated perfused rat and mouse liver and in mouse lung slices, *Biochem. Pharmacol. 28*, 1051–1056 (1979).
67. K. Kawajiri, H. Yonekawa, N. Harada, M. Noshiro, T. Omura, and Y. Tagashira, Immunochemical study on the role of different types of microsomal cytochrome P-450 in mutagenesis by chemical carcinogens, *Cancer Res. 40*, 1652–1657 (1980).
68. H. W. S. King, M. H. Thompson, and P. Brookes, The role of 9-hydroxybenzo(*a*)pyrene in the microsome mediated binding of benzo(*a*)pyrene to DNA, *Int. J. Cancer 18*, 339–344 (1976).

69. P. Kraus and H.-D. Kloft, The activity of glutathione *S*-transferases in various organs of the rat, *Enzyme 25*, 158–160 (1980).
70. E. J. LaVoie, L. Tulley, V. Bedenko, and D. Hoffmann, Mutagenicity of methylated fluorenes and benzofluorenes, *Mutat. Res. 91*, 167–176 (1981).
71. E. J. LaVoie, D. T. Coleman, R. L. Tonne, and D. Hoffmann, Mutagenicity, tumor initiating activity and metabolism of methylated anthracenes, in: *Polynuclear Aromatic Hydrocarbons: Formation, Metabolism and Measurement* (M. Cooke and A. J. Dennis, eds.), pp. 785–798, Battelle Press, Columbus, Ohio (1983).
72. E. J. LaVoie, L. Tulley-Freiler, V. Bedenko, and D. Hoffmann, Mutagenicity of substituted phenanthrenes in *Salmonella typhimurium*, *Mutat. Res. 116*, 91–102 (1983).
73. R. E. Lehr, A. W. Wood, W. Levin, A. H. Conney, D. R. Thakker, H. Yagi, and D. M. Jerina, The bay region theory: History and current perspectives, in: *Polynuclear Aromatic Hydrocarbons: Physical and Biological Properties* (M. Cooke, A. J. Dennis, and G. L. Fisher, eds.), pp. 21–37, Battelle Press, Columbus, Ohio (1982).
74. W. Levin, A. Y. H. Lu, D. Ryan, A. W. Wood, J. Kapitulnik, S. West, M.-T. Huang, A. H. Conney, D. R. Thakker, G. Holder, H. Yagi, and D. M. Jerina, Properties of the liver microsomal monooxygenase system and epoxide hydrase: Factors influencing the metabolism and mutagenicity of benzo(*a*)pyrene, in: *Origins of Human Cancer* (H. H. Hiatt, J. D. Watson, and J. A. Winsten, eds.), Book B, pp. 659–682, Cold Spring Harbor Laboratory, Cold Spring Harbor, New York (1977).
75. W. Levin, M. K. Buening, A. W. Wood, R. L. Chang, B. Kedzierski, D. R. Thakker, D. R. Boyd, G. S. Gadaginamath, R. N. Armstrong, H. Yagi, J. M. Karle, T. J. Slaga, D. M. Jerina, and A. H. Conney, An enantiomeric interaction in the metabolism and tumorigenicity of (+)- and (−)-benzo(*a*)pyrene 7,8-oxide, *J. Biol. Chem. 255*, 9067–9074 (1980).
76. P. E. Lombardi, J. W. Mayhew, B. R. Goldin, M. E. Gregory, M. A. Lynch, C. E. Sullivan, and S. L. Gorbach, Enzymatic methylation of microsomal metabolites of benzo(*a*)pyrene, *Cancer Res. 41*, 4415–4419 (1981).
77. R. J. Lorentzen and P. O. P. Ts'o, Benzo(*a*)pyrenedione/benzo(*a*)pyrene-diol oxidation–reduction couples and the generation of reactive reduced molecular oxygen, *Biochemistry 16*, 1467–1473 (1977).
78. R. J. Lorentzen, W. J. Caspary, S. A. Lesko, and P. O. P. Ts'o, The autoxidation of 6-hydroxybenzo(*a*)pyrene and 6-oxobenzo(*a*)pyrene radical, reactive metabolites of benzo(*a*)pyrene, *Biochemistry 14*, 3970–3977 (1975).
79. R. A. Lubet, J. Capdevila, and R. A. Prough, The metabolic activation of benzo(*a*)pyrene and 9-hydroxybenzo(*a*)pyrene by liver microsomal fractions, *Int. J. Cancer 23*, 353–357 (1979).
80. A. D. MacNicoll, C. S. Cooper, O. Ribeiro, K. Pal, A. Hewer, P. L. Grover, and P. Sims, The metabolic activation of benz(*a*)anthracene in three biological systems, *Cancer Lett. 11*, 243–249 (1981).
81. C. Malaveille, T. Kuroki, P. Sims, P. L. Grover, and H. Bartsch, Mutagenicity of isomeric diol-epoxides of benzo(*a*)pyrene and benz(*a*)anthracene in *S. typhimurium* TA98 and TA100 and in V79 Chinese hamster cells, *Mutat. Res. 44*, 313–326 (1977).
82. L. Marnett, J. Reed, A. Gregory, and J. T. Johnson, Prostaglandin synthetase dependent benzo(*a*)pyrene oxidation: Products of the oxidation and inhibition of their formation by antioxidants, *Biochem. Biophys. Res. Commun. 79*, 569–576 (1977).
83. H. Marquardt, S. Baker, P. L. Grover, and P. Sims, Malignant transformation and mutagenesis in mammalian cells induced by vicinal diol-epoxides derived from benzo(*a*)pyrene, *Cancer Lett. 3*, 31–36 (1977).
84. J. McCann, E. Choi, E. Yamasaki, and B. N. Ames, Detection of carcinogens as

mutagens in the *Salmonella*/microsome test: Assay of 300 chemicals, *Proc. Natl. Acad. Sci. USA 72*, 5135–5139 (1975).

85. J. Meijer, J. W. DePierre, and U. Rannug, Measurement of drug-metabolizing systems in *Salmonella typhimurium* strains G46, TA1535, TA100, TA1538 and TA98, *Chem. Biol. Interact. 31*, 247–254 (1980).
86. R. Morgenstern, C. Guthenberg, B. Mannervik, J. W. DePierre, and L. Ernster, Benzo(*a*)pyrene metabolism by rat liver microsomes: Effects of adding purified glutathione *S*-transferases A, B and C, *Cancer Res. 42*, 4215–4221 (1982).
87. M. Nagao, T. Sugimura, S. K. Yang, and H. V. Gelboin, Mutagenicity of optically pure (−)*trans*-7,8-dihydroxydihydrobenzo(*a*)pyrene, *Mutat. Res. 58*, 361–365 (1978).
88. C. Nagata, M. Kodama, T. Kimura, and M. Aida, Metabolically generated free radicals from many types of chemical carcinogens and binding of the radicals with nucleic acid bases, in: *Carcinogenesis: Fundamental Mechanisms and Environmental Effects* (B. Pullman, P. O. P. Ts'o, and H. V. Gelboin, eds.), pp. 43–54, D. Reidel, Dordrecht, Holland (1980).
89. N. Nemoto and H. V. Gelboin, Enzymatic conjugation of benzo(*a*)pyrene oxides, phenols and dihydrodiols with UDP-glucuronic acid, *Biochem. Pharmacol. 25*, 1221–1226 (1976).
90. N. Nemoto, S. Takayama, and H. V. Gelboin, Enzymatic conversion of benzo(*a*)pyrene phenols, dihydrodiols and quinones to sulfate conjugates, *Biochem. Pharmacol. 26*, 1825–1829 (1977).
91. R. F. Newbold and P. Brookes, Exceptional mutagenicity of benzo(*a*)pyrene diol epoxide in cultured mammalian cells, *Nature 261*, 52–54 (1976).
92. R. F. Newbold, P. Brookes, and R. G. Harvey, A quantitative comparison of the mutagenicity of carcinogenic polycyclic hydrocarbon derivatives in cultured mammalian cells, *Int. J. Cancer 24*, 203–209 (1979).
93. R. L. Norman, U. Muller-Eberhard, and E. F. Johnson, The role of cytochrome P-450 forms in 2-aminoanthracene and benzo(*a*)pyrene mutagenesis, *Biochem. Biophys. Res. Commun. 89*, 195–201 (1979).
94. F. Oesch, Epoxide hydratase, in: *Progress in Drug Metabolism* (J. W. Bridges and L. F. Chasseaud, eds.) Vol. 3, pp. 253–301, Wiley, London (1979).
95. F. Oesch and H. R. Glatt, Evaluation of the relative importance of various enzymes involved in the control of mutagenic and cytotoxic metabolites, in: *Screening Tests in Chemical Carcinogenesis* (R. Montesano, H. Bartsch, and L. Tomatis, eds.), IARC Scientific Publications No. 12 pp. 255–274, International Agency for Research on Cancer, Lyon (1976).
96. F. Oesch and M. Golan, Specificity of mouse liver cytosolic epoxide hydrolase for K-region epoxides derived from polycyclic aromatic hydrocarbons, *Cancer Lett. 9*, 169–175 (1980).
97. F. Oesch, P. Bentley, and H. R. Glatt, Prevention of benzo(*a*)pyrene-induced mutagenicity by homogeneous epoxide hydratase, *Int. J. Cancer 18*, 448–452 (1976).
98. F. Oesch, H. R. Glatt, and H. U. Schmassmann, The apparent ubiquity of epoxide hydratase in rat organs, *Biochem. Pharmacol. 26*, 603–607 (1977).
99. F. Oesch, D. Raphael, H. Schwind, and H. R. Glatt, Species differences in activating and inactivating enzymes related to the control of mutagenic metabolites, *Arch. Toxicol. 39*, 97–108 (1977).
100. F. Oesch, M. Bücker, and H. R. Glatt, Activation of phenanthrene to mutagenic metabolites and evidence for at least two different activation pathways, *Mutat. Res. 81*, 1–10 (1981).
101. F. Oesch, H. R. Glatt, K. Vogel, A. Seidel, P. Petrovic, and K. L. Platt, Dihydrodiol

dehydrogenase: A new level of control by both sequestration of proximate and inactivation of ultimate carcinogens, in: *Biochemical Basis of Chemical Carcinogenesis* (H. Greim, R. Jung, M. Kramer, H. Marquardt, and F. Oesch, eds.), pp. 23–31, Raven Press, New York (1984).

102. H. Okuda, Y. Kitahara, K. Shudo, and T. Okamoto, Identification of an ultimate mutagen of 10-azabenzo(*a*)pyrene: Microsomal oxidation of 10-azabenzo(*a*)pyrene to 10-azabenzo(*a*)pyrene-4,5-oxide, *Chem. Pharm. Bull. 27*, 2547–2549 (1979).
103. I. S. Owens and P. I. Makenzie, A comparison of the isoelectric points of mouse liver UDP-glucuronosyltransferase enzymes conjugating the twelve benzo(*a*)pyrene phenols, *Biochem. Biophys. Res. Commun. 109*, 1075–1082 (1982).
104. I. S. Owens, G. M. Koteen, and C. Legraverend, Mutagenesis of certain benzo(*a*)pyrene phenols *in vitro* following further metabolism by mouse liver, *Biochem. Pharmacol. 28*, 1615–1622 (1979).
105. I. S. Owens, C. Legraverend, and O. Pelkonen, Deoxyribonucleic acid binding of 3-hydroxy- and 9-hydroxybenzo(*a*)pyrene following further metabolism by mouse liver microsomal cytochrome P_1-450, *Biochem. Pharmacol. 28*, 1623–1629 (1979).
106. O. Pelkonen, A. R. Boobis, H. Yagi, D. M. Jerina, and D. W. Nebert, The tentative identification of benzo(*a*)pyrene metabolite–nucleoside complexes produced *in vitro* by mouse liver microsomes, *Mol. Pharmacol. 14*, 306–322 (1978).
107. B. W. Penman, D. A. Kaden, H. L. Liber, T. R. Skopek, and W. G. Thilly, Perylene is a more potent mutagen than benzo(*a*)pyrene for *S. typhimurium*, *Mutat. Res. 77*, 271–277 (1980).
108. J. N. Pitts, Jr., K. A. Van Cauwenberghe, D. Grosjean, J. P. Schmid, D. R. Fitz, W. L. Belser, Jr., G. B. Knudson, and P. M. Hynds, Atmospheric reactions of polycyclic aromatic hydrocarbons: Facile formation of mutagenic nitro derivatives, *Science 202*, 515–519 (1978).
109. A. Pullman and B. Pullman, Electronic structure and carcinogenic activity of aromatic molecules: New developments, *Adv. Cancer Res. 3*, 117–169 (1955).
110. W. H. Rastetter, R. B. Nachbar, Jr., S. Russo-Rodriguez, R. V. Wattley, W. G. Thilly, B. M. Andon, W. L. Jorgensen, and M. Ibrahim, Fluoranthene: Synthesis and mutagenicity of four diol epoxides, *J. Org. Chem. 47*, 4873–4878 (1982).
111. E. G. Rogan, R. W. Roth, P. A. Katomski-Beck, J. R. Laubscher, and E. L. Cavalieri, Non-enzymatic ATP-mediated binding of hydroxymethyl derivatives of aromatic hydrocarbons to DNA, *Chem. Biol. Interact. 31*, 51–63 (1980).
112. H. S. Rosenkranz and R. Mermelstein, Mutagenicity and genotoxicity of nitroarenes: All nitro-containing chemicals were not created equal, *Mutat. Res. 114*, 217–267 (1983).
113. R. M. Santella, D. Grunberger, and I. B. Weinstein, DNA–benzo(*a*)pyrene adducts formed in a *Salmonella typhimurium* mutagenesis assay system, *Mutat. Res. 61*, 181–189 (1979).
114. I. Mertes, H. R. Glatt, and F. Oesch, Activities of drug-metabolizing enzymes in cultured mammalian cells used for the detection of mutagens, *Mutat. Res. 147*, 125–126 (1985).
115. J. K. Selkirk, R. G. Croy, and H. V. Gelboin, High-pressure liquid chromatographic separation of 10 benzo(*a*)pyrene phenols and the identification of 1-phenol and 7-phenol as new metabolites, *Cancer Res. 36*, 922–926 (1976).
116. K. Shinohara and P. A. Cerutti, Excision repair of benzo(*a*)pyrene–deoxyguanosine adducts in baby hamster kidney $21/C_{13}$ cells and in secondary mouse embryo fibroblasts, *Proc. Natl. Acad. Sci. USA 74*, 979–983 (1977).
117. P. Sims, P. L. Grover, A. Swaisland, K. Pal, and A. Hewer, Metabolic activation of benzo(*a*)pyrene proceeds by a diol-epoxide, *Nature 252*, 326–328 (1974).

118. K. Sivarajah, J. M. Lasker, and T. E. Ehling, Prostaglandin synthetase-dependent cooxidation of (±)-benzo(*a*)pyrene-7,8-dihydrodiol by human lung and other mammalian tissues, *Cancer Res. 41*, 1834–1839 (1981).
119. N. H. Sloane and T. K. Davis, Hydroxymethylation of the benzene ring: Microsomal hydroxymethylation of benzo(*a*)pyrene to 6-hydroxymethylbenzo(*a*)pyrene, *Arch. Biochem. Biophys. 163*, 46–52 (1974).
120. M. Sugiura, Y. Yamazoe, T. Kamataki, and R. Kato, Reduction of epoxy derivatives of benzo(*a*)pyrene by microsomal cytochrome P-450, *Cancer Res. 40*, 2910–2914 (1980).
121. K.-H. Summer, W. Göggelmann, and H. Greim, Glutathione and glutathione *S*-transferases in the *Salmonella* mammalian-microsome mutagenicity test, *Mutat. Res. 70*, 269–278 (1980).
122. D. R. Thakker, H. Yagi, H. Akagi, M. Koreeda, A. Y. H. Lu, W. Levin, A. W. Wood, A. H. Conney, and D. M. Jerina, Metabolism of benzo(*a*)pyrene: VI. Stereoselective metabolism of benzo(*a*)pyrene and benzo(*a*)pyrene 7,8-dihydrodiol to diol epoxides, *Chem. Biol. Interact. 16*, 281–300 (1977).
123. D. R. Thakker, H. Yagi, R. E. Lehr, W. Levin, M. Buening, A. Y. H. Lu, R. L. Chang, A. W. Wood, A. H. Conney, and D. M. Jerina, Metabolism of *trans*-9,10-dihydroxy-9,10-dihydrobenzo(*a*)pyrene occurs primarily by arylhydroxylation rather than formation of a diol epoxide, *Mol. Pharmacol. 14*, 502–513 (1978).
124. D. R. Thakker, W. Levin, M. Buening, H. Yagi, R. E. Lehr, A. W. Wood, A. H. Conney, and D. M. Jerina, Species-specific enhancement by 7,8-benzoflavone of hepatic microsomal metabolism of benzo(*e*)pyrene 9,10-dihydrodiol to bay-region diol epoxides, *Cancer Res. 41*, 1389–1396 (1981).
125. G. Theall, M. Eisinger, and D. Grunberger, Metabolism of benzo(*a*)pyrene and DNA adduct formation in cultured human epiderminal keratinocytes, *Carcinogenesis 2*, 581–587 (1981).
126. M.-H. Tu, D. Perry, and C. Chen, Mutagenic effect of 7,12-dimethylbenz(*a*)anthracene-epidioxide on *Salmonella typhimurium, Mol. Pharmacol. 15*, 189–191 (1979).
127. J. Van Cantfort, L. Manil, J. E. Gielen, H. R. Glatt, and F. Oesch, A new assay for glutathione *S*-transferase using (^{3}H)-benzo(*a*)pyrene 4,5-oxide as substrate: Inducibility by various chemicals in different rat tissues compared to that of aryl hydrocarbon hydroxylase and epoxide hydratase, *Biochem. Pharmacol. 28*, 455–460 (1979).
128. T. Watabe, T., Ishizuka, M. Isobe, and N. Ozawa, A 7-hydroxymethyl sulfate ester as an active metabolite of 7,12-dimethylbenz(*a*)anthracene, *Science 215*, 403–405 (1982).
129. A. Weston, P. L. Grover, and P. Sims, Formation of the 11,12-diol as a metabolite of benzo(*a*)pyrene by rat skin *in vivo, Biochem. Biophys. Res. Commun. 105*, 935–941 (1982).
130. D. L. Whalen, A. M. Ross, J. A. Montemarano, D. R. Thakker, H. Yagi, and D. M. Jerina, General acid catalysis in the hydrolysis of benzo(*a*)pyrene 7,8-diol 9,10-epoxides, *J. Am. Chem. Soc. 101*, 5086–5088 (1979).
131. F. J. Wiebel, J. K. Selkirk, H. V. Gelboin, D. A. Haugen, T. A. Van Der Hoeven, and M. J. Coon, Position-specific oxygenation of benzo(*a*)pyrene by different forms of purified cytochrome P-450 from rabbit liver, *Proc. Natl. Acad. Sci. USA 72*, 3917–3920 (1975).
132. F. J. Wiebel, S. Brown, H. L. Waters, and J. K. Selkirk, Activation of xenobiotics by monooxygenases: Cultures of mammalian cells as analytical tool, *Arch. Toxicol. 39*, 133–148 (1977).

133. P. G. Wislocki, A. W. Wood, R. L. Chang, W. Levin, H. Yagi, O. Hernandez, P. M. Dansette, D. M. Jerina, and A. H. Conney, Mutagenicity and cytotoxicity of benzo(*a*)pyrene arene oxides, phenols, quinones, and dihydrodiols in bacterial and mammalian cells, *Cancer Res. 36*, 3350–3357 (1976).
134. A. W. Wood, R. L. Goode, R. L. Chang, W. Levin, A. H. Conney, H. Yagi, P. M. Dansette, and D. M. Jerina, Mutagenic and cytotoxic activity of benzo(*a*)pyrene 4,5-, 7,8-, and 9,10-oxides and the six corresponding phenols, *Proc. Natl. Acad. Sci. USA 72*, 3176–3180 (1975).
135. A. W. Wood, W. Levin, A. Y. H. Lu, H. Yagi, O. Hernandez, D. M. Jerina, and A. H. Conney, Metabolism of benzo(*a*)pyrene and benzo(*a*)pyrene derivatives to mutagenic products by highly purified hepatic microsomal enzymes, *J. Biol. Chem. 251*, 4882–4890 (1976).
136. A. W. Wood, P. G. Wislocki, R. L. Chang, W. Levin, A. Y. H. Lu, H. Yagi, O. Hernandez, D. M. Jerina, and A. H. Conney, Mutagenicity and cytotoxicity of benzo(*a*)pyrene benzoring epoxides, *Cancer Res. 36*, 3358–3366 (1976).
137. A. W. Wood, R. L. Chang, W. Levin, R. E. Lehr, M. Schaefer-Ridder, J. M. Karle, D. M. Jerina, and A. H. Conney, Mutagenicity and cytotoxicity of benz(*a*)anthracene diol epoxides and tetrahydro-epoxides: Exceptional activity of the bay region 1,2-epoxides, *Proc. Nat. Acad. Sci. USA 74*, 2746–2750 (1977).
138. A. W. Wood, R. L. Chang, W. Levin, H. Yagi, D. R. Thakker, D. M. Jerina, and A. H. Conney, Differences in mutagenicity of the optical enantiomers of the diastereomeric benzo(*a*)pyrene 7,8-diol-9,10-epoxides, *Biochem. Biophys. Res. Commun. 77*, 1389–1396 (1977).
139. A. W. Wood, W. Levin, A. Y. H. Lu, D. Ryan, S. B. West, H. Yagi, H. D. Mah, D. M. Jerina, and A. H. Conney, Structural requirements for the metabolic activation of benzo(*a*)pyrene to mutagenic products: Effects of modifications in the 4,5-, 7,8-, and 9,10-positions, *Mol. Pharmacol. 13*, 1116–1125 (1977).
140. A. W. Wood, W. Levin, P. E. Thomas, D. Ryan, J. M. Karle, H. Yagi, D. M. Jerina, and A. H. Conney, Metabolic activation of dibenz(*a,h*)anthracene and its dihydrodiols to bacterial mutagens, *Cancer Res. 38*, 1967–1973 (1978).
141. A. W. Wood, R. L. Chang, W. Levin, D. E. Ryan, P. E. Thomas, H. D. Mah, J. M. Karle, H. Yagi, D. M. Jerina, and A. H. Conney, Mutagenicity and tumorigenicity of phenanthrene and chrysene epoxides and diol epoxides, *Cancer Res. 39*, 4069–4077 (1979).
142. A. W. Wood, W. Levin, D. R. Thakker, H. Yagi, R. L. Chang, D. E. Ryan, P. E. Thomas, P. M. Dansette, N. Whittaker, S. Turujman, R. E. Lehr, S. Kumar, D. M. Jerina, and A. H. Conney, Biological activity of benzo(*e*)pyrene: An assessment based on mutagenic activities and metabolic profiles of the polycyclic hydrocarbon and its derivatives, *J. Biol. Chem. 254*, 4408–4415 (1979).
143. A. W. Wood, R. L. Chang, M.-T. Huang, W. Levin, R. E. Lehr, S. Kumar, D. R. Thakker, H. Yagi, D. M. Jerina, and A. H. Conney, Mutagenicity of benzo(*e*)pyrene and triphenylene tetrahydroepoxides and diol-epoxides in bacterial and mammalian cells, *Cancer Res. 40*, 1985–1989 (1980).
144. A. W. Wood, R. L. Chang, W. Levin, D. E. Ryan, P. E. Thomas, M. Croisy-Delcey, Y. Ittah, H. Yagi, D. M. Jerina, and A. H. Conney, Mutagenicity of the dihydrodiols and bay-region diol-epoxides of benzo(*c*)phenanthrene in bacterial and mammalian cells, *Cancer Res. 40*, 2876–2883 (1980).
145. A. W. Wood, M.-T. Huang, R. L. Chang, H. L. Newmark, R. E. Lehr, H. Yagi, J. M. Sayer, D. M. Jerina, and A. H. Conney, Inhibition of the mutagenicity of bay-region diol epoxides of polycyclic aromatic hydrocarbons by naturally occurring

plant phenols: Exceptional activity of ellagic acid, *Proc. Natl. Acad. Sci. USA 79*, 5513–5517 (1982).

146. A. W. Wood, W. Levin, R. L. Chang, M.-T. Huang, D. E. Ryan, P. E. Thomas, R. E. Lehr, S. Kumar, M. Koreeda, H. Akagi, Y. Ittah, P. Dansette, H. Yagi, D. M. Jerina, and A. H. Conney, Mutagenicity and tumor-initiating activity of cyclopenta(*c,d*)pyrene and structurally related compounds, *Cancer Res. 40*, 642–649 (1980).
147. A. W. Wood, R. L. Chang, W. Levin, D. E. Ryan, P. E. Thomas, R. E. Lehr, S. Kumar, D. J. Sardella, E. Boger, H. Yagi, J. M. Sayer, D. M. Jerina, and A. H. Conney, Mutagenicity of the bay-region diol-epoxides and other benzo-ring derivatives of dibenzo(*a,h*)pyrene and dibenzo(*a,i*)pyrene, *Cancer Res. 41*, 2589–2597 (1981).
148. A. W. Wood, R. L. Chang, W. Levin, H. Yagi, M. Tada, K. P. Vyas, D. M. Jerina, and A. H. Conney, Mutagenicity of the optical isomers of the diastereomeric bay-region chrysene 1,2-diol-3,4-epoxides in bacterial and mammalian cells, *Cancer Res. 42*, 2972–2976 (1982).
149. A. W. Wood, J. M. Sayer, H. L. Newmark, H. Yagi, D. P. Michaud, D. M. Jerina, and A. H. Conney, Mechanism of the inhibition of mutagenicity of benzo(*a*)pyrene 7,8-diol 9,10-epoxide by riboflavin 5′-phosphate, *Proc. Natl. Acad. Sci. USA 79*, 5122–5126 (1982).
150. S. K. Yang and H. V. Gelboin, Nonenzymatic reduction of benzo(*a*)pyrene diol-epoxides to trihydroxypentahydrobenzo(*a*)pyrenes by reduced nicotinamide adenine dinucleotide phosphate, *Cancer Res. 36*, 4185–4189 (1976).
151. S. K. Yang, P. P. Roller, P. P. Fu, R. G. Harvey, and H. V. Gelboin, Evidence for a 2,3-epoxide as an intermediate in the microsomal metabolism of benzo(*a*)pyrene to 3-hydroxybenzo(*a*)pyrene, *Biochem. Biophys. Res. Commun. 77*, 1176–1182 (1977).
152. S. K. Yang, P. P. Roller, and H. V. Gelboin, Enzymatic mechanism of benzo(*a*)pyrene conversion to phenols and diols and an improved high-pressure liquid chromatographic separation of benzo(*a*)pyrene derivatives, *Biochemistry 16*, 3680–3687 (1977).
153. P. L. Chesis, D. E. Levin, M. T. Smith, L. Ernster, and B. N. Ames, Mutagenicity of quinones: Pathways of metabolic activation and detoxification, *Proc. Natl. Acad. Sci. USA 81*, 1696–1700 (1984).
154. H. R. Glatt, A. Seidel, W. Bochnitschek, H. Marquardt, H. Marquardt, R. M. Hodgson, P. L. Grover, and F. Oesch, Mutagenicity in bacterial and mammalian cells of diol-epxoides, triol-epoxides and other metabolites of chrysene, in: *Polynuclear Aromatic Hydrocarbons: Chemistry, Characterization and Carcinogenesis* (M. Cooke and A. J. Dennis, eds.), Battelle Press, Columbus, Ohio, in press.
155. R. M. Hodgson, A. Weston, and P. L. Grover, Metabolic activation of chrysene in mouse skin: Evidence for the involvement of a triol-epoxide, *Carcinogenesis 3*, 1639–1643 (1984).
156. R. M. Hodgson, A. Weston, A. Seidel, W. Bochnitschek, H. R. Glatt, F. Oesch, and P. L. Grover, Metabolism of chrysene to triols and a triol-epoxide in mouse skin and rat liver preparations, in: *Polynuclear Aromatic Hydrocarbons: Chemistry, Characterization and Carcinogenesis* (M. Cooke and A. J. Dennis, eds.), Battelle Press, Columbus, Ohio, in press.
157. S. Nesnow, S. Leavitt, R. Easterling, R. Watts, S. H. Toney, L. Claxton, R. Sangaiah, G. E. Toney, J. Wiley, P. Fraher, and A. Gold, Mutagenicity of cyclopenta-fused isomers of benz(*a*)anthracene in bacterial and rodent cells and identification of the major rat liver microsomal metabolites, *Cancer Res. 44*, 4993–5003 (1984).

158. M. Sakai, D. Yoshida, and S. Mizusaki, Mutagenicity of polycyclic aromatic hydrocarbons and quinones on *Salmonella typhimurium* TA97, *Mutat. Res. 156*, 61–67 (1985).
159. J. M. Sayer, H. Yagi, P. J. van Bladeren, W. Levin, and D. M. Jerina, Stereoselectivity of microsomal epoxide hydrolase toward diol epoxides and tetrahydroepoxides derived from benz(*a*)anthracene, *J. Biol. Chem. 260*, 1630–1640 (1985).

CHAPTER 3

Chromosomal Mutations: The Genetic Approach

Antony G. Searle

1. Introduction

The latest information from cytogenetic surveys in many countries[59] suggests that about 0.6% of newborns suffer from numerical or structural chromosomal abnormalities, with serious consequences for themselves or (in the case of some structural rearrangements) for their descendants. A much large proportion of abortions and stillbirths also show chromosomal abnormalities. Thus the human social load from chromosomal diseases is already very high. Since a large number of different chemicals have already been shown to induce chromosome aberrations in mammals,[51] it is important to have a clear idea of the types of anomalies that may be induced and how they can be detected and studied. This chapter will be particularly concerned with genetic methods for transmitted anomalies, rather than cytological ones, which are often used to detect anomalies in germ cells before transmission. Of course, the genetic approach may have to be followed up with cytological confirmation, but it has the advantage of dealing with the actual inherited anomaly and of obviating the need for too much skilled but laborious microscopic scanning. I first review what is known about

Antony G. Searle • Medical Research Council, Radiobiology Unit, Chilton, Didcot, Oxon OX11 ORD, England.

the various classes of anomaly in experimental mammals (especially the mouse) as compared with the human situation, then consider in detail methods of detection and analysis.

2. The Main Categories of Chromosomal Mutation

2.1. Numerical Anomalies

In this section I deal only with changes in the number of whole chromosomes, not with those alterations in chromosome number that arise as a secondary result of translocation induction and lead to gains or losses of parts of chromosomes (partial or tertiary monosomies and trisomies). Numerical anomalies can be split into two categories, polyploidy and aneuploidy, of which the latter is by far the most important in mutagenesis studies and general hazard evaluation.

2.1.1. Polyploids

These are lethal *in utero*. The commonest type, and the most studied, is triploidy, comprising about 16% of specific chromosome anomalies in human abortions,[(49)] but tetraploidy is about one-third as frequent. In the mouse, some triploids survive to the day 10 of embryonic life, but usually show gross abnormalities; most tetraploids die at the egg cylinder stage. The incidence of triploidy can be increased by various treatments, such as heat shock, around the time of fertilization or by fertilization *in vitro*, while the treatment of embryos approaching second cleavage division with cytochalasin B increases the frequency of tetraploidy. The genetic background is also important in determining the incidence and manifestation of triploidy, which is frequently the result of dispermy.

2.1.2. Aneuploids

These arise mainly by meiotic nondisjunction, but nullosomic gametes (giving monosomy) can also arise from chromosome damage and loss at haploid postmeiotic stages in the male and at the pronuclear stage. Chromosome losses at premeiotic stages appear to be cell-lethal.

2.1.2a. Monosomies. The XO condition is the only monosomy compatible with viability, although some XO mice and nearly all human XOs die *in utero*. Autosomal monosomies are eliminated in early embryonic life (soon after implantation at the latest) in both human and mouse development.

2.1.2b. Trisomies. The XXY and XYY conditions are viable in the mouse, although the former is sterile (as in man), being more or less completely devoid of spermatogenic cells. XYY mice are usually sterile with impaired spermatogenesis, but may be subfertile when young. Thus, they seem to differ from human XYYs, who have normal testes as a rule, apart from a few with hypogonadism.[27] The triple-X condition has not been described in the mouse, although it is well known in human females. XXX women show no congenital malformations and may have offspring, although they do not transmit the condition.

Most autosomal trisomies in the mouse lead to death in mid- to late pregnancy and may be associated with specific malformations, e.g., exencephaly with trisomy 12 and less frequently with trisomy 14.[25] Mice with trisomy 19 may survive for some days beyond birth, while it is now known that some trisomies for chromosomes 16 and 18 may also survive to birth, though the genetic background may be important.[25] This trisomy 16 survival is just one of the reasons for believing that there may be homologies between mouse chromosome 16 and human chromosome 21, trisomics for which may survive to show Down syndrome.[15,39] A small proportion of human trisomies for chromosomes 13 and 18 may also survive to birth, but with serious malformations which are usually fatal within the first few months of life (Patau and Edwards syndromes, respectively).

2.2. Structural Anomalies

Structural anomalies must be considered from two different points of view: (1) their immediate phenotypic effects on individuals carrying them, e.g., complete sterility in males heterozygous for certain translocations, and (2) their effects in subsequent generations, especially as a result of meiotic segregation. Although the latter have been concentrated on in the past, there is now a growing realization that deleterious, immediate effects may be an important source of detriment in the offspring of those exposed to certain clastogens.

2.2.1. Translocations

These involve the transfer of a chromosomal segment from one position to another. Usually this is in the form of a reciprocal exchange (reciprocal translocation) involving the formation of lesions in two separate, usually nonhomologous, chromosomes followed by an interchange of segments, but nonreciprocal insertion of material from one chromosome into another is also possible. However, this involves the formation of three lesions and is therefore much less common.

Robertsonian translocations involve whole chromosome arms and lead to the formation of a metacentric chromosome from two acrocentric ones or vice versa. Recently, a special category of reciprocal translocation known as an "intercentric translocation" has been described[(12)] in which breakpoints are very close to the centromere, so that effectively whole arms are exchanged.

2.2.1a. Reciprocal Translocations (RT). Homologous pairing at meiosis leads to the formation of cross-shaped configurations when the germ cells concerned are heterozygous for an RT, which can be recognized cytologically. In this situation, alternate and adjacent-1 disjunction are normally equivalent[(17)] and lead to a ratio of one balanced normal:one balanced and translocated:two unbalanced gametes, with duplications and deficiencies. In addition, there is some adjacent-2 disjunction (nondisjunction of homologous centromeres), which will give unbalanced gametes only. Since, for most translocations studied in the mouse, the unbalanced gametes give rise to zygotes that die mainly around the time of implantation, a translocation heterozygote generally produces litters that are about half or somewhat less than half normal size. This phenomenon is known as "semisterility" or "partial sterility." True sterility is found in males heterozygous for X-autosome translocations and for a number of wholly autosomal translocations[(31)] in which one breakpoint is in or near centromeric heterochromatin and the other more distal. This male sterility results from spermatogenic impairment and is dealt with in Chapter 13.

It seems probable that the unbalanced zygotes produced by most human heterozygotes also lead to very early embryonic death, so are largely unnoticed and do not affect the total number of offspring. However, some unbalanced structural aberrations (including RTs) are compatible with survival to birth, but usually show severe malformations. In surveys of newborn infants their frequency has been 0.06%.[(49)]

Phenotypic effects have been reported for some mouse RTs when in balanced form and apart from the male sterility already mentioned. Thus some have phenotypic effects associated with loci at their breakpoints, such as nonagouti, short-ear, and steel,[(50)] while some are associated with skeletal defects.[(58)] A number of recent cytogenetic surveys of patients in mental institutions or of mentally retarded children have shown higher than expected frequencies of balanced structural rearrangements, especially translocations,[(49)] so this aspect should be included in future risk estimates.

2.2.1b. Insertions. These involve the placing within one chromosome A of a nonterminal segment of another chromosome B; thus in the balanced form there is another chromosome B with a complementary deficiency of chromosome material. As with reciprocal translocations, about half the resultant gametes from an insertion heterozygote will be unbalanced, with duplications or deficiencies, which will normally result in early embryonic lethality. The most studied insertion in the mouse is Cattanach's translocation Is(In7;X)1Ct, in which an inverted part of chromosome 7 is inserted into the X chromosome.[50] The balanced form is viable, but soon becomes sterile in the male, while about one-quarter of males with the transposed part of chromosome 7 present as a duplication are viable and fertile. Male-sterile autosomal insertions have also been discovered.[57]

2.2.1c. Robertsonian Translocations. Large numbers of these translocations (otherwise known as centric fusions or whole arm translocations) have now been recovered from wild and laboratory mouse populations, although first reported in the mouse only 15 years ago. All the autosomes are involved in at least one existing Robertsonian, but not the sex chromosomes. When wild-derived Robertsonians are placed on a laboratory mouse genetic background they may show quite a high frequency of nondisjunction. This may also be shown by compounds of two Robertsonians with a common arm (i.e., with monobrachial homology), which are frequently sterile in the male.

The commonest human Robertsonian translocations are between acrocentric chromosomes 13 and 14 to give a metacentric, and similarly between chromosomes 14 and 21. Nondisjunction in heterozygotes for the latter can lead to trisomy 21 with Down syndrome. The risk of this in female carriers is about 10%.[49]

2.2.1d. Intercentric Translocations. These comprise a special type of reciprocal translocation in which both breakpoints are near the centromere, so that virtually whole arms are exchanged. They resemble Robertsonian translocations in the involvement of whole arms, but this does not lead to the formation of metacentrics from acrocentrics or vice versa. Crocker and Cattanach[12] have shown how this class of translocation can be detected by the use of stocks containing metacentric rather than acrocentric chromosomes. Crocker[11] distinguished two types of intercentric translocation, namely (1) those with breaks in the euchromatin near the centromeres, in which case the chromosomally unbalanced progeny will either die *in utero* or be phenotypically abnormal, or (2) those with breaks in the centromeric heterochromatin, in which case the chromosomally unbalanced progeny are viable and fertile. He reported that the only translocations recovered after treatment of mice with mitomycin C and ethylnitrosourea (ENU) were of the intercentric type, and deduced that preferential induction of this type of exchange would generally occur after treatment of mouse spermatogonia with S-dependent mutagens, which appear to produce aberrations mainly in heterochromatic regions. This is in line with Adler's observation[1] that the chromosome aberrations induced by the S-dependent mutagen mitomycin C were in centromeric regions only.

2.2.2. Inversions

These involve the inversion of part of a single chromosome. There are two main types: (1) paracentric, in which the centromere is not included, and (2) pericentric, in which it is.

2.2.2a. Paracentric Inversions. Heterozygotes for these show meiotic anaphase bridges and acentric fragments[41] and the most efficient way of detecting such inversions has proved to be the cytological identification of these bridges. Male heterozygotes for such inversions show very little reduction in litter size[42] and little extra embryonic mortality[18] in progeny. Such males also have normal viability, although the homozygous condition may be lethal. Ford *et al.*[18] showed that female heterozygotes for one particular inversion, In(2)5Rk, showed high postimplantation intrauterine mortality when 11–12 weeks old and even higher pre- and postimplantation mortality when 8–9 months old. Evans and Phillips[16] showed that a long paracentric inversion in the mouse X chromosome was responsible for the high frequency of XO progeny from carrier females and that this is what would be expected because of the formation of dicentric chromatids after single chiasma formation within the inversion loop.

Most of the inversions studied in man have been pericentric, but Fryns and van den Berghe[20] surveyed information on 12 paracentric ones, which suggested that heterozygotes tended to show some pathological features, ranging from slight retardation to some more serious conditions.

Roderick and Hawes[43] have shown that paracentric inversions can be induced by the chemicals ethyl methanesulfonate (EMS) and triethylenemelamine (TEM) as well as by radiation.

2.2.2b. Pericentric Inversions. Heterozygotes for these inversions, in which breaks are located each side of the centromere, do not produce anaphase bridges or acentric fragments, because the centromere is within the inversion loop. Instead, they generate duplication-deficiency gametes, like reciprocal translocations, when there is a chiasma within the inversion loop. Thus the frequency of their unbalanced gametes depends on the length of the inversion.

The frequency of inversions found in human newborn infants is only about 20% that of reciprocal or of Robertsonian translocations.[49] On the whole they seem to present similar risks to reciprocal translocations, in that congenital malformations may arise from zygotic imbalance, while balanced heterozygotes may be more liable to mental retardation or other disorders.

2.2.3. Deletions and Duplications

2.2.3a. Deletions. Studies on deletions in the mouse have concentrated on those induced by irradiation that affect the albino (*c*) locus and the closely linked *d* and *se* loci. Homozygotes are lethal, usually *in utero*, although those for several *c* locus deletions die around the time of birth[24] with biochemical, renal, and other defects. Larger deletions (up to about 5 cM long) can be detected cytologically, by effects on G bands or by the presence of diagnostic loops at pachytene.[28] Heterozygotes sometimes show reduced transmission,[46] with growth retardation and possibly some effects on viability. Tertiary monosomies have been recovered from certain male-sterile translocations; these are deficient for the centromeric end of one chromosome and the distal end of another. They are also male-sterile, tend to be retarded, and show other abnormalities.[4,31]

Human chromosome deletions may be manifest as pathological conditions with dominant inheritance, e.g., retinoblastoma and deletion of band 13q14,[60] while many viable partial monosomies with deleterious effects are now known, e.g., the *cri-du-chat* syndrome, with mental and physical retardation and minor malformations, which is associated with deficiencies in the short arm of human chromosome 5. This has an incidence of about 1.5 per 1000 among mentally retarded individuals.[35]

2.2.3b. Duplications. Most duplications seem to arise as a secondary consequence of translocation induction, especially of those translocations (usually male-sterile) with breaks near the proximal and distal ends of chromosomes, which thereby generate tertiary trisomies with duplication of the proximal and distal regions concerned. These trisomies may be viable, but are usually aspermic or oligospermic and may have specific malformations.[50] However, a tandem duplication of part of mouse chromosome 7, involving the *Hbb* locus, has been described by Russell *et al.*[47] Growth of heterozygotes was retarded, while homozygotes died prenatally. The unbalanced form of the mouse insertion known as Cattanach's translocation is also a duplication Dp(7;X)Ct of part of chromosome 7. Only about 25% of male carriers are viable, but these are usually fertile.[50]

Human partial trisomies are more frequent than partial monosomies and tend to have less deleterious effects, since deficiencies of chromosomal material seem more serious than duplications. There are some exceptions, however, and factors other than length are involved in determining the severity of chromosomal disorders. There is increasing evidence that duplications or deletions of late-replicating regions are less harmful than those involving earlier replicating regions.[49]

3. Genetic Detection of Aneuploidy in the Mouse

So far, direct genetic tests for aneuploidy have dealt only with the X chromosome. This is because autosomal monosomies, in contrast to XO, are all lethal *in utero*, while only a few autosomal trisomies survive for a limited time after birth. However, a number of mutants (especially those involving eye pigmentation or gross morphological effects on limbs or tail as well as biochemical effects) can be detected in late fetal life, while certain trisomies, e.g., those from chromosomes 13, 14, 16, 18, and 19,[25] usually survive to this late fetal period. Intrauterine genetic detection systems would be possible, therefore, if mutants could be found that manifested themselves in late fetal life and also showed a clear-cut dosage effect, such that two doses of the mutant and one of the wild type allele or vice versa would be phenotypically different from a heterozygote. Of course, specific abnormalities associated with particular trisomies would be an aid to detection.

3.1. Sex-Chromosomal Aneuploidy

Russell[45] has described in detail the methodology and the results of using X-linked markers to determine the frequencies of the XO and XXY conditions with and without mutagenic treatment of male and female mice. XO detection rests on the observation of offspring with homozygous phenotypes where heterozygotes would be expected, while the rare XXY can be picked up where a hemizygous male is expected but a heterozygous one observed. The most used genetic markers are tabby (*Ta*), greasy (*Gs*), and blotchy (Mo^{blo}), all located centrally on the X chromosomes, and all with very distinct phenotypic effects in the heterozygotes. Other possible markers are harlequin (*Hq*), in which heterozygotes have bald patches while homo- and hemizygotes are almost completely bald,[24] and the terminal marker cream (*Crm*), in which all three genotypes can be distinguished on an albino background under UV light, since the cream pigment fluoresces, giving a mosaic pattern in heterozygous females.[38] The existence of *Crm* also raises the possibility of detecting terminal deficiencies of the X chromosomes, especially of seeing if these lead to nonrandom inactivation of the deficient X chromosome, as would be expected on the basis of human cytogenetic experience.

3.2. Autosomal Aneuploidy

3.2.1. Tests for Dominant Lethals and Congenital Malformations

As described earlier, most of the trisomic products of nondisjunction die in the second half of pregnancy in the mouse, although a few survive to be born, usually retarded and sometimes with specific malformations. In general, therefore, their time of death differs from that characteristic of dominant lethals, most of which die around the time of implantation, as deciduomata or "small moles." Thus, treatment with an agent that induces high levels of nondisjunction should lead to a shift in the pattern of intrauterine lethality when the germ cells sampled are at the right stage (e.g., meiosis) to undergo nondisjunction. The most efficient way to detect the shift would be to autopsy the females concerned very late in pregnancy (see ref. 30 for use of this procedure on rats) and to make careful counts of small moles, live normal embryos, large moles (placentas), dead embryos, and live malformed (including retarded) embryos. The last three categories would include most trisomies, presence of which could be confirmed by chromosomal examination of the last category.

X rays and urethane can increase the frequency of congenital malformations (e.g., open eyelids, kinky tails, cleft palates, dwarfism) in 19-day mouse fetuses after parental treatment.[29,36,37] Urethane produced no chromosome rearrangements or dominant lethals, but did induce heritable tumors in the F_1 as well as malformations,[36,37] which suggested that gene mutations or small deletions were responsible in this instance. However, mouse trisomies are also associated with a number of malformations, e.g., exencephaly, cleft palate, many cardiac defects, and retardation. Thus, increased frequencies of nondisjunction should lead to increased frequencies of malformations, although many of the trisomies will be lethal before the optimum age for detection of malformations, i.e., just before birth.

3.2.2. Use of Genetic Markers

When heterozygotes for translocations are intercrossed the fusion of gametes with complementary duplications and deficiencies usually leads to a fully viable zygote, though there are exceptions.[54] The complementation products can be recognized by the presence of appropriate genetic markers on the chromosomes concerned. Thus Lyon *et al.*[34] intercrossed heterozygotes for the wild-derived Robertsonian Rb(4.6)2Bnr in which one sex was homozygous for the recessive chromosome 4 markers brown (*b*) and misty (*m*) while the other was homozygous for the wild-type alleles of these. Twenty-one of 1997 or about 1% of offspring showed the brown misty phenotype instead of the expected wild type. Clearly, as a result of nondisjunction in both parents, gametes from the brown misty parent that were disomic for chromosome 4 had complemented gametes from the wild-type parent nullisomic for that chromosome, to give brown misty offspring. The authors calculated the nondisjunction frequency in each parent for the chromosome concerned to be 15%. However, in similar crosses involving the laboratory-derived Robertsonian Rb(9.19)163H and the chromosome 19 recessive marker ruby (*ru*), no ruby offspring were recovered in 5260 mice. This suggested that the nondisjunction frequency was very low, as seems usual in laboratory-derived Robertsonians.

There are some doubts on whether the mechanisms of nondisjunction in these Robertsonian heterozygotes giving high frequencies of it are the same as those found in normal homozygous stocks. Therefore it would seem advisable to use such stocks giving high frequencies of nondisjunction as tester stocks, in the same way as with the specific locus test. Cattanach *et al.*[10] and Searle and Beechey[55] have discussed various approaches to the problem of constructing a stock with a sufficiently high frequency of nondisjunction to make its use feasible. These include (1) genetically marking both arms of a Robertsonian, (2) using markers with heterozygous expression if possible, so that both types of homozygote could be recognized, (3) combining two or more independent Robertsonians, and (4) combining two Robertsonians with an arm in common (monobrachial homology), since such compounds have higher ratios of nondisjunction than individual translocation heterozygotes.[26] Unfortunately, a number of these compounds are known to be sterile in the male (on certain genetic backgrounds at least), and so could not be used for studies on nondisjunction in female germ cells. However, Cattanach *et al.*[10] have reported promising results from the use of such systems after irradiation of male and female germ cells.

A final possibility discussed by Searle and Beechey[55] is the combination of three genetically marked Robertsonians with tribrachial homology so as to give a compound that would produce only nullisomic and disomic gametes. Thus this would resemble the well-known isochromosome stocks of *Drosophila*. The three translocations being used for this purpose are Rb(4.6)2Bnr, Rb(4.15)4Rma, and Rb(6.15)1Ald with light (B^{lt}), white (Mi^{wh}), and belted (*bt*) as genetic markers. Production of such a triple compound has now been achieved (Searle, unpublished results) but it remains to be seen whether a stock can be established. Thus, good genetic alternatives to the cytological approach for nondisjunction are now being developed.

4. Reciprocal Translocations: Detection and Analysis

The genetic detection of reciprocal translocations (RTs) rests largely on the cytological behavior of heterozygotes at meiosis, since the pairing of homologous segments which takes place then leads to the formation of unbalanced gametes and resultant semisterility, as described earlier. However, RTs also can have other phenotypic effects.[50]

4.1. Translocations in Specific Locus Tests

One very interesting group of RTs comprises those that can cause coat color variegation in heterozygous females. Nearly all have arisen in specific locus experiments in which irradiated meiotic or postmeiotic germ cells have been sampled after the treated mice were crossed to homozygotes for a number of recessive coat color genes. All are associated with X-autosome translocations,[50] since it is the spread of inactivation from the translocated X to the attached autosome that leads to the variegation if the female translocation heterozygote is heterozygous also for a suitable recessive coat color marker, with the recessive allele on the intact autosome. A small proportion of wholly autosomal translocations also have visible phenotypic effects, either in combination with a recessive allele at a specific lucus (*a, se*) or at some other locus (e.g., the steeloid series at the *S1* locus), where they behave like a dominant mutant with semisterility.[6] However, all these special categories of translocation are rare and there seems to be no obvious way in which their detection could be incorporated into a genetic test for translocations, although clearly they should be borne in mind. Their existence makes the combination of a specific locus test with an F_1

TABLE 1. Semisterility Test Criteria Used at Harwell[7]

Number of dead implants	Result for given total number of implantations[a]						
	1–5	6	7	8	9	10	>10
>4	S	S	S	S	S	S	S
4	S	S	S	S	S	S	S
3	I	I	I	I	I	I	I
2	I	I	I	I	I	I	F
<2	I	F	F	F	F	F	F

[a] F, Fertile; I, test inconclusive; S, semisterile.

translocation test especially valuable, because it would allow further examples of these valuable translocations to be identified.

4.2. Heritable Translocation Assays

This name is usually (but not always) applied to an assay in which F_1 offspring from treated male or female mice are tested to see if they are heterozygous for a reciprocal translocation, mainly on the basis of litter size.[21] It is misleading to apply it to cytological tests on metaphase spermatocytes of the exposed animals themselves (as has been done), since translocations identified at this stage may not all be heritable, because of the direct effects they may have on spermatogenesis.

A protocol for detecting translocation heterozygosity in the progeny of treated mice was first clearly set out by Carter *et al.*[7] It is routinely used at Harwell for linkage and fertility tests since it has proved very satisfactory with our stocks. F_1 males are mated to two females, which are killed in midpregnancy (around 14½ days post coitum), with counts of live and dead implantantations after the uterine horns have been carefully dissected open. The criteria given in Table 1 are used to decide whether the male can be diagnosed immediately as semisterile (both females giving the S criterion, or one S and one I) or fertile, or whether two more females must be added because the result is inconclusive (both I or one S and one F). Clearly, this type of test cannot be carried out on F_1 females if they are required for breeding. Instead, Carter and co-workers[7] proposed a diagnosis of semisterility if the female's first three litters by a normal male totalled 12 young or fewer, with confirmation by progeny test wherever possible.

No attempt has been made to modify these criteria for use with stocks in which average litter sizes are unusually low or high, but clearly they will be less efficient under these circumstances. They will also tend to miss those interesting but apparently rare RTs in which one of the main unbalanced products survives to late pregnancy or to birth, unless such survival is accompanied by gross malformation, which can be regarded as equivalent to intrauterine death and thus put in the "dead implant" category. Otherwise, a series of inconclusive results may indicate a translocation of this type.

Cattanach[9] has discussed the various protocols that have been used in the heritable translocation test in mice. The most usual method is the sequential one in which F_1 males are successively eliminated as fully fertile on the basis of the sizes of the first and, if necessary, the second and third litters. Those still suspect are tested further by midpregnancy autopsies similar to those of Carter *et al.*[7] In addition, Generoso *et al.*[22] have compared the efficiency of the sequential method and cytological screening for translocations[2] and have also shown how to estimate the misclassification errors involved in both methods. Adler[3] has also compared two modifications of the test, namely (1) the mating together for fertility testing of F_1 males and females, so that two offspring from the treated parent are tested at once, and (2) omission of the fertility testing, with cytological analysis of each F_1 male instead. In view of all these publications on different versions of the test it seems unnecessary to deal in detail with the various protocols suggested, but a few general points may be helpful:

1. The only male germ cells worth testing are those in late meiotic or postmeiotic stages, since those in early meiotic or premeiotic (spermatogonial) stages are best examined in the treated males themselves at meiotic metaphase I by cytological methods. Treated males should be mated for the following 3 weeks to cover the relevant stages.

2. Much more work should be done on the effects of treating oocytes, especially with matings up to 6 weeks afterward (sampling mature and maturing oocytes).

3. If translocations are being induced, quite a high proportion of these will lead to sterility in the male. Such male sterility is the result of spermatogenic impairment and associated with reduced testis size. This will help to differentiate it from some other types of male sterility (e.g., loss of libido), but cytological examination of the testis is very desirable, to look for translocation configurations at metaphase I. If possible, this should be combined with cytological examination of somatic tissues (e.g., bone marrow or cornea), since affected spermatocytes may not survive to metaphase I but are very likely to carry somatic marker chromosomes if a male-sterile translocation is present.

4. If the intention is to detect only reciprocal translocations, then

it is advisable to test only F_1 males. Both genetic and cytological confirmation of translocation heterozygosity are more difficult in F_1 females. Moreover, the use of males only will prevent the possible presence of XO mice, in which litter sizes are reduced to much the same extent as in translocation heterozygotes.

5. Use of S-dependent mutagens will lead to the induction of intercentric translocations, as described earlier. In a normal mouse stock with acrocentric chromosomes, heterozygotes for these would be expected to show semisterility, with extra embryonic lethality, just like other RTs. However, no multivalent configurations will be formed at metaphase I, so they can only be diagnosed by the existence of heritable semisterility.

6. Another important type of reciprocal translocation associated with few or no multivalent configurations is one with distal breakpoints on both chromosomes concerned, so that only short segments are interchanged. Most unbalanced products, therefore, will have relatively small duplications and deficiencies, which may be compatible with postnatal survival but may well lead to malformations or other abnormalities. Such translocations might be suspected if matings of particular F_1 progeny consistently lead to extra intrauterine lethality at late stages of pregnancy, with similar effects in about half the surviving offspring but without cytological confirmation of the presence of a translocation.

7. It is important to rule out the presence of preexisting translocations that have arisen spontaneously, so both parents of a translocation carrier should be examined with respect to breeding performance and (preferably) meiotic metaphase chromosome constitution.

8. It would be very useful if a large historical control could be built up, as with specific locus mutations.

4.3. Genetic Analysis of Reciprocal Translocations and Their Products

It may sometimes be feasible to carry out further genetic studies on the translocations detected, except for those found in sterile males. Such studies are desirable for two reasons: (1) much remains unknown about the properties of reciprocal translocations, especially about the extent of spermatogenic impairment and of other possible phenotypic effects in the heterozygote, and (2) translocations are very useful tools in, for example, the more accurate location of genes on particular chromosomes or in the study of position effects, so the existence of further translocation stocks is very desirable. Therefore, some brief notes are given on analytical methods.

TABLE 2. Linkage-Testing Stocks Used at Harwell[a]

Stock name	Chromosome tested	Markers[b]
LI	2, 5, 11	*Sd* A^w *Ra*, W^e, *Re*
LII	3, 13, 15, 17	*Va*, *Xt*, *Ca*, *T*
LIII	1, 2, 10, 19	*fz ln*, *a*, *v*, *ep*
LIVA	2, 4, 6, 7, 9, 14	*a*, *b*, *wa-1*, c^{ch}, *d se*, *s*
LVC	4, 11, 13	*je*, *wa-2*, pe^H
LVI	8, 10, 18	*Os* E^{so}, $Sl^{gb\text{-}H}$, *Tw*

[a] For meanings of locus symbols and positions of these loci on the linkage map see Green.[24]

[b] Where chromosomes carry more than one marker they are given in order from the centromeric end.

4.3.1. Linkage Tests

In general, heterozygotes for mouse RTs behave as if a dominant gene for semisterility was located at each of the two breakpoints concerned. Therefore the same linkage-testing stocks can be used as for ordinary dominant and recessive mutants, e.g., those shown in Table 2. The translocation heterozygote (or homozygote if available) is crossed with a member of the linkage-testing stock. If this is a recessive stock, semisterile F_1 mice are backcrossed to it and progeny are classified for the recessive markers and for sensitivity. If dominant, semisterile F_1's carrying the dominant markers are outcrossed to wild type, with classification of progeny as before.

In practice, assignment of translocation breakpoints to chromosomes will be normally be done by cytological methods (G or Q banding), which are outside the scope of this chapter. However, genetic methods are needed to locate the breakpoint more accurately on the linkage map of each of the chromosomes involved. A three-point linkage test (i.e., involving the translocation breakpoint and two known loci) will be needed to allow identification of double recombinant classes of progeny. These will be less frequent than the single recombinant classes and thus will allow the order of loci and breakpoint to be determined. However, it should be noted that some translocations tend to suppress crossing-over in particular regions, so that accurate positions on the linkage map may be difficult to determine. Examples of position-testing stocks are given in Table 3; these can be used in essentially the same way as the linkage-testing stocks described above. Due to a dearth of good genetic markers on chromosomes 12, 16, and 18, no satisfactory position-testing stocks are available yet for these chromosomes, while there are diffi-

TABLE 3. Examples of Position-Testing Stocks Used at Harwell[a]

Stock name	Chromosome	Markers
MA	3	*ma Va*
PS	4	*b m*
PM	5	W^{sh} *go bf*
SIG	6	*Sig Lc* Mi^{wh}
FR	7	*p* c^{ch} *sh-1 fr*
MT	9	*cw se tk*
VGR	10	*v gr* Sl^{con}
PCE	11	*wa-2 vt*
PM	13	*sa mu* pe^{H}
FK	14	*hr s* Rn^{fkl}
UBC	15	*uw bt Ca*
TFH	17	*T tf*
BM	19	*bm ep ru*

[a] Markers(24) are given in order from the centromeric end.

culties also with the X chromosome, because of hemizygous lethality and poor breeding of heterozygotes. Further details on linkage testing can be found in Burdette,(5) Robinson,(40) and Green.(23)

4.3.2. Phenotypic Effects

The extent to which balanced reciprocal translocations cause deleterious effects in the carriers themselves in still far from clear. For instance, many lead to sterility in the male, but it is not known what proportion of the rest have some spermatogenic defect. Therefore a comparison of sperm counts in male carriers of particular chemically induced translocations and in their brothers would be desirable, (see ref. 53 for method). Measurements of body weight at birth and weaning would also be desirable, while possible skeletal defects can be studied easily by the use of the alizarin or papain techniques.

4.3.3. Tertiary Monosomics and Trisomics

These are unlikely to be encountered unless female as well as male progeny of mutagen-treated mice are tested for the presence of translocations, since they are mainly associated with the type of translocation (having a proximal breakpoint on one chromosome and a distal on the other) that leads to male infertility. The only known viable and fertile exception so far is Ts(1^{13})70H.[13] Such partial aneusomies are very useful for the study of dosage effects and for duplication-deficiency mapping.[14]

4.3.4. Use of Genetic Markers

Genetic markers can be useful both in maintenance and study of particular RTs. The former use applies where the translocation is lethal in the homozygote, so will normally be kept by outcrossing the heterozygote. If there is a closely linked recessive marker, then heterozygotes for the RT and the recessive, with the latter on one of the intact chromosomes concerned, can be backcrossed to the homozygous recessive in each generation, so that wild-type offspring have a high probability of carrying the RT. Thus T(1;2)5Ca, with a breakpoint near the *a* locus on chromosome 2, is maintained by the cross T5Ca +/+ *a* × + *a*/+ *a*. If a dominant marker is obtained close to the breakpoint on one of the translocated chromosomes, then it can be used in the same way by outcrossing to wild type; if it is on one of the intact chromosomes (i.e., T17H +/+ *Os*), then it can only be used to identify the RT carrier in alternate generations unless homozygotes for the dominant are viable (which is not the case with *Os*) and easily distinguished from heterozygotes.

Genetic markers proximal to the translocation breakpoints can help in the study of the RTs concerned because they allow the extent of adjacent-2 disjunction (i.e., nondisjunction of homologous centromeres) of the translocated chromosomes to be estimated.[56] This is done by intercrossing RT heterozygotes, one sex of which is homozygous for the proximal marker and the other homozygous wild type. Only gametic products of adjacent-2 disjuncton can complement each other so as to produce offspring homozygous for the proximal marker concerned instead of heterozygous as expected. However, the expected complementation does not always occur[54] and translocations with genetic markers are very useful in exploring this noncomplementation phenomenon and finding out which chromosomes are involved.

5. Detection and Study of Other Translocations and of Inversions

5.1. Robertsonian Translocations

Although heterozygotes for Robertsonian translocations tend to show increased frequencies of nondisjunction, leading to extra fetal mortality, the increase is generally small in laboratory-derived ones. Therefore, it seems insufficient to form the basis of a genetic test for Robertsonian induction, nor has any other genetic method for their detection been proposed. There is good evidence[19] that ionizing radiations do not induce Robertsonian translocations in any significant numbers in male mice, but an entirely different situation might hold for certain chemical mutagens. If and when such translocations are induced by chemicals, it will be interesting to see how their nondisjunctional properties compare with those arising spontaneously.[50] Genetic as well as cytological methods can be used to ascertain frequencies of nondisjunction[34] in similar fashion to those used for RTs.

5.2. Insertions

The two insertional translocations maintained in the mouse[50] are both male-sterile in the balanced form. However, the well-known Cattanach or "flecked" translocation Is(In7;X)1Ct, which involves an inverted insertion of one-third of chromosome 7 into the X chromosome, is male-fertile when this insertion is duplicated (type II males). It has been much studied genetically because spread of inactivation from the X into the insertion leads to coat-color variegation when recessive alleles at the *c* and/or *p* are on the intact chromosome.[8] This insertion was recovered after treatment of male mice with TEM followed by mating to the seven-locus tester stock; no doubt more X-autosome reciprocal and insertional translocations may be recovered in the same way. There seems to be no simple genetic method for differentiating autosomal insertions from RTs. However, since single crossover events within the insertion lead to unbalanced gametes and zygotic lethality, linkage tests on insertion heterozygotes will reveal no recombination between the insertion and loci within the inserted chromosome segment.[57]

5.3. Inversions

As already discussed, the most efficient method of detecting paracentric inversions is a cytological one, namely the scoring of meiotic anaphases for bridges and acentric fragments.[41] With the advent of stocks of mice that carry a number of metacentric chromosomes, like the human karyotype, there are greater possibilities of scoring pericentric inversions, for which this cytological method does not apply. Instead, some may be detected by the same methods as for heritable translocations because of the extra embryonic lethality that heterozygotes for such inversions generate. The amount of this lethality will depend on the length of the inverted segment, since this will determine the frequency of single chiasmata within the segment, which generate duplication-deficiency gametes. Thus, if the inverted segment is 20 map units long, then the inversion heterozygote would be expected to produce almost 20% extra embryonic lethality. Double chiasmata (which restore the balanced condition if two-strand) will become more frequent as the length of the inverted segment increases, so that the approximately 50% embryonic lethality ("semisterility") found with RT heterozygotes would not be expected unless the inverted segment was more than 50 units long. Half the surviving progeny of an inversion heterozygote would be expected to carry the inversion. Thus the induction of a fairly large pericentric inversion would be suspected after treatment of a metacentric stock and a heritable translocation assay on the F_1 if (1) there was increased embryonic lethality but no cytological evidence for an RT (i.e., no quadrivalent configurations) and (2) about half the progeny of the propositus showed the same phenomenon. This is a similar picture to that found with intercentric translocations between acrocentric chromosomes. Cytological confirmation of the condition might come from an altered position of the centromere (if the inversion was asymmetric), from the presence of an inversion loop at pachytene, or from a change in the G-banding pattern. Because of the elimination of most crossover products within the loop, linkage tests would suggest that genetic markers proximal and distal to the inversion were more closely linked than was really the case, while there would be very little crossing-over between markers within the loop.

6. Detection and Study of Deletions

6.1. X-Chromosomal Deletions

As mentioned earlier, it seems likely that most X-chromosomal deletions will lead to nonrandom inactivation in somatic cells of the X chromosome carrying them, so will behave genetically like the XO condition but have 40 chromosomes. If a deficiency of a particular locus (e.g., *Crm* at the distal end of the X chromosome) is compatible with random inactivation, then this can be detected by (1) the occurrence of a *Crm*/*Crm* phenotype among the progeny of a +/+ × *Crm* Y or *Crm*/*Crm* × +/Y cross, (2) the production of *Ta*/+ rather than *Ta*/O daughters when such a female is crossed to a *Ta* male (to show that the chromosome concerned is not inactivated or missing), with determination of a chromosome number of 40, or (3) cytological confirmation of a deficiency, if possible.

6.2. Autosomal Deletions

Genetic detection of autosomal deletions mainly depends on the discovery of closely linked pairs of genes which are suitable for test stocks because (1) mutant alleles are obtainable in doubly homozygous form, and (2) they breed well and have clearly visible effects. The *d* and *se* specific loci are only 0.2 unit apart and have been successfully used for this purpose.[46]

6.2.1. Use of Dilute Short-ear

Control or treated wild-type mice are crossed to homozygotes for *d, se* (and usually the other five specific-locus mutant alleles); progeny are classified at weaning age. Very occasionally, mice of the dilute short-ear phenotype are found. To see if they carry a deficiency Df, they are crossed to heterozygotes for the lethal allele of dilute d^1, since Df(d)/d^1 shows the typical opisthotonic d^1 phenotype. If no such dilute lethals appear in the progeny, then the dilute short-ear mouse is probably *d se*/*d se* and thought to be the result of complementation of gametes monosomic and disomic for chromosome 9.[46] However, it could be the result of nondisjunction in + +/*d se* F_1 zygotes at an early cleavage division, with only the *d se* homozygous products forming the embryo proper.

Deleterious effects of *d se* deficiencies can be studied by (1) intercrossing and outcrossing heterozygotes, with dissections of females in late pregnancy, to establish the time of death of homozygotes and to determine whether heterozygosity is also associated with extra intrauterine lethality, and (2) crosses to d^1 heterozygotes to check for reduced transmission of the deficiency, when less than 25% of the offspring would have the dilute lethal phenotype. Various other tests for deleterious effects (e.g., body weight, sperm count) would seem desirable also.

6.2.2. Use of Other Pairs of Markers

Table 4 shows all the pairs of closely linked genes that seem suitable for use in trying to induce more specific-locus deficiencies with a phenotypic effect that is immediately visible in F_1 progeny. Some of

TABLE 4. Closely Linked Pairs of Genes with Visible Effects in the Mouse[a]

Chromosome	Symbols	Names	Map units apart (approx.)
1	*ln . . . th*	Leaden, tilted head	1.9
2	*a . . . bp*	Nonagouti, brachypodism	0.4
2	*ro . . . pa*	Rough, pallid	0.6
2	*we . . . mg*	Wellhaarig, mahogany	0.1
7	*p . . . ru-2*	Pink-eyed dilution, ruby-2	2.5
9	*d . . . se*	Dilute, short ear	0.2
9	*ash . . . se*	Ashen, short ear	0.2
11	*sh-2 . . . vt*	Shaker-2, vestigial tail	1.7
13	*cr . . . bg*	Crinkled, beige	0.2
19	*ru . . . ab*	Ruby, asebia	1.5
19	*ru . . . ep*	Ruby, pale ear	1.5

[a] For further details, see Green.[24]

these pairs are not available yet in homozygous form, but a stock homozygous for *a bp*, *d se*, and *ru ep* has been constructed,[52] to which *p ru-2* has now been added. The *a bp* pair has been used already in the Harwell specific locus stock,[32,33] but no deletions covering these two loci were found. The same sorts of tests for deleterious effects as suggested for *d se* would be useful.

6.2.3. Another Approach: Complementation Tests

If several of the mutations induced by a chemical mutagen at a specific locus turn out to be lethal in the homozygote, then it is possible that some of them are intergenic deletions of the locus concerned rather than base-pair or other intragenic changes. This is especially true if gene action at the locus concerned would not be expected to have a lethal effect even if completely abolished. Suspected deletions may be detected and studied genetically either by (1) crossing heterozygotes for the lethal mutations to mutant homozygotes for neighboring genes to see if the suspected deletions overlap these, or (2) complementation tests. These involve crossing together heterozygotes for different lethal "alleles" at the locus concerned. If they are true alleles, the compound is likely to be lethal also, but if they are deletions, they may partially complement each other to give a viable product. In this way, as Russell and colleagues[44,48] have shown, complementation maps can be built up for various loci showing the position and extent of deletions.

7. Conclusions

It is clear that genetic tests can be of considerable value in the task of detecting the induction of chromosome aberrations by chemicals and of studying their properties, although at some stage or other they must be followed up by cytological examination. Some indications have been given of the kinds of genetic stocks needed. Careful planning is needed to determine the most efficient test system, which will give the maximum amount of information with a given level of staffing and of cage space. If the main purpose of the tests is to gain a better idea of the genetic hazards from particular chemicals, then treatment of a multimetacentric mouse stock would seem very desirable in view of its greater similarity to the human karyotype. The use of test stocks containing pairs of closely linked recessives will allow the detection of both deletions and specific locus point mutations in the F_1, while the assessment of late fetal lethality and malformation in F_1 progeny can give much useful information. It is difficult to justify large-scale and expensive *in vivo* tests with particular chemicals unless they yield a commensurate amount of information on the induction of both point mutations and chromosome aberrations.

8. References

1. I.-D. Adler, Comparative cytogenetic study after treatment of mouse spermatogonia with mitomycin C, *Mutat. Res. 23*, 369–379 (1974).
2. I.-D. Adler, The cytogenetic heritable translocation test, *Biol. Zentralbl. 97*, 441–451 (1978).
3. I.-D. Adler, New approaches to mutagenicity studies in animals for carcinogenic and mutagen agents. 1. Modification of the heritable translocation test, *Teratogen. Carcinogen. Mutagen. 1*, 75–86 (1980).
4. C. V. Beechey, M. Kirk, and A. G. Searle, A reciprocal translocation induced in an oocyte and affecting fertility in male mice, *Cytogenet. Cell Genet. 27*, 129–146 (1980).
5. W. J. Burdette, *Methodology in Mammalian Genetics*, Holden-Day, San Francisco (1963).
6. N. L. A. Cacheiro and L. B. Russell, Evidence that linkage group IV as well as linkage group X are in chromosome 10, *Genet. Res. (Camb.) 25*, 193–195 (1975).
7. T. C. Carter, M. F. Lyon, and R. J. S. Phillips, Gene-tagged chromosome translocations in eleven stocks of mice, *J. Genet. 53*, 154–166 (1955).
8. B. M. Cattanach, Position effect variegation in the mouse, *Genet. Res. (Camb.) 23*, 291–206 (1974).
9. B. M. Cattanach, The heritable translocation test in mice, in: *Cytogenetic Assays of Environmental Mutagens* (T. C. Hsu, ed.), pp. 289–323, Allanheld, Osmun, Totowa, New Jersey (1981).
10. B. M. Cattanach, D. Papworth, and M. Kirk, Genetic tests for autosomal nondisjunction and chromosome loss in mice, *Mutation Res. 126*, 189–204 (1984).

11. A. J. M. Crocker, Mutation Studies in Mice Carrying Robertsonian Translocations Ph.D. Thesis, Council for National Academic Awards (1982).
12. M. Crocker and B. M. Cattanach, X-ray induction of translocations in mice carrying metacentrics (Robertsonian fusions); detection of whole arm chromosome exchanges, *Mutat. Res. 91*, 353–357 (1981).
13. P. de Boer, Fertile tertiary trisomy in the mouse, *Cytogenet. Cell Genet. 12*, 435–442, (1973).
14. E. M. Eicher and L. L. Washburn, Assignment of genes to regions of mouse chromosomes, *Proc. Natl. Acad. Sci. USA 75*, 946–950 (1978).
15. C. J. Epstein, Mouse monosomies and trisomies as experimental systems for studying mammalian aneuploidy, *Trends in Genetics 1*, 129–134 (1985).
16. E. P. Evans and R. J. S. Phillips, Inversion heterozygosity and the origin of XO daughters of *Bpa*/+ female mice, *Nature 256*, 40–41 (1975).
17. C. E. Ford, Meiosis in mammals, in: *Comparative Mammalian Cytogenetics* (K. Benirschke, ed.), pp. 91–106, Springer, New York (1969).
18. C. E. Ford, E. P. Evans, and M. D. Burtenshaw, in: *Current Chromosome Research* (K. Jones and P. E. Brandham, eds.), pp. 123–132, Elsevier/North-Holland, Amsterdam (1976).
19. C. E. Ford, E. P. Evans, and A. G. Searle, Failure of irradiation to induce Robertsonian translocations in germ cells of male mice, in: *Conference on Mutations: Their Origin, Nature and Potential Relevance to Genetic Risk in Man*, pp. 102–108, Deutsche Forschungsgemeinschaft, Harald Boldt, Boppard (1978).
20. J. P. Fryns and H. van den Berghe, Paracentric inversion in man: Personal experience and review of the literature, *Hum. Genet. 54*, 413–416 (1980).
21. W. M. Generoso, K. T. Cain, S. W. Huff, and D. G. Gosslee, Heritable-translocation test in mice, in: *Chemical Mutagens: Principles and Methods for Their Detection* (A. Hollaender, ed.), Vol. 5, pp. 55–77, Plenum Press, New York (1978).
22. W. M. Generoso, M. Krishna, C. T. Cain, and C. W. Sheu, Comparison of two methods for detecting translocation heterozygotes in mice, *Mutat. Res. 81*, 177–186 (1981).
23. E. L. Green, *Genetics and Probability in Animal Breeding Experiments*, MacMillan, London, (1981).
24. M. C. Green, *Genetic Variants and Strains of the Laboratory Mouse*, Gustav Fischer, Stuttgart (1981).
25. A. Gropp, Value of an animal model for trisomy, *Virchows Arch. Pathol. Anat. 395*, 117–131 (1982).
26. A. Gropp, U. Kolbus, and D. Giers, Systematic approach to the study of trisomy in the mouse. II, *Cytogenet. Cell Genet. 14*, 42–62 (1975).
27. J. L. Hamerton, *Human Cytogenetics*, Vol. II, Academic Press, New York (1971).
28. G. Jagiello, J. S. Fang, H. A. Turchin, S. E. Lewis, and S. Gluecksohn-Waelsch, Cytological observations of deletions in pachytene stages of oogenesis and spermatogenesis in the mouse, *Chromosoma 58*, 377–386 (1976).
29. M. Kirk and M. F. Lyon, Induction of congenital anomalies in offspring of female mice exposed to varying doses of X-rays, *Mutat. Res. 106*, 73–83 (1982).
30. I. Knudsen, E. V. Hansen, O. A. Meyer, and E. Poulsen, A proposed method for the simultaneous detection of germ-cell mutations leading to fetal death (dominant lethality) and of malformations (male teratogenicity) in mammals, *Mutat. Res. 48*, 267–270 (1977).
31. M. F. Lyon and R. Meredith, Autosomal translocations causing male sterility and viable aneuploidy in the mouse, *Cytogenet. 5*, 335–354 (1966).

32. M. F. Lyon and T. Morris, Mutation rates at a new set of specific loci in the mouse, *Genet. Res.* (*Camb.*) 7, 12–17 (1966).
33. M. F. Lyon and T. Morris, Gene and chromosome mutation after large fractionated or unfractionated doses to mouse spermatogonia, *Mutat. Res. 8*, 191–198 (1969).
34. M. F. Lyon, H. C. Ward, and G. M. Simpson, A genetic method for measuring non-disjunction in mice with Robertsonian translocations. *Genet. Res.* (*Camb.*) *26*, 283–295 (1976).
35. E. Niebuhr, The *cri du chat* syndrome, epidemiology, cytogenetics and clinical features, *Hum. Genet. 44*, 227–275 (1978).
36. T. Nomura, Induction of heritable tumours and anomalies in mice by parental exposure to X-rays and chemicals, *Nature 296*, 575–577 (1982).
37. T. Nomura, Quantitative studies on mutagenesis, teratogenesis and carcinogens in mice, in: *Problems of Threshold in Chemical Mutagenesis* (Y. Tazima, ed.), pp. 27–34, Environmental Mutagen Society of Japan, Tokyo (1984).
38. J. Peters and A. G. Searle, Personal communication, *Mouse News Letter 61*, 37 (1979).
39. P. C. Polani and M. Adinolfi, Chromosome 21 of man, 22 of the great apes and 16 of the mouse, *Dev. Med. Child Neurol. 22*, 223–233 (1980).
40. R. Robinson, *Gene Mapping in Laboratory Mammals*, Plenum Press, London (1971).
41. T. H. Roderick, Producing and detecting paracentric chromosomal inversions in mice, *Mutat. Res. 11*, 59–69 (1971).
42. T. H. Roderick and N. L. Hawes, Two radiation-induced chromosomal inversions in mice, *Proc. Natl. Acad. Sci. USA 67*, 961–967 (1970).
43. T. H. Roderick and N. L. Hawes, Nineteen paracentric chromosomal inversions in mice, *Genetics 76*, 109–117 (1974).
44. L. B. Russell, Definition of functional units in a small chromosome segment of the mouse and its use in interpreting the nature of radiation-induced mutations, *Mutat. Res. 11*, 107–123 (1971).
45. L. B. Russell, Numerical sex-chromosome anomalies in mammals: Their spontaneous occurrence and use in mutagenesis studies, in: *Chemical Mutagens: Principles and Methods for Detection* (A. Hollaender, ed.), Vol. 4, pp. 55–91 Plenum Press, New York (1976).
46. L. B. Russell and W. L. Russell, Genetic analysis of induced deletions and of spontaneous non-disjunction involving chromosome 2 of the mouse, *J. Cell. Comp. Physiol. 55*(Suppl. 1), 169–188 (1960).
47. L. B. Russell, W. L. Russell, N. L. A. Cacheiro, C. M. Vaughan, R. A. Popp, and K. B. Jacobson, A tandem duplication in the mouse, *Genetics 80*, s71 (1975).
48. L. B. Russell, C. S. Montgomery, and G. D. Raymer, Analysis of the albino-locus region of the mouse: IV. Characterization of 34 deficiencies, *Genetics 100*, 427–453 (1982).
49. K. Sankaranaryanan, *Genetic Effects of Ionizing Radiation in Multicellular Eukaryotes and the Assessment of Genetic Radiation Hazards in Man*, Elsevier, Amsterdam (1982).
50. A. G. Searle, Numerical variants and structural rearrangements, in: *Genetic Strains and Variants of the Laboratory Mouse* (M. C. Green, ed.), pp. 324–356, Gustav Fischer, Stuttgart (1981).
51. A. G. Searle, Germ-cell sensitivity in the mouse: A comparison of radiation and chemical mutagens, in: *Environmental Mutagens and Carcinogens* (T. Sugimura, S. Kondo, and H. Takebe, eds.), pp. 169–177, University of Tokyo Press, Tokyo and Liss, New York (1982).
52. A. G. Searle, Some ideas on future test systems, in: *Utilization of Mammalian Specific-Locus Studies in Hazard Evaluation and Estimation of Genetic Risk* (F. J. de Serres and W. Sheridan, eds.), Vol. 28, pp. 279–288, Plenum Press, New York (1983).

53. A. G. Searle and C. V. Beechey, Sperm count, egg-fertilization and dominant lethality after X-irradiation of mice, *Mutat. Res. 22*, 63–72 (1974).
54. A. G. Searle and C. V. Beechey, Complementation studies with mouse translocations, *Cytogenet. Cell Genet. 20*, 282–303 (1978).
55. A. G. Searle and C. V. Beechey, The use of Robertsonian translocation in the mouse for studies on non-disjunction, *Cytogenet. Cell. Genet. 33*, 81–87 (1982).
56. A. G. Searle, C. E. Ford, and C. V. Beechey, Meiotic disjunction in mouse translocations and the determination of centromere position, *Genet. Res.* (*Camb.*) *18*, 215–235 (1971).
57. A. G. Searle, C. V. Beechey, P. de Boer, D. G. de Rooij, E. P. Evans, and M. Kirk, A male-sterile insertion in the mouse, *Cytogenet. Cell Genet. 36*, 617–626 (1983).
58. P. B. Selby, Greater genetic hazard from radiation results from balanced reciprocal translocations than from segmental aneuploids, *Environ. Mutagen. 1*, 158–159 (1979).
59. United Nations Scientific Committee on the Effects of Atomic Radiation, *Ionizing Radiation: Sources and Biological Effects*, Report to the General Assembly, United Nations, New York (1982).
60. J. J. Yunis and N. Ramsey, Retinoblastoma and sub-band deletion of chromosome 13, *Am. J. Dis. Child. 132*, 161–163 (1978).

CHAPTER 4

Cytogenetic Assays for Mitotic Poisons Using Somatic Animal Cells

T. C. Hsu, Jan C. Liang, and Kanithahalli L. Satya-Prakash

1. Introduction

It is well known that most carcinogens are mutagens. They interact with cellular DNA via a variety of mechanisms resulting in gene mutation (mutagens) and chromosome breakage (clastogens). A number of short-term assay systems are available to detect mutagenic and/or clastogenic potential of environmental agents. However, some carcinogens, such as diethylstilbestrol (DES), halothane, and benzene, showed negative results in tests for mutations and for chromosome aberrations. Sawada and Ishidate[24] reported that DES induced a very high frequency of polyploidy in cultured Chinese hamster cells. Kusyk and Hsu,[11] using cell cultures and chick embryos, showed that halothane caused mitotic accumulation as well as mitotic anomalies such as multipolar anaphases. Liang *et al.*,[14] using grasshopper embryos exposed to vapors of volatile liquids, confirmed that halothane is a mitotic arrestant, but added data on benzene, which not only is capable of arresting dividing cells at metaphase, but also can cause multipolar divisions and lagging chro-

T. C. Hsu, Jan C. Liang, and Kanithahalli L. Satya-Prakash • Department of Genetics, University of Texas System Cancer Center, M.D. Anderson Hospital and Tumor Institute at Houston, Houston, Texas 77030.

mosomes. As will be discussed later, mitotic arrest alone would only induce polyploidy, not aneuploidy. Since a large number of neoplastic tissues are aneuploid, numerical chromosomal changes constitute a type of mutation that drastically alters the genome instead of single genes. Representing a committee, Crow[5] proposed the term "genomic mutation" (as opposed to gene mutation and chromosomal mutation) for aneuploidy.

Although cell division is a complex process necessary to ensure precise distribution of chromosomes to daughter cells, chromosomes (and DNA) are only passively involved in mitosis and meiosis. The major actors on the stages of mitosis and meiosis should include the spindle, the centrioles, the kinetochores, and the cytokinetic mechanism. The major molecules of all of these structure or processes are protein in nature. Therefore, it is not unexpected that agents that exclusively interact with DNA do not interfere with structures such as microtubules and that agents that can interact with such molecules (e.g., tubulin) do not induce mutations or chromosome breaks.

Aneuploidy is a cytogenetic phenomenon, but the prevailing assay systems for aneuploidy employ genetic markers in fungi,[7,17] yeast,[19] *Drosophila*,[21] and the laboratory mouse.[23] A few direct cytogenetic assays, such as observing aneuploidy in cleaving eggs, are laborious and expensive. The reason for the lack of cytogenetic test systems to estimate aneuploidy induction by environmental agents was primarily a technical one. Most cell lines are aneuploid with a range of chromosome numbers within a cell population. This poses a serious problem because the frequency of induced aneuploidy may be masked by existing aneuploidy. Furthermore, artifactual aneuploidy, the result of cell rupture during slide preparation, is unavoidable. In most cytogenetic analyses, it is easy to concentrate on metaphases with the modal chromosome number, but in estimating aneuploid frequency, aneuploid cells (both real and artifactual), must be recorded. Again, this technical defect hinders the progress in this area of research. In the present chapter, we present descriptions of protocols for three quick and inexpensive test systems using mammalian cells *in vitro*, mouse bone marrow *in vivo*, and grasshopper embryos *in ovo*.

2. Mitotic Poisons and Their Genomic Effects

The term mitotic poison may be loosely defined as any agent that can cause disturbances in the orderly process of cell division, which requires synchrony and coordination of the following cellular structures and events: separation of centrioles at the beginning of mitosis, formation of the spindle apparatus, attachment of kinetochores at the spindle, separation of sister chromatids, and cytokinesis to divide one cell into two. Conceivably, any disturbance of this orderly process of cell division could lead to abnormalities resulting in possible genomic alterations in the daughter cells. However, the bulk of information on damage and disturbances of these structures has been related to the spindle apparatus. Therefore, our discussion necessarily emphasizes this organelle also.

2.1. The Spindle Apparatus

The spindle apparatus is the most dominant cellular structure of cell division. It is well known that the major component of the spindle is the microtubule, a polymerized product of tubulin. Although spindle microtubules are the same as the microtubules of interphase cells, the cell does not use existing microtubules to form a spindle. Normally, an equilibrium between microtubule depolymerization and tubulin polymerization exists in interphase cells, but the mitotic spindle must be formed by new polymerization. Thus, cell division cannot proceed when the formation of spindle microtubules in mitosis is inhibited by one mechanism or another. The cell is arrested at metaphase (c-metaphase). Agents that can prevent spindle formation are known as *mitotic arrestants.*

When a cell is arrested at metaphase for a long period of time, the chromatids of each chromosome separate (c-anaphase), and the chromosomes eventually decondense as if they were entering telophase (c-telophase). However, the chromosomes may be scattered because of the absence of anaphase; therefore, when the nuclear envelope is formed, it is formed haphazardly, which may result in a number of micronuclei. The cell then enters interphase as a polyploid.

Whether a blocked metaphase can recover (reorganizing the spindle apparatus) and proceed with anaphase movement depends upon a number of factors:

1. The duration of arrest. When the cell is already in c-anaphase or c-telophase, there will be no recovery.
2. Mechanism of arrest. This will be discussed in a later section.
3. Type of cells. HeLa cells appear to be more susceptible to Colcemid than do the Chinese hamster cells.
4. Concentration of the arrestant. Probably recovery will always occur when the drug concentration is at borderline effective-

ness; however, at higher concentrations, some agents will inhibit recovery.

2.1.1. Mechanisms of Mitotic Arrest

Metaphase arrest is a cytological phenomenon, but the mechanisms that lead to metaphase arrest are different. At present, several mechanisms are known. These are briefly described below, each with a model agent as an example.

2.1.1a. Prevention of Tubulin Polymerization. The best known mitotic arrestant is colchicine.[6] The basic mechanism for mitotic arrest by colchicine (or its synthetic derivative, Colcemid) is the prevention of polymerization of tubulin. Colchicine does not interfere with the microtubules already formed. Therefore, colchicine does not arrest cells already at metaphase. A detailed electron microscopic study of Colcemid action on mitosis in cultured Chinese hamster cells[4] revealed that in arrested metaphase the two sets of centrioles are situated in the center of the cell. The chromosomes are arranged such that one chromatid of each chromosome faces the centrioles (with a scanty amount of microtubules connecting the centrioles and the kinetochore) and the other sister chromatid faces the plasma membrane with no microtubules.

Colchicine-induced metaphase arrest may be easily reversed by removing the agent from the culture medium or the environment in which the cells were exposed. In cultured Chinese hamster cells, the reversal process takes only a matter of minutes.[4] The two sets of centrioles move apart and the chromosomes start to reorient themselves. As the centrioles proceed to reach the poles, a typical metaphase plate is formed with a normal spindle apparatus. Anaphase ensues and the cell divides. The recovery process takes no more than 30 min. By 1 hr after Colcemid removal, most arrested mitotic cells have finished division.

Whether this series of events, typical of the colchicine reaction, can be used to explain a similar mechanism for other mitotic arrestants whose effect can be readily reversed remains to be verified.

2.1.1b. Crystalization of Microtubules and Tubulin. Two major compounds, vincristine and vinblastine, are currently used for cancer chemotherapy. The unique reaction of these alkaloids is that at high concentrations they aggregate microtubules and tubulin molecules to form crystals.[2] Cells entering mitosis have no tubulin available to form spindle microtubules, hence there is a metaphase block. At high concentrations, e.g., 10^{-6} M or higher, there is no recovery even if the agent is removed, because all available tubulin molecules have been crystallized. The cells at metaphase have no transcriptional and little translational activities to synthesize new tubulin molecules.

2.1.1c. Depolymerization of Microtubules. It is well known that cold shock destroys the spindle. This is somewhat different from the colchicine action, because colchicine does not interfere with spindles already in existence. The action of hypothermia on spindles is reversible. Cells can reorganize microtubules when proper temperature returns. However, no information is available on the recovery capability after prolonged hypothermia.

2.1.1d. Stabilization of Microtubules. Thus far, only one compound, taxol, an alkaloid extracted from the plant *Taxus*, behaves in this fashion. As mentioned previously, microtubules normally depolymerize and polymerize to maintain an equilibrium. When the cell enters mitosis, most existing microtubules dissociate to return to tubulin dimers or small peptides. The cell then reutilizes these molecules to organize its spindle. Taxol allows polymerization but prevents depolymerization; in other words, it stabilizes microtubules. Therefore, under the influence of taxol, all available tubulin molecules are eventually used to form microtubules. Since a mitotic cell cannot use existing microtubules to form its spindle, the cell is unable to organize a spindle apparatus in the presence of taxol, because the supply of tubulin molecules has been exhausted. Thus, mitosis is blocked despite the network of microtubules in the cell. At high concentrations of taxol, there is no reversal (T. C. Hsu, J. C. Liang, and K. L. Satya-Prakash, unpublished data).

2.1.2. Consequences of Mitotic Arrest

If arrested metaphases either became polyploid or recovered to divide normally, then mitotic arrest would be a relatively harmless episode. However, normalcy of mitotic recovery appears to be inversely correlated with the duration of arrest. In cinematographic records, Stubblefield[26] found that mitotic recovery from Colcemid arrest appeared essentially normal when the treatment time was relatively short (2 hr). With prolonged treatment time, more and more anaphases became multipolar. At metaphase there are two sets of centrioles, each consisting of a mother centriole and a daughter centriole (or procentriole). As mentioned, both sets of centrioles are situated in the center of the cell when mitosis is arrested by Colcemid. When the arrest period is short, the two sets of centrioles separate following reversal and move to the poles as in normal metaphase. However, when the arresting period is prolonged, the mother and the daughter centrioles of one or both sets may separate. Thus, when anaphase resumes, there may be three or four spindle organization centers, resulting in multipolar anaphases.[27]

Multipolar division, with its severe imbalance in chromosome distribution, will most likely lead to the death of daughter cells. Therefore, the anomaly, though drastic, may be inconsequential as far as its fate is concerned. However, prolonged metaphase arrest may result in other mitotic abnormalities, such as nondisjunction and lagging chromosomes, when the cell eventually recovers. When two sister chromatids move to the same pole instead of to opposite poles (nondisjunction), the result is aneuploid daughter cells ($2n + 1$ and $2n - 1$). This could occur when one chromatid attaches to the spindle apparatus while the sister fails to do so. Unfortunately, direct observation of nondisjunction in anaphase is extremely difficult; therefore, estimation of the frequency of nondisjunction must await data from DNA measurement of daughter nuclei or actual chromosome counts of the next metaphase. However, nondisjunction is not the only mitotic disturbance for aneupoloid production. Lagging chromosomes constitute another (probably major) mechanism for aneuploid production.

When both chromatids of a chromosome fail to attach to the spindle, the chromosome lags during anaphase movement. The laggard(s) may be eventually included in one of the daughter nuclei or may form micronuclei. Since the sister chromatids do not separate, aneuploidy may result in either situation. Cytologically, lagging chromosome(s) are easily detected in late anaphase or early telophase. When the dose of the arrestant is at borderline effectiveness, a portion of the mitotic cell population may be blocked at metaphase while the rest may proceed to undergo anaphase. Ostergren and Levan[18] observed that under such conditions, lagging chromosomes were common. Liang *et al.*[14] found the same phenomenon in grasshopper embryo cells exposed to benzene.

Another possibility that can produce a high percentage of lagging chromosomes is a prolonged metaphase arrest and subsequent recovery. The degree of mitotic disturbance appears to be dependent on the duration of arrest, a phenomenon already observed by Stubblefield[26] on multipolar spindles. In observing living cells at a low magnification, Stubblefield could not detect laggards. Hsu *et al.*[9] reported that diazepam is a mitotic arrestant at high concentrations, but in the absence of the drug, blocked metaphase can quickly recover. In squashed preparations, cells recovering from a 2-hr mitotic block showed no significant increase of laggards compared with the control samples, but cells recovered from a 7-hr block exhibited more than 10% of anaphases with laggards. The data of Satya-Prakash *et al.*[23] showed that this may be a generalized phenomenon.

The percentage of aneuploid cells that can survive and enter the next mitotic division conceivably depends upon many factors. However, some of the aneuploid cells may be able to survive the next cycle or even to perpetuate. Kato and Yosida[10] noted a significantly high frequency of aneuploids in cultured Chinese hamster cells recovering from Colcemid block. Similar findings were reported by Hsu *et al.*[8]

2.2. The Centriole

In animal cells, centrioles are spindle organization centers. When mitosis begins, the two sets of centrioles migrate apart and finally reach the poles at prometaphase. Formation of the spindle is complete at this stage. After cell division, each daughter cell receives one set of centrioles, and during the ensuing interphase, a duplication process takes place. The cell then once again possesses two sets of centrioles.

Centrioles are highly specialized and organized structures of microtubules. There is no definite information on what happens to cell division when centrioles are damaged or their duplication is inhibited. We strongly suspect that centriolar damage may be one cause of endoreduplication, which represents two DNA replication cycles without an intervening mitosis. McGill *et al.*[15] found a relatively high frequency of endoreduplication in Chinese hamster cells following an exposure to ethidium bromide. Electron microscopic observations revealed structural abnormalities of centrioles in cells treated with ethidium bromide.[16] Regardless of the mechanism, it is conceivable that damage to centrioles could lead to abnormalities in cell division.

2.3. Kinetochores

Electron microscopic studies suggested that the kinetochore is a proteinaceous plate to which the spindle microtubules are attached.[3] If the development of the kinetochore structure is not synchronous with the rest of the mitotic process or if a kinetochore is damaged, then it is possible that a chromosome or a chromatid will not be able to move normally in anaphase. Unfortunately, no direct electron microscopic data on kinetochore damage are available. Whether immunofluorescence microscopy using antikinetochore antiserum can offer some clue to this process also remains to be explored.

2.4. Cytokinesis

Without question, DNA and chromosomes are not involved in cytokinesis. Since cytokinesis takes place after karyokinesis, abnormality in cytokinesis means the failure of cell separation, resulting in a binucleated (polyploid) cell. Since syncytes (cells with more than one nucleus each, such as binucleated cells) are common in normal tissues, binucleated cells should not pose a serious problem to genomic balance.

3. Assays with Diploid Mammalian Cells *in Vitro*

For preliminary information, it is a good policy to test all water- and dimethylsulfoxide (DMSO)-soluble compounds with cells *in vitro* because cell culture systems offer many advantages, such as versatility, rapidity, and economy. It takes only a few hours to find out, for example, whether an agent is capable of causing mitotic disturbances. Recovery experiments can also be performed with relative ease. Therefore, this system will be described in more detail than others.

3.1. Choice of Cell Lines

If one merely wishes to find out whether an agent can cause mitotic disturbances, any relatively stable cell line may be employed because the experiments do not demand a diploid constitution. Pertinent criteria for determining a candidate cell line include (1) fast growth—the generation time should be no more than 24 hr, preferably shorter, because the shorter the cell cycle, the higher the frequency of mitotic divisions at a given time, thus making analyses easier—and (2) relative stability—the cells should not exhibit frequent spontaneous mitotic disturbances such as multipolar spindles, lagging chromosomes, and chromatin bridge formation. Many Chinese hamster cell lines are suitable for such experiments. The popular Chinese hamster ovary (CHO) line, though aneuploid, may be good for such use.

Nevertheless, the ultimate test is aneuploidy induction. If the stemline cell of an aneuploid line is constant at one number, then this line may be used for aneuploidy enumeration. Deviating chromosome numbers would indicate induced changes. Unfortunately, most aneuploid cell lines have a range of chromosome numbers, and even a range of only two to three chromosomes becomes undesirable. Therefore, diploid cell lines are preferred, even for preliminary experiments, since

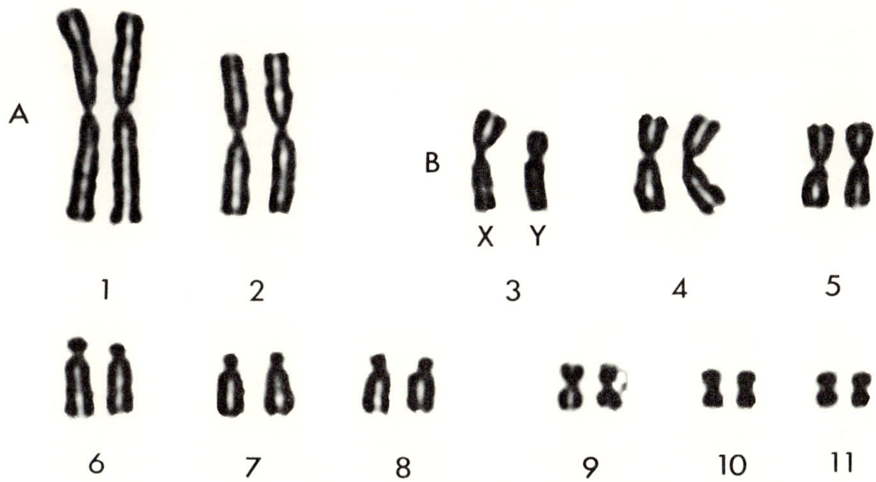

FIGURE 1. A karyotype of the male Chinese hamster.

there is no reason to use one line for preliminary experiments and another line for aneuploidy enumeration.

Most human fibroblast lines are predominantly diploid. The chromosome number of 46 is not very high. An experienced cytogeneticist can accurately count chromosomes of a human metaphase plate and construct a rough karyotype by microscopic examination in less than 1 min. Therefore, human fibroblast lines may be seriously considered as alternative candidates. However, the disadvantage of using human fibroblast lines is the phenomenon of senescence. Periods with vigorous growth potential are limited, so it becomes necessary to initiate new lines from time to time. Nonsenescent lymphoid lines probably can be used as test materials.

Diploid cell lines derived from other mammals may also be considered, as long as (1) they maintain vigorous growth for long periods of time, (2) the chromosome number is not exceedingly high, and (3) the cell lines remain predominantly (more than 95%) diploid. Avian cells, with a large number of microchromosomes, are not useful unless one concentrates only on the macrochromosomes.

In our laboratory, we have been using the diploid male Chinese hamster line Don. The diploid number is low ($2n = 22$) and the karyotype is one of the best with conventional staining (Figure 1). Many chromosome pairs are identifiable without banding. The Chinese hamster chromosomes are excellent for analyzing chromosome(s) that most likely will contribute to aneuploidy. However, most rodent species have one drawback, namely, the cells do not remain diploid for many generations. Sooner or later they become aneuploid. The Don line is no exception. In our experience, the Don cells maintain a very high frequency of diploidy (95–98%) for long periods of time if subcultured 1:4 every day. Although this may be a nuisance in laboratory management, the cultures are excellent. However, a laboratory does not perform experiments on aneuploidy at all times. Seed stocks can be frozen and revived when experiments demand diploid cells. In the following description, we present the protocols and techniques pertinent to diploid Chinese hamster cells. Some modifications should be made when other materials are used.

3.2. Mitotic Arrest

Available data are not sufficient to answer two key questions: (1) whether all mitotic poisons cause mitotic arrest and (2) whether mitotic arrest always results in aneuploidy formation. If we use our definition for mitotic poisons, then some of them, e.g., those that cause a disturbance in cytokinesis, will not cause mitotic anomaly. As for the second question, we believe the answer will be forthcoming soon.

To test the ability of a given agent to induce mitotic arrest is relatively simple. Test cells are seeded in T-25 flasks with an inoculum to assure an exponential growth in 24 hr (for slow-growing cells, 48 hr). The test compound is prepared in concentrate, either in saline or in DMSO. No more than 0.2 ml of the drug solution should be added to each flask, but the DMSO concentration should be calculated to be less than 2% of the final dilution and preferably lower. Only a saline control is necessary for water-soluble compounds, but for DMSO-soluble compounds, a DMSO control is necessary in addition to the saline control.

For preliminary experiments, we suggest the use of four concentrations for each compound: 10^{-4}, 10^{-5}, 10^{-6}, and 10^{-7} M. The culture flasks are inoculated with the chemical solution and the cultures are reincubated for 3–5 hr. For Chinese hamster cells, a 3-hr treatment should be sufficient. For slower growing cells, 5 hr is more desirable.

If no metaphase arrest is recorded at all these concentrations, it is safe to conclude that the agent is not a mitotic arrestant. If mitotic arrest is noted at the lowest concentration (10^{-7} M), the dose must then be reduced to find the minimum effective concentration.

As demonstrated by Hsu *et al.*,[9] the mitotic-arresting ability of diazepam was mitigated by the addition of S9 liver microsome mix to the culture medium. Therefore, it may be a prudent move to conduct experiments on mitotic arrestant with metabolic activation (or inactivation) by adding the S9 mix or by cocultivation with hepatocytes.

3.2.1. Cell Harvest and Slide Preparation

In our laboratory, we prefer to use squash preparations for analyzing mitotic arrest (or the lack of it) because the procedure is simple and because the slides are very easy to read with phase-contrast optics. The treated cells as well as the control cells are dislodged from the flask and centrifuged to form a pellet, and the supernatant is decanted. Hypotonic solution treatment should not be applied in these samples because strong hypotonicity will distort anaphases. However, we found it helpful to treat the cells with a very mild hypotonic solution (four parts growth medium: one part distilled water). In mammalian cells, following breakdown of nucleoli, the nucleolar material tends to adhere to the chromosomes in lumps, thus making the chromosome morphology unclear. A treatment with a mild hypotonic solution for 10 min is adequate to disperse the nucleolar particles but does not spread the chromosomes. The cells are recentrifuged and approximately 3 ml of 45% glacial acetic acid should be added to each cell pellet. The pellet should not be disturbed for at least 10 min. The cells can be stored in the fixative almost indefinitely. When slides are made, the fixative is then decanted. There will always be a small amount of fixative runback, but this does not interfere with staining.

A stock acetic orcein staining solution should be prepared by dissolving 2 g of natural orcein in 100 ml of 50% acetic acid. The solution should be boiled, allowed to cool, and filtered into an amber bottle. This solution does not require refrigeration and can be stored indefinitely. Two to three drops of acetic orcein should be added into the tube with the cell pellet, and the cells are gently suspended with a Pasteur pipette. The combination of stain and the runback fixative yields approximately 1% orcein instead of the original 2%. The precipitates, if any, will be redissolved when the stain concentration is reduced. The cells should be as thoroughly suspended in the staining solution as is practicable with no large lumps. One small drop of the cell suspension is placed on a clean slide, and a 22×30 mm coverslip is placed over the suspension. Excess stain is absorbed by a blotting paper. Gentle pressure is then applied by the thumb over the blotting paper, rolling the thumb from the center of the coverslip toward the edges. Tapping should be avoided. Thumb pressure should gradually increase to press out more stain, moving the blotting paper each time the pressure is applied. Two important precautions for squash preparation are: (1) the cell suspension should not be too thick (if it is, add one or two drops of additional stain) and (2) the coverslip should not slide or slip when thumb pressure is applied. If this occurs, the slide should be discarded and another one tried.

When no more stain can be squeezed out, the edges of the coverslip should be sealed with Kroenig wax or other sealants. The preparations are temporary, but under refrigeration the slides can be kept in good condition for up to 1 year. Two slides of this type are sufficient for each sample.

3.2.2. Data Collection

Phase contrast optics should be used to read the acetic orcein squash slides. For data collection of this type, medium magnification (40×) is best. An area should be selected containing good, flattened cells with complete cellular outlines (Figure 2), not fields with round and overcrowded cells (Figure 3) or broken cells.

Two parameters, the mitotic index (MI) and the anaphase/metaphase ratio A/M, can be effectively used to estimate the potency of mitotic arrest. Mitotic index can be calculated in several ways, but we believe that the method suggested here is the easiest and the least problematic. We count approximately 2000 random cells and record only the frequencies of metaphase, anaphase, and interphase. Prophase and telophase are counted as interphase. Mitotic index is the combined percentage of metaphase and anaphase. For the purpose of estimating potency of mitotic arrest of a given compound, this method has the advantages of avoiding having to identify prophases and telophases, thus saving time in recording, and avoiding having to judge when prophase begins and when telophase ends.

3.2.3. Anaphase/Metaphase Ratio

We usually count 100 metaphase and anaphase figures to obtain the A/M ratio. Since the duration of anaphase is invariably shorter than that of metaphase, the A/M ratio in control samples rarely exceeds 0.4. When there is a total mitotic block, the A/M ratio becomes zero. When the A/M ratio is between 0 and the control value, the drug concentration is regarded as one with incomplete mitotic arrest. It should be added that the A/M ratio in control samples varies considerably. For an individual set of experiments, it is necessary to use control cultures of that set as the standard.

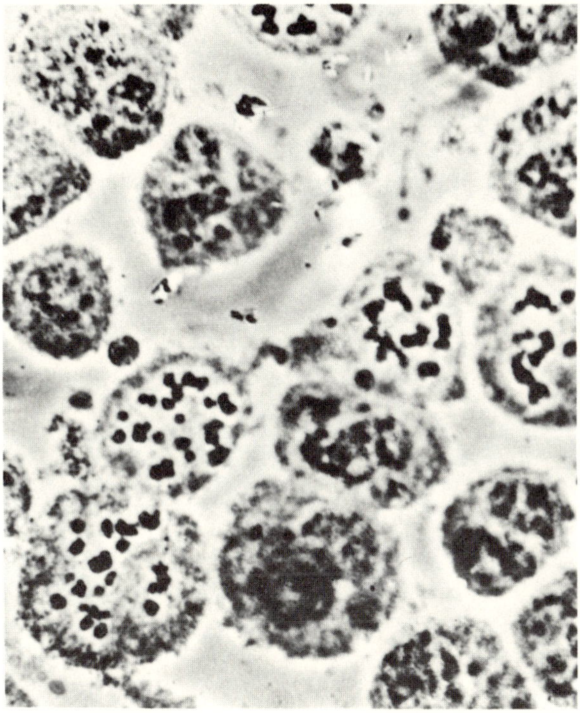

FIGURE 2. Don cells treated with taxol for 3 hr: a low magnification field from a squash preparation showing proper flattening of cells and intact cell outlines.

3.2.4. Frequency of Abnormal Anaphases

As observed decades ago by Ostergren and Levan,[18] the frequency of abnormal anaphases increases when mitotic block is incomplete. Common anaphase abnormalities include multipolar spindles and lagging chromosomes. Chromatin bridges should not be included, since bridge and fragment formation suggest clastogenic effects. To estimate the frequency of abnormal anaphases, only late anaphase figures should be used because laggards are not evident in early anaphases. Whenever possible, at least 100 late anaphase figures should be scored.

3.3. Recovery Experiments

Depending upon the mechanism of action and the concentrations employed, some mitotic arrestants allow reversal (resumption of ana-

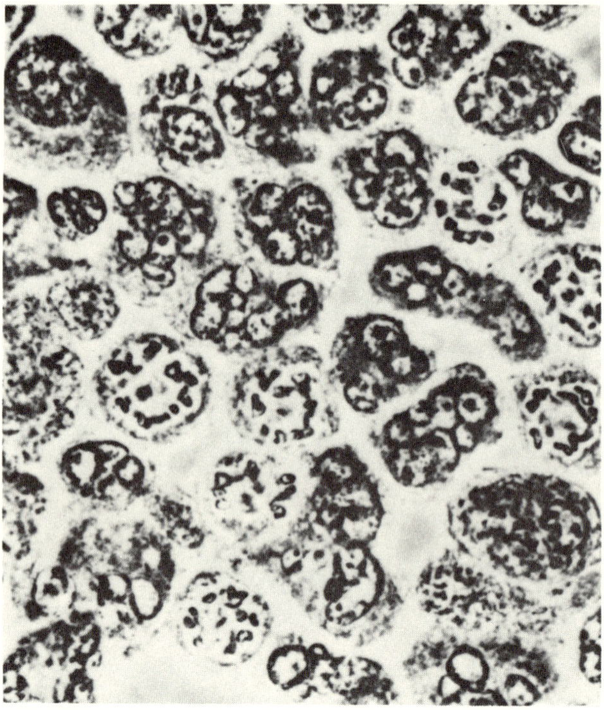

FIGURE 3. Same preparation as in Figure 2, showing crowded cells unsuitable for microscopic observations. Note mitotic cells with poor spreads.

phase after metaphase block) when they are removed from the culture medium, while others do not. For example, Colcemid permits recovery when it is removed.[8] Quick recovery was also noted when diazepam was used to arrest mitosis. On the other hand, no recovery was registered at higher concentrations (e.g., 10^{-6} M) of vinblastine and taxol. The frequency of metaphases continued to increase and the chromosomes continued to condense. At lower concentrations, e.g., 10^{-7} M, anaphase recovery was achieved.[23]

We suggest the following procedures. The minimum effective concentration should be used at which a total (or near total) mitotic block is achieved, along with a higher concentration (100× or 1000×). Cells should be treated for 2–3 hr and 7 hr. At the termination of the treatment period, the cells should be rinsed twice with growth medium for 2 min each, and reincubated for 30, 60, and 120 min. Squash preparations are then made for each sample and the A/M ratio is

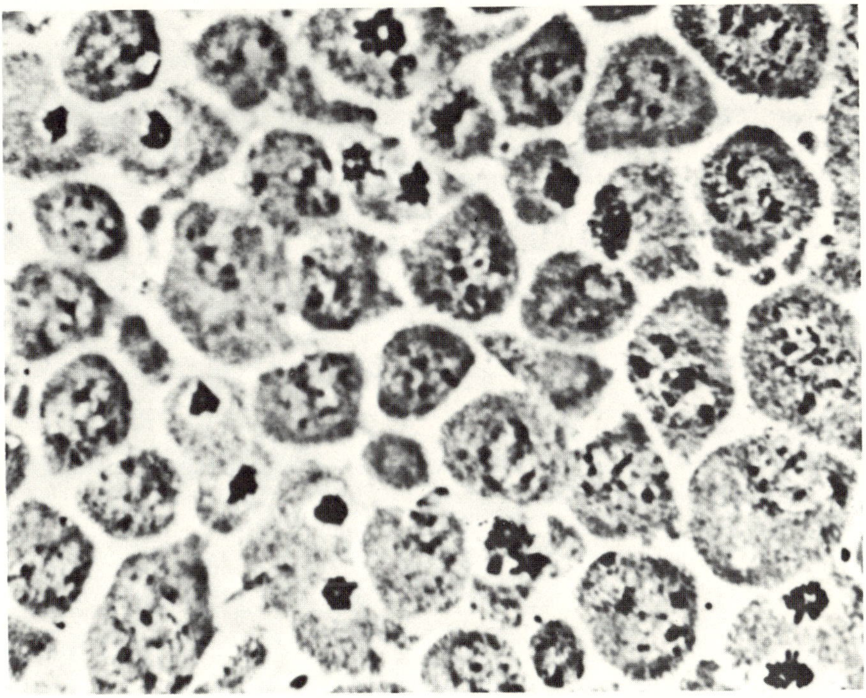

FIGURE 4. A low-magnification field showing Don cells rushing to anaphase after a DES block (10^{-4} M) for 3 hr and reversal with regular medium for 30 min.

recorded. If recovery is prompt, the A/M ratio may be extremely high at 30 min recovery (Figure 4) and may reach the control level after 60 min. If the recovery is slow, the A/M ratio will require a longer time to become high. If no recovery takes place, the A/M ratio will remain 0. There is no point in recording mitotic index in the recovery experiments, but it is most important to record the frequencies of abnormal anaphases. At least 100 late anaphases should be read and all categories of anomalies should be recorded. Control cultures (both saline and DMSO) should always be performed in conjunction with each experiment.

3.4. Aneuploidy Enumeration

At the present time, we have no firm conclusions regarding (1) whether mitotic arrest invariably leads to aneuploidy in the future generations, and (2) whether prolonged arrest invariably induces more aneuploidy. From our experiments with a few agents (Colcemid, diazepam, vinblastine, and DES), the answer is positive.[23]

When a mitotic arrestant behaves in a way similar to colchicine, i.e., by preventing polymerization of microtubules, then most likely the arrested cells will quickly recover when the agent is removed. In such cases, the interphase cells should receive no injury. When such cells subsequently enter mitosis, cell division is expected to be normal. Thus, in experiments designed to detect aneuploid induction, the uninjured interphase cells should remain diploid such that the large quantity of uninjured cells will dilute the frequency of aneuploids. Therefore, for detection of aneuploid cells in the recovering cell populations, the best starting cell populations would be one containing only arrested mitotic cells. This can easily be accomplished by disloding mitotic cells (mitotic shakeoff) with trypsin or trypsin–EDTA mixture.[28] Since only a fraction of the cell population is used for seeding, the initial populations should be much larger than those cultivated in T-25 or T-30 flasks. We recommend the use of T-75 (or even larger) flasks for monolayer cultures. If mitotic division is blocked for 5–7 hr, the shakeoff mitotic cells from each T-75 flask should be sufficient to seed a T-25 flask. For cultures arrested for a short period (e.g., 2 hr), one may have to pool dislodged mitotic cells from several large flasks to start a small flask. This system is not applicable to suspension cell cultures.

If the purpose of the experiment is merely to determine whether aneuploidy is generated, then a 7-hr arrest is most desirable. If the experiment is designed to compare aneuploidy induction in relation to the duration of arrest, it is necessary to perform recovery experiments with several intervals.

If an agent allows quick anaphase recovery, the arrested cells soon enter interphase and start a new cycle. In such a case, mitotic stages of the very next cell generation should be approximately one normal cell generation following the reversal. For Chinese hamster cells, a convenient experimental schedule is to reverse the drug treatment late in the afternoon and to harvest the cells the next morning (approximately 18 hr later). For cell lines with a longer cell cycle, the harvest time should be delayed accordingly. In our experience, the frequency of aneuploids for Chinese hamster cells is almost the same when the recovering cells are harvested at 18 and 24 hr. For safety, it may be desirable to obtain a 36-hr or 48-hr sample.

It should be noted that if the agent does not allow mitotic recovery (e.g., vinblastine and taxol at higher concentrations), then the mitotic shakeoff method should not be applied because the great majority of arrested cells will either die or become tetraploid. In such cases, all treated cells should be used for aneuploid evaluation, in the hope that some interphase cells will recover and will enter the next cell generation. The harvest time may have to be extended to 72 hr and the 18-hr sample omitted.

Since this series of experiments is for the enumeration of aneuploidy, the cultures should be treated with colchicine or Colcemid for 1 hr to arrest mitoses prior to harvest, and the cells, after disloding, should be treated with a hypotonic solution (0.075 M KCl or 1% sodium citrate) for 10–15 min. Either squash preparations or air-dried preparations may be made. For acetin orcein squash, the cells should be fixed with 45% acetic acid and stained as described previously. For air-dried preparations, the cells should be fixed with Carnoy's mixture (three parts methanol: one part acetic acid) and conventional air-dried slides should be made. Slides may be stained with Giemsa (4% in 0.01 M phosphate buffer) for 7–8 min.

3.5. Slide Reading

A minimum of 100 well-spread metaphases should be recorded for each sample. In our laboratory, we prefer to use squash slides lightly stained with acetic orcein because under phase-contrast optics, the intactness of the cellular outline can be determined. In air-dried preparations, only metaphases with undistorted metaphase plates should be used. However, despite such a precaution, occasional recording of artifactual aneuploids is unavoidable. Since cell rupture during slide preparation tends to result in the loss of chromosomes rather than a gain of stray chromosomes, a compromise method may be used by disregarding all the hypodiploid counts and recording only diploids and hyperdiploids. Theoretically, generation of a hyperdiploid cell by nondisjunction or lagging chromosomes should be accompanied by a hypodiploid cell. Thus, the frequency of aneuploids, if only hyperdiploids are counted, should roughly double that of hyperdiploids alone. In reality this is a conservative estimate because anaphase anomaly may give rise to two hypodiploid daughter cells when a lost chromosome in the cytoplasm is not incorporated in either of the daughter genomes.

4. Assays with Bone Marrow Cells *in Vivo*

The environments of cells in culture and tissues *in vivo* differ in a number of respects. (1) All cells in culture, monolayer or suspension, are directly exposed to any agent contained in the culture medium, whereas drugs ingested by or injected into an animal do not directly make contact with the tissues or cells until they are circulated. (2) A drug may be metabolically activated or inactivated *in vivo* but not *in vitro*. Even addition of liver microsomal fractions to the culture medium or cocultivation with hepatocytes cannot completely simulate the *in vivo* condition. (3) The chemicals in the culture medium remain in the medium until the medium is changed; however, in the *in vivo* condition, they may be excreted at various rates. If an agent capable of blocking cell division is of the colchicine type, allowing quick recovery when it is removed, and if this agent is easily excreted, then the duration of metaphase block may be relatively short. In such cases, a less harmful effect on the cell genome may be expected. For these reasons, results obtained from tests performed in cell culture must be confirmed or modified by experiments *in vivo*.

Only a few adult tissues constantly renew themselves, and some of them (e.g., epidermis, intestinal mucosa) are not suitable for cytogenetic analysis. Bone marrow seems to be the most workable tissue. In young adult laboratory mice (3–4 months of age), a certain proportion of mitotic cells is usually present in the bone marrow. In older animals, stimulation of mitotic activity may be performed.[12] We have not used the stimulation method in conjunction with experiments on mitotic poisons, but we believe this can be done when necessary.

Since we have already described technical aspects in some detail in the preceding section, the presentation on the protocol *in vivo* will be limited to specific requirements.

4.1. Animals

The recommended test animal is the laboratory mouse, which affords several advantages. (1) The animals are small and easy to handle. When the quantity of the test chemical is limited, small animals are particularly desirable. (2) They are comparatively inexpensive. There is no compelling reason to use inbred stains for routine studies. The non-inbred Swiss strain suffices. (3) The diploid number (40) is not too difficult for counting. (4) For special problems requiring certain genetic markers, numerous inbred strains are available.

Other animals (e.g., rats, Syrian hamsters, Chinese hamsters, *Peromyscus*) may also be used, particularly in special cases. The Chinese hamster, with its excellent cytogenetic characteristics and small size, should be ideal for this purpose; unfortunately, their supply is very limited. It is practically impossible to purchase these animals for experimentation. If a testing laboratory can raise a large colony of Chinese hamsters, then this species is the material of choice.

Syrian hamsters ($2n = 44$) are useful in some respects; however, they are extremely resistant to colchicine. *Peromyscus* ($2n = 48$) may be another suitable material, since it can be bred in the laboratory with relative ease and can produce several litters per year.

The present description of the procedure deals with the laboratory mouse, but the principles can be applied to any rodent species.

4.1.1. Treatment Protocol

The most convenient route of drug treatment is intraperitoneal injection, although intravenous injection is preferred. The test compound is dissolved in physiological saline (if it is water-soluble) or DMSO, then mixed with saline (if it is DMSO-soluble). Preliminary experiments should be done to determine the lethal and LD_{50} doses. In actual experiments, the highest dose should be near or below the LD_{50} level; three additional doses, at one log each step, are tested simultaneously.

We recommend the use of three instead of five animals per harvest; when needed, the experiments are repeated. Two separate sets of experiments should be performed.

4.1.2. Short-Term Experiment

The short-term experiment is to determine the capability of mitotic arrest of a given compound. A 5-hr treatment should be sufficient. If mitoses are arrested, MI should increase and A/M ratio should decrease.

4.1.3. Long-Term Experiment for Aneuploidy Evaluation

Since the chemical, once injected, cannot be removed as in the *in vitro* system, recovery experiments as described in the previous section cannot be performed. Mitotic recovery, if any, must wait until the agent is metabolically altered or until it is eliminated by excretion. Therefore, recovery may not be as synchronous as it is in the cell culture system. However, as far as aneuploidy enumeration is concerned, this disadvantage is not serious. A series of harvest times may be designed, e.g., 18, 24, 48, 72, and 96 hr after injection of the drug.[13] Again, each time spot should consist of three animals and the bone marrow processed for chromosome counts.

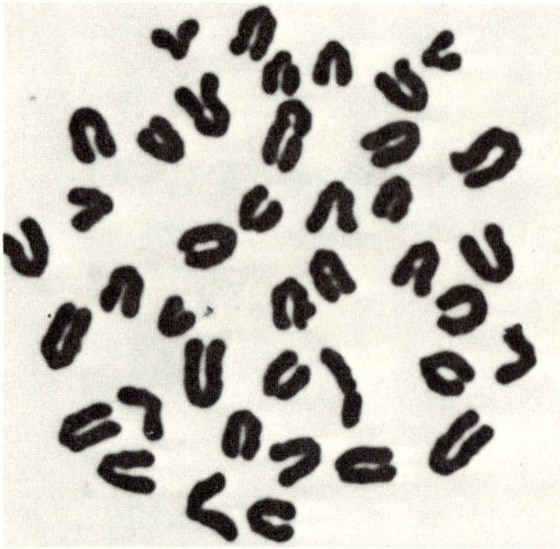

FIGURE 5. A diploid metaphase (40 chromosomes) from bone marrow of a mouse.

4.2. Techniques

At the time of harvesting bone marrow cells, the animals should be killed either with neck dislocation or anesthesia. The epiphyses of the femurs are trimmed off, and a small syringe (2–3 ml capacity) equipped with a small needle is used to flush out the marrow with either saline (for squash preparations) or a hypotonic solution such as 1% sodium citrate (for air-dried chromosome spreads).

For estimating MI and A/M, the marrow cells are suspended in saline and centrifuged. The pellet is then fixed with 45% acetic acid and later squashed in acetic orcein as described previously.

Although cytogeneticists have been using bone marrow cells to analyze karyotypes of vertebrates and blood diseases and to estimate the degree of chromosome damage induced by mutagens, the existing technique is not entirely satisfactory for evaulation of aneuploidy. Conventionally, a mitotic arrestant such as Colcemid is injected into the animals to accumulate mitoses, and air-dried or flame-dried preparations are made from aspirated bone marrow 1 or 2 hr later. Such preparations are usually adequate for the purposes just mentioned, even though they may have defects, such as insufficient spreading and overspreading of metaphase chromosomes. In the former case, chromosomes may be so crowded that counting of chromosome numbers become inaccurate; in the latter case, chromosomes may be lost during the slide-drying process, so that chromosome counts may be artifactual. Since the frequencies of artifactual counts vary among preparations, no standard artifact frequency can be established. For determining karyotypes, one can always select a few good spreads with the modal

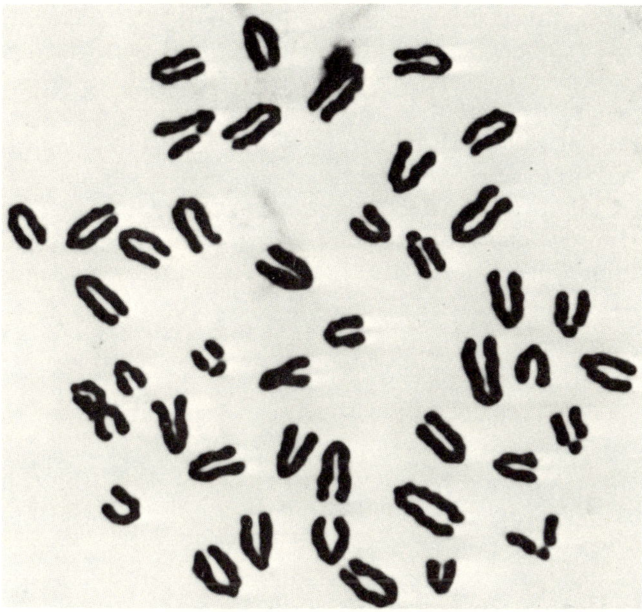

FIGURE 6. An aneuploid metaphase (42 chromosomes) from bone marrow of a mouse treated with vinblastine for 96 hr.

chromosome number and disregard poor metaphase plates. In analyzing chromosome breakage, moderately crowded metaphases and overspread metaphases can still be used. However, for estimating aneuploidy frequencies, such defects are intolerable. Ideally, every metaphase should be countable without artifactual loss or gain of chromosomes. Because of the technical defects already mentioned, cytogenetic data on aneuploidy induction in bone marrow have been deficient, and a number of published reports are not completely reliable.

In order to analyze effectively aneuploidy in bone marrow preparations, two technical goals must be reached: (1) to increase the yield of mitotic divisions and (2) to produce good metaphase spreads with a minimum number of cells that have artifactual chromosome loss or gain. These goals have been achieved by Liang and Satya-Prakash.[13] Briefly, to arrest mitosis prior to sacrifice, the mice are injected intraperitoneally with vinblastine (9 mg/kg) for 1 hr. The marrow is flushed out into centrifuge tubes with fetal calf serum[25] and centrifuged, and the supernatant is removed. The cell pellet is resuspended in 10 ml of prewarmed 1% sodium citrate and is incubated at 37°C for

25 min. At the end of this period, 1 ml of methanol–acetic acid fixative (3:1) is added to each tube, the suspension is centrifuged, and the supernatant is removed, leaving 0.5–1 ml of the solution in the tube. The cells are resuspended in the remaining solution to break up cell clumps, and 10 ml of the fixative is added. Cells are washed with fixative three times and finally are suspended in a methanol–acetic acid mixture (5:2) before air-dried preparations are made. The 5:2 mixture seems to produce better chromosome spreads than does the 3:1 mixture. Typical diploid and aneuploid metaphase figures are illustrated in Figures 5 and 6.

5. Assays with Grasshopper Embryos

People are exposed to one class of chemicals through inhalation. Most of these chemicals are volatile liquids, and several of them are already known to be mitotic posions.[11,14]

Neither cell cultures nor intact animals are suitable as test material for volatile liquids. In cell culture, the liquid medium is a barrier between the gaseous compounds and the cells. If the liquid phase of these agents is added to the culture medium, most agents form globules because they are immiscible with water. Therefore, exposure to the cells is never uniform. It is feasible to design and construct an apparatus suitable for such experiments, but for preliminary information, the expense appears not worthwhile.

Many volatile liquids are anesthetics and others are highly toxic to the central nervous system. Intact animals exposed to such agents, especially when the doses are relatively high, succumb to anesthesia or other toxic effects prior to expressing any cytogenetic toxicity. One may use sophisticated inhalation chambers to conduct experiments with animals, but the doses must be exceedingly low. These experiments should be performed for low-dose, chronic, and long-term exposures. To ascertain whether an agent is a mitotic poison, test materials without a central nervous system should be advantageous.

5.1. Test Materials

Embryos *in ovo*, prior to their development of a central nervous system, are considered suitable. Chick embryos at 3 or 4 days of incubation may be utilized for such experiments; however, as previously mentioned, the high diploid number with 60-odd microchromosomes make this material a poor candidate for aneuploidy enumeration.

It appears that insect embryos, especially those of orthopteran insects, offer many advantages. The diploid number of most orthopteran insects (grasshoppers, crickets, etc.) is low (in the 20s) and the chromosomes are large. Grasshoppers are classic experimental material for cytologists and for demonstration in cytology classes. During early stages of embryonic development, mitoses of neuroblasts and other somatic cells are abundant. The eggs can be directly exposed to the vapors of volatile liquids in a sealed container.

One disadvantage of using grasshopper eggs as test material is that in nature these insects have only two generations per year. Thus, the supply of eggs becomes seasonal. In our laboratory, we circumvent this drawback by using a strain of *Melanoplus sanguinipes* that has no diapause.[20] Five to six generations per year can be obtained. The eggs can be stored at low temperature (10–13°C) to retard their embryonic growth, and rapid growth will resume in a matter of hours at room temperature.

In case there is difficulty in rearing grasshoppers, a substitute species, the domestic cricket, *Gryllus domesticus*, may be tried because, at least in the U.S., crickets can be purchased from cricket farms. Adult crickets readily mate and the female insects lay copious amounts of eggs. A moistened black paper may be used to collect the white eggs for experimentation. Unfortunately, the cricket eggs (and embryos) are considerably smaller than the grasshopper eggs, so they are more difficult to handle when making cytological preparations. The following procedure is mainly for grasshopper embryos. Laboratories using cricket embryos must do their groundwork.

5.2. General Information about Grasshopper Embryos

A number of grasshopper eggs are usually embedded in an egg case. Individual eggs can be isolated from the egg case by careful dissection. The ideal age of the embryos for cytological preparations is about 12–14 days, but after cold storage, estimation of the age of the embryos by examining intact eggs is not accurate. However, there is always a sufficient number of mitotic divisions in the embryos, regardless of the exact age. The age of embryos may be determined more accurately after the egg shell is removed. Grasshopper eggs can be incubated at 37°C with a cell cycle time of 4 hr or less. At room temperature, the cell cycle time is approximately double that at 37°C.

5.3. Procedure

5.3.1. Mitotic Arrest and Anomalies

Three to five eggs may be placed in a paper boat or a bottle cap and the container suspended in a Coplin jar or any medium-sized jar that can be tightly capped. The volume of the jar should be measured. A measured amount of the volatile liquid to be tested (using micropipettes or syringes) is placed in the bottom of the jar and the cap should be sealed with petroleum jelly. The liquid will soon evaporate and enter the embryos through respiration. Several doses, e.g., 0.5, 0.1, 0.05, and 0.01 ml for 100 ml of air space, should be tried. Control samples are exposed to jars with ambient air similarly capped. For determining mitotic arrest, an exposure time of 5 hr should be sufficient.

After the exposure, shells of individual eggs should be removed and the embryos fixed in 45 or 50% acetic acid. It is preferable to fix the embryos *in toto*. After fixation, the fixative is changed to acetic orcein, and one embryo is placed on a clean slide and cut with a needle into several smaller pieces, and the tissue pieces spread to minimize cell crowding. A coverslip is then placed over the tissue pieces and squash preparations are made.

The slides are read to obtain MI, A/M, and the percent abnormal anaphases. If mitotic arrest is complete at all the doses tested, the experiment should be repeated with lower doses, always overlapping the lower doses of the previous experiment, until no metaphase arrest is recorded. At borderline doses, special attention should be paid to anaphase anomalies.

5.3.2. Recovery Experiments

From experiments on mitotic arrest, at least two doses should be chosen: one that gave nearly complete arrest, and another, ten times higher. The eggs should be exposed to the vapor for 5 hr as described for the previous experiment.

5.3.2a. Short-Term Recovery. Exposed eggs should be moved to ambient air and embryos fixed 1 and 2 hr after removal. Squash preparations should be made as described previously. The MI and A/M values should be obtained to determine whether there is recovery and, if so, the rate of recovery. If recovery is fast, the MI and A/M values of the 1-hr sample should approach those of controls. If by 2 hr the MI is still high and the A/M ratio is still close to 0, it may be concluded that there is no recovery.

5.3.2b. Long-Term Recovery for Aneuploidy Evaluation. It should be kept in mind that the first cell generation after recovery is probably the best cell population for evaluating aneuploid frequency. Counting in the 5-hr exposure time and time for recovery, harvest time for aneuploidy evaluation falls late in the evening, which is not suitable for routine work. One way to adjust work schedules yet still obtain cell populations at or near the peak of mitoses of the first recovering cell generation is to store the eggs after drug exposure at 10–13°C overnight to retard the cell cycle.

Since this series of experiments requires accurate chromosome counting, a short mitotic arrest is necessary to accumulate metaphases and to condense the chromosomes. The egg shell should be removed and the embryo immersed in Mark's insect culture medium containing Colcemid. The embryos should be incubated at 37°C for 1 hr. It is important that if an embryo is broken, the tissue pieces should not be mixed with other embryos. Because of the cytological dimorphism between the sexes ($2n$ = F 24 XX, M 23 XO), a male cell with $2n + 1$ constitution has the same chromosome number as does a normal female cell. The best way to avoid embryo breakage is to practice liberating embryos without damage. Embryos should be treated individually with 1% sodium citrate for 10 min and fixed again individually with 50% acetic acid. Acetic orcein squash preparations should be made, one embryo per slide. Since the cells are softer after a hypotonic solution treatment, the chances of cellular breakage as the result of squash pressure increase. It is advisable to cut the embryos into smaller pieces and to squash less heavily than usual. Slides should be monitored to adjust pressure. There will be a large number of mitoses. One can always avoid counting chromosomes of crowded metaphase plates and use only good spread with undistorted metaphase plates. At least 100 good metaphases should be counted for each embryo.

6. Conclusions

Admittedly, cytogenetic assays for aneuploidy induction by environmental agents using animal cells are still in their formative stage; however, information seems to support two tentative conclusions: (1) anaphase abnormalities are the major cause of aneuploid production and (2) prolonged metaphase arrest increases the frequency of anaphase abnormalities. If these conclusions can be substantiated by carefully planned experiments with a number of mitotic arrestants, then routine determination of agents that may induce genomic mutations can be simplified by verifying the capability of mitotic arrest using MI and A/M ratio as criteria without resorting to actual chromosome counts in the subsequent cell generations.

We believe that with intensive investigations into refining the procedures and by drawing more concrete conclusions, cytogenetic assays for mitotic poisons should soon become routine for somatic cells. However, scanty cytogenetic data have been available on meiosis, although genetic and epidemiologic information suggests that aneuploidy can occur in germinal lines. The next important phase of research should be in gametogenesis. Since a great deal of basic cytological information on spermatogenesis is still sketchy, efforts should also be made to study many fundamental aspects of this field.

ACKNOWLEDGMENT

This work was supported in part by research grant R-808582 from the U.S. Environmental Protection Agency.

7. References

1. L. C. Andersson, V.-P. Lehto, S. Stenman, A. Badley, and I. Vintanen, Diazepam induces mitotic arrest at prometaphase by inhibiting centriolar separation, *Nature 291,* 247–248 (1981).
2. K. G. Bensch and S. E. Malawista, Microtubule crystals in mammalian cells, *J. Cell Biol. 40,* 95–107 (1969).
3. B. R. Brinkley and E. Stubblefield, The fine structure of the kinetochore of a mammalian cell *in vitro, Chromosoma 19,* 28–43 (1966).
4. B. R. Brinkley, E. Stubblefield, and T. C. Hsu, The effects of Colcemid inhibition and reversal on the fine structure of the mitotic apparatus of Chinese hamster cells *in vitro, J. Ultrastruct. Res. 19,* 1–18 (1967).
5. J. F. Crow, Chemical mutagen testing: A committee report, *Environ. Mutagen. 5,* 255–261 (1983).
6. O. J. Eigsti and P. Dustin, Jr., *Colchicine—In Agriculture, Medicine, Biology, and Chemistry,* Iowa State College Press, Ames, Iowa (1955).
7. A. J. F. Griffith, *Neurospora* prototroph selection system for studying aneuploid production, *Environ. Health Perspect. 31,* 75–80 (1979).
8. T. C. Hsu, L. R. Shirley, and H. Takanari, Cytogenetic assays for mitotic poisons: The diploid Chinese hamster cell system, *Anticancer Res. 3,* 155–160 (1983).
9. T. C. Hsu, J. C. Liang, and L. R. Shirley, Aneuploidy induction by mitotic arrestants: Effects of diazepam on diploid Chinese hamster cells, *Mutat. Res. 122,* 201–209 (1983).
10. H. Kato and T. H. Yosida, Nondisjunction of chromosomes in a synchronized cell population initiated by reversal of Colcemid inhibition, *Exp. Cell. Res. 60,* 459–464 (1970).
11. C. J. Kusyk and T. C. Hsu, Mitotic anomalies induced by three inhalation halogenated anesthetics, *Environ. Res. 12,* 366–370 (1976).

12. M. R. Lee and F. F. B. Elder, Yeast stimulation of bone marrow mitosis for cytogenetic investigations, *Cytogenet. Cell Genet. 26,* 36–40 (1980).
13. J. C. Liang and K. L. Satya-Prakash, Induction of aneuploidy by mitotic arrestants in mouse bone marrow, *Mutat. Res. 155,* 61–70 (1985).
14. J. C. Liang, T. C. Hsu, and J. E. Henry, Cytogenetic assay for mitotic poisons: The grasshopper embryo system for volatile liquids, *Mutat. Res. 113,* 467–479 (1983).
15. M. McGill, S. Pathak, and T. C. Hsu, Effects of ethidium bromide on mitosis and chromosomes: A possible material basis for chromosome stickiness, *Chromosoma 47,* 157–167 (1974).
16. M. McGill, D. P. Highfield, T. M. Monahan, and B. R. Brinkley, Effects of nucleic acid specific dyes on centrioles of mammalian cells, *J. Ultrastruct Res. 57,* 43–53 (1976).
17. G. Morpargo, D. Bellincampi, G. Gualandi, L. Baldinelli, and O. Crescenzi, Analysis of mitotic nondisjunction with *Aspergillus nidulans, Environ. Health Perspect 31,* 81–95 (1979).
18. G. Ostergren and A. Levan, The connection between c-mitotić activity and water solubility in some monocyclic compounds, *Hereditas 29,* 496–498 (1943).
19. J. M. Parry, D. Sharp, and E. M. Parry, Detection of mitotic and meiotic aneuploidy in the yeast *Saccharomyces cerevisiae, Environ. Health Perspect. 31,* 97–111 (1979).
20. R. Pickford and R. L. Randell, A non-diapause strain of the migratory grasshopper, *Melanoplus sanguinipes* (Orthoptera: Acrididae), *Can. J. Entomol. 101,* 894–896 (1969).
21. C. Ramel and J. Magnusson, Chemical induction of nondisjunction in *Drosophila, Environ. Health Perspect. 31,* 59–66 (1979).
22. L. B. Russell, Meiotic nondisjunction in the mouse: Methodology for genetic testing and comparison with other methods, *Environ. Health Perspect. 31,* 113–128 (1979).
23. K. L. Satya-Prakash, T. C. Hsu, and W. J. Wheeler, Metaphase arrest, anaphase recovery and aneuploidy induction in cultured chinese hamster cells following exposure to mitotic arrestants, *Anticancer Res. 4,* 351–356 (1984).
24. M. Sawada and M. Ishidate, Colchicine-like effect of diethylstilbestrol (DES) on mammalian cells *in vitro, Mutat. Res. 57,* 175–182 (1978).
25. W. Schmid, The micronucleus test, *Mutat. Res. 31,* 9–15 (1975).
26. E. Stubblefield, DNA synthesis and chromosome morphology of Chinese hamster cells cultured in media containing *N*-desacetyl-*N*-methylcolchicine (Colcemid), in: *Cytogenetics of Cells in Culture* (R. J. C. Harris, ed. pp. 223–248), Academic Press, New York (1964).
27. E. Stubblefield, Centriole replication in a mammalian cell, in: *The Proliferation and Spread of Neoplastic Cells,* pp. 175–193, Williams and Wilkins, Baltimore, (1968).
28. E. Stubblefield and R. Klevecz, Synchronization of Chinese hamster cells by reversal of Colcemid inhibition, *Exp. Cell Res. 40,* 660–664 (1965).

CHAPTER 5

Detection of Aneuploidy in *Drosophila*

Polly A. Foureman

1. Introduction

Aneuploidy, a numerical change in the chromosome complement, is a major contributor to spontaneous abortions and birth defects in human populations. Approximately 15% of recognized pregnancies result in spontaneous abortion.(14,117) Of these, approximately 30–50% are due to chromosome abnormalities.(48,94) In addition, roughly 0.4% of live-births suffer from an aneuploid defect.(15) These are manifested primarily as one of three autosomal trisomies (Down, Edward, and Patau syndromes), as sex-chromosome trisomies (Klinefelter syndrome), sex-chromosome monosomy (Turner syndrome), or as polyploids, including triploids and tetraploids.(102) Of conceptions that continue long enough to be recognized, approximately 5–8% are aneuploid. It is believed that because certain types of aneuploidy are underrepresented among recognized pregnancies, the rate of aneuploidy among conceptuses is actually much higher than this.(9)

It is not known how much, if any, of the enormous genetic burden associated with aneuploidy is due to exposure to environmental agents. X rays are the most extensively studied agent in this regard. Results obtained from epidemiologic studies are ambiguous with respect to the role of X rays in the production of nondisjunction.(26,113) Thus, there

Polly A. Foureman • Department of Zoology and Physiology, Louisiana State University, Baton Rouge, Louisiana 70803.

are no known human aneuploidy-inducing agents. Because the aneuploidy-inducing activity of various agents and chemicals has been demonstrated in other species, however, it may be assumed that agents capable of inducing aneuploidy in humans do indeed exist. There is a need, therefore, for large-scale screening of environmental agents for their ability to induce aneuploidy.[20]

1.1. Role of *Drosophila*

Screening systems for aneuploidy must, by necessity, utilize eukaryotic organisms. These systems should be simple to use, quick, and inexpensive. Such systems are being developed in such organisms as fungi,[7,43,44,77,87,88] silkworms,[101] *Drosophila*,[31,41,96,109] and rodents.[92,98,100] Those systems with greatest relevance to the human germ line probably cannot be determined until more is known about environmentally induced aneuploidy in humans. Interspecies comparisons between plant, insect, and mammalian systems, however, would be useful for determining how universal the critical lesions are in the aneuploidy-inducing process. There is, for instance, evidence that tubulin differs not only between species[66] but within a species.[52,53] Moreover, although the universality of mitotic and meiotic processes in eukaryotic organisms is considered dogma, the details of these processes differ between species and between sexes of the same species.[6,51] Testing for human hazard assessment, then, should be done in those systems most closely approximating the human situation. The mouse generally fills this role in mutagenicity testing. Systems utilizing the mouse, however, tend to be time-consuming and expensive. If lower eukaryotic systems could be found that could reliably predict aneuploidy-inducing ability in the mouse, these quicker and more economical systems could be substituted for the mouse in screening for aneuploidy-inducing agents. It is crucial for human hazard evaluation, therefore, that those lower eukaroytic systems that best predict aneuploidy-inducing ability in man's testing surrogate, the mouse, be identified.

Drosophila is a logical choice for inclusion as one of the lower eukaryotes for several reasons. It has been shown to have metabolic activating and deactivating systems similar to those found in humans.[3] In addition, insect systems have an advantage over plant systems in that (1) the gametogenesis process of insects is similar to that of humans and (2) the chromosome makeup of insects is similar to that in humans with respect to the presence of a pair of heterologous sex chromosomes and a general intolerance of aneuploidy. Finally, the genetics of *Drosophila* has been extensively studied and a variety of genetic markers and special chromosomes are available to maximize the efficiency of testing systems. For reviews detailing the advantages of *Drosophila* in mutagen testing, see Abrahamson and Lewis,[1] Auerbach,[2] Sobels,[97] Valencia,[114], Vogel,[116] and Würgler *et al.*[119]

Although aneuploidy was one of the first mutational responses to be studied in *Drosophila melanogaster*,[79] little is known of the various processes or interactions with the cell that produce changes of number in the chromosome complement. As early as 1921, Mavor demonstrated the effects of X-ray exposure on the induction of aneuploidy in *Drosophilia* in experiments involving irradiation of female germ cells.[73,74] Work has continued since that time to identify conditions that influence spontaneously occurring nondisjunction in *Drosophila* males and females (e.g., Dobzhansky,[27] Cooper,[21] Oksala,[82] Grell,[35,36] Novitski,[80] Merriam and Frost,[76] Baker and Carpenter,[4] and Baker *et al.*[6]) It was not until the 1960s, however, after the demonstration by Lejeune *et al.*[62] in 1959 that aneuploidy constitutes a significant genetic hazard to humans, that the effects of chemical and physical agents on changes in chromosome number began to be studied in earnest. Since that time, reports of the induction of aneuploidy by environmental agents have appeared, but these are much rarer than reports on the induction of gene mutation (see reviews by Lee *et al.*[57] and Valencia *et al.*[115]). Work is now beginning with serious attempts to develop *Drosophila* assays capable of large-scale screening of chemicals for aneuploidy-inducing ability.

1.2. Definition of Aneuploidy

Aneuploidy will be defined in this chapter as a quantitative change in the chromosome complement. This includes the gain or loss of (1) an entire chromosome, or (2) a major portion of a chromosome, or (3) a set of chromosomes, so as to produce a genome with a chromosome number uncharacteristic of the species. Although this definition includes as aneuploid some conditions, such as triploidy, which would classically be considered euploid, it is useful within the confines of aneuploidy testing in *Drosophila*.

Events that generate aneuploidy in the germ line, the only type with which I will be concerned, include nondisjunction, chromosome breakage, and polyploidization. Nondisjunction is defined as the failure of homologous centromeres (in the case of reduction division) or sister centromeres (in the case of equational or mitotic division) to segregate to opposite poles during meiosis or mitosis. This results in the gain or loss of one entire chromosome. Chromosome breakage can lead either to the loss of all or part of a chromosome or to the gain of all or part of a chromosome by attaching it to a homologous (in the case of attached-X's or X–Y interchange) or nonhomologous (in the case of translocation) chromosome. Polyploidization, which leads to the gain of an entire set of chromosomes, can be caused by a failure to undergo either a mitotic or meiotic segregation, or by polyfertilization, or fusion of gametes.

Thus, aneuploidy could be induced by disrupting any of a large number of genetic and epigenetic targets within the cell. For example, disrupting centromere function, pairing processes (e.g., synaptonemal complex), segregation processes (e.g., microtubules or microtubule-associated proteins), nuclear membrane function, or the timing of critical events could theoretically produce aneuploidy.[60,83]

It is important, therefore that tests designed to detect aneuploidy be capable of detecting damage to a variety of endpoints. This is particularly true in light of the fact that chromosome gain, chromosome loss, interchanges or translocations, and polyploidization all give rise to liveborn human aneuploids. In testing for aneuploidy, chromosome breakage is generally considered to be a less critical endpoint because other *Drosophila* assays, such as the sex-linked recessive lethal test[1,57] or Zimmering's chromosome loss test,[124] would detect agents inducing this sort of damage. Thus, chromosome gain is considered to be the best indicator of the nondisjunctional events that would be missed in other *Drosophila* tests and therefore remains the critical endpoint for aneuploidy testing. In an ideal test, the chromosome gain detected could result from nondisjunction in either the first or second meiotic division, from nondisjunction in a mitotic division, or from polyploidization.

2. Issues

There are several important and as yet unresolved issues that must be considered in the use of *Drosophila* for aneuploidy testing. These include determining (1) which sex is more sensitive to damage and/or is more relevant to humans, (2) which germ cell stages are most sensitive to damage and/or more relevant to humans, and (3) what other factors affect the sensitivity of a *Drosophila* assay system. This last issue involves determination of spontaneous rates of aneuploidy and determination of the effects of the ambient environmental conditions, as well as the effects of various factors such as the age, chromosome constitution, and genetic background of the parents on the spontaneous rate of nondisjunction.

2.1. Sex

The meiotic process differs in males and females of *Drosophila melanogaster* and studies with meiotic mutants have in fact shown that the first meiotic division is under separate genetic control in the two sexes.[5] One manifestation of this is the "aberrant" sequence of events occurring in prophase I of the male, in which no synaptonemal complex is present and crossing-over between homologues does not regularly occur.[22,49] A consequence of this for screening is that any chemical capable of inducing nondisjunction only by virtue of a disruption of processes involved in crossing-over would not be detectable in the male. Although gene mutations have been recovered that affect this process,[5] it does not necessarily follow that chemicals will be found that have this effect.

Drosophila females, on the other hand, exhibit a feature in addition to normal chiasma formation that ensures the regular segregation of homologues.[38,81] The universality of this "distributive pairing" process is not known. Nonhomologous pairing similar to that postulated to occur as part of the distributive pairing process has been observed in mice,[78] but studies of the frequency of chiasma formation indicate that in human males such a process would not be necessary and probably does not occur.[50] In addition, the large cytoplasmic volume of cells undergoing meiosis in females and the fact that the meiotic divisions are not complete until after fertilization could reduce the probability of successful exposure of critical targets.

It is therefore unclear whether the male or female of *Drosophila* is the more suitable for testing. That sex that is more sensitive would be better for making hazard evaluations, but the sex that most closely mimics the meiotic processes found in humans would be better for risk estimation. At this point, one cannot *a priori* select the more appropriate sex for exposure, particularly in the absence of knowledge of the general pattern of damage induced by various chemicals and without a detailed understanding of meiosis in human males and females.

Among those few aneuploidy-inducing agents that have been studied in both males and females, no clear pattern emerges. X rays,[30–33,59,60,85,86,121–123] heat shock,[31–33,37] and cold shock[103–105] are apparently effective in both sexes, although in each case the effects are germ cell stage-specific. Colchicine induces large increases in nondisjunction[109,112] in the female. In the male, however, colchicine induces a large amount of sterility,[45] but is ineffective in inducing increases in nondisjunction.[31–33] Finally, although more work needs to be done with this compound, studies indicate that dimethylsulfoxide increases nondisjunction in the male[31,33] but not in the female.[110]

2.2. Germ Cell Stage

It appears that aneuploidy-inducing agents acting in different ways are effective at different times in gametogenesis. Thus, a second critical issue in detection of aneuploidy is the exposure and sampling of the proper germ cell stages. This necessitates a knowledge of oogenesis and spermatogenesis. A brief description of each is presented below. For further information the reader is referred to reviews by Mahowald and Kambysellis[72] and Lindsley and Tokuyasu.[65]

2.2.1. Female

Oogenesis in *Drosophila* has been divided into 14 morphological stages[55] during which the oocyte undergoes a 100,000-fold increase in volume. The 14 stages are thought to correspond roughly to stages of meiosis from late or postpachytene I[13] through metaphase I.[89] The first oocytes appear in developing females at approximately the time of pupation.[39] These germ cells will have matured to stage 7 oocytes by the time of imago emergence.[39] A minimum of approximately 12 hr is required for the stage 7 oocytes to mature to stage 14 oocytes.[55] Meiosis is usually arrested at this point until after fertilization, when the two meiotic divisions occur.

In adult females, it takes approximately 3.5 days for an oocyte to go from stage 1 to stage 14.[24] The first cells undergoing premeiotic DNA synthesis at the beginning of an exposure of adult females will first appear after approximately 5–6 days of egg laying.[16] Thus, roughly 1.5 days are required between the time of DNA synthesis and oocyte stage 1. Because meiosis is not complete until after fertilization, exposure of virtually any germ cell in the female could lead to recovery of aneuploids.

Following chemical exposures, particular germ cell stages are sampled by collecting successive samples of fertilized eggs. One of the easiest stages to sample reliably following chemical exposure is the stage 14 oocytes from females exposed as mature adults. Virgin females tend to accumulate oocytes in this stage, so the first eggs collected from females exposed as adults will have been exposed only in stage 14.[120] A second fairly straightforward stage to sample is the late oogonia/early oocytes from females exposed as third-instar larvae. Again, these are collected as the first eggs produced by exposed females. Possible persistence of chemicals in the germ line, however, makes definite assignments to particular stages impossible during larval exposure. Because of variability due to such factors as persistence of chemicals, strain differences, environmental conditions, and exposure effects, precise stage assignments to germ cells of females is not possible following chemical exposure.

Physical agents such as X rays and temperature shocks can be delivered in acute exposures to permit more reliable sampling of germ cell stages. No pattern emerges concerning the most sensitive stage in the female. Cold shock is only effective following exposure of stage 14 oocytes(30,61,103,104); X rays are effective in immature oocytes and oogonia,(19,42,95) but not in stage 14 oocytes(56,106); and heat shock is only effective during a short period occurring near the time of oocyte formation.(37,40) Thus, exposure of any germ cell stage in the female can result in an increase in aneuploidy and exposure of all stages is necessary to rule out aneuploidy-inducing activity.

2.2.2. Male

In *Drosophila melanogaster* the first spermatocytes appear in the first instar larvae(21) approximately one day after hatching (25°C). These primary spermatocytes enter a growth phase lasting approximately 90 hr during which cell volume increases 25-fold.(65) At the end of the growth period, and corresponding approximately to the prepupal stage, the two meiotic divisions occur in rapid succession. The first of these divisions corresponds with the prepupal stage, occurring at approximately 115 hr of development (25°C). The pupae, then, contain germ cells in all stages up to and including spermatids.

Exposure of the spermatocyte stage can be accomplished by exposing larvae and collecting the first sperm ejaculated. Subsequent matings will sample stages first exposed as spermatogonia or stem cells. Appearance of large clusters of, for instance, interchange events, would herald the appearance of these germ cells.

To sample germ cells exposed as spermatocytes, spermatogonia, or stem cells from exposure of adult males, one must collect sperm from sequential matings to fresh virgin females. Valencia *et al.*(115) recommend three to four matings each for 2–3 days, at a ratio of three females per male. When this protocol is followed, sperm exposed at meiosis generally are ejaculated 6–9 days postmating.(17) Those sperm sampled before this time would have been exposed only as spermatozoa or spermatids, and no exposure-induced chromosome gain could be recovered from these.

That the germ cell stage exposed is crucial to detection of certain types of aneuploid events is illustrated by acute exposures to physical agents. Tokunaga found a short critical period of sensitivity to nondisjunction induced by cold shock exposure of males.[105] This period correlates with germ cells that were in the spermatocyte stage at the time of exposure. In addition, Foureman[31] determined that a 24-hr heat shock delivered to the early spermatocytes found in first-instar larvae failed to induce nondisjunction, while the same exposure delivered to the older spermatocytes present in second-instar larvae was effective in inducing nondisjunction.

2.3. Other Factors

Although some agents, such as X rays, cold shock, and colchicine, produce tremendous increases in aneuploidy, most agents appear to increase aneuploidy only three- to four-fold over the control.[91] The fact that many agents are germ cell stage-specific and the likelihood that there is a delicate balance between the production of recoverable nondisjunction and cell death can make the choice of brooding pattern, exposure level, and experimental sample size critically important in successfully detecting aneuploidy. Brood length must be short enough that an increase in aneuploidy occurring from a particular critical germ cell stage will not be swamped by germ cells that were in an insensitive phase at the time of exposure. On the other hand, the brooding time must be long enough that the labor involved is not prohibitive to performing the test. The exposure level must be high enough to induce aneuploidy without completely sterilizing the exposed flies. In addition, the experimental size should be large enough to detect at least a two- to fivefold increase in events. This size will vary with the spontaneous rate of aneuploidy of a particular stock in a particular set of environmental conditions as well as with the particular assay system used. As a general rule, one needs to monitor tens of thousands of germ cells rather than thousands of germ cells.

In 1920, Safir[93] reported that different genetic lines of *Drosophila* display marked differences in rates of primary nondisjunction. Differences of this sort continue to be noted for various stocks.[91,115] An elegant series experiments performed by Lüning[68–71] demonstrates the genetic and cytoplasmic effects on nondisjunction and chromosome loss in inbred *Drosophila melanogaster* strains. The proper choice of strain can be important to the sensitivity of a test system and therefore it is worthwhile to test any strain for sensitivity to known aneuploidy-inducing agents before routinely using it in an assay system.

The effects of environmental conditions on the rate of nondisjunction are largely unknown. Enormous temperature shocks (±10–15°C) have marked effects on nondisjunction frequencies, but the effects of more subtle temperature changes have not been studied. Gethmann[34] mentioned anecdotally that a meiotic mutant, *mei-G87*, failed to show the mutant phenotype when the laboratory humidity was high. Traut[108] has shown that the aging of mature oocytes in females results in increased nondisjunction following storage. Therefore, although little is understood about these effects, temperature, humidity, and the "age" of germ cells at the time of exposure must all be considered as possible variables that can influence the induction of nondisjunction in *Drosophila*.

3. Assay Systems

Several assays have been developed and are available for use in aneuploidy testing with *Drosophila*. Tests available include nonselective, semiselective, and selective types and among them they can detect the induction of sex-chromosome or autosomal aneuploidy in exposed males and females. Taken together, it is possible to test for increases in reductional or equational division, nondisjunction, and polyploidization. There is no single test available, however, that would enable an investigator to test all pertinent endpoints in both sexes. In choosing a testing system, then, one must weigh advantages and disadvantages of the various systems in order to choose the most efficacious one for the problem at hand.

All of the tests described below have three features critical to a *Drosophila* aneuploidy assay. First, the stocks employed carry genetic markers that permit one to distinguish and discard aneuploid parents that could produce offspring from secondary nondisjunction that would otherwise obfuscate the results. Second, the parental stocks also have markers that preclude prefertilization of the parental females (i.e., "nonvirginity") as a possible origin of those offspring phenotypically described as aneuploid. Finally, the genetic markers used to distinguish aneuploid exceptional offspring from normal offspring are easy to detect and nonambiguous in phenotype.

3.1. Classical Test

One of the most straightforward tests for aneuploidy is a variation of the test first employed by Bridges[10] to demonstrate the chromosome

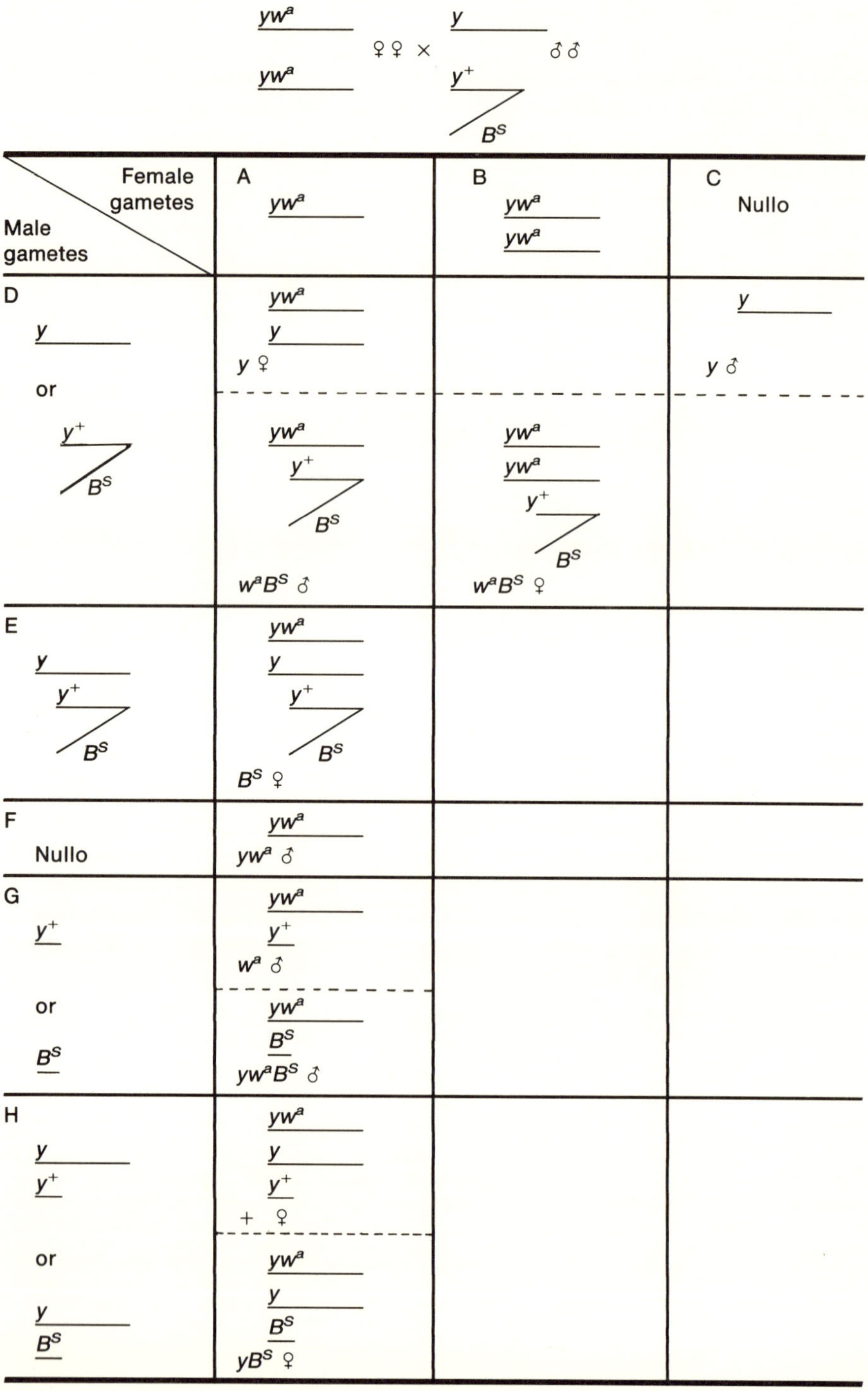

yw^a
yw^a
♀♀ ×
y
y^+
B^S
♂♂
Female gametes
Male gametes
A
B
C
Nullo
D
or
E
F
Nullo
G
or
H
or
y ♀
y ♂
w^aB^S ♂
w^aB^S ♀
B^S ♀
yw^a ♂
w^a ♂
yw^aB^S ♂
$+$ ♀
yB^S ♀

theory of inheritance and by Mavor[73–75] to demonstrate the aneuploidy-inducing effects of X rays. The modern version of this test uses a female homozygous for the recessive marker yellow (*y*) and one other X-linked marker (e.g., white-apricot, w^a) mated to a male with an X chromosome marked with yellow and a Y chromosome (such as the one developed by Brosseau[12] in 1958) doubly marked with the wild-type allele of yellow (y^+) on the long arm (Y^L) and the dominant eye shape marker Bar of Stone (B^S) on the short arm (Y^S). (See Lindsley and Grell[64] for a description of the genetic markers mentioned in this section.) All regular progeny as well as a series of aneuploid exceptions survive from this cross. Figure 1 presents the major classes of aneuploid progeny that can be scored, including both male and female exceptions. From the male, (1) chromosome gain resulting from first division nondisjunction, (2) chromosome loss from nondisjunction in either meiotic division, from chromosome breakage or chromosome lagging, and (3) partial losses and interchanges can be detected and distinguished. From the female, chromosome gain classes include products resulting from first division nondisjunction, second division nondisjunction, or attachment of X chromosomes (or other translocations). Chromosome loss could result from nondisjunction at either meiotic division, from chromosome breakage, or from chromosome lagging. Because the X chromosome used in this test contains no marker on its right end, chromosome gain resulting from nondisjunction in the female cannot, without further testing, be distinguished from gain resulting from breakage events. In addition, most triploid exceptions could be easily overlooked because they are distinguishable from normal progeny only by the triploid, intersex, or metamale phenotypes. Finally, gene mutations or chromosome breakage resulting in the deletion of all but the most distal part of the X chromosome could lead to an occasional misclassification of exceptional progeny.

It should also be noted that in order to recover exceptional ova (gain or loss events from the female), fertilization with the complementary gamete is necessary. Therefore, only one-half of such ova will be

←

FIGURE 1. Progeny of interest recovered from the classical nonselective test. Sex-chromosome constitutions of gametes and F_1 progeny are shown. Either sex can be exposed. Female gametes include those produced from (A) normal disjunction; (B) chromosome gain resulting from first division nondisjunction, breakage events, or triploidization; or (C) chromosome loss. Male gametes include those produced from (D) normal disjunction; (E) chromosome gain resulting from first division nondisjunction, (F) chromosome loss, (G) partial loss of Y^L (recovery of y^+) or Y^S (recovery of B^S); or (H) partial loss of Y or X–Y interchange leading to the recovery of the y^+ or B^S marker.

recovered. Exceptional sperm, however, will almost certainly fertilize a single X-bearing ovum, and therefore, at least theoretically, all such sperm could be recovered. These differences must be taken into account when rates of exceptional events are determined.

The major advantages of this test are its conceptual simplicity, the fact that normal chromosomes are used, the fact that either sex can be exposed, and the fact that nearly every endpoint of interest can be scored. Rates of induced events can be easily determined and compared to the occurrence of spontaneous events recovered from the nonexposed parent. In addition, the majority of chemicals that have been assayed in *Drosophila* have been tested using this system. Both parents could be exposed in a single experiment, but this is discouraged because effects in one parent (e.g., decreased fertility) could affect rates obtained for events from the other parent.

The major disadvantage of the test is that all progeny must be meticulously sorted and counted, thereby making scoring a tedious and time-consuming procedure.

3.2. Selective and Semiselective Tests

Tests that address the major disadvantage of the classical test, that of sorting and counting thousands of progeny in order to find the rare aneuploid exception, have been developed. Tests of this type use balanced lethal situations or specialized chromosome stocks to kill the normal progeny of one or both sexes. Most of these tests are designed to detect efficiently exceptional progeny following exposure of either one sex or the other, but not both.

3.2.1. Tests for the Detection of Exceptions Occurring in the Female

3.2.1a. Semiselective Test for Sex-Chromosome Nondisjunction. Grell[41] has proposed a test, which has been used by Brodberg *et al.*[11] In this test (Figure 2), the X chromosomes of the female contain balanced lethals so that all regular male progeny die. Chromosome gain occurring from first division nondisjunction can be recovered, as can gain resulting from breakage, although these two classes cannot be distinguished from each other without further testing. Triploid ova fertilized by a Y-bearing sperm can be detected as intersex progeny. Chromosome loss can also be detected.

The major advantages of this test are the use of "normal" chromosomes and the semiselective nature of the test. Having only female

$\frac{y w^a\ f\ l(1)Q56}{y w^a\ v\ f\ mal\ l(1)20}$ ♀♀ × $\frac{y^2\ v\ B\ car}{y^+}$ ♂♂

Female gametes \ Male gametes	y^2 v B car	y^+	Phenotype
A yw^a f l(1)Q56	y^2 v B car / yw^af l(1)Q56		y^2 B ♀
or yw^a v f mal l(1)20	y^2 v B car / yw^a v f mal l(1)20		y^2 v B ♀
B yw^a f l(1)Q56 / yw^a v f mal l(1)20		yw^a f l(1)Q56 / yw^a v f mal l(1)20 / y^+	w^a f ♀
C Nullo	y^2 v B car		y^2 v B car ♂

FIGURE 2. Progeny of interest obtained from Grell's semiselective test. The sex-chromosome constitution of gametes and F_1 progeny are presented. Male gametes shown result from normal disjunction. First division nondisjunction in the male would result in phenotypically "*B*" or "*vB*" F_1 females. Female gametes shown include those produced from (A) normal disjunction; (B) nondisjunction in the first meiotic division, or (C) chromosome loss.

progeny recovered from normal meiocytes greatly simplifies both sorting and counting and enables the use of weighing procedures instead of manual counting. (Male and female *Drosophila* differ in weight and therefore estimates based on the weight of a mixture of males and females are less accurate than those based on single-sex weights.)

The major disadvantage encountered with this test is the fact that individuals resulting from chromosome gain in the second division cannot be recovered. Only one compound [*cis*-platinum (II) diamminodichloride (*cis*-PDD), reported to be negative] has been tested with this system to date.[11] Further work is necessary in order to determine if the lack of recovery of second division nondisjunction damages the overall power of this test.

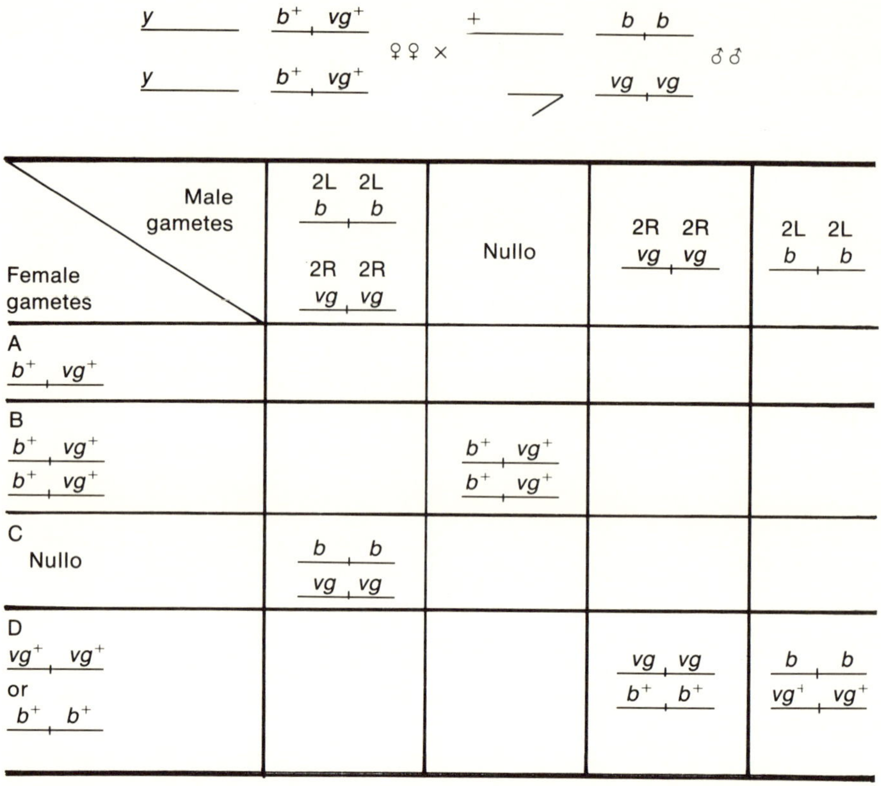

FIGURE 3. Progeny of interest recovered from Traut's selective test. The chromosome 2 constitutions of gametes and F_1 progeny are shown. The four regular types of male gametes are produced in approximately equal numbers. Female gametes represent the following events: (A) normal disjunction, (B) nondisjunction resulting in gain of chromosome 2, (C) loss of chromosome 2 (due either to nondisjunction or breakage), (D) isochromosome 2 formation resulting from breakage or interchange.

3.2.1b. Selective Test for Sex-Chromosome and Autosomal Nondisjunction. Traut[109] has proposed a selective scheme for the detection of aneuploidy in females. This test (Figure 3) uses a male with compound chromosome 2 autosomes marked with black (*b*) and vestigial (*vg*) mated to a femal · with structurally normal, unmarked second chromosomes and X chromosomes homozygous for the marker yellow. The principle is that in males, the two compound 2 chromosomes segregate approximately at random from each other,[47] producing, in roughly equal numbers, sperm that (1) have two copies of chromosome 2 (diplo-2), (2) are lacking chromosome 2 (nullo-2), (3) have two copies of the right arm of chromosome 2 (2R) and no copies of the left arm, or (4) have two copies of the left arm of chromosome 2 (2L) and no copy of the right arm. When any of these sperm fertilize a normal haplo-2 ovum, the resulting offspring is aneuploid and consequently dies at an early developmental stage. An aneuploid ovum with the proper chromosome constitution, however, will complement the aneuploid sperm to, in essence, "rescue" the resulting individual. Thus, the only survivors are those resulting from aneuploid exceptions in the females, or rare spontaneous chromosome rearrangements in the untreated male. Both nondisjunction and chromosome breakage occurring in the female can be recovered and distinguished. A nondisjunction of the second chromosomes in the female will, of course, result in gametes that are either diplo-2 or nullo-2. Because the female also has marked X chromosomes, those diplo-2 or nullo-2 exceptional gametes can be additionally classified as being nullo, haplo, or diplo with respect to the X chromosomes (chromosome 1 in *Drosophila*). Traut therefore lists six classes of exceptions with respect to nondisjunction of chromosome 2 (Figure 4). Note that because the male's X and Y chromosomes are unmarked, sex-chromosome exceptions from the sperm, provided that they fertilize a haplo-1 ovum, would not be detected. In those rare cases in which a diplo-1 sperm fertilizes a nullo-1 ovum, the resulting progeny could be misclassified as resulting from a haplo-1 ovum. All other combinations would be expected to be lethal.

In addition to detecting nondisjunction for chromosome 2, Traut's scheme can detect the presence of presumed breakage events leading to production of isochromosomes in the female. These ova are, of course, of two possible types with respect to chromosome 2, i.e., compound for the left arm, designated C(2L); or compound for the right arm, designated C(2R). Again, these can be further divided according to the chromosome 2 constitution of the compound 2 exceptional gametes. Because the events resulting in compound chromosomes are probably not mishaps of segregation, the other chromosomes in these meiocytes would be less likely to be affected (i.e., less likely to missegregate) than in those meiocytes undergoing nondisjunction of chromosome 2.[38] Thus, Traut identifies as his two classes of isochromosomal exceptional gametes those with normal haplo constitution with respect to chromosome 1 (class VII) and those with abnormal (nullo or diplo) constitution with respect to chromosome 1 (designated class VIII).

In this system, analysis is done by performing a χ^2 comparison of the mean values of the eight recovered classes in the control series

	Chromosome 2	Chromosome 1	Recovered phenotype	Class
Diplo-2	b^+ vg^+ / b^+ vg^+	Nullo	+ ♂	I
		Haplo *y*	+ ♀ *y* ♂	II
		Diplo *y* / *y*	*y* ♀	III
Nullo-2		Nullo	*b vg* ♂	IV
		Haplo *y*	*b vg* ♀ *y*; *b vg* ♂	V
		Diplo *y* / *y*	*y*; *b vg* ♀	VI
Haplo-2	b^+ b^+ or vg^+ vg^+	Haplo *y*	*vg* ♀ *y*; *vg* ♂	VII
		Haplo *y*	*b* ♀ *y*; *b* ♂	
	b^+ b^+ or vg^+ vg^+	Nullo	*vg* ♂	VIII
		Diplo *y* / *y*	*y*;*vg* ♀	
		Nullo	*b* ♂	
		Diplo *y* / *y*	*y*; *b* ♀	

FIGURE 4. The first and second chromosome constitutions of recovered female gametes and the phenotypes and classes of the surviving progeny recovered from Traut's selective test.

compared to the mean values of the eight recovered classes in the exposed group.

Traut has used this test successfully with two known aneuploidy-inducing agents, colchicine and X rays. In addition, one compound (dimethylsulfoxide) with unknown aneuploidy-inducing capacity has been tested. The results obtained were similar (negative) to those obtained with a classical test.[110] No concurrent control was performed with either test, however, so this result must be viewed with caution.

The major advantage of this test is the fact that enormous numbers of gametes can be tested with relatively little time and effort. Traut[109] estimates that his 15 control experiments, in which 617 spontaneous aneuploid exceptions were recovered, sampled 3.2×10^6 gametes. Other advantages include the fact that nondisjunction of structurally normal chromosomes are assayed and that nondisjunction arising from either the first or second meiotic division can be recovered and distinguished on the basis of phenotype from isochromatid production. Finally, Traut's system would be the most suitable for the detection of agents that cause nondisjunction by specifically inhibiting chromosome pairing, because with this test, those gametes with both chromosomes 1 and 2 nondisjoining would result in viable offspring.

Disadvantages include the fact that rates of aneuploidy cannot be determined, that rare events occurring in the untreated male would be confused with presumed female nondisjunction (although further testing of F_1 progeny could resolve this), and that triploid exceptions would not be recovered. In addition, there is no internal control. The effect of this is that any agent that does not affect the relative ratios of events, i.e., that increases or decreases all endpoints to the same degree, would not be scored as positive in this test system.

3.2.1c. Optionally Selective Test. A third type of test has been developed by Smith.[96] This test (Figure 5) depends on the hypersensitivity to killing by treatments with aqueous purine solutions of mutants with the X-linked marker maroonlike (*mal*). The test involves mating a female heterozygous for complementing *mal* alleles to a male bearing a *mal* allele that is noncomplementary to both *mal* alleles carried by the female. All regular F_1 males are hemizygous for *mal* and therefore are subject to killing by purine exposure. In addition, all regular F_1 females are heterozygous for noncomplementing *mal* alleles and are likewise subject to killing by purine exposure. XXY progeny arising from first division nondisjunction in the parental female, however, are heterozygous for complementing alleles of *mal* and are not subject to killing by the purine exposure. Thus, the selective system in this test can be turned on by adding a purine solution to any culture. Alternatively, if it is preferred, no selection need be used.

This test can detect chromosome gain arising from first division nondisjunction, from the attachment of X chromosomes, or from triploidization. Further testing of exceptional females would distinguish among these possibilities. Smith[96] has successfully used this system to detect chromosome gain following exposures of adult females to X-irradiation and 9°C cold shock.

The major advantage of this system is the fact that true rates of nondisjunction can be determined for a sample of exposed females in the absence of selection, and enormous numbers of progeny can easily be scored from the remaining females by utilizing the selective system.

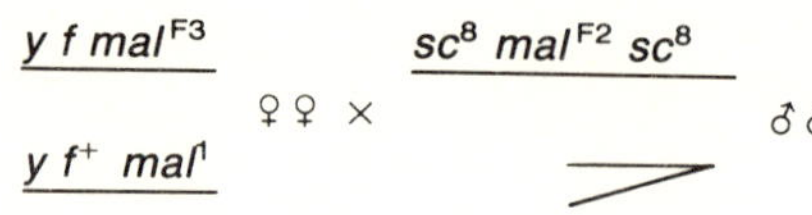

Female gametes \ Male gametes	$sc^8\ mal^{F2}\ sc^8$	Y	Phenotype: + Purine	Phenotype: − Purine
A $y\ f\ mal^{F3}$	$sc^8\ mal^{F2}\ sc^8$ / $y\ f\ mal^{F3}$	$y\ f\ mal^{F3}$ / Y	Dead	mal ♀ y f mal ♂
or $y\ f^+\ mal^{f1}$	$sc^8\ mal^{F2}\ sc^8$ / $y\ f^+\ mal^{f1}$	$y\ f^+\ mal^{f1}$ / Y	Dead	mal ♀ y mal ♂
B $y\ f\ mal^{F3}$ / $y\ f^+\ mal^{f1}$		$y\ f\ mal^{F3}$ / $y\ f^+\ mal^{f1}$ / Y	y mal ♀	y mal ♀
C Nullo	$sc^8\ mal^{F2}\ sc^8$		Dead	mal ♂

FIGURE 5. Progeny of interest obtained from Smith's optionally selective system. The sex-chromosome constitution of gametes and F_1 progeny and the associated phenotypes in the presence (+ purine) and absence (− purine) of selective conditions are shown. Male gametes include only those produced from normal disjunction (aneuploid sperm cannot be detected). Female gametes include those produced from (A) normal disjunction, (B) nondisjunction in the first meiotic division, attachment of X chromosomes, or triploidization, or (C) chromosome loss.

The major disadvantage of the system is the fact that chromosome gain arising from second division nondisjunction cannot be detected, nor can chromosome loss of any kind. The effect of this in terms of screening efficiency is not known.

3.2.2. Tests for the Detection of Exceptions Occurring in the Male

3.2.2a. Semiselective Test. Foureman[31,32] has described a semiselective test for detection of sex-chromosome aneuploidy induction in males (Figure 6). An attached X-female marked with yellow and prune (*pn*) is mated with a Killer of prune (*K-pn*) male with an X chromosome marked with yellow and a Y chromosome marked with the wild-type allele of yellow on the short arm and the dominant marker Bar of Stone on the long arm. All females resulting from normal disjunction

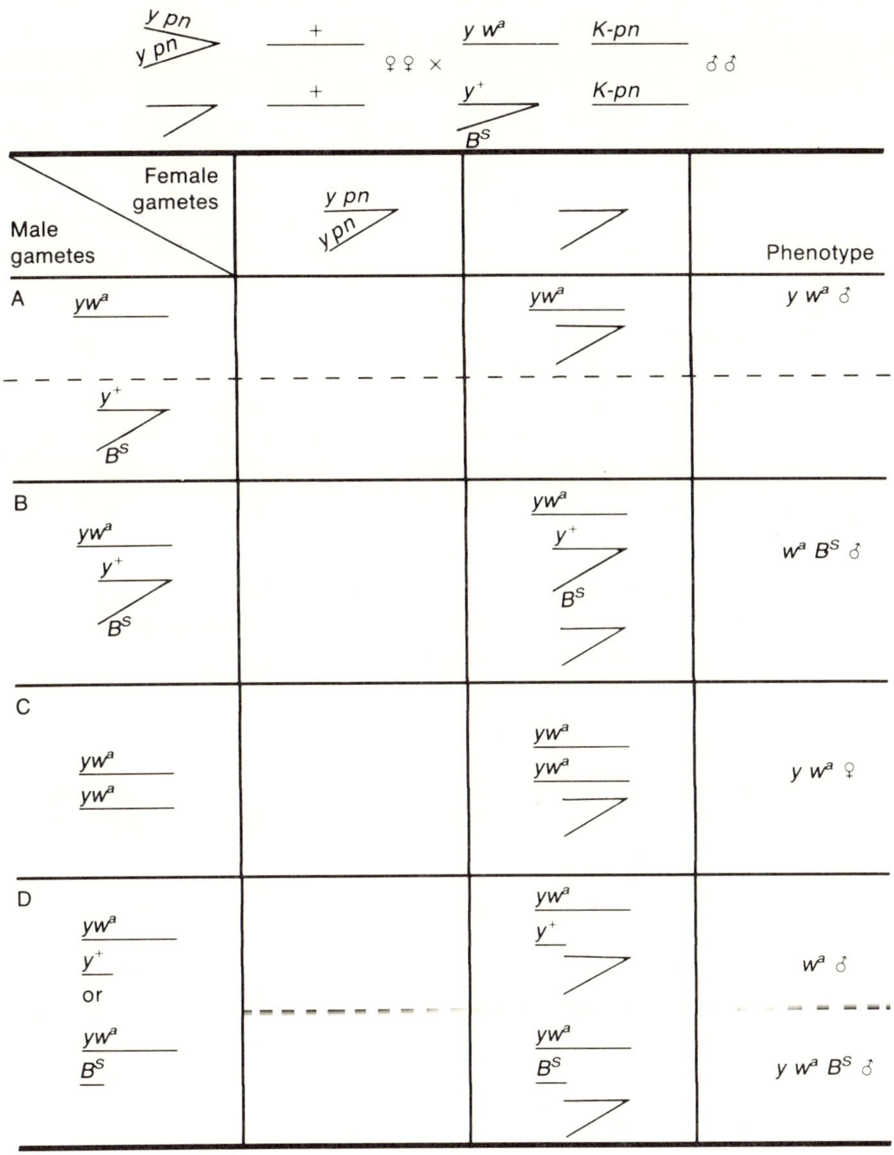

FIGURE 6. Progeny of interest recovered from the Killer of prune test. The sex-chromosome constitution of male and female gametes is shown. Male gametes include those produced from (A) normal X–Y disjunction, (B) nondisjunction at the first meiotic division, (C) nondisjunction at the second meiotic division, and (D) partial recovery of the Y following breakage or X–Y recombination. All progeny homozygous for *pn* are killed by the action of *K-pn*.

die either from genetic imbalance resulting from trisomy of the X chromosome (the 1% or so that survive are readily distinguishable as yellow females), or are killed prior to the second larval instar due to the interaction of Killer of prune with prune.[99] All males arising from normal disjunction survive. Exceptional progeny resulting from first division nondisjunction, second division nondisjunction, or partial recovery of the Y chromosome can be recovered and distinguished using this test. In addition, triploid progeny can be recovered, although further testing (or an experienced observer) is necessary to distinguish these from other possible exceptional types. Chromosome loss events cannot be detected using this test.

Advantages of this system include the fact that rates of events can be determined, that scoring is simplified, and that large numbers of progeny can be assayed. Because only male progeny survive, weighing procedures can be used in lieu of counting. In addition, first and second division nondisjunctional events in the male can be distinguished from each other and from breakage and interchange events.

The major disadvantage of this system is the fact that whole chromosome loss events cannot be detected.

3.2.2b. Selective Test. The TX;Y test was first proposed by Craymer[23] and has been used by Foureman.[29–31,33] This system (Figure 7) uses a male with an X–Y translocation mated to a female with normal X chromosomes marked with the recessive alleles yellow and white-apricot. All regular progeny die prior to eclosion as a result of genetic imbalance. The only survivors result from various events occurring in either the male or female to produce aneuploid gametes that restore genetic balance in the resulting progeny. These events include chromosome gain resulting from first division nondisjunction in the male, chromosome loss in the male, partial chromosome loss in the male, X–Y interchange events in the male, or X-chromosome gain due to nondisjunction (first and second division), breakage, or triploidization in the female. The etiology of the female events cannot be determined, since they all produce the same sterile female phenotype in this system.

One important feature of this system is the presence of an internal control class. When the male is the exposed parent, those progeny classes arising only from events in the female serve as an internal control. When the female is the exposed parent, those progeny classes arising only from events in the male serve as the internal control. Thus, statistical analysis can be performed separately on each class of interest by comparing the ratio of the class of interest and the internal control class in the exposed group to the ratio of the class of interest and the internal control class in the unexposed group.

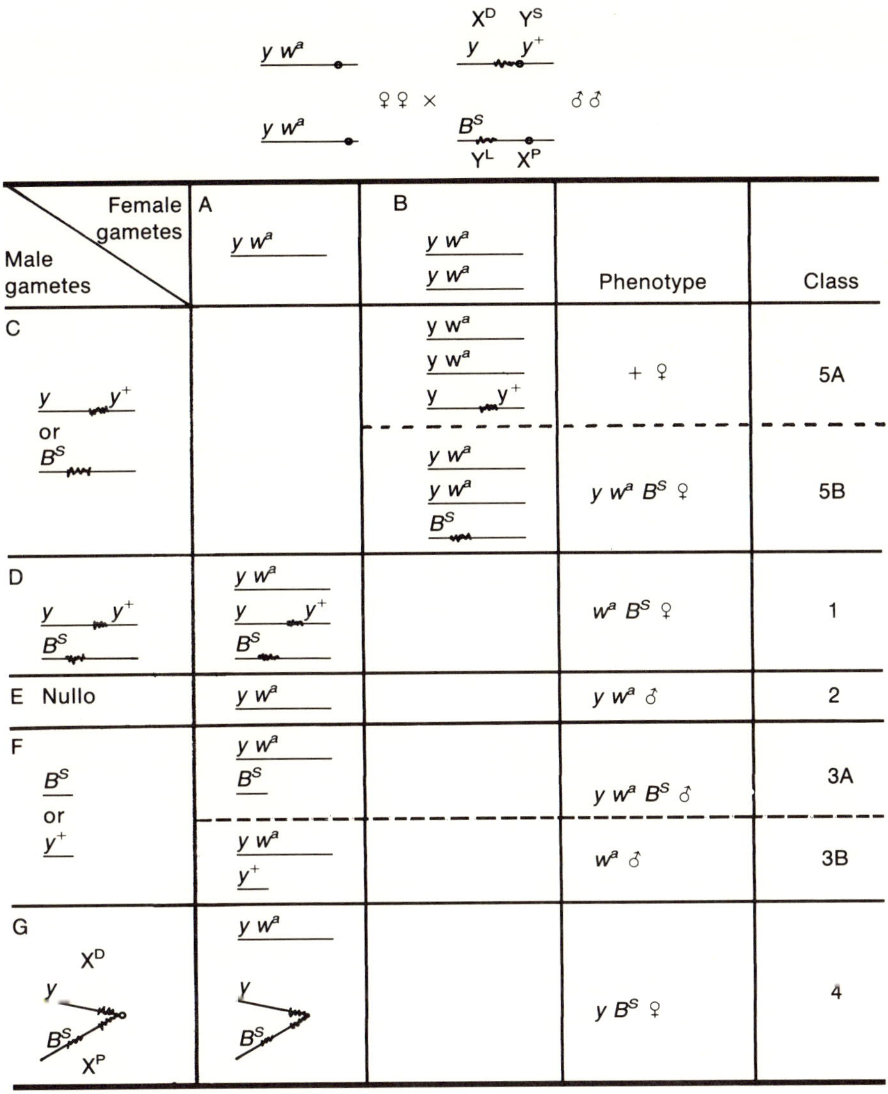

FIGURE 7. Progeny of interest recovered from the TX;Y test. Chromosome 1 constitutions of gametes and surviving progeny are shown. Female gametes presented arise either from (A) normal disjunction or (B) chromosome gain resulting from nondisjunction, attachment of X chromosomes, or triploidization. Male gametes arise from (C) normal disjunction, (D) nondisjunction in division I, (E) whole chromosome loss, (F) partial chromosome loss resulting either from breakage (classes 3A and 3B) or interchange (class 3B), or (G) interchange.

Because it has been found[31] that recovery of survivors is nonrandom in some of the phenotypic classes, including the internal control class, the χ^2 statistic is not adequate to use with these data. The nonparametric rank sum test[58] is used instead. In this procedure, each culture bottle is considered to be an independent test. For each bottle in the control and exposed groups, the number of individuals in the class of interest is divided by the total number of individuals in the class of interest plus the internal control class in that bottle. This corrects for differences in fertility between the bottles. The control and exposed groups are combined and ranked in ascending order. The two sets (control and exposed) are then separated and the ranks for each test are summed and compared.[31,33]

The major advantage of this system is that because it is selective, it is very quick and inexpensive to use. Hundreds of thousands of gametes can be screened in a single test. The genetic markers used are very easy to distinguish, so little experience is necessary in order to do the scoring. The presence of an internal control class enables changes in fertility following exposure to be monitored. In addition, a variety of endpoints can be recovered and distinguished, and although the male is usually the exposed parent, events occurring in the female can be recovered as well.[30,31] Thus, only one stock need be kept to screen both male and female events.

Disadvantages of the system include the fact that rates of events cannot be determined and that the exposed males have structurally abnormal chromosomes. The effect of this is not known. In addition, control results from different experiments are not always homogeneous,[33] so a concurrent control series is always required.

4. Known Positives

One important aspect of the development of assays for aneuploidy in *Drosophila* is the accumulation of a data base of known positive and negative chemical compounds. This will permit comparisons of the various testing schemes described above, help determine which sex is more appropriate for testing and which germ cell stages are most important, and aid in the development of new testing assays for aneuploidy-inducing agents. A solid data base is also important for interspecies comparisons.

Although a recent review of the literature by Valencia *et al.*[115] lists 15 chemicals that are considered positive for aneuploidy-inducing ability in *Drosophila*, only a few of these have been adequately tested in more than one system. Several of these will be discussed here to demonstrate the differences in sensitivity found according to endpoint scored, sex exposed, and germ cell stage exposed.

4.1. Colchicine

Colchicine and its closely related derivative Colcemid are among the most extensively studied aneuploidy-inducing agents in *Drosophila*. Colchicine is known to interfere with the spindle structure by binding to the tubulin subunit of microtubules.[8,67] Studies by Traut and Sommer[112] and by Traut and Scheid[111] demonstrated that feeding exposures of adult females result in an increase in both X-chromosome gain and loss with exposure concentrations of colchicine as low as 3 ppm and of Colcemid as low as 1 ppm. Similar results were obtained by Ramel and Magnusson[91] following larval feeding exposures and by Foureman[30,31] following adult or larval feeding exposures. Traut[109] has also demonstrated an increase in nondisjunction of the second chromosome as well as of the X chromosome following feeding exposures of adult females. Finally, Held[46] has induced triploidy in oogonia of females exposed as larvae to Colcemid. No effective increase in nondisjunction from mature stage 14 oocytes has been described. Thus, colchicine effectively induces aneuploidy in both early oocytes and oogonia, but is probably ineffective in mature oocytes.

In males, however, despite the fact that both colchicine and Colcemid have been shown to reach the critical germ cells,[54,118] colchicine is ineffectivve in inducing X–Y nondisjunction[31–33] following feeding exposure of larval males. In addition, Held[46] reported an inability to induce triploidy following exposure of male larvae to Colcemid. Thus, in males, although critical germ cells are exposed and fertility is reduced by colchicine exposure,[31,45] aneuploid endpoints are not affected by colchicine or Colcemid exposure.

Colchicine, then, is an example of a chemical that induces a sex-specific (females only) and germ cell stage-specific increase in aneuploidy in *Drosophila*. Insufficient data are available to demonstrate the existence of chemicals that induce aneuploidy exclusively in males or exclusively in mature oocytes of females.

4.2. Methyl Mercury Hydroxide

Ramel and Magnusson[90,91] have extensively studied the effects of methyl mercury hydroxide, a potent inhibitor of the spindle fiber mechanism.[65] Following exposures throughout the larval period of females with structurally homozygous X chromosomes, they observed the puzzling result of an increase in chromosome-gain (XXY) progeny with no concomitant increase in chromosome-loss (XO) progeny.[90] Comparable exposures of larval females with structurally heterozygous chromosomes, however, gave the opposite effect, with an increase in XO progeny and no increase in XXY progeny.[91] No adequate explanation for this effect has been proposed. Exposures of adult males indicate[90] that methyl mercury hydroxide is also capable of inducing chromosome gain and chromosome loss during spermatogenesis.

Other heavy metal compounds have also been tested following larval exposures of both males and females. Ramel and Magnusson[91] report positive results for induction of X-chromosome gain induced by phenyl mercury hydroxide, diethyl lead dichloride, triethyl lead chloride, and trimethyl tin chloride.

4.3. Inhibitors of Nucleic Acid Synthesis

There is a small body of evidence to support the notion that inhibitors of nucleic acid synthesis such as 5-fluorodeoxyuridine, (FUdR), azathioprine, and actinomycin D induce nondisjunction in *Drosophila*. Traut[107] has observed a significant increase in both X-chromosome gain and loss (XXY and XO) progeny resulting from germ cells first exposed to FUdR as early oocytes and oogonia in adult females. Clark[18] found a significant increase in X–Y nondisjunction (both the gain and loss classes) in spermatocytes sampled from adult males exposed to azathioprine. In neither of these cases was more than one exposure level used and neither result has been corroborated in a second *Drosophila* system.

Félix and Rodriguez[28] and de la Rosa and Félix[25] report a slight non-dose-related increase in X-chromosome gain and loss from oocytes in adult females injected with 10 or 100 μg/ml actinomycin D. A much larger experiment reported by Ostertag and Haake[84] in which larval females were exposed to feeding, however, failed to detect any increase in nondisjunction following 400 μg/ml exposure. Ramel and Magnusson[91] do report a marginally significant increase in chromosome loss after a larval feeding exposure (10 μg/ml) of actinomycin D to both males and females. Clearly, more work is needed with this compound before an assessment of its aneuploidy-inducing ability can be made.

5. Summary

Numerous assay systems are available in *Drosophila* for the large-scale screening of chemicals for their aneuploidy-inducing ability.

Future work must focus on using these assays to screen a large number of chemicals in order to increase the data base of known positive and negative chemicals. In addition, more studies are needed on the environmental and genetic conditions that affect spontaneous rates of aneuploidy. The ultimate goal of this work is twofold. First, it will improve hazard assessment for aneuploidy in humans, and second, it will increase our knowledge of the intricacies of meiosis itself.

ACKNOWLEDGMENTS

The author is grateful to Dr. S. Abrahamson for critical readings of the manuscript. Work performed by the author and reported here was supported in part by Reproductive Hazards in the Workplace Grant No. 15-13 from March of Dimes Birth Defects Foundation.

6. References

1. S. Abrahamson and E. B. Lewis, The detection of mutations in *Drosophila melanogaster*, in: *Chemical Mutagens: Principles and Methods for Their Detection* (A. Hollaender, ed.), Vol. 2, pp. 461–487, Plenum Press, New York (1971).
2. C. Auerbach, The role of *Drosophila* in mutagen testing, in: *Mutagenesis in Sub-Mammalian Systems* (G. E. Paget, ed.), pp. 13–25, MTP Press, Lancaster, England (1979).
3. A. J. Baars, J. A. Zijlstra, E. Vogel, and D. D. Breimer, The occurrence of cytochrome P-450 and aryl hydrocarbon hydroxylase activity in *Drosophila melanogaster* microsomes, and the importance of this metabolizing capacity for the screening of carcinogenic and mutagenic properties of foreign compounds, *Mutat. Res. 44*, 257–268 (1977).
4. B. S. Baker and A. T. C. Carpenter, Genetic analysis of sex chromosomal meiotic mutants in *Drosophila melanogaster, Genetics 71*, 255–286 (1972).
5. B. S. Baker and J. C. Hall, Meiotic mutants: Genic control of meiotic recombination and chromosome segregation, in: *Genetics and Biology of Drosophila* (E. Novitski and M. Ashburner, eds.), Vol. 1a, pp. 351–434, Academic Press, New York (1976)
6. B. S. Baker, A. T. C. Carpenter, M. S. Esposito, R. E. Esposito, and L. Sandler, The genetic control of meiosis, *Annu. Rev. Genet. 10*, 53–134 (1976).
7. D. J. Bond and L. McMillan, Meiotic aneuploidy: Its origins and induction following chemical treatment in *Sordaria brevicollis, Environ. Health Perspect. 31*, 67–74 (1979).
8. G. G. Borisy and E. W. Taylor, The mechanism of action of colchicine: Binding of colchicine-^{3}H to cellular protein, *J. Cell Biol. 34*, 525–533 (1967).
9. J. Boué, A. Boué, and P. Lazar, Retrospective and prospective epidemiological studies of 1500 karyotyped spontaneous human abortions, *Teratology 12*, 11–26 (1975).
10. C. B. Bridges, Non-disjunction as proof of the chromosome theory of heredity, *Genetics 1*, 1–52 (1916).
11. R. K. Brodberg, R. F. Lyman, and R. C. Woodruff, The induction of chromosome

aberrations by *cis*-platinum (II) diamminodichloride in *Drosophila melanogaster, Environ. Mutagen. 5*, 285–297 (1983).

12. G. E. Brosseau, Jr., Crossing over between Y chromosomes in male *Drosophila, Drosophila Inform. Serv. 32*, 115–116 (1958).
13. A. T. C. Carpenter, Electron microscopy of meiosis in *Drosophila melanogaster* females. I. Structure, arrangement, and temporal change of synaptonemal complex in wild-type, *Chromosoma 51*, 157–182 (1975).
14. D. H. Carr and M. Gedeon, Population cytogenetics of human abortuses, in: *Population Cytogenetics: Studies in Humans* (E. B. Hook and I. H. Porter, eds.), pp. 1–9, Academic Press, New York (1977).
15. C. O. Carter, The relative contribution of mutant genes and chromosome abnormalities to genetic ill-health in man. in: *Progress in Genetic Toxicology* (D. Scott, B. A. Bridges, and F. H. Sobels, eds.), pp. 1–14, Elsevier/North-Holland, New York (1977).
16. A. C. Chandley, The effect of X-rays on female germ cells of *Drosophila melanogaster.* III. A comparison with heat-treatment on crossing-over in the X-chromosome, *Mutat. Res. 5*, 93–107 (1968).
17. A. C. Chandley and A. J. Bateman, Timing of spermatogenesis in *Drosophila melanogaster* using tritiated thymidine, *Nature 193*, 299–300 (1962).
18. J. M. Clark, The mutagenicity of azathioprine in mice, *Drosophila melanogaster* and *Neurospora crassa, Mutat. Res. 28*, 87–99 (1975).
19. A. M. Clark and F. H. Sobels, Studies on non-disjunction of the major autosomes in *Drosophila melanogaster.* I. Methodology and rate of induction by X-rays for the compound second chromosome, *Mutat. Res. 18*, 47–61 (1973).
20. Committee on the Biological Effects of Ionizing Radiations (BEIR), *The Effects on Populations of Exposure to Low Levels of Ionizing Radiation: 1980*, National Academy Press, National Academy of Sciences, Washington, D.C. (1980).
21. K. W. Cooper, Normal segregation without chiasmata in female *Drosophila melanogaster, Genetics 30*, 472–484 (1945).
22. K. W. Cooper, Normal spermatogenesis in *Drosophila*, in: *Biology of Drosophila* (M. Demerec, ed.), pp. 1–61, Wiley, New York (1950).
23. L. Craymer, A new genetic testing procedure for potential mutagens, *Drosophila Inform. Serv. 51*, 62 (1974).
24. J. David and J. Merle, A reevaluation of the duration of egg chamber stages in oogenesis of *Drosophila melanogaster, Drosophila Inform. Serv. 43*, 122–123 (1968).
25. M. E. de la Rosa and R. Félix, Chromosome X loss and non-disjunction in oocytes of *D. melanogaster*, induced by treatment with actinomycin D, *Drosophila Inform. Serv. 50*, 36–39 (1973).
26. C. Denniston, Low level radiation and genetic risk estimation in man, *Annu. Rev. Genet. 16*, 329–355 (1982).
27. T. Dobzhansky, Studies on chromosome conjugation. II. The relation between crossing-over and disjunction of chromosomes, *Z. Indukt. Abstammungs. Vererbungsl. 64*, 269–309 (1933).
28. R. Félix and R. Rodriguez, Actinomycin D effects on the frequency of X-chromosome loss and non-disjunction in *Drosophila melanogaster* females, *An. Inst. Biol. Nal. Autón. México, Ser. Biol. Exp. 39(1)*, 13–21 (1968).
29. P. A. Foureman, A translocation X;Y system for detecting meiotic nondisjunction and chromosome breakage in males of *Drosophila melanogaster, Environ. Health Perspect. 31*, 53–58 (1979).
30. P. A. Foureman, Use of the translocation X;Y test for the detection of sex chromosome aneuploidy induced in *Drosophila melanogaster* females, *Environ. Mutagen. 5*, 376 (1983).

31. P. A. Foureman, The Translocation X;Y Test: A Selective System for the Detection of Sex Chromosome Aneuploidy in *Drosophila melanogaster*, Ph.D Thesis, University of Wisconsin at Madison (1983).
32. P. A. Foureman, A semi-selective test for the detection of first and second division nondisjunction in *Drosophila melanogaster* males, *Environ. Mutagen*, in press.
33. P. A. Foureman, The TX; Y test for the detection of nondisjunction and chromosome breakage in *Drosophila melanogaster*. I. Exposure of males, *Mutat. Res.*, submitted for publication.
34. R. C. Gethmann, The genetic analysis of a chromosome-specific meiotic mutant that permits a premature separation of sister chromatids in *Drosophila melanogaster, Genetics 107*, 65–77 (1984).
35. R. F. Grell, A new hypothesis on the nature and sequence of meiotic events in the female of *Drosophila melanogaster, Proc. Natl. Acad. Sci. USA 48*, 165–172 (1962).
36. R. F. Grell, A new model for secondary nondisjunction: The role of distributive pairing, *Genetics 47*, 1737–1754 (1962).
37. R. F. Grell, Induction of sex chromosome nondisjunction by elevated temperature, *Mutat. Res. 11*, 347–349 (1971).
38. R. F. Grell, Distributive pairing, in: *The Genetics and Biology of Drosophila* (M. Ashburner and E. Novitski, eds.), Vol. 1a, pp. 435–486, Academic Press, New York (1976).
39. R. F. Grell, Time of recombination in the *Drosophila melanogaster* oocyte: Evidence from a temperature-sensitive recombination-deficient mutant, *Proc. Natl. Acad. Sci. USA 75*, 3351–3354 (1978).
40. R. F. Grell, Origin of meiotic nondisjunction in *Drosophila* females, *Environ. Health Perspect. 31*, 33–39 (1979).
41. R. F. Grell, A test for nondisjunction in *Drosophila, Environ. Mutagen. 3*, 365 (1981).
42. R. F. Grell, E. R. Muñoz, and W. R. Kirschbaum, Radiation-induced non-disjunction and loss of chromosomes in *Drosophila melanogaster* females. I. The effect of chromosome size, *Mutat. Res. 3*, 494–502 (1966).
43. A. J. F. Griffiths, *Neurospora* prototroph selection system for studying aneuploid production, *Environ. Health Perspect. 31*, 75–80 (1979).
44. A. J. F. Griffiths, Short-term tests for chemicals that promote aneuploidy, in: *Chemical Mutagens: Principles and Methods for Their Detection* (F. E. de Serres and A. Hollaender, eds.), Vol. 7, pp. 189–210, Plenum Press, New York (1982).
45. S. M. Hannah, Effect of colchicine on lethality and sterility, *Drosophila Inform. Serv. 52*, 122 (1977).
46. L. I. Held, Jr., Polyploidy and aneuploidy induced by Colcemid in *Drosophila melanogaster, Mutat. Res. 94*, 87–101 (1982).
47. D. G. Holm, Compound autosomes, in: *The Genetics and Biology of Drosophila* (M. Ashburner and E. Novitski, eds.), Vol. 1b, pp. 529–561, Academic Press, New York (1976).
48. E. B. Hook, Perspectives in mutation epidemiology: 3. Contribution of chromosome abnormalities to human morbidity and mortality and some comments upon surveillance of chromosome mutation rates, *Mutat. Res. 114*, 389–423 (1983).
49. A. F. Huettner, The spermatogenesis of *Drosophila melanogaster, Z. Mikrosk. Anat. Forsch. 11*, 615–637 (1930).
50. M. Hultén, Chiasma distribution at diakinesis in the normal human male, *Hereditas 76*, 55–78 (1974).
51. B. John, Myths and mechanisms of meiosis, *Chromosoma 54*, 295–325 (1976).
52. K. J. Kemphues, R. A. Raff, T. C. Kaufman, and E. C. Raff, Mutation in a structural

gene for a β-tubulin specific to testis in *Drosophila melanogaster, Proc. Natl. Acad. Sci. USA 76*, 3991–3995 (1979).
53. K. J. Kemphues, E. C. Raff, R. A. Raff, and T. C. Kaufman, Mutation in a testis-specific β-tubulin in *Drosophila*: Analysis of its effects on meiosis and map location of the gene, *Cell 21*, 445–451 (1980).
54. K. J. Kemphues, T. C. Kaufman, R. A Raff, and E. C. Raff, The testis-specific β-tubulin subunit in *Drosophila melanogaster* has multiple functions in spermatogenesis, *Cell 31*, 655–670 (1982).
55. R. C. King, A. C. Rubinson, and R. F. Smith, Oogenesis in adult *Drosophila melanogaster, Growth 20*, 121–157 (1956).
56. W. C. Kiriazis, The Effect of Varying Doses of X-rays in the Production of Chromosome Loss and Non-disjunction of the X-Chromosome and the Fourth Chromosome in Stage-14 Oocytes of *Drosophila melanogaster*, Master's Thesis, University of Wisconsin at Madison (1969).
57. W. R. Lee, S. Abrahamson, R. Valencia, E. S. von Halle, F. E. Würgler, and S. Zimmering, The sex-linked recessive lethal test for mutagenesis in *Drosophila melanogaster*: A report of the U. S. Environmental Protection Agency Gene-Tox Program, *Mutat. Res. 123*, 183–279 (1983).
58. E. L. Lehman, *Nonparametrics: Statistical Methods Based on Ranks*, Holden-Day, San Francisco (1975).
59. B. Leigh, Induced nondisjunction in *Drosophila* oocytes, *Mutat. Res. 61*, 65–68 (1979).
60. B. Leigh, Mechanisms of nondisjunction induction in *Drosophila* oocytes, *Environ. Health Perspect. 31*, 41–43 (1979).
61. B. Leigh and D. R. Parker, Meiotic disturbances caused by cold treatment of mature oocytes of *Drosophila melanogaster, Genen. Phaenen. 17*, 141–142 (1974).
62. J. Lejeune, M. Gautier, and R. Turpin, Étude des chromosomes somatiques de neuf enfants mongoliens, *C. R. Acad. Sci. Paris 248*, 1721–1722 (1959).
63. A. Léonard, P. Jacquet, and R. R. Lauwerys, Mutagenicity and teratogenicity of mercury compounds, *Mutat. Res. 114*, 1–18 (1983).
64. D. L. Lindsley and E. H. Grell, *Genetic Variations of Drosophila melanogaster*, Carnegie Institution of Washington (1968).
65. D. L. Lindsley and K. T. Tokuyasu, Spermatogenesis, in: *Genetics and Biology of Drosophila* (M. Ashburner and T. R. F. Wright, eds.), Vol. 2d, pp. 225–294, Academic Press, New York (1980).
66. M. A. Lopata, J. C. Havorcroft, L. T. Chow, and D. W. Cleveland, Four unique genes required for β tubulin expression in vertebrates, *Cell 32*, 713–724 (1983).
67. R. F. Ludueña, Biochemistry of tubulin, in: *Microtubules* (K. Roberts and J. S. Hyams, eds.), pp. 65–116, Academic Press, New York (1979).
68. K. G. Lüning, Genetics of inbred *Drosophila melanogaster*. II. Primary non-disjunction and chromosome loss, *Hereditas 95*, 189–200 (1981).
69. K. G. Lüning, Genetics of inbred *Drosophila melanogaster*. IV. Tests of paternal non-disjunction, chromosome loss and X.Y exchange, *Hereditas 96*, 97–99 (1982).
70. K. C. Lüning, Genetics of inbred *Drosophila melanogaster*. V. Genetic and cytoplasmic effects on primary non-disjunction and chromosome loss, *Hereditas 96*, 101–104 (1982).
71. K. G. Lüning, Genetics of inbred *Drosophila melanogaster*. IX. Interference of genotype and cytoplasm on primary non-disjunction and chromosome loss, *Hereditas 99*, 49–56 (1983).
72. A. P. Mahowald and M. P. Kambysellis, Oogenesis, in: *Genetics and Biology of Drosophila* (M. Ashburner and T. R. F. Wright, eds.), Vol. 2d, pp. 141–224, Academic Press, New York (1980).

73. J. W. Mavor, On the elimination of the X-chromosome from the egg of *Drosophila melanogaster* by X-rays, *Science 54*, 277–279 (1921).
74. J. W. Mavor, The production of non-disjunction by X-rays, *Science 55*, 295–297 (1922).
75. J. W. Mavor, The production of nondisjunction by X-rays, *J. Exp. Zool. 39*, 381–432 (1924).
76. J. R. Merriam and J. N. Frost, Exchange and nondisjunction of the X chromosomes in female *Drosophila melanogaster, Genetics 49*, 109–122 (1964).
77. G. Morpurgo, D. Bellincampi, G. Gualandi, L. Baldinelli, and O. S. Crescenzi, Analysis of mitotic nondisjunction with *Aspergillus nidulans, Environ. Health Perspect. 31*, 81–95 (1979).
78. M. J. Moses and P. A. Poorman, Synaptonemal complex analysis of mouse chromosomal rearrangements. II. Synaptic adjustment in a tandem duplication, *Chromosoma 81*, 519–535 (1981).
79. H. J. Muller, Artificial transmutation of the gene, *Science 66*, 84–87 (1927).
80. E. Novitski, An alternative to the distributive pairing hypothesis in *Drosophila, Genetics 50*, 1449–1451 (1964).
81. E. Novitski, The relation of exchange to nondisjunction in heterologous chromosome pairs in the *Drosophila* female, *Genetics 88*, 499–503 (1978).
82. T. Oksala, Chromosome pairing, crossing over, and segregation in meiosis in *Drosophila melanogaster* females, *Cold Spring Harbor Symp. Quant. Biol. 23*, 197–210 (1958).
83. A. Önfelt and C. Ramel, Some aspects on the organization of microfilaments and microtubules in relation to nondisjunction, *Environ. Health Perspect. 31*, 45–52 (1979).
84. W. Ostertag and J. Haake, The mutagenicity in *Drosophila melanogaster* of caffeine and of other compounds which produce chromosome breakage in human cells in culture, *Z. Vererbungsl. 98*, 299–308 (1966).
85. D. R. Parker and N. Busby, Observations concerning the effects of radiations on the segregation of chromosomes, *Mutat. Res. 18*, 33–46 (1973).
86. D. R. Parker and J. H. Williamson, Aberration induction and segregation in oocytes, in: *Genetics and Biology of Drosophila* (M. Ashburner and E. Novitski, eds.), Vol. 1c, pp. 1251–1268, Academic Press, New York (1976).
87. J. M. Parry, D. Sharp, and E. M. Parry, Detection of mitotic and meiotic aneuploidy in the yeast *Saccharomyces cerevisiae, Environ. Health Perspect. 31*, 97–111 (1979).
88. J. M. Parry, E. M. Parry, and J. C. Barrett, Tumor promoters induce mitotic aneuploidy in yeast, *Nature 294*, 263–265 (1981).
89. J. Puro and S. Nokkala, Meiotic segregation of chromosomes in *Drosophila melanogaster* oocytes: A cytological approach, *Chromosoma 63*, 273–286 (1977).
90. C. Ramel and J. Magnusson, Genetic effects of organic mercury compounds: II. Chromosome segregation in *Drosophila melanogaster, Hereditas 61*, 231–254 (1968).
91. C. Ramel and J. Magnusson, Chemical induction of nondisjunction in *Drosophila, Environ. Health Perspect. 31*, 59–66 (1979).
92. L. B. Russell, Meiotic nondisjunction in the mouse: Methodology for genetic testing and comparison with other methods, *Environ. Health Perspect. 31*, 113–128 (1979).
93. S. R. Safir, Genetic and cytological examination of the phenomena of primary non-disjunction in *Drosophila melanogaster, Genetics 5*, 459–487 (1920).
94. K. Sankaranarayanan, The role of non-disjunction in aneuploidy in man: An overview, *Mutat. Res. 61*, 1–28 (1979).
95. M.-L. Savontaus, Relationship between effects of X-rays on nondisjunction and crossing over in *Drosophila melanogaster, Hereditas 80*, 195–204 (1975).

96. P. D. Smith, A rapid selection technique for detecting meiotic X-chromosomal nondisjunction in *Drosophila melanogaster, Mutat. Res. 108*, 169–174 (1983).
97. F. H. Sobels, The advantages of *Drosophila* for mutation studies, *Mutat. Res. 26*, 277–284 (1974).
98. R. M. Speed and A. C. Chandley, The response of germ cells of the mouse to the induction of non-disjunction by X-rays, *Mutat. Res. 84*, 409–418 (1981).
99. A. H. Sturtevant, A highly specific complementary lethal system in *Drosophila melanogaster, Genetics 41*, 118–123 (1956).
100. A. D. Tates and N. de Vogel, Further studies on effects of X-irradiation on prespermatid stages of the northern vole *Microtus oeconomus*: Low induction of sex-chromosomal nondisjunction and very high induction of diploid spermatids, *Mutat. Res. 82*, 323–330 (1981).
101. Y. Tazima, Chemical mutagenesis in the silkworm, in: *Chemical Mutagens: Principles and Methods for Their Detection* (F. J. de Serres and A. Hollaender, eds.), Vol. 6, pp. 203–238, Plenum Press, New York (1980).
102. E. Therman, *Human Chromosomes*, Springer, New York (1980).
103. C. Tokunaga, The effects of low temperature and aging on nondisjunction in *Drosophila, Genetics 65*, 75–94 (1970).
104. C. Tokunaga, Aspects of low-temperature-induced meiotic nondisjunction in *Drosophila* females, *Genetics 66*, 653–661 (1970).
105. C. Tokunaga, The effects of temperature and aging of *Drosophila* males on the frequencies of XXY and XO progeny, *Mutat. Res. 13*, 155–161 (1971).
106. H. Traut, The resistance of mature oocytes of *Drosophila melanogaster* to the induction of nondisjunction by X-rays, *Mutat. Res. 10*, 156–158 (1970).
107. H. Traut, The induction of X-chromosomal aneuploidy by 5-fluorodeoxyuridine (FUdR) fed to *Drosophila melanogaster* females, *Can. J. Genet. Cytol. 20*, 259–263 (1978).
108. H. Traut, X-Chromosomal nondisjunction induced by aging oocytes of *Drosophila melanogaster*: The special susceptibility of mature eggs, *Can. J. Genet. Cytol. 22*, 433–437 (1980).
109. H. Traut, Aneuploidy patterns in *Drosophila melanogaster, Environ. Mutagen. 3*, 275–286 (1981).
110. H. Traut, The solvent dimethylsulfoxide (DMSO) does not induce aneuploidy in oocytes of *Drosophila melanogaster, Environ. Mutagen. 5*, 273–277 (1983).
111. H. Traut and W. Scheid, The induction of aneuploidy by Colcemid fed to *Drosophila melanogaster* females, *Mutat. Res. 23*, 179–188 (1974).
112. H. Traut and U. Sommer, Die induktion von chromosomenverlust und -gewinn durch Kolchizin, *Münch. Med. Wochenschr. 118*, 1113–1116 (1976).
113. United Nations Scientific Committee on the Effects of Atomic Radiation, *Ionizing Radiation: Sources and Biological Effects*, Report to the General Assembly, United Nations, New York (1982).
114. R. Valencia, The versatility of *Drosophila melanogaster* for mutagenicity testing, *Ann. N. Y. Acad. Sci. 407*, 197–207 (1983).
115. R. Valencia, S. Abrahamson, W. R. Lee, E. S. von Halle, R. C. Woodruff, F. E. Würgler, and S. Zimmering, Chromosome mutation tests for mutagenesis in *Drosophila melanogaster, Mutat. Res. 134*, 61–88 (1984).
116. E. Vogel, Recent achievements with *Drosophila* as an assay system for carcinogens, in: *Short-term Tests for Chemical Carcinogens* (H. F. Stich and R. H. C. San, eds.), pp. 379–398, Springer, New York (1981).
117. D. Warburton and F. C. Fraser, Spontaneous abortion risks in man: Data from reproductive histories collected in a medical genetics unit, *Am. J. Hum. Genet. 16*, 1–25 (1964).

118. R. F. Wilkinson, H. P. Stanley, and J. T. Bowman, Genetic control of spermiogenesis in *Drosophila melanogaster*: The effects of abnormal cytoplasmic microtubule populations in mutant ms(3)10R and its Colcemid-induced phenocopy, *J. Ultrastruct. Res. 48*, 242–258 (1974).
119. F. E. Würgler, F. H. Sobels, and E. Vogel, *Drosophila* as an assay system for detecting genetic changes, in: *Handbook of Mutagenicity Test Procedures* (B. J. Kilbey, ed.), pp. 335–373, Elsevier, New York (1977).
120. S. Zimmering and J. Scott, Measurements of X-ray induced mutational damage in stage-14 oocytes of *Drosophila, Mutat. Res. 6*, 179–180 (1968).
121. S. Zimmering and C. K. Wu, Radiation induced X–Y exchange and nondisjunction in spermatocytes of the immature testis of *Drosophila, Genetics 48*, 1619–1623 (1963).
122. S. Zimmering and C. K. Wu, X–Y nondisjunction and exchange induced by X-rays in primary spermatocytes of the adult *Drosophila, Genetics 49*, 499–504 (1964).
123. S. Zimmering and C. K. Wu, Meiotic X–Y exchange and nondisjunction induced by irradiation in the *Drosophila* male, *Genetics 50*, 633–638 (1964).
124. S. Zimmering, A. W. Hartmann, and S. F. Cooper, Evidence that the repair deficient *mei-9*^{a} female in *Drosophila melanogaster* is a strong potentiator of chromosome loss induced in the paternal genome by dimethylnitrosamine, *Environ. Mutagen. 2*, 187–190 (1980).

CHAPTER 6

Mammalian Cytogenetic and Genetic Tests for Nondisjunction

Charles Tease and Bruce M. Cattanach

1. Introduction

Meiotic chromosome nondisjunction generates gametes that are chromosomally unbalanced (aneuploid). Two types of aneuploid gamete are produced by meiotic nondisjunction, one containing fewer than the haploid number of chromosomes (hypoploid), and one containing chromosomes additional to the haploid complement (hyperploid). Upon fertilization, these gametes give rise to monosomic and trisomic embryos, respectively. Such aneuploid embryos are developmentally impaired in mammals and most do not survive to term.[35,45] In man, the aneuploid fetuses that survive to birth show, with a few exceptions, severe congenital abnormalities. Recent reviews provide estimates of the frequencies of aneuploid conception in humans and describe the fate of these embryos.[11,36,106]

A number of epidemiologic studies have investigated the intrinsic and extrinsic factors believed to be potentially capable of influencing the incidence of nondisjunction, and thereby of aneuploidy, in human populations. Various factors have been studied: maternal and paternal age, the use of or exposure to chemicals and drugs (including contra-

Charles Tease and Bruce M. Cattanach • Medical Research Council, Radiobiology Unit, Chilton, Didcot, Oxon OXll ORD, England.

ceptive hormones), and history of exposure to radiation. The strength of evidence linking particular factors to fluctuations in the frequency of aneuploidy varies. Although, for example, a maternal age effect upon nondisjunction has consistently been found, the importance of paternal age is contentious. Stene *et al.*,[118] in an analysis of the incidence of prenatally recognized Down syndrome fetuses, described a strong paternal age effect, whereas Hook and Cross,[61] in a similar type of study, concluded that the incidence was independent of paternal age. Similar problems have been encountered in determining unequivocally the effect on nondisjuction of exposure to environmental insults such as X rays.[107]

The use of animal models in nondisjunction studies overcomes many of the problems associated with the epidemiologic approach. Specifically, the relative importance of any single factor, such as age, radiation treatment, or exposure to chemical mutagens, can more easily be investigated. A battery of test systems have been devised using laboratory mammals to examine both the induction of aneuploidy and the mechanisms of meiotic nondisjunction. The ease of carrying out the different tests and the quantity and quality of data each generates all vary. It is the purpose of this review to attempt an assessment of the relative merit of each method.

2. Cytogenetic Tests

A variety of cytogenetic tests are available for assaying mutagen-induced nondisjunction in mammals. These systems entail determinations of chromosome number in cell types ranging from metaphase II gametocytes to postimplantation embryos. Further assays have been devised which exploit morphological features of sex chromosomes, such as the presence of large blocks of heterochromatin, to screen nondisjuction affecting these elements. In addition to these quantitative approaches, it has been suggested that examination of primary gametocytes for meiotic anomalies may prove of value. First, they may lead to a better understanding of the mechanisms of nondisjunction and, second, they may be of use for predicting potential nondisjunction-inducing effects of environmental insults.[60]

The different test systems are described below in a temporal sequence from meiosis to postimplantation embryos. An evaluation of the systems is presented in Section 4.

2.1. Methods Using Primary and Secondary Gametocytes

The principal features of the meiotic cell cycle of relevance to nondisjunction are the pairing of homologous chromosomes at pachytene, crossing-over (chiasma formation) between the paired homologues, the coorientation and segregation of homologous chromosomes at metaphase I/anaphase I, and chromatid segregation at the second meiotic division. Disturbances of these processes, whether spontaneous or induced by environmental insults, may cause nondisjunction. Examination of primary gametocytes (pachytene to metaphase I in this instance) can provide information on the effects of mutagens on chromosome pairing and chiasma formation, while screening of secondary gametocytes (metaphase II cells) provides quantitative evaluation of nondisjunction at anaphase I. Nondisjunction at the second meiotic division is determined in postmeiotic cells, zygotes, or embryos.

2.1.1. Synaptonemal Complexes

The intimate pairing of homologous chromosomes at pachytene is mediated by a proteinaceous structure, the synaptonemal complex.[41,143] Homologous chromosome pairing is an essential prerequisite of chiasma formation, which in turn is critical for chromosome segregation at anaphase I. Any interruptions or anomalies of the normal sequence of chromosome pairing may therefore result indirectly in nondisjunction. On this basis, analysis of chromosome pairing via the synaptonemal complexes has been suggested to be of value both to the elucidation of suspected mechanisms of nondisjunction and to an assessment of the potential nondisjunction-inducing effect of environmental insults.[60]

2.1.1a. Methods. There are two methods for examining synaptonemal complexes (SCs). First, fixed ovarian[25] or testicular[59] material embedded in plastic is serially sectioned using an ultramicrotome and three-dimensional reconstructions of SCs are made from serial electron micrographs of suitable pachytene gametocytes. This method provides highly detailed information on the fine structure of SCs and their spatial relationships within pachytene nuclei. Second, whole-mount spreads are made of ovarian[115] or testicular[83] cell suspensions to provide two-dimensional images, by light or electron microscopy, of flattened gametocytes in which complete SC complements are visible.

2.1.1b. Results. SCs have been used to investigate aspects of spontaneous nondisjunction, as, for example, by Gropp and colleagues in male mice heterozygous for Robertsonian translocations[46] and by

Speed and Chandley[117] in oocytes of fetal mice. However, no clear link has been established between pairing behavior at pachytene and subsequent nondisjunction at anaphase I. No studies have been carried out on SCs in laboratory mammals exposed to a radiation or chemical regimen known to induce nondisjunction.

2.1.1c. Assessment. Critical assessment of the usefulness of SCs in nondisjunction studies is hampered by the lack of experimental results. However, the difficulty of extrapolation from spontaneous or induced anomalies in pachytene gametocytes to presumptive effects on chromosome segregation makes the use of SCs impractical for screening the nondisjunction-inducing potential of environmental agents. Nevertheless, SCs may ultimately provide valuable information on the mechanisms of nondisjunction.

2.1.2. Diakinesis/Metaphase I Gametocytes

Gametocytes at the diakinesis/metaphase I (D/MI) stages of development may be analyzed to determine (1) the frequency and distribution of chiasmata and (2) the incidence of univalent (unpaired) chromosomes. Although the distribution of chiasmata may influence chromosome segregation, at least in certain translocation heterozygotes of the mouse,[26] most interest in D/MI cells has centered on the phenomenon of univalence. In the absence of a chiasma to hold homologues together, the resulting univalents may segregate independently to cause aneuploidy in a proportion of the secondary gametocytes. The phenomenon of univalence has therefore been examined in a number of studies to evaluate its relevance both as a mechanism of aneuploidy and as a test for potential nondisjunction.

2.1.2a. Methods. Three routines are currently in use to produce D/MI oocytes:

1. Germinal vesicle stage oocytes are collected from excised ovaries and cultured *in vitro* to D/MI. The length of the culture period varies with species, but the method has been used successfully with mice,[58] hamsters,[122] rabbits,[82] and humans.[29]
2. Ovulation is induced by injection of pregnant mare's serum (PMS) followed 48 hr later (in the case of the mouse) by human chorionic gonadotrophin (HCG). A few hours after HCG application, oocytes at D/MI are present in the ovaries.[30]
3. Ovulation is induced as in routine 2, but a dilute solution of colchicine is also injected at the same time as the HCG. The colchicine treatment arrests oocyte development at metaphase

I and the oocytes are ovulated at this stage. The cells are collected from the oviduct.[88]

Tarkowski[129] described a hypotonic and fixation schedule for oocyte preparation which has since been widely adopted and is applicable to all three methods. Kamiguchi *et al.*[64] and Tanaka[127] have recently developed alternative schedules which are claimed to increase the proportions of analyzable cells.

Spermatocyte preparations are generally made using the technique of Evans *et al.*[32] or by modifications of it.[2,12]

2.1.2b. Results. Various attempts have been made to determine the possible relationship between aneuploidy at metaphase II and univalence at D/MI. De Boer and van der Hoeven,[27] for example, compared chromosome behavior in oocytes of young and old female mice homozygous for the T(1;13)70H reciprocal translocation. They observed an age-related increase in the incidence of univalence at D/MI. This was particularly evident for the small marker chromosome 1,[11] which displayed an eightfold increase in univalence in old females. Parallel increases in the rates of chromosome nondisjunction were found at metaphase II, the small marker chromosome showing a ninefold increment with age. A similar positive relationship between univalence and aneuploidy has been recorded by Wauben-Penris and van Buul-Offers.[141] Male hypopituitary Snell dwarf mice have a higher incidence of univalence than phenotypically normal male mice of the same population. The frequency of aneuploidy at metaphase II is also higher in the dwarf males compared to normal males.[141] Although the results from these two studies favor the possibility that univalence is an indicator of subsequent aneuploidy, other observations dispute this. Polani and Jagiello[92] were unable to detect a correlation between univalence and aneuploidy in their study of age effects on meiosis in male and female mice. Old male mice had no greater incidence of univalents than young male mice and at metaphase II the incidence of aneuploidy was again similar in both age groups. In contrast, female mice displayed an age-related increase of univalents and also a decrease in chiasma frequencies. However, old female mice did not show a concomitantly increased level of aneuploidy.

Sugawara and Mikamo[122] recently surveyed meiotic chromosome behavior in primary and secondary ooctytes of the Chinese hamster. They found a significant increase in the incidence of cells with univalents in old females. The proportion of aneuploid metaphase II oocytes was also larger in old females. However, while univalence at metaphase I was confined mainly to bivalents of the smallest metacentric chromosome group, nondisjunction affected all chromosome groups to a similar extent. This observation argues against a correlation between univalence at diakinesis/metaphase I and aneuploidy at metaphase II.

2.1.2c. Assessment. The principal objection to the use of D/MI cells to study nondisjunction is the lack of a consistent correlation between the phenomenon of univalence and aneuploidy. A number of confounding factors may, in part at least, serve to explain the discrepant conclusions drawn from the different experiments described in Section 2.1.2b. First, the fate of cells carrying univalents is uncertain. X/Y dissociation is believed to be cell-lethal in mammalian spermatogenesis.[10] This may also be true for autosomal univalence, since in some mouse translocations failure of chiasma formation in small marker chromosomes tends to be associated with a breakdown of spermatogenesis.[110] Second, univalents may encounter difficulties of orientation on the meiotic spindle. This may result in either the loss of univalents, for example, by lagging, or the premature separation of the centromere and segregation of single chromatids. Single chromatids are occasionally observed at metaphase II,[50] but the subsequent fate of these elements is not known. Third, unambiguous identification of univalents is hampered by the occurrence of apparently achiasmate homologue associations, which have been variously named "false," "opposite," or "ambiguous" univalence.[92,114,122] Whether such associations are in reality chiasmate and their appearance due to fixation and spreading artifacts is unknown.

Overall, the lack of an established relationship between univalence and aneuploidy and the uncertainties over identification and fate of univalents all diminish the usefulness of this system as a means of investigating spontaneous or induced nondisjunction.

2.1.3. The Metaphase II Gametocyte Method

Nondisjunction at anaphase I generates metaphase II cells with either additional (hyperploid) or fewer (hypoploid) chromosomes than expected. The proportion of such aneuploid products can be used to measure the frequency of nondisjunction.

2.1.3a. Methods. Three routines are currently in use to obtain metaphase II oocytes in the mouse:

1. Germinal vesicle stage oocytes are released from excised ovaries and cultured *in vitro* for a period of 18–24 hr.[42,92]
2. Naturally ovulated eggs are collected from the oviduct on the morning of estrus. These are normally arrested at metaphase

II. The timing of ovulation is determined either by following the stages of estrus by examination of vaginal smears or by mating the females to sterile males and checking daily for vaginal plugs.

3. Ovulation is induced by application of gonadotrophic hormones (see Section 2.1.2a). The number of oocytes ovulated varies with hormone doses, mouse strain, and maternal age.

These techniques are generally applicable to other mammalian species and have been used in Chinese and Syrian hamsters[51,121] and the rat.[24] However, there is some evidence that the hormone doses used to induce ovulation in the Djungarian hamster may affect the level of nondisjunction (I. Hansmann, personal communication).

Metaphase II spermatocytes are prepared using the same methods as for primary spermatocytes.[2,12,32] The number of metaphase II spermatocytes may be increased by a colchicine pretreatment.[105]

2.1.3b. Results. Nondisjunction at anaphase I should generate equal proportions of hyperploid and hypoploid metaphase II cells. However, it is a common experience to record significantly higher incidences of hypoploid cells in the samples from both control and treated animals. This deviation from the expected ratio is usually ascribed to the artifactual loss of chromosomes by cell breakage during the fixation and spreading procedures. Loss of mutagen-damaged chromosomes may also account for part of the hypoploid excess in treated animals. Whatever the mechanism producing a higher than expected proportion of hypoploid cells, the usefulness of this class of cells for nondisjunction testing is obviously diminished. Since hyperploidy can only result from chromosome nondisjunction, such cells provide a more reliable measure. An estimate of the actual level of chromosome nondisjunction is therefore simply obtained by doubling the incidence of hyperploidy. In all studies described below, the effects of environmental agents on nondisjunction have been determined from the proportions of hyperploid cells.

The variation in data presentation in different reports hampers both comparison and evaluation of these experiments. For example, Gates *et al.*[40] include all cells analyzed (haploid, hyperploid and hypoploid) in their tabulated results, whereas Hansmann *et al.*[56] omit hypoploid cells, since these could be artifacts of the preparation procedure. Of least value for this purpose of comparison are those studies that report only the proportion of aneuploid cells and do not detail the separate frequencies of hyperploidy and hypoploidy,[7,97] thereby including artifacts, nondisjunction, and other mechanisms of

TABLE 1. Aneuploidy Induction Detected in Metaphase II Oocytes of Irradiated Female Mice

Treatment[a]	Interval to sampling, days	Maternal age	Number of cells with $n \geqslant 20$	Number of hyperhaploid cells	Percent hyperhaploidy	Ref.
Control	—	3–6 months	1023	0	0	137
γ rays, 10–30 R	1–7	3–6 months	1115	6	0.54	
Control	—	12 months	1248	4	0.32	136
γ rays, 10–30 R	1–7	12 months	1063	15	1.41	
Control	—	10–12 weeks	135	0	0	93
X rays, 22.2–200 R, 3 hr post-HCG	0.5	10–12 weeks	181	6	3.32	
Control	—	8–12 weeks	846	1	0.12	56
X rays, 5–80 cGy, 2 hr post-HCG	0.5	8–12 weeks	1188	5	0.42	

[a] HCG, Human chorionic gonadotrophin.

chromosome loss within the category of aneuploidy. In order to provide a consistent basis for a comparison of different experiments, only those giving numbers of haploid and hyperploid cells are presented (Tables 1–4). The incidence of hyperploidy is given as the proportion of hyperploid cells among the total haploid plus hyperploid number. It should be realized, however, that this method of analysis provides a slight overestimate of the incidence of hyperploidy as a consequence of eliminating hypoploid cells from the sample.

Only four studies have been carried out using the metaphase II oocyte test to characterize the effects of radiation on chromosome nondisjunction in female mice (Table 1). The studies of Uchida and Lee[(137)] and Uchida and Freeman[(136)] adopted the *in vitro* culture method, while Reichert *et al.*[(93)] and Hansmann *et al.*[(56)] induced ovulation by means of gonadotrophic hormones. Uchida and Lee[(137)] reported that exposure of maturing oocytes to low doses of γ-irradiation resulted in a significant increase in the level of hyperploidy in oocytes from young female mice (Table 1), but no indication of a dose-related response was detected. In a subsequent experiment to investigate the influence of maternal age on response to radiation, Uchida and Freeman[(136)] recorded a significant induction of nondisjunction after exposure of old females to γ rays. Once again no evidence of a dose-related response was found. However, on comparing their results with those of the earlier study[(137)] they concluded that older females had a higher sensitivity to radiation-induced nondisjunction than younger females. These two studies provide the only evidence to date that low doses of radiation are capable of inducing nondisjunction in maturing oocytes. Subsequent experiments using different nondisjunction tests have not substantiated these results and have occasionally reached opposing conclusions (see Sections 2.3.1b and 2.3.3b).

Reichert *et al.*[93] concluded that exposure of immediately preovulatory oocytes to X rays did not result in a significant increase in the incidence of hyperploidy at metaphase II. However, reanalysis of their data (Table 1) indicates that a significant effect ($p = 0.034$, Fisher's exact test) is present if a comparison between control and combined irradiated groups is made with hypoploid cells excluded. The low cell sample sizes per radiation dose preclude analysis for a dose-related response. Hansmann *et al.*[56] likewise investigated the effects of X rays on nondisjunction in immediately preovulatory oocytes, although the protocol differed slightly from that of Reichert *et al.*[93] Although Hansmann *et al.*[56] reported a significant induction of hyperploidy after the highest radiation dose (80 cGy), reanalysis of their data indicates that this conclusion was incorrect. No significant difference in the relative numbers of hyperploid cells was present between controls and females given 80 cGy of X rays ($p = 0.12$, Fisher's exact test). Similarly, there was no difference in the levels of hyperploidy (Table 1) between controls and combined irradiated group ($p = 0.21$, Fisher's exact test). Overall, however, the results of these separate experiments suggest that irradiation of immediately preovulatory oocytes increases the likelihood of nondisjunction, although the data are far from conclusive.

Although there have been relatively few studies on the effects of radiation on nondisjunction in male mice, the results so far reported have been reasonably consistent (Table 2). Significantly increased levels of nondisjunction were detected following X-irradiation of preleptotene through to metaphase I spermatocyte stages.[105,125] Russo *et al.*[105] also recorded a significant induction of nondisjunction after exposing spermatogonial stages to fission neutrons, but neither they nor Szemere and Chandley[125] could achieve such an effect after X-irradiation. The lack of any response from spermatogonial cells following irradiation may be due, in part at least, to cell selection.

Nijhoff and de Boer[86] carried out a radiation study in male mice heterozygous for the Robertsonian translocation Rb(11.13)4Bnr, which causes a high spontaneous frequency of nondisjunction of chromosomes 11 and 13. Metaphase II spermatocytes were examined 2–3 hr after exposure of the animals either to fission neutrons or X rays. They

TABLE 2. Aneuploidy Induction Detected in Metaphase II Spermatocytes of Irradiated Male Mice

Absorbed dose, cGy	Cell stage irradiated[a]	Number of cells with $n \geq 20$	Number of hyperhaploid cells	Percent hyperhaploidy	Ref.
Control	—	200	0	0	125
X rays, 100	Pachytene	200	2	1.0	
	Preleptotene	197	6	3.05	
X rays, 200	Pachytene	198	3	1.52	
	Preleptotene	199	11	5.53	
	Spermatogonia	199	1	0.5	
Control	—	509	89	17.09	86
γ rays, 15	D/MI	481	57	11.85	
X rays, 60	D/MI	500	74	14.8	
Control	—	402	3	0.75	105
X rays, 200	D/MI	219	42	19.18	
	Diplotene	200	18	9.0	
	Pachytene	637	31	4.87	
	Zygotene	200	8	4.0	
	Preleptotene	232	12	5.17	
	Spermatogonia	220	5	2.27	
γ rays, 56	D/MI	231	39	16.88	
	Diplotene	200	38	19.0	
	Pachytene	412	53	12.86	
	Zygotene	210	11	5.24	
	Preleptotene	200	11	5.5	
	Spermatogonia	218	7	3.21	

[a] D/MI, Diakinesis/metaphase I.

found the nondisjunction frequency to be reduced rather than increased (Table 2) and suggested this was the result of selective cell-killing effects.

Very little work has been carried out using the metaphase II gametocyte test to investigate the effects of radiation on nondisjunction in experimental mammals other than the mouse. Cawood and Breckon[22] gave Syrian hamsters an acute dose of 260 cGy of X rays. Cell preparations were made at intervals between 5 and 15 weeks postirradiation to sample effects on spermatogonial cell stages. All metaphase II spermatocytes examined from control and irradiated animals had the correct haploid chromosome number. That is, exposure of spermatogonial stages to X-irradiation did not result in an increased incidence of nondisjunction at anaphase I in spermatocytes.

A number of chemicals have been examined for their effects on nondisjunction in female mice using the metaphase II oocyte test (Table 3). Trenimon, an antineoplastic agent which acts by alkylation, has been demonstrated to be a potent inducer of nondisjunction in immediately preovulatory oocytes.[9,95] No information is available on its effects in earlier oocyte maturation stages. Extensive data are available for two other antineoplastic agents, cyclophosphamide and methotrexate, although the conclusions reached in different experiments are occasionally at variance (Table 3). Röhrborn and Hansmann[95] and Hansmann[49] reported that cyclophosphamide induced nondisjunction when immediately preovulatory oocytes were treated. However, Becker and Schöneich[9] were unable to confirm this conclusion in a similarly designed experiment. Although they recorded a rise in the incidence of hyperploidy following cyclophosphamide treatment (Table 3), this increase was not significant. Hansmann[49] also found that transplacental exposure to cyclophosphamide did not subsequently result in a significantly higher level of nondisjunction in female mice 11–13 weeks postpartum.

Methotrexate was shown to induce nondisjunction to a significant extent only in one experiment. Röhrborn and Hansmann[95] treated hormone-stimulated oocytes with methotrexate (200 mg/kg) 3 hr after HCG application and recorded a large rise in the incidence of hyperploid metaphase II ooctyes (Table 3). However, Gates *et al.*[40] found no induction of hyperploidy by methotrexate when they carried out an experiment of similar design (Table 3). This disparity in result may possibly be due to the use of different mouse strains in the two experiments and/or to the use of mature female mice in one instance[95] and prepubertal females in the other.[40] Hansmann[49] carried out a series of experiments to determine the effects of varying the concentration of methotrexate, exposure of different oocyte maturation stages, and of chronic treatment with the agent (Table 3). None of these experiments resulted in the significant induction of nondisjunction by methotrexate.

Also listed in Table 3 are compounds that have been tested on only one occasion. These chemicals range from a heavy metal (cadmium chloride),[27] to a contraceptive hormone (norethisterone acetate),[96] to a suspected carcinogen (4-chloromethylbiphenyl).[13] None of these chemicals proved to have any significant effect on nondisjunction in mouse oocytes (Table 3). In addition to these chemicals, the antitubercular drug isoniazid and the bactericidal agent trypaflavin have also been investigated. The reports of these experiments do not provide sufficient detail to determine whether or not the chemicals affect the incidence of hyperploid metaphase II oocytes. They do, however, state

TABLE 3. Aneuploidy Induction Detected in Metaphase II Oocytes of Female Mice Treated with Chemicals

Chemical	Treatment[a]	Dose, mg/kg	Number of cells with $n \geqslant 20$	Number of hyperhaploid cells	Percent hyperhaploidy	Ref.
Trenimon	Control	—	67	5	7.46	95
	1. Acute, 8 hr pre-HCG	0.25	6	3	50.0	
	2. Acute, with HCG	0.25	21	11	52.38	
	3. Acute, 3 hr post-HCG	0.25–1.0	26	6	23.08	
	Control	—	136	1	0.74	9
	Acute, to preovulatory oocytes	1.0	75	4	5.33	
	Control	—	140	3	2.14	9
	Acute, 3 hr post-HCG	1.0	116	6	5.17	
Cyclophosphamide	Control	—	89	1	1.12	95
	Acute, 3 hr post-HCG	145	34	4	11.76	
	Control	—	140	3	2.14	9
	Acute, 3 hr post-HCG	150	52	3	5.77	
	Control	—	28	1	3.57	49
	Acute, 3 hr post-HCG	30–50	44	2	4.55	
	Control	—	35	0	0	49
	Acute, 3 hr post-HCG	145	24	4	16.67	
	Control	—	38	1	2.63	49
	Transplacental	5–100	38	2	5.26	

Methotrexate	Control	—	89	1	1.12	95
	Acute, 3 hr post-HCG	200	37	16	43.24	
	Control	—	146	6	4.11	49
	Acute, 3 hr post-HCG	15–100	109	10	9.17	
	Control	—	18	0	0	49
	Acute, 8.5 hr post-HCG	200	30	0	0	
	Control	—	204	0	0	40
	Acute, 3 hr post-HCG	200	180	0	0	
	Control	—	38	1	2.63	49
	Transplacental	15–50	110	11	10	
	Control	—	20	0	0	49
	Chronic, 7× 15 mg/kg	15	39	3	7.69	
Cadmium chloride	Control	—	198	0	0	112
	Acute, 3 hr post-HCG	3–6	276	3	1.09	
Norethisterone acetate	Control	—	118	3	2.54	96
	1. Chronic, 28× 1–10 mg/kg, cells sampled 1–2 weeks later	1–10	68	5	7.35	
	2. Chronic, 28× 1–10 mg/kg, cells sampled 7–9 weeks later	1–10	125	8	6.4	
4-Chloromethylbiphenyl	Control	—	71	0	0	13
	Acute, 3 hr post-HCG	50–100	140	0	0	

[a] HCG, Human chorionic gonadotropin.

that no increased frequencies of aneuploid methaphase II oocytes were recorded after chronic exposure to either compound.[7,98]

Evaluation of the published data on chemical effects on nondisjunction in female mice is hampered by the low cell sample sizes reported in many investigations (Table 3). Indeed, such small sample sizes reduce the likelihood of detecting low levels of induced nondisjunction. It is notable that in several studies in which significant induction of nondisjunction did not occur, the samples from treated females displayed higher incidences of hyperploidy than their corresponding controls. Possibly, significant nondisjunction-inducing effects would have been detected had the samples been larger. That is, in such experiments, the results so far are best regarded as inconclusive rather than negative.

The range of spontaneous nondisjunction in control animals is also surprisingly wide. The control values for hyperploidy range from 0% in a sample of 204 oocytes[40] to 7.46% in a sample of 67 oocytes.[95] This variation does not accord with the reports by Hansmann and El-Nahass[50] and Golbus,[42] which indicated that neither the mouse strain used nor the hormonal induction of ovulation influenced significantly the level of spontaneous chromosome nondisjunction. Moreover, the range of hyperploidy in control females reported in studies of radiation effects on nondisjunction (Table 1) is quite small by comparison with the chemical studies. Notably, the sample sizes in the former studies are generally much larger than those in the latter group.

Very few chemicals have been tested for their abilities to induce nondisjunction in spermatocytes of male mice using the metaphase II test (Table 4). Pacchierotti *et al.*[87] showed that a low dose (30 mg/kg) of the drug cyclophosphamide caused increased frequencies of hyperploid metaphase II spermatocytes following treatment of spermatogonial and spermatocyte stage (Table 4). Higher concentrations of the chemical had no such effect. Since testis mass was found to be reduced substantially at these higher doses, Pacchierotti *et al.*[87] suggested that a selective cell killing was responsible for the lack of a nondisjunction effect at such doses. The alkylating agents ethyl methanesulfonate and methyl methanesulfonate have been tested for their effects on preleptotene spermatocytes of the mouse.[126] Although no concurrent controls were screened in this experiment, it was suggested that each chemical induced a significant increase in hyperploidy compared to a previously ascertained spontaneous incidence (Table 4). Brook[13] treated metaphase I stage spermatocytes with the chemical 4-chloromethylbiphenyl and examined secondary spermatocytes 4 hr later. The chemical had no effect on chromosome segregation, no hyperploid cells being found

TABLE 4. Aneuploidy Induction Detected in Metaphase II Spermatocytes of Male Mice Treated with Chemicals

Chemical	Treatment	Cell treated	Dose mg/kg	Number of cells with $n \geq 20$	Number of hyperhaploid cells	Percent hyperhaploidy	Ref.
Ethyl methane-sulfonate	Acute[a]	Preleptotene spermatocytes	240	412	7	1.7	126
Methyl methane-sulfonate	Acute[a]	Preleptotene spermatocytes	50	391	11	2.81	126
Cyclophosphamide	Control	—	—	750	7	0.93	87
	Acute	Spermatogonia to metaphase I spermatocytes	30	1397	42	3.01	
			50	1627	14	0.86	
			145	1498	18	1.2	
4-Chloromethyl-biphenyl	Control	—	—	301	0	0	13
	Acute	Metaphase I spermatocytes	100	406	0	0	

[a] No concurrent controls described.

either in treated or control animals (Table 4). A limited amount of information has been reported on the effects of both chronic and acute exposures to the drug isoniazid of male mice and Chinese hamsters.[3] The drug did not increase the frequencies of aneuploid metaphase II spermatocytes in either species.

A number of studies have been carried out in experimental mammals other than the mouse to examine the effects of various chemicals on nondisjunction using the metaphase II test. The work of Adler *et al.*[3] described briefly above is the only experiment to have involved males, the remaining reports being on female Chinese and Syrian hamsters and female rats (Table 5). Trenimon has been shown to induce nondisjunction in immediately preovulatory oocytes of the Chinese hamster when given to the female 8 hr before HCG application.[54] However, two other preovulatory oocyte stages examined showed no response to this drug (Table 5). This is in contrast to the situation in the female mouse, in which trenimon is a potent inducer of nondisjunction in all preovulatory oocyte stages tested (Table 3).

Cadmium chloride has been screened for its effects on nondisjunction in oocytes of Syrian hamster females.[139] Although an increased incidence of hyperploidy was found in oocytes of females treated 5 hr before the estimated time of ovulation compared to the corresponding controls, this difference was not significant (Table 5). A similar result has been described in the mouse (Table 3). The results reported from experiments using mercuric compounds provide an interesting example of a difference between species in their responses to treatment with a mutagen. Watanabe *et al.*[140] found that exposure of immediately preovulatory oocytes of the Chinese hamster to either mercuric chloride or methylmercuric chloride had no effect in the frequency of hyperploidy at metaphase II (Table 5). In contrast, Mailhes[74] reported that exposure to methylmercuric chloride increased significantly the frequency of hyperploidy in Syrian hamster oocytes (Table 5). The reason for this difference between species is unknown and awaits further investigation.

Chebotar[24] reported that acute exposure of preovulatory rat oocytes either to the antimalarial drug chlorodin or to the pesticide 2,4,5-trichlorophenoxyacetic acid caused significant increases in the frequencies of hyperploid metaphase II oocytes (Table 5). No information is available on the effects of these agents on nondisjunction after treatment of other oocyte maturation stages. A low concentration of colchicine has been found to disrupt the meiotic spindle in oocytes of the Chinese hamster, thereby impairing the correct segregation of homologous chromosomes.[121] A highly significant increase in the

TABLE 5. Aneuploidy Induction Detected in Metaphase II Oocytes of Various Mammalian Species (Other Than the Mouse) Treated with Chemicals

Chemical	Treatment[a]	Dose, mg/kg	Species	Number of cells with $n \geq 20$	Number of hyperhaploid cells	Percent hyperhaploidy	Ref.
Trenimon	Control	—	Chinese hamster	219	0	0	54
	1. Acute, 8 hr pre-HCG	0.25–1.0		48	3	6.25	
	2. Acute, with HCG	0.25		31	0	0	
	3. Acute, 3 hr post-HCG	0.25–1.0		66	0	0	
Cadmium chloride	Control	—	Syrian hamster	140	1	0.71	139
	Acute	1–4		384	7	1.82	
Methylmercuric chloride	Control	—	Syrian hamster	136	0	0	74
	Acute, 6 hr pre-HCG	10		122	6	4.92	
	Control	—	Chinese hamster	149	0	0	140
	1. Acute, to preovulatory oocytes	6.4–12.8		249	0	0	
	2. Acute, 5 days before sampling	6.4–12.8		250	3	1.2	
Mercuric chloride	1. Acute, to preovulatory oocytes	12.8		120	0	0	140
	2. Acute, 5 days before sampling	6.4–12.8		44	0	0	
Chlorodin	Control	—	Rat	801	4	0.49	24
	Acute, with HCG	2–25		830	15	1.81	
2,4,5-Trichlorophenoxyacetic acid	Acute, with HCG	200–400		398	10	2.51	24
Colchicine	Control	—		1721	14	0.81	121
	Acute, to preovulatory oocytes	3		316	32	10.31	

[a] HCG, Human chorionic gonadotrophin.

proportion of hyperploid metaphase II oocytes was recorded after an acute exposure (Table 5).

2.1.3c. Assessment. As is evident from the foregoing descriptions of the effects of mutagen exposure on nondisjunction in experimental mammals, the metaphase II gametocyte test has been widely used and more importantly has identified agents capable of inducing significantly increased levels of aneuploidy. Metaphase II analysis therefore provides a validated system for nondisjunction testing. There are limitations to the system, however. First, only nondisjunction at anaphase I can be screened. Second, the morphology of metaphase II chromsomes can make accurate assessment of numbers difficult. This difficulty is increased if the agent under investigation causes chromosome breakage, but the use of C banding to identify centromeres partially overcomes the problem. Third, the conventional methods of oocyte and spermatocyte preparation cause cell breakage and concomitant chromosome loss in a substantial proportion of cells. Therefore hypoploid cells do not provide a reliable indicator of nondisjunction and only hyperploid cells are informative for this purpose. Kamiguchi *et al.*(64) recently described methodologies claimed to reduce drastically the relative numbers of hypoploid oocytes arising through cell breakage and chromosome loss. Certainly, improved preparation techniques which allow identification of both hypoploid and hyperploid products of nondisjunction would greatly increase the usefulness of the metaphase II system.

The importance of screening a large sample size per treatment group has been mentioned earlier and cannot be stressed too often. Given the low rate of spontaneous nondisjunction in laboratory mammals, large sample sizes are required to provide unequivocal evidence of a nondisjunction-inducing effect. The use of small sample sizes, for example, less than 200 cells per treatment group, makes it difficult (if not impossible) to detect even such an important effect as the doubling of the rate of nondisjunction. Only large increases in the rate of nondisjunction could therefore be demonstrated. However, as Bond and Chandley(11) have pointed out, the nondisjunction-inducing effects of chemicals are seldom very great, a conclusion that emphasizes the importance of screening sufficiently large numbers of cells.

The metaphase II system has the advantage of being technically the least demanding of the validated cytogenetic approaches to nondisjunction testing. The other tests described below each have their strengths and limitations regarding the quality and quantity of data generated and ease of performance. Overall, however, in comparison to such other systems, metaphase II gametocytes provide the most useful cytogenetic approach to routine screening of environmental mutagens for effects on chromosome disjunction.

2.2. Methods Using Spermatids and Spermatozoa

2.2.1. Sex-Chromosome Aneuploidy in *Microtus oeconomus*

The X and Y chromosomes of *Microtus oeconomus* have large C-band-positive (heterochromatic) regions, which distinguish them from the autosomes, which do not possess large heterochromatic blocks. The Y chromosome is completely C-band-positive, while the X has a large block of centric heterochromatin. These patterns of heterochromatin distribution can be used to identify X-bearing and Y-bearing spermatids. Nondisjunction involving the sex chromosomes at either anaphase I or II should produce spermatids containing either no large heterochromatic blocks or two such bodies. Spermatids can thus be screened to assay spontaneous or induced chromosome nondisjunction involving the sex chromosomes.

2.2.1a. Method. A full description of the cytological technique has recently been given by Tates.[130] In brief, a testicular cell suspension is made and is fixed in 3:1 ethanol–acetic acid and the fixed cells spread on glass slides. Differentiation of heterochromatic bodies is achieved by barium hydroxide and 2× SSC (saline sodium citrate) treatments followed by Giemsa staining.

2.2.1b. Results. Initial results indicated a significant induction of sex-chromosome nondisjunction and diploid spermatid formation after exposure to X-irradiation.[132] However, more extensive recent experiments have not confirmed in full the earlier observations. Although the incidence of diploid spermatids showed a linear response to doses of X rays, as observed in the first experiments, significant induction of sex-chromosome aneuploidy could no longer be detected.[131] A change in the genetic constitution of the *Microtus* colony since its inception has been suggested as a possible reason for the discrepant responses to X-irradiation. Further work is underway to determine whether this explanation is correct.

A number of chemical mutagens have also been tested using this system.[130] Preliminary results indicate that methyl methanesulfonate, procarbazine, 4-fluor-d-p-phenylalanine, vincristine, bleomycin, carbendazine, and halothane are each able to induce sex-chromosome nondisjunction. No significant effect was observed after treatment with colcemid, nocodazol, benomyl, or 6-mercaptopurine.

2.2.1c. Assessment. Large numbers of cells can be screened rapidly using this method. Although the assay is limited to sex-chromosome nondisjunction, the effects of environmental agents on chromosome segregation are screened in both meiotic divisions. However, the problems encountered in the X-irradiation experiments indicate that further development of the system is required to enable consistent and comparable results to be obtained.

2.2.2. Fluorescent (F) Bodies in Human Spermatozoa.

A brightly fluorescent region is present after quinacrine staining on the human Y chromosome, which can thereby be distinguished in all cell stages from spermatocytes through to spermatozoa.[6,90,124] The potential of identifying Y-bearing sperm by the presence of a fluorescent (F) body stimulated the investigation of the feasibility of using this feature to detect Y-chromosome nondisjunction. Preliminary studies indicated an incidence of 1–3% of spermatozoa with two F bodies.[89] This level of presumed Y chromosome nondisjunction was higher than anticipated and led to the suggestion that not all F bodies actually represent Y chromosomes.[8]

2.2.2a. Method. A detailed description of the routine used to fix and stain spermatozoa for fluorescence microscopy can be found in Kapp.[65]

2.2.2b. Results. The F-body system has been used to monitor genotoxic effects in men exposed to environmental mutagens.[66,67,145] Significantly increased proportions of spermatozoa with two F bodies were recorded after exposure to dibromochloropropane, adriamycin, and X rays in comparison to comparable unexposed individuals.[66,67] Exposure to carbaryl was not found to increase the proportions of spermatozoa with two F bodies.[145]

2.2.2c. Assessment. The F-body system is obviously limited in scope, being concerned only with Y-chromosome nondisjunction at anaphase II. Its potential value is limited to screening selected populations of men suspected of being at risk of exposure to environmental mutagens. There are also uncertainties concerning the validity of the method itself. Beatty[8] has detailed a number of criticisms of the assumption of equality between F bodies and Y chromosomes. These criticisms allied to the unexpectedly high estimates of spontaneous Y nondisjunction provided by this method caution against its use as a nondisjunction test.

2.2.3. Sperm DNA Content Estimation

The rationale behind this approach is that since chromosome nondisjunction generates gametes containing either additional or fewer chromosomes than the haploid number, such gametes must possess either a larger or a smaller DNA content, respectively, than the expected haploid (IC) amount and can be identified on this basis. Various partly and fully automated systems may be used to estimate sperm DNA contents. These function by measuring either the nuclear dry mass of fixed or unfixed material or the amount of stain bound to nuclear DNA of fixed material stained with stoichiometric dyes.

2.2.3a. Methods. Stolla and Gropp[119] measured DNA contents of unstained mouse sperm heads by UV microspectrophotometry. The cells are prepared by first washing in saline, smearing them onto a glass slide, and covering with glycerol.

Sumner and Robinson[123] compared DNA content estimations from fixed and unfixed human spermatozoa and concluded that the former routine provided more consistent data. Washed spermatozoa were fixed in methanol–acetic acid (3:1) and spread on glass slides. Nuclear dry mass was measured by integrating microinterferometry. They were able to demonstrate a positive relationship between nuclear dry mass and DNA amount.

Flow cytometry is a fully automated approach to estimating cell DNA contents. Single-cell suspensions of fixed spermatids/spermatozoa are prepared. The cells are stained with a fluorescent dye, e.g., ethidium bromide, and passed individually through a beam of exciting light. The amount of fluorescence emitted is recorded and is proportional to the nuclear DNA content (e.g., ref. 80).

2.2.3b. Results. Stolla and Gropp[119] examined spermatozoa of mouse hybrids heterozygous for Robertsonian translocations. Such animals have high levels of chromosome nondisjunction and thus produce a large proportion of aneuploid spermatozoa. Comparison of the spermatozoan DNA amounts of hybrid males and parental species revealed a greater range in the former. This increased variation was ascribed to the presence of significantly higher proportions of anueploid spermatozoa in the hybrid males. A similar study with comparable results was carried out by Meistrich *et al.* [79] using the flow cytometry assay.

Pinkel *et al.* [91] used the flow cytometry assay to investigate the effects of chemical and physical agents on DNA content variability of mouse spermatozoa. The two chemicals used, mitomycin C and benzo(*a*)pyrene, did not produce an increased spermatozoan DNA content variation after treatment of spermatogonial cells. In contrast, exposure of spermatogonial cells to a range (100–600 cGy) of X-irradiation doses resulted in an increased variation of spermatozoan DNA contents upon sampling 35 days after treatment. However, Pinkel *et al.*[91] concluded that the cause of the variation could not be attributed with certainty to either nondisjunction or a finer fragmentation of the genome.

2.2.3c. Assessment. Flow cytometry has largely superseded the other approaches to DNA content estimation. This is due in large part to the fact that it is a fully automated system, capable of analyzing extremely rapidly large numbers of cells.

None of these assay systems is capable of identifying individual cells that are aneuploid. Increases in aneuploidy, such as in the Robertsonian heterozygotes described by Stolla and Gropp,[119] can only be demonstrated by comparison of cell populations. Obviously, technical variables have to be reduced to a minimum to enhance the sensitivity of the assay system. However, the conclusion reached by Pinkel *et al.*[91] that "mouse sperm DNA contents are not sensitive to small amounts of aneuploidy and as such will only be useful in detecting agents that produce substantial DNA content variability" indicates that at present DNA content assays have a very limited role to play in routine nondisjunction testing.

2.3. Methods Using Pre- and Postimplantation Embryos

2.3.1. Zygotes at the First Cleavage Division

Zygotes arrested at metaphase of the first cleavage division offer the first postfertilization opportunity to assess the incidence of spontaneous or induced aneuploidy. Maternally and paternally derived pronuclei often remain separate and this enables determination of the parental origin of chromosome anomalies. The pronuclei are asynchronous in development, with the maternal pronucleus more advanced in the cell cycle, thus allowing their distinction through differences in chromosome contraction and staining.[28] The use of chromosome markers can further aid identification of the pronuclei.[38,69]

2.3.1a. Methods. A number of routines are available in the mouse for collecting zygotes at the first cleavage metaphase. Natural or gonadotrophin-induced ovulation can be used:

1. Approximately 16 hr after natural ovulation[128] or 30 hr after HCG treatment,[76] mated females are given an injection of colchicine. The optimum time for injection may vary with mouse strain used. Four hours later the females are killed and the zygotes recovered from the oviduct.
2. Fertilized eggs are recovered from mated females during the afternoon following observation of the plug. The cells are

cultured *in vitro* in the presence of a mitotic inhibitor to the first cleavage metaphase.(38)

3. Unfertilized, ovulated eggs are collected from the oviducts and put into a dilute suspension of capacitated sperm. After 6 hr they are transferred to a medium containing vinblastine and cultured to the first cleavage division.(38) The increased technical complexity of this approach in comparison to methods 1 and 2 has mitigated against its adoption for systematic nondisjunction screening.

Zygotes at the first cleavage division can also be obtained from mammalian species other than the mouse using routines similar to those described above (e.g., ref. 81).

Preparation of the cells for microscopy is generally carried out using Tarkowski's(129) method. Alternatively, the more recently devised schedules of Kamiguchi *et al.* (64) or Tanaka(127) may be adopted.

2.3.1b. Results. Low doses of X rays have been found to induce nondisjunction in oocytes at diakinesis (Table 6). Tease(133) described significant increases in the incidence of hyperploid pronuclei of young and old female mice after X-irradiation. The increases were dose-related and similar in both young and old animals. This result is in contrast to that of Uchida and Freeman,(136) who found a higher sensitivity to radiation in old females (Section 2.1.3b), but is consistent with the results reported from postimplantation embryo tests on radiation effects (Section 2.3.3b). The zygote system has also been used to examine the effects of X-irradiation on nondisjunction in other mouse oocyte stages (Table 6). Tease(133) found that exposure of dictyate oocytes 9.5 days before ovulation to low doses of X rays did not increase the level of hyperploidy in maternal pronuclei compared to controls. Further investigation using larger X-ray doses has shown that oocytes 3.5 days from ovulation are relatively insensitive to radiation-induced nondisjunction.(150) These observations are difficult to reconcile with those of Uchida and Lee,(137) who described a significant induction of hyperploidy in metaphase II oocytes of female mice given low doses of γ rays 1–7 days before sampling (see Section 2.1.3b).

Tease(150) has shown that relatively large doses (100 to 600 cGy) of X-rays to various dictyate oocyte stages induced nondisjunction (Table 6). Moreover, the incidence of induced nondisjunction was dose-related in all oocyte maturation stages sampled. The most mature stage, 3.5 days before ovulation, was less responsive than other, younger maturation stages.

The dose of gonadotrophic hormones used to induce superovula-

TABLE 6. Aneuploidy Induction Detected in First Cleavage Zygotes from Female Mice Exposed Either to Radiation or Chemicals

Agent	Treatment[a]	Dose	Number of cells	Number of hyperdiploid cells	Percent hyperdiploidy	Ref.
X rays	Control (young)	—	523	1	0.19	133
	Acute, 3 hr post-HCG	10–50 cGy	1516	35	2.31	
	Control (old)	—	324	5	1.54	133
	Acute, 3 hr post-HCG	10–50 cGy	835	27	3.23	
	Control	—	256	2	0.78	133
	Acute, 9.5 days before ovulation	10–50 cGy	1212	5	0.41	
	Control (low hormone dose)	—	241	3	1.25	134
	Acute, 3 hr post-HCG	50 cGy	242	15	6.2	
	Control (high hormone dose)	—	440	4	0.91	134
	Acute, 3 hr post-HCG	50 cGy	293	17	5.8	
	Control	—	236	2	0.85	150
	Acute, 3.5 to 23.5 days preovulation	100–600 cGy	3127	184	5.88	
Trenimon	Control	—	76	1	1.32	9
	Acute, 3 hr post-HCG	0.25 mg/kg	101	9	8.91	
Methotrexate	Control	—	32	2	6.25	48
	1. Acute, 3 hr post-HCG	200 mg/kg	15	1	6.67	
	2. Acute, 8.5 hr post-HCG	200 mg/kg	21	1	4.76	
Methyl methanesulfonate	Control	—	354	3	0.85	127
	Acute, 0–20.5 hr postmating	100 mg/kg	293	3	1.0	
Isopropyl methanesulfonate	Acute, 0–20.5 hr postmating	75 mg/kg	323	2	0.62	127
Mitomycin C	Acute, 0–20.5 hr postmating	5 mg/kg	341	3	0.88	127
Ethanol	Control	—	12	0	0	69
	Acute, 13.5 hr post-HCG	1 ml, 10–15% ethanol	180	10	5.56	

[a] HCG, Human chorionic gonadotrophin.

tion does not alter significantly mouse oocyte sensitivity to radiation-induced nondisjunction [134] (Table 6).

Chemical induction of nondisjunction in mouse oocytes has been investigated on a number of occasions using the zygote test system. Trenimon, an antineoplastic agent, caused a significant increase in the frequency of hyperploid zygotes when oocytes at diakinesis were treated[9] (Table 6). Similar results have been reported from other nondisjunction tests (Sections 2.1.3b and 2.3.3b). A second antineoplastic drug, methotrexate, did not cause a significant increase in the incidence of hyperploid zygotes following treatment of two oocyte maturation stages[48] (Table 6), although this apparent lack of effect may have resulted because not enough embryos were screened. The results from the metaphase II oocyte test were somewhat equivocal on the ability of methotrexate to induce nondisjunction (Section 2.1.3b). Tanaka[127] tested the effects of methyl methanesulfonate, isopropyl methanesulfonate, and mitomycin C on oocyte nondisjunction using the zygote assay. None of these compounds had a significant effect on the incidence of hyperploidy (Table 6).

In a recent experiment, Kaufman[69] administered ethanol to mated female mice. The oocytes of these females were at metaphase II at the time of treatment. The incidence of hyperploidy was found to be significantly higher in the alcohol-treated females than in controls, indicating that ethanol induced nondisjunction at anaphase II (Table 6). Fraser and Maudlin[39] investigated the hormones progesterone and estradiol. Neither caused an increase of aneuploidy in first cleavage zygotes after treatment of female mice.

Although the zygote test can be used to screen nondisjunction in spermatogenesis, no experiments have yet been reported using this system to investigate mutagen-induced aneuploidy in male mice. There is a similar dearth of information on nondisjunction in both sexes of other laboratory mammals.

2.3.1c. Assessment. Nondisjunction at both anaphase I and II can contribute to the observed aneuploidy in zygotes. The relative contributions of each division to the aneuploid total cannot, however, be distinguished by zygote analysis. Theoretically, screening chromosome numbers in zygotes should provide an estimate of the primary rate of aneuploidy immediately postfertilization. However, as has been discussed earlier, for technical reasons only hyperploid cells are unambiguous indicators of nondisjunction. An estimate of the initial level of aneuploidy has to be derived from that of hyperploidy. The important advantage of the zygote test over other cytogenetic assays is the ability to identify maternally and paternally derived pronuclei in a large proportion of cells. This permits an accounting of the maternal and paternal contributions to total aneuploidy. A major disadvantage of the test (which applies equally to the metaphase II test) is the difficulty of distinguishing primary and tertiary trisomies. Although karyotyping using G or Q banding is feasible,[68,75,85] in practice the proportion of cells that can be unambiguously analyzed is limited for technical reasons.[68]

The necessity of screening large numbers of cells to ensure unequivocal identification of mutagen-induced nondisjunction applies equally well to this test as to the metaphase II gametocyte assay (Section 2.1.3c).

2.3.2. Other Preimplantation Embryo Stages

Chromosome preparations can be made from all preimplantation embryo stages from two-cell through to blastocyst.

2.3.2a. Methods. Two routines are available to obtain chromosome preparations from preimplantation embryos: (1) the embryos are flushed from the oviduct or the uterine horn into a culture medium containing a mitotic arrestant, e.g., colchicine or vinblastine. After a few hours, chromosome preparations are made in the conventional manner;[64,129] (2) colchicine is administered to pregnant females and 4–5 hr later the embryos are flushed from the oviduct or uterine horn and processed as above.

In the mouse, embryos are at the two-cell stage approximately 52 hr after HCG, at the eight-cell stage after 72 hr and at the blastocyst stage after 100 hr.[85]

2.3.2b. Results. As far as we are aware, this test has not been used in any study of radiation-induced nondisjunction in laboratory mammals. However, a variety of chemicals have been tested for their effects on nondisjunction in mouse and rat oocytes (Table 7). Treatment of mouse diakinesis oocytes with trenimon resulted in a significant increase in the proportions of hyperploid metaphases in two-cell embryos and morulae compared to controls[53,97] (Table 7). Similar descriptions of trenimon-induced nondisjunction in mouse oocytes using other cytogenetic tests have been described earlier (Sections 2.1.3b and 2.3.1b). The antimalarial preparation pyrimethamine has been found to induce nondisjunction in rat oocytes, as evidenced by an increased frequency of hyperploid blastocysts[5] (Table 7). Neither of the remaining compounds listed in Table 7, methotrexate and cadmium chloride, caused an increased frequency of aneuploid preimplantation embryos following treatment of oocytes.[48,138] However, since the numbers of embryos

TABLE 7. Aneuploidy Induction Detected in Various Preimplantation Embryo Stages (Other Than First Cleavage Zygotes) from Female Mammals Treated with Chemicals

Chemical	Treatment[a]	Dose mg/kg	Species	Embryo stage	Number of embryos[b]	Number of hyperdiploid embryos	Percent hyperdiploidy	Ref.
Trenimon	1. Control	—	Mouse	Two to four cell	35 (26)	3	8.57	97
	Acute, 3 hr post-HCG	0.25			22 (18)	3	13.64	
	2. Control	—			29 (24)	1	3.45	
	Acute, 3 hr post-HCG	0.25			46 (40)	7	15.22	
	Control	—	Mouse	Morulae	28	0	0	53
	Acute, 3 hr post-HCG	0.25			56	5	8.93	
Pyrimethamine	Control	—	Rat	Blastocyst	226	0	0	5
	Acute, to preovulatory oocytes	2–5			145	6	4.14	
Methotrexate	Control	—	Mouse	Two cell	21	1	4.76	48
	1. Acute, 3 hr post-HCG	200			38	2	5.26	
	2. Acute, 6 hr after vaginal plug	200			14	0	0	
Cadmium chloride	Control	—	Mouse	Blastocyst	138	0	0	138
	Acute, 3 hr post-HCG	1.5–3.0			298	7	2.35	

[a] HCG, Human chorionic gonadotrophin.
[b] The figure given is the number of metaphases analyzed; the parentheses give the number of embryos from which these were derived.

scored in these latter experiments were very low, these results should be regarded with caution.

2.3.2c. Assessment. There seems little merit in the use of two- to four-cell embryos for nondisjunction testing. On most occasions, each embryo provides but a single karyotype for analysis, that is, equivalent to a zygote at first cleavage. Given the advantage of the zygote test in ascertaining maternally and paternally derived aneuploidy, two- to four-cell embryos provide on the whole less information than zygotes. The latter embryo stages (morulae/blastocysts), in contrast, have the advantage of providing regularly more than a single karyotype for analysis. Hypoploid embryos, arising through mutagen-induced chromosome loss or nondisjunction, can therefore be identified, whereas the metaphase II and zygote tests cannot distinguish artifactual chromosome loss from those induced by mutagen treatment. Although detailed karyotyping by G or Q banding is feasible in morulae/blastocysts, it is still less reliable than karyotyping of postimplantation embryos, which offer so much more material for study (Section 2.3.3).

2.3.3. Postimplantation Stages

Since most trisomic embryos survive until midgestation,[35,37,47] karyotypic analysis of postimplantation embryos provides a means of screening for nondisjunction. Embryo analysis is usually done at about 9–10 days of gestation, by which time most monosomic embryos have degenerated. There is therefore considerable loss of information compared with data from earlier fetal stages. This, however, is compensated by the availability of relatively quick and simple techniques for making good-quality metaphase spreads, and with the large number of cells available for karyotyping at these embryonic stages exact knowledge of the chromosome constitution of each embryo studied can be assured. G-banding techniques can also be applied, so that the identity of the extra, or missing, chromosomes can be established.

2.3.3a. Methods. The basic procedures are simple: treated/control males or females are caged together and the females thereafter checked daily for vaginal plugs (taken as evidence of mating). The mated females are killed at about day 9–10 of gestation and the embryos removed for examination and cytological analysis. Whole embryos, living or, in some circumstances, those that have recently died, and their membranes are then cytologically studied. The cytological techniques of Evans *et al.*[33] and Wróblewska[144] or modifications of them have been used for making chromosome preparations.

Most experiments upon postimplantation embryos have scored five or more cells/embryo. Such a number is to be recommended, since it reduces artifactual error and may demonstrate true chromosome mosaicism attributable to mitotic chromosome loss or nondisjunction in the zygotes. Hansmann and co-workers[52,55] scored only a single cell/embryo, however.

2.3.3b. Results. Comparison of the data is difficult because different investigators have used different criteria as measures of effect. Thus, Yamamoto *et al.*[148,149] included both monosomies and trisomies plus mosaic monosomies and trisomies in their aneuploid totals. Speed and Chandley,[116] on the other hand, recognized that mosaics may derive from zygotic chromosome losses (artifactual or real) and that most monosomies will have died and degenerated prior to analysis. They therefore based their conclusions upon estimates of the nondisjunction frequencies in the paternal germ lines calculated according to the method of Ford and Evans:[37] monosomic embryos are excluded; the number of trisomic embryos is doubled and expressed as a percentage of the number of diploid embryos plus the doubled number of trisomics; the mosaics are allotted to the normal and trisomic classes according to their presumed original zygotic constitutions. Yet another criterion for aneuploidy was used by Max,[78] who made the valid point that 40/41-chromosome mosaics may not necessarily derive from 41-chromosome zygotes as Speed and Chandley[116] presumed, but may originate from normal, 40-chromosome zygotes by mitotic nondisjunction to give 39/40/41 mosaics in which the 39-chromosome cell line has degenerated. Table 8 summarizes the data recalculated according to the method of Max,[78] i.e., aneuploid frequency $= 41 \times 2/40 + 41 \times 2$, with all mosaics excluded. This procedure has very little influence upon the overall results and conclusions from each experiment. Mosaic frequencies are shown separately and, since it has been suggested that triploidy may also be induced by paternal meiotic germ cell treatment,[23] the triploid frequencies are also given.

Most experimental work with female mice has been stimulated by reports based on human epidemiologic data that low-level maternal irradiation may increase the risk of aneuploidy in children, especially those born to older women.[4] The mouse data are summarized in Table 8 and it may be noted that a general problem is the low incidence of aneuploidy and relatively low numbers of embryos scored.

The paper of Yamamoto *et al.*[148] clearly suggests that aneuploidy is higher among postimplantation embryos of older females than among those of younger females. The claim was made on the basis of combined total of monosomies, trisomies, and mosaics, but it still evident with the corrected estimates (Table 8). Some further data (not shown) were

TABLE 8. Aneuploid Induction Detected in Postimplantation Embryos following X-Irradiation of Female Mice

Strain of mice (age)	X-ray dose	Treatment–mating interval	Number of embryos	Percent aneuploid embryos[a] (number)	Percent mosaics (number)	Percent triploids (number)	Ref.
CF (3–5 months)	—	—	149	0	1.3 (2)	0	148
CF (11–16 months)	—	—	156	4.0 (4)	3.9 (6)	0	148
CF (3–5 months)	Control	—	149	0	1.3 (2)	0	149
	5 R	3–14 days	111	0	3.6 (4)	0	
CF (11–16 months)	—	—	156	4.0 (4)	3.9 (6)	0	149
	5 R	3–14 days	43	21.7 (5)	4.7 (2)	0	
CBA (6 weeks)	Control	26–29 weeks	78	0	?	2.6 (2)	78
	2 R		44	0		0	
	4 R		83	2.4 (1)		0	
	8 R		78	2.5 (1)		0	
	16 R		84	0		1.2 (1)	
CBA (15 weeks)	Control	9 weeks	108	0	?	0	78
	4 R		103	0		1.9 (2)	

	8 R		108	0		1.9 (2)	
	16 R		96	0		0	
CBA (45–47 weeks)	Control	1–? days	27	0	?	7.4 (2)	78
	4 R		28	0		0	
	8 R		18	0		0	
C57BL (11 months)	Control	5.5–9.5 days	496	1.6 (4)	1.2 (6)	0.4 (2)	120
	4 R		99	0	0	1.0 (1)	
	8 R		246	2.4 (3)	0	1.2 (3)	
	16 R		223	1.8 (2)	0.5 (1)	0.9 (2)	
Q (6–8 weeks)	Control	4–8 days	571	0.7 (2)	0.2 (1)	0.2 (1)	23, 116
	5 rad		131	1.5 (1)	0.8 (1)	0	
Q (9 months)	Control	4–8 days	168	1.2 (2)	0.6 (1)[b]	0.6 (1)	23, 116
	5 rad + aged		145	1.4 (2)	0.7 (1)	0	
	Aged + 5 rad		168	3.6 (3)	0	1.8 (3)	
NMRI (8–12 weeks)	Control	1–	90	0	?	0	52
	20 rad		95	0		0	
	200 rad		22	9.1 (1)		0	

[a] Calculated according to the method of Max[78].
[b] Excluding one 38, X/40, X deleted X mosaic.

added in a later report.[147] The age effect was also evident in the data of Max[78] and Speed and Chandley.[116] The data obtained from studies upon postimplantation embryos are therefore consistent with other reports based on chromosome counts in metaphase II oocytes (Section 2.1.3) and preimplantation embryos (Section 2.3.1 and 2.3.2), which showed the incidence of aneuploidy to increase with maternal age.[27,43,77,133]

More controversial data were provided in a further paper by Yamamoto *et al.*[149] This was interpreted by the authors to show that low-level X-irradiation (5 R) of female mice increased the aneuploid frequency among their young and that the effect was greater in older females than in younger females. This conclusion was strongly criticized by Gosden and Walters,[44] however, on the grounds that (1)mosaics were included among the aneuploids, (2) the statistical analyses used were inappropriate, and (3) earlier published data appeared to have been used as controls. After reanalyzing the data, (with mosaics still included) Gosden and Walters[44] considered that the differential effect of X-irradiation on old and young females was not proven. Calculation of the estimated aneuploid frequencies (Table 8) provides better evidence for Yamamoto *et al.*'s[149] claim, but the validity of the control data remains questionable.

Max,[78] Strausmanis *et al.*,[120] and Speed and Chandley[116] attempted to repeat Yamamoto *et al.*'s[149] X-ray result, but without success. Max[78] was unable to find any increase in aneuploidy following low-dose X-irradiation of female mice. In a large experiment employing a range of X-ray exposures (2–16 R) and different treatment–mating intervals, she found only two trisomic embryos. It is interesting that both of these derived from the group of females that had been irradiated when young and investigated after aging, a test regime appropriate for events liable to occur in human populations. No evidence of X-ray-induced aneuploidy was evident in Strausmanis *et al.*'s[120] equally large experiment upon aged females, although several monosomies, trisomies, and mosaics were found in both the control and irradiated groups given 4–16 R of X rays. Similar negative results were obtained by Speed and Chandley,[116] who, like Max,[78] checked for an interaction between age and radiation (5 rad). No such interaction was detected, although, as indicated earlier, an age effect on aneuploidy was found. Speed and Chandley[116] G-banded selected cells from each embryo and were thus able to identify the additional or missing chromosomes in their aneuploids. The nondisjunctional events detected appeared to have affected chromosomes of all sizes, and autosomal and sex-chromosome aneuploids were thought to be equally represented. A strange feature of the data, however, was the detection of three XXY embryos, two of which derived from the young control group. XXY mice are thought to be fully viable (Section 3.1), but the frequency with which they occur in adult mouse populations is very low (0.01%[102]).

Only one other study with female mice has been reported and this was mentioned only in brief in a review on nondisjunction by Hansmann and Probeck.[52] One group of females was given a far higher X-ray dose (200 rad) than used in any other such experiment and among 22 fetuses examined, two were aneuploid. Banding studies showed one to be 39,XO and the other trisomic for chromosome 17. No aneuploids were found among the far larger number of embryos from a low-dose (20 cGy) group and the control, so it is likely that the two aneuploids derived from the radiation dose given. No information on mosaicism was obtained in this experiment, perhaps because only a single cell was examined from each embryo unless a chromosome abnormality was detected in the cell studied.

Triploid embryos were detected in most of these experiments, but occurred in control as well as treatment groups. There is little indication overall to indicate that the incidence of triploidy is increased by the radiation exposure.

It may be concluded from these experiments that the karyotypic screening of postimplantation embryos has provided reasonable support for the belief that nondisjunction in female mice increases with age. It has presented less substantial evidence that X-irradiation increases the frequency of nondisjunction, although Hansmann and Probeck's[52] finding of two aneuploid embryos among a very small number of embryos from females given 200 rad of X rays might indicate that higher doses of X rays than used in most of these experiments may induce nondisjunction.

Most studies with males have screened for aneuploidy in embryos derived from X-irradiated meiotic or premeiotic germ cells (spermatocytes or spermatogonia) in which the numerical chromosome changes could have been caused by nondisjunction. Hansmann *et al.*,[55] however, also screened for aneuploidy in treated postmeiotic stages (spermatozoa, spermatids) in which any induced changes could be attributed only to chromosome loss events.

The earliest experiment upon males was that of Szemere and Chandley,[125] who found a higher frequency of chromosome anomalies in embryos from X-irradiated (100 rad) spermatocytes and spermatogonia than in those of their control series (Table 9). However, all their anomalies were either 39/40- or 40/41-chromosome mosaics or triploids. They suggested that triploidy may be induced by spermatocyte irradia-

TABLE 9. Aneuploid Induction Detected in Postimplantation Embryos following Mutagen Treatment of Male Mice

Strain of mice	Treatment[a]	Dose	Germ cell stage treated (treatment–mating interval)	Number of embryos	Percent aneuploid embryos[b] (number)	Percent mosaics (number)	Percent triploids (number)	Ref.
Q	Control	—	—	64	0	6.3 (4)	0	125
	X rays	100 rad	Spermatocytes (week 5) + spermatogonia (week 6)	164	0	3.7 (6)	1.2 (2)	
Q	Control	—	—	571	0.7 (2)	0.5 (3)	0.2 (1)	23, 116
	X rays	100 rad	Spermatocytes (week 5)	1024	0.4 (4)	0.7 (7)	0.5 (5)	
	X rays	100 rad	Spermatogonia (week 6)	708	2.0 (9)	1.1 (8)	0.4 (1)	
NMRI	Control	—	—	211	0	?	0	55
	X rays	20 rad	Mostly spermatozoa/spermatids (days 1–22)	138	2.9 (2)		0.7 (1)	
	X rays	200 rad		246	1.6 (4)		1.2 (3)	
	X rays	20 rad	Spermatocytes (days 23–36)	78	2.5 (1)	?	0	
	X rays	200 rad		139	2.9 (3)		0	
Random-bred	EMS	240 mg/kg	Spermatogonia (days 43–50)	108	0	0	0	126
Q	Control	—	—	561	0.7 (2)	0.7 (2)	0.2 (1)	23
	MC	2 mg/kg	Spermatocytes (week 5)	325	0	0	0	

[a] EMS, Ethyl methanesulfonate; MC, mitomycin C.
[b] Calculated according to the method of Max.[78]

tion, but this view has not yet found support from the data of other investigators.[55] Szemere and Chandley[125] attributed their failure to detect any monosomies or trisomies to gametic and zygotic selection, i.e., loss, but further studies carried out later in the same laboratory[23,116] using virtually the same procedures were more successful. Several autosomal trisomies and one autosomal monosomy were found, and two cases of the rare XXY class additional to one found in their female study and the two in the common control. Mosaics and triploids were also detected. The incidence of aneuploidy was not significantly higher in the treatment group than in the control. An unexpected feature of the data was the appearance of aneuploid embryos following spermatogonial irradiation. Although it is possible that these may all have been of spontaneous origin, their occurrence might suggest that mitotic nondisjunction in the spermatogonia was the cause.[23] However, since extra chromosomes have only rarely been observed in the many studies on mouse spermatocytes at metaphase I or II following X-irradiation of spermatogonia,[105,108] this explanation would seem unlikely. A more credible explanation is that the aneuploidy derived from irradiated spermatocytes which, because of the oligospermy resulting from cell killing, were retained longer as spermatozoa in the testis and released at the time when spermatozoa derived from the irradiated spermatogonia would normally have been present in the ejaculate.[116]

Hansmann *et al.*[55] screened for aneuploid embryos in G-banded preparations following the X-irradiation (20 and 200 cGy) of male mice. The data as presented do not satisfactorily distinguish which anomalies derived from post- and premeiotic germ cells, but the approximate groupings are shown in Table 9. No aneuploidy was detected in the control group, but two autosomal trisomies were found following the exposure of spermatozoa/spermatids to 200-rad X rays. Since meiotic nondisjunction cannot have been responsible for these aneuploidies, they may have represented undetected mosaics or they could have been of spontaneous origin. Two further autosomal trisomics were also recovered from the 20-rad, 1- to 22-day-interval group, but these could have originated from irradiated spermatocytes because they were conceived on day 21 (not indicated in Table 9). Spermatocyte irradiation yielded three trisomies (one from the 20-rad group) and an X monosomy. The uncertainty as to the origin of the trisomies precludes any conclusion on whether X-irradiation increases the frequency of nondisjunction in male mouse spermatocytes.

Cytogenetic analysis of postimplantation embryos has not to our knowledge been used to investigate the chemical induction of nondisjunction in female mice, but two such studies have been carried out in males (Table 9). Szemere and Marczinovits[126] failed to find any evidence of trisomy among the progeny of males treated with ethyl methanesulfonate (EMS). However, because the germ cells treated were spermatogonia, the lack of effect is not surprising. The other study was carred out by Chandley and Speed[23] at the same time as most of their X-ray work and used essentially the same control. Mitomycin C (MC) was chosen for test because it is thought to produce chromosome aberrations chiefly in the centric heterochromatin of homologous chromosomes. Only two aneuploid embryos were found, both X monosomies, and of two mosaics, one was a 40/41 mosaic and the other possessed three cell lines (39/40/41). No increase in nondisjunction over the control levels was therefore indicated.

Chinese hamsters have been used to investigate the incidence of chromosome anomalies in postimplantation embryos following procedures to cause hypoxia and alter tissue pH in the mothers prior to conception[62,113] to delay fertilization and cause ovum aging,[1,146] Maternal age effects have also been studied,[63] but since the main effect was the production of triploidies and tetraploidies, this work is beyond the scope of this chapter.

2.3.3c. Assessment. The results obtained with this method confirm that many trisomies and some monosomies can be detected among postimplantation embryos.[35,37,47] As such, the method has proved effective in detecting the age effect upon nondisjunction in females that was observed using other cytogenetic assays. It has the advantage over metaphase II (Section 2.1.3) and first cleavage division (Section 2.3.1) studies that many cells are available for scoring. Scoring errors are therefore reduced and, with G banding, not only is it practicable to identify additional or missing chromosomes, but tertiary monosomies and trisomies are potentially distinguishable. Although the methodology is relatively quick and simple, it is still sufficiently laborious that the number of embryos that investigators choose to score is low. This, together with the low incidence of aneuploid embryos detected, reduces the value of the method.

3. Genetic Test Systems with the Mouse

Until recently there have been no established genetic assays for nondisjunction in the mouse such as exist, for example, in *Drosophila*.[107] The zygotic lethality of most mouse trisomies and monosomies[35,37,47] was the major problem, but, in addition, stocks that would allow nondisjunction products to be recovered in chromosomally balanced viable forms had not been constructed. The viability problem is overcome when only the sex chromosomes are considered, however; most sex-chromosome trisomies and monosomies are viable in the mouse[16,101] and those not yet detected, e.g., XXX, might reasonably be expected to be so on the basis of observations upon other mammalian species.[57,94] Thus, the numerical sex-chromosome anomaly (NSA) method, originally developed by Russell[99] as a means of assessing chromosome damage additional to the dominant lethal and heritable translocation tests, may be considered a feasible assay for nondisjunction and has been projected as such by Russell in a recent review.[102]

The discovery that Robertsonian translocations (biarmed metacentric chromosomes) derived from feral mouse populations cause high rates of nondisjunction when heterozygous with their homologous laboratory mouse acrocentric chromosomes has also offered the prospect of methods for assaying autosomal nondisjunction. Three methods are currently being investigated—the Robertsonian (Rb) complementation method[72] and the Rb tester and the monobrachial homology (MBH) methods[21]—and a fourth is being developed.[111] Each relies on the high rate of nondisjunction that typifies certain Rb heterozygotes or Rb compounds so that chromosomal gains (or losses) in one parent may be complemented in reasonably high numbers by losses (or gains) in the other. Chromosomally balanced, and therefore viable, nondisjunction products can then be recovered among the progeny.

Since very little data have yet been obtained with any of these genetic tests, evaluation has to be limited to the technical procedures involved and the general potential applications of the test systems. For this reason some detail of the methodologies and the background of these genetic tests will be presented.

3.1. The Numerical Sex-Chromosome Anomaly (NSA) Method

For full details of the methodology and for reviews of the historical data on the spontaneous frequencies of sex-chromosome loss and nondisjunction reference should be made to Russell's[101,102] extensive reviews. Here we present only sufficient information to allow comparison of the method with other genetic test systems.

The test relies upon the viability of most of the numerical sex chromosome anomalies (NSAs) in the mouse and the availability of gene markers that permit their detection simply by brief examination of the phenotype.

Simple nondisjunction can produce the following sex-chromosome constitutions: OY, XO, XXY, XYY, and XXX; however, as can be seen

TABLE 10. Viability and Detectability of Mice Having Numerical Sex-Chromosome Anomalies

Sex-chromosome constitution[a]	Sex	Viability	Detectability with X-linked genes	Ref.
OY	—	Preimplantation lethal	−	142
X^MO and X^PO	F	Normal as adults; retarded as preimplantation embryos, near-normal to normal by birth[b]	+	14
X^MX^PY and X^MX^MY	M	Probably normal (cf. X/X *Sxr* ♂♂); X^MX^MY not yet detected in the mouse	+	73
X^MYY	M	Probably normal	−	16, 101
$X^MX^MX^P$	F	Not yet detected in the mouse	+/−; may require three marker genes	—

[a] X^M, maternally derived X; X^P, paternally derived X.
[b] P. Burgoyne, personal communication.

from the comments on each anomaly in Table 10, only the XO, XXY, and possibly the XXX conditions have any application for the test. Successive nondisjunction events that can cause additional sex-chromosome constitutions are thought to be very rare and are not considered.[102]

The NSAs listed in Table 10 may arise from meiotic nondisjunction (meiosis I or II) or, in theory, from mitotic nondisjunction. However, when nondisjunction occurs in the mitotically dividing spermatogonia of the male none of the resulting cells (XO, XYY, XXY, or OY) may be expected to survive to form functional gametes.[109] The XXX and XO cells that arise from mitotic nondisjunction in the oogonia of the female probably can survive, but may be less efficient than XX germ cells in producing gametes. Oogonial nondisjunction could be detected in the NSA test by clustering of the rare X^PO females or as clusters of the as yet undetected XXX female or X^MX^MY male classes (Table 10).

To be of value for detecting NSAs, X-chromosome markers should (1) be compatible with viability in mice of all possible chromosome constitutions at least up to the age of weaning and (2) show distinguishable phenotypes in the hemizygous and heterozygous conditions. Examples of genes that have been or could be used are given in Table 11, but from the comments on each, it can be concluded that most of

TABLE 11. X-Chromosomal Marker Genes for the NSA Test[a]

Gene	Viability, fertility, and ease of classification		Location on X
	Hemizygote/homozygote	Heterozygote	
Tabby (*Ta*)	Obvious phenotype; near-normal fertility in male, variable in female; because of tooth abnormalities may require supplementary feeding; reduced fitness	Normal viability and fertility; variable phenotype[b] occasional overlaps with normal; compounds with *Gs* and *Hq* difficult to distinguish	Central
Greasy (*Gs*)	Obvious phenotype; normal viability and near-normal fertility, especially in male	Normal viability and fertility; variable phenotype,[b] occasional overlaps with normal; compounds with *Hg* possibly not distinguishable from Gs^+ or *Hq*+	1 cM from *Ta*
Blotchy (Mo^{blo})	Obvious phenotype; variable viability when adult; variable fertility in both sexes; reduced fitness	Normal viability and fertility; variable phenotype,[b] classification may be difficult to distinguish from wild type; compounds with other genes regularly difficult to distinguish	4 cM distal to *Ta*
Eye–ear reduction (*Ie*)	Presumed viable[c]	Presumed classifiable[c]	—
Harlequin (*Hq*)[d]	Obvious phenotype; near-normal fertility in male, variable in female	Normal viability and fertility; variable phenotype[b]	21 cM proximal to *Ta*

[a] For other possible genes, reference should be made to recent issues of the *Mouse News Letter*.
[b] Variable phenotypes in heterozygotes attributable to X-chromosome inactivation.
[c] Not described in detail (used at Oak Ridge).
[d] *Hq* not yet used in NSA test.

them do not fulfill these conditions. Several have deleterious effects in hemizygotes and homozygotes, all show a wide range of heterozygous phenotypes, and it is debatable whether any three could be reliably used to detect XXX females. However, although these effects limit the range of crosses, enough possibilities remain to ensure that XO females and XXY males are readily detectable.

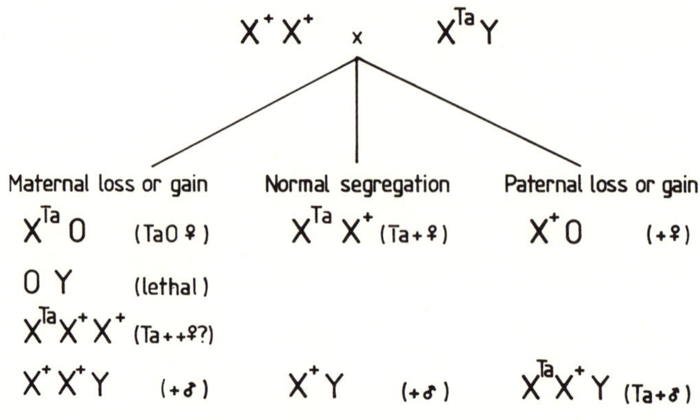

FIGURE 1. Simple mating scheme for detecting induced or spontaneous paternal sex-chromosome loss or nondisjunction. With further testing (see test) 50% of X^M loss can also be recognized, but the $X^MX^MX^P$ and X^MX^MY classes deriving from maternal nondisjunction cannot be distinguished.

3.1.1. Methods

The marker genes can be used in a variety of ways, but the principal feature of two main recommended schemes are outlined in Figures 1 and 2. With the simplest scheme the maternal and paternal X chromosomes are distinguished by the use of a single marker, as in the cross of +/+ ♀♀ with *Ta*/Y ♂♂. The procedure is effective for scoring paternal losses and nondisjunction, but only 50% of maternal losses are recoverable (the OY zygote being lethal, Figure 1) and in order to verify that X^PO progeny arise from *de novo* events the possibility of a preexisting XO condition in the mother has to be ruled out by cytological analysis. Maternal nondisjunction cannot be detected with this scheme.

A modified version of the scheme uses two markers to distinguish the maternal X chromosomes, as in the cross Mo^{blo} +/+ *Ta* ♀♀ with + +/Y ♂♂. This arrangement obviates the need to test for a preexisting XO condition in the mothers, but recombination between the two markers increases the risk of a failure to detect of X^MX^PY progeny. For this reason the two markers must be closely linked to be of value.

With the more complex scheme (Figure 2) each of the parental chromosomes is marked, as in the cross of Mo^{blo} + +/+ *Ta* + ♀♀ with + + *Ie*/Y ♂♂. In theory, this scheme should allow the detection of all XO females and all XXX females and XXY males except for those that

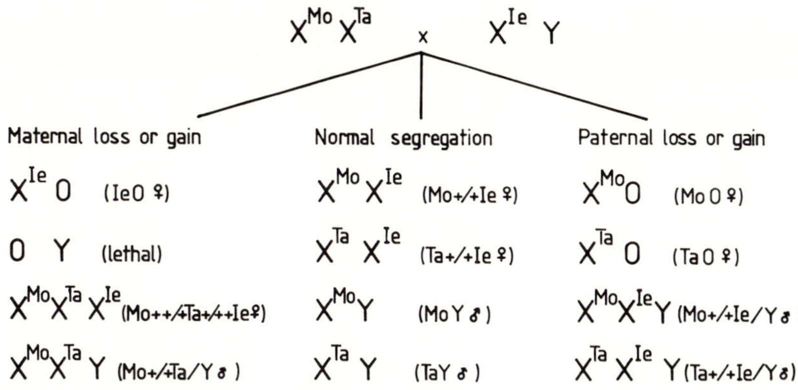

FIGURE 2. More complex mating scheme for detecting induced or spontaneous maternal and paternal sex-chromosome loss and nondisjunction. This is the method of choice for experiments involving the treatment of female germ cells, but does not offer any advantage over the simpler mating scheme (Figure 1) for experiments upon treated male germ cells. It is not known whether the markers could distinguish the triple X female.

receive two unmarked X chromosomes from the mother as a result of recombination between the maternal markers.

To operate the test, at least two stocks need to be maintained, one carrying one or more markers and the other perhaps homozygous wild type. Thereafter, the breeding procedures are in principle very simple. After the treatments have been given and the matings set up all that is needed is regular checking for litters and classification of progeny when they are old enough for the phenotypes to be distinguished. At weaning, only the few exceptional progeny may be retained for further testing; the remainder of the litters can be discarded and the untreated parents used again if necessary. A problem in practice is that it is usually appropriate to score for nondisjunction in certain germ cell stages; this requires timed matings. Hence a laborious feature of the method is the need to set up new matings frequently with newly treated animals. This, however, is a problem of all genetic tests.

A further problem with the NSA test is that gene mutations and chromosome rearrangements can give rise to mimics of the phenotype of XO females and XXY males. Verification of the chromosomal constitutions of exceptional progeny is therefore advisable. If the animals are to be kept alive, short-term blood cultures can provide material for study. Otherwise, chromosome preparations can be made within a matter of minutes from corneal epithelium or other tissues.

Counts of 39 and 41 reliably confirm the XO and XXY conditions, respectively.

3.1.2. Results

Almost all the results using the NSA method derive from Oak Ridge National Laboratory and, reflecting the original use of the method as an assay for chromosome damage, most of the data obtained provide information on sex-chromosome loss rather than specifically nondisjunction. The radiation data on females have been published only in summary in Russell's[101] review and therefore details of the markers used, etc., cannot be presented. The summarized data are given in Table 12. It is evident that X-irradiation of dictyate oocytes of adult females can bring about increased frequencies of X^PO females, but in none of these experiments was the arrangement of markers such that X^MX^MY males (or $X^MX^MX^P$ females) could have been detected. Only in a smaller study upon preleptotene to diplotene oocytes of the late fetus to neonate could such XXY males have been detected, and none were found.

The Oak Ridge data on males (Table 13) derive partly from crosses using doubly marked (Mo^{blo} +/+ *Ta*) females and partly from singly marked females. These tests therefore had the potential for detecting some of any X^MX^PY males produced. However, although X^MO females were recovered following spermatocyte X-irradiation, only a single X^MX^PY male was detected and this could have been of spontaneous origin. The discordance between the recovered XO and XXY frequencies strongly suggests that most of the former class derives from events other than nondisjunction.

Of the two experiments upon spermatogonia, one was carried out at Oak Ridge and used crosses involving marked females.[103] The other was carried out by Léonard and Schröder[70] and used marked (*Ta*) males. In both experiments there was a failure of the spermatogonial irradiation to yield a excess of X^MO females over control values or any X^MX^PY males. As previously indicated (Section 3.1.), this could be understood in terms of the lethality of XO and XXY cells in the testis.

The NSA test also has been used to assess sex-chromosome loss following X-irradiation of male postmeiotic cells and pronuclear stages of the fertilized egg. The data are not reviewed here since they do not concern nondisjunction, but a summary of this work may be found in Russell's[101] review.

Isopropyl methanesulfonate (IMS) and hycanthone have been investigated for NSA induction in female mice. The results have not

TABLE 12. Recovery of Numerical Sex-Chromosome Anomalies (NSAs) following X-Irradiation of Female Mice[a]

Treatment–mating interval, days	Cell stage sampled	X-ray dose, R	Dose rate, R/min	Number of progeny classified	X^PO Percent[b]	X^PO Adjusted induced frequency/R × 10^{-5b}	X^MX^MY,[c] %
Control	—	0	—	391	0	0	0
>42	Preleptotene–diplotene	221	~80	1,221	0.08–0.33	0.13–1.24	0
Control	—	0	—	11,528	0.06	—	—
1–42	Dictyate	400	66–78	15,006	0.61–0.77	1.33–1.78	—
1–42	Dictyate	50	66–78	12,256	0.02–0.08	0	—
>42	Dictyate	50	66–78	10,172	0.09	0	—
1–42	Dictyate	400	0.6	10,900	0.36–0.40	0.76–0.88	—
Control	—	0	—	12,163	0.04–0.05	—	—
0–7	Dictyate	400	0.006	1,906	0.26	0.52	—
8–42	Dictyate	400	0.006	1,779	0.0–0.06	0	—
>42	Dictyate	400	0.006	6,444	0.08	0	—
Control	—	0	—	5,154	0.04	—	—
1–49	Dictyate	400	65–70	5,170	0.77–0.87	1.80–2.04	—
1–49	Dictyate	8 × 50[d]	65–70	6,099	0.46–0.51	1.02–1.14	—

[a] Data from Russell.[101]
[b] The lower number represents only genetically/cytologically verified cases; the higher number represents cases scored by phenotype only.
[c] A dash indicates that the cross made did not allow detection of X^MX^MY.
[d] Fractionated dose; 75 min interval between fractions.

TABLE 13. Induction of Numerical Sex-Chromosome Anomalies (NSAs) following X-Irradiation of Male Mouse Premeiotic Germ Cells

Treatment–mating interval, days	Cell stage sampled	X-ray dose, R	Number of progeny classified	Number of exceptionals		Adjusted induced frequency/R × 10^{-5}		Ref.
				X^MO	X^MX^PY	X^MO	X^MX^PY	
Control	—	—	3423	0 + 1[a]	0	—	—	100
22–28	Spermatocytes	200	2118	2 + 1[a]	0	1.0	0	
29–35	Spermatocytes	200	2370	8 + 1[a]	1	3.3	0.4	
36–42	Spermatocytes	200	1726	5 + 1[a]	0	2.9	0	
Control	—	—	4994	11	1	—	—	103
83–>700	Spermatogonia	600	4173	9	0	0	0	
		100 + 500[b]	3982	6 + 1[a]	0	0	0	
Control	—	—	1508	2	0	—	—	70
"Poststerile"[c]	Spermatogonia	600	1347	1	0	0	0	

[a] The lower number represents only the genetically/cytologically verified cases; the higher number represents cases scored by phenotype only.
[b] Twenty-four-hour fractionation.
[c] Greater than 42 days.

been formally published, but summaries are given in Russell's[101] review. In both experiments the arrangement of gene markers was such that X^M loss could be detected, but any X^MX^MY males would not have been found. With IMS, Generoso and colleagues (see Russell[101]) obtained significant increases in X^PO female frequencies following treatment of dictyate oocytes (weeks 1–6). The doses used were 50 and 75 mg/kg and the scale of the experiment was large, over 9000 female progeny being screened. In an even larger experiment Russell and Hunsicker (see Russell[101]) found that hycanthone at a dose of 150 mg/kg induced X^M loss but only in mature oocytes (week 1) and they suggested that because this pattern of response differed from that with X rays, as found also for dominant lethals,[101] the two mutagens may bring about chromosome loss by different mechanisms. Interestingly, with a lower hycanthone dose (75 mg/kg) the X^PO frequency detected was considerably below the linear interpolation from the 150 mg/kg point and was not significantly above the control level. A nonlinear dose–response curve was therefore indicated.

Triethylenemelamine (TEM) and methane methylsulfonate (MMS) have been investigated for NSA induction in male mice, but since the X^MO females detected in the TEM study[15] derived from treated postmeiotic germ cells and therefore could not have arisen from nondisjunctional events, the data will not be considered here. All germ cell stages were screened in the MMS study[84] (dose 60 mg/kg) and despite the small scale of the experiment and the fact that the arrangement of gene markers allowed the detection of only half of the exceptional progeny, some evidence of X^P loss and possible evidence of paternal sex-chromosome nondisjunction was obtained. Two presumptive X^MO females were found among female progeny derived from treated meiotic and premeiotic cells and five suspected X^MX^PY males. However, since the latter exhibited features that are not seen in XXY mice and cytological evidence was not sought, the occurrence of these mice does not provide adequate proof for the induction of nondisjunction by MMS.

3.1.3. Assessment

The NSA method should be a simple, effective test for sex-chromosome nondisjunction. Of the five possible sex-chromosome nondisjunction products, four are likely to be viable and three of these should be detectable with the use of appropriate X-chromosome marker genes. In the absence of any known selection against these genotypes, the frequencies of occurrence of exceptional progeny should directly represent the frequencies of the responsible chromosomal events in the germ cells and in some cases the phenotypes of the exceptional young may distinguish the germ cell types in which the events occurred.

However, a weakness of the method is the limited range of markers available; most have some degree of deleterious effect in the hemizygote or when doubly heterozygous, making stock maintenance difficult. Furthermore, classification is not always easy, so that experienced staff are required. Of yet greater importance, the available markers do not facilitate the detection of XXX or X^MX^MY animals, and this is probably the main reason why these classes have not yet been found in the mouse. This leaves only the XO and X^MX^PY classes, which can be easily recognized and, from the experimental data, it is clear that most of the former derive from chromosome loss events other than nondisjunction. Therefore, at present the X^MX^PY is the only genotypic class that in practice can serve as an indicator of sex-chromosome nondisjunction with the NSA method. The spontaneous frequency of this class is very low (0.01%)[101] and it has to be accepted that very large numbers of progeny need to be scored to give any certainty that a treatment is not causing nondisjunction. However, the converse is also true; the detection of only a few XXY progeny provides strong evidence that a treatment is causing nondisjunction. Only one XXY male has so far been detected following mutagen treatment and this could have been of spontaneous origin. Therefore positive evidence for the induction of nondisjunction by any treatment has not been obtained with the NSA method.

A further limitation of the method is that it only scans for nondisjunction in one of 20 pairs of chromosomes and, this being the sex-chromosome pair, the processes leading to regular disjunction at meiosis may not be typical of autosomal bivalents (see Sections 3.2.2 and 3.2.3). This may be especially true in the male, where there is a big size difference between the X and Y chromosomes and the postulated homologous region they share must be very small.[34]

3.2. Tests Using Robertsonian Translocations

Three methods of screening for nondisjunction with the use of Robertsonian translocations (Rbs) have been developed and subjected to some experimentation, and a further application is being investigated. Before presenting these, however, a brief summary of the properties and characteristics of Rbs will be given so that their potential and limitations for nondisjunction testing may be recognized.

Rbs comprise two acrocentric chromosomes joined at their centric ends to form biarmed metacentric or submetacentric chromosomes.

Several years ago they were thought to be very rare in mice, but large numbers are now known and these involve many different combinations of chromosome arms.[46] Most Rbs derive from wild mouse populations, but several have arisen spontaneously in laboratory stocks. The key property shared by the wild-derived forms is that when heterozygous with the homologous acrocentric chromosomes of the laboratory mouse meiotic disjunction is irregular such that a significant proportion of the gametes are chromosomally unbalanced; monosomic and trisomic zygotes therefore result.[135] Studies in different laboratories have shown that each Rb tends to cause a characteristic frequency of nondisjunction. Frequencies of unbalanced gametes or zygotes have been found to range from about 4% to over 30%, with good general agreement being obtained with different methods of assessment.[21,35,47,71] Such high levels of nondisjunction are not usually found with Rbs of laboratory mouse origin, although some examples have been found.[46,71] For this reason it has been postulated that the irregular segregation of the metacentric and homologous acrocentric chromosomes is at least in part attributable to other genic or minor chromosome differences between the wild and laboratory mouse chromosomes additional to the Robertsonian changes.[21] Some evidence supporting this comes from the observation that Rb heterozygotes produced by crossing wild mouse populations from neighboring geographic areas may not show the high levels of nondisjunction found in wild–laboratory mouse hybrids.[46] On two grounds, therefore—major structural change and minor nonhomologies—the nondisjunction associated with Rbs may not be representative of the process that occurs in like, normal chromosomes within a species or subspecies.

It is evident from a number of studies[21,35,47] that the nondisjunction in Rb heterozygotes is a meiotic event. Although, in principle, the nondisjunction could occur at meiosis I and/or meiosis II, two findings suggest that it occurs only, or principally, at meiosis I: (1) the nondisjunction frequencies at meiosis I and detected at metaphase II generally agree well with the frequencies detected with methods that assess total meiosis I and meiosis II events; (2) with the Rb complementation method, analysis of the progeny derived from nondisjunction in both parents has shown that most nondisjunctional events must have occurred at meiosis I. For further discussion on this point reference should be made to Lyon *et al.*[72]

Although the diverse data are relatively consistent in showing which Rbs produce the highest rates of nondisjunction, they have also drawn attention to a number of factors that could create technical difficulties for any assay system using these chromosome changes. These include:

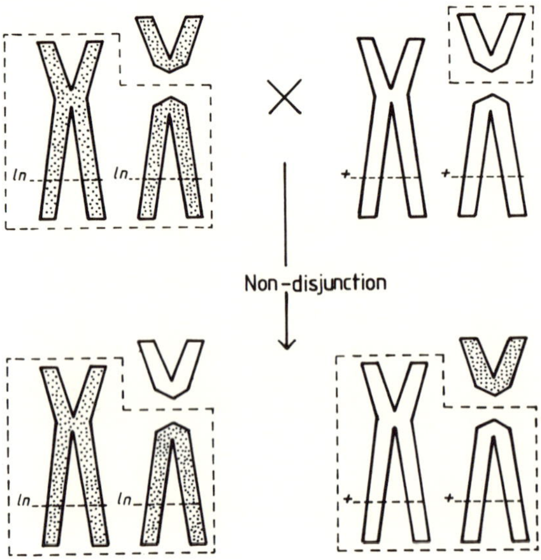

FIGURE 3. Diagrammatic representation of the cross between heterozygotes for an Rb translocation. One parent is shown homozygous for the gene marker *ln*, carried on one arm of the Rb chromosome and on the homologous acrocentric. The other parent is homozygous for the dominant wild-type allele. With normal segregation of the Rb and the acrocentrics all the progeny will be phenotypically wild type. However, if the segregation in both parents is as illustrated and complementary gametes unite, a proportion of the progeny will be homozygous for the gene marker and show the *ln* phenotype.

(1) variability between experiments, which could be attributable to genetic background effects;[21,71] (2) variability between repeat experiments, which suggests that there is interanimal variation;[71] (3) sex differences, since nondisjunction tends to be higher in females than males;[46] (4) age effects—where found, the nondisjunction has increased with maternal age.[72]

3.2.1. The Rb Complementation Test

The problem of the lethality of autosomal monosomies and trisomies is surmounted when gains (or losses) in one parent are complemented by losses (or gains) in the other. This is achieved in crosses between heterozygotes for one or more Rbs having high frequencies of nondisjunction. With the Rb complementation (RBC) test the parents are made homozygous for different alleles of suitable marker genes and, as a consequence, when nondisjunction occurs in both parents and the uniting gametes carry complementary abnormalities (gains/losses) the resulting chromosomally balanced offspring may be recognized by their phenotype with respect to the marker genes.[72] Thus, the offspring formed when disjunction is normal will be heterozygous for the marker; but any homozygotes must have received both homologues from one parent and therefore be the product of simultaneous, complementary nondisjunction in both parents (Figure 3). The frequency of nondisjunction can be calculated from the incidence of young homozygous for the marker, although to do this it has to be accepted that a number of artificial assumptions must be made.[72]

The main premise underlying the method is that any increase in nondisjunction brought about by a given treatment will be detected by appropriate increases in the nondisjunction frequencies associated with Rb heterozygosity.

3.2.1a. Methods. In principle, any gene marker can be used so long as it (1) affects the external phenotype to permit immediate recognition, (2) gives the homozygote a phenotype distinguishable from the heterozygote, and (3) is viable at least up to weaning age in the homozygous condition. Most chromosomes have a number of suitable markers. Hence, the problems encountered with the NSA test (Section 3.1) need not cause problems for tests using Rbs.

All the markers so far used have been recessives (see Table 14), but semidominants could also be used, as could dominants when the recessive wild-type alleles then serve as the markers.

The markers can be used in a variety of ways: when recessive markers are used and are carried by the treated parent, chromosome gains (but not losses) attributable to nondisjunction may be detected; marking the untreated parent of the cross screens for chromosome losses (but not gains) and these may result from events other than nondisjunction. The former procedure is therefore to be recommended for nondisjunction screening. The use of semidominants confers an advantage over the use of recessives, permitting the detection of both gains and losses within any one cross, whichever parent is treated.

To operate the test, four stocks need to be maintained: (1) a homozygous Rb stock, homozygous also for the markers; (2) a wild-type homozygous Rb stock; (3) a chromosomally normal stock, homozygous for the markers; and (4) a chromosomally normal wild-type stock. Test crosses are set up from animals derived from matings of stocks 1 × 3 and stocks 2 × 4. Thereafter, according to the germ cell stage being investigated, the basic procedures only involve the regular checking of litters and classification of progeny. Normally only the marked progeny are retained for cytological analysis and perhaps also for breeding tests.

3.2.1b. Results. The Rb complementation test has been investigated with seven different Rbs. These are listed in Table 14 together with the individual results obtained. The six Bnr Rbs, Rb1Bnr–Rb6Bnr, are of wild mouse origin deriving from the tobacco mouse (*Mus poschiavinus*), and it may be seen that all give a high level of nondisjunction as indicated by the frequencies of marked young. Rb163H is of laboratory

TABLE 14. Marker Genes Used and Results Obtained with the Rb Complementation Test

Experiment number	Translocation	Chromosomes	Marker[a]	Marked parent	Offspring: Marked	Offspring: Total	Offspring: Percent marked	Ref.
1	Rb(1.3)1Bnr	1	*ln*	♂	28	1156	2.54	20
2a				♀	11	631	1.74	17
2b				♂	7	512	1.37	17
3a				♀	12	589	2.04	17
3b				♂	1	540	0.19	17
4a	Rb(4.6)2Bnr	4	*b, m*	♀	6	728	0.82	72
4b				♂	15	1269	1.18	72
5			*b*	♂	15	733	2.05	20
6			*m*	♂	4	921	0.43	20
7		6	*wa-1*	♂	8	698	1.15	20
8a		4/6	*b/wa-1*	♀	3/0	411	0.73/0.0	18
8b				♂	12/3	481	2.50/0.62	18
9a				♀	13/0	537	2.42/0.0	18
9b				♂	14/7	632	2.22/1.11	18
10	Rb(5.15)3Bnr	5	*bf*	♂	5	1090	0.46	20
11			*go*	♂	3	471	0.64	20

12		15	*bt*	♂	8	1321	0.61	20
13a		4/15	*bf*/*bt*	♀	3/3	580	0.52/0.52	17
13b				♂	9/7	654	1.38/1.07	17
14a				♀	2/1	380	0.53/0.26	17
14b				♂	12/2	522	2.29/0.38	17
15	Rb(11.13)4Bnr	11	*vt*	♂	10	1146	0.87	20
16		13	*pe*	♂	9	1592	0.57	20
17a		11/13	*vt*/*sa*	♀	11/3	464/444	2.37/0.68	[b]
17b				♂	14/9	546/526	2.56/1.71	[b]
18	Rb(8.12)5Bnr	8	*e*	♂	6	2298	0.26	20
19a	Rb(9.14)6Bnr	14	*hr*, *s*	♀	5	469	1.07	72
19b				♂	14	1229	1.14	72
20a		9/14	*d*, *se*/*hr*, *s*	♀	4/2	811	0.49/0.25	71
20b				♂	15/15	917	1.64/1.64	71
21		9	*d*, *se*	♂	9	710	1.26	20
22		9/14	*d*, *se*/*s*	♂	15/22	1219	1.23/1.81	20
23		14	*s*	♂	1	409	0.27	20
24a	Rb(9.19)163H	19	*ru*	♀	0	2077	0.0	72
24b				♂	0	3183	0.0	72

[a] *b*, brown; *bf*, buff; *bt*, belted; *d*, dilute; *e*, recessive yellow; *go*, angora; *hr*, hairless; *ln*, leaden; *m*, misty; *pe*, pearl; *ru*, ruby; *s*, recessive spotting; *sa* satin; *se*, short ear; *vt*, vestigial tail; *wa-1*, waved-1.

[b] B. M. Cattanach, unpublished results.

mouse origin and clearly gives low or zero levels of nondisjuncton on the basis of the test results obtained. Other characteristics of Rbs described earlier are also evident. Thus, very different frequencies of marked young were obtained in cross 5 versus 6 and in 19a versus 20a, in which different or additional genetic markers were used. The manipulation of the markers had necessitated crosses with different strains of mice and hence genetic background influences could be suspected as the cause. The differing results obtained in some repeat experiments (e.g., in crosses 2b versus 3b, 8a versus 9a, 13b versus 14b) also suggest that interanimal variation exists. There is clear evidence of reciprocal cross differences with some Rbs (e.g., in crosses 8a and 9a versus 8b and 9b) and markedly different results for each chromosome arm were also obtained with Rb2Bnr (crosses 8a, 8b, 9a, and 9b). Of yet greater significance for the application of the Rb complementation test as an assay for nondisjunction was the finding of a maternal age effect with Rb2Bnr (not detailed in Table 14). Older matings produced significantly higher yields of marked young than younger matings[72] and in some of these matings substitution of the young females for older ones established that it is increasing maternal, rather than paternal, age that is responsible (B. M. Cattanach, unpublished results). No such effect was found with Rb6Bnr; data from the other crosses have not yet been sufficiently analyzed.

Estimates of the frequency of nondisjunction in each of the crosses have been calculated; these range from about 8 to 24%, with most values being above 10%.

The Rb complementation test has only been applied as an assay for nondisjunction in females following X-irradiation.[71] An effect upon females was investigated following doses of 0-, 50-, 100-, and 200-cGy X rays. The females were placed with males in trio matings (two females:one male) immediately following treatment and progeny recovered throughout their fertile life span were screened for marked young. Since fertility lasted only a few weeks following the higher doses, more matings had to be set up in the higher dose groups than in the zero and lower dose groups. The numbers of matings per group ranged from six to about 30. In the smaller groups at least, interanimal variation could have influenced the overall results. Two Rbs were used, Rb(4.6)2Bnr with *b* and *m* being the markers for one chromosome arm (chromosome 4), and Rb(9.14)6Bnr with *hr* and *s* marking one of its chromosome arms (chromosome 14), and these gave different responses (Table 15). With Rb2Bnr the sterilizing effect was severe, so that only a few young were obtained and, among these, there was no sign of any increase in the frequency of marked young with radiation dose. Rather,

TABLE 15. Recovery of Products of Complementary Nondisjunction following X-Irradiation of Females Using the Rb Complementation Test[a]

Dose, rad	Marked female: Marked young	Marked female: Total young	Marked male: Marked young	Marked male: Total young	Total: Marked young	Total: Total young	Percent marked young
Rb2Bnr (*b m*)							
0	1	136	4	155	5	291	1.72
50	1	108	7	195	8	303	2.42
100	0	84	1	89	1	173	0.58
200	0	62	0	64	0	126	0.0
Rb6Bnr (*hr s*)							
0	3	194	4	527	7	721	0.97
50	2	260	4	357	8	617	1.30
100	1	178	4	262	5	440	1.14
200	6	144	5	243	11	387	2.84

[a] From Lyon.[71]

the incidence appeared to fall with increasing dose and this was explained in terms of the maternal age effect previously found with this Rb. Thus, due to the sterilizing effect of the radiation, the females that received the higher doses produced fewer litters at lower maternal ages, whereas the control and lower dose groups produced more litters at higher maternal ages, when more marked young normally appear. While this effect would indeed be expected, another explanation might be that the X-irradiation did in fact reduce the nondisjunction frequency. This would be consistent with the cytological finding made by Nijhoff and de Boer[86] with Rb4Bnr, the explanations for which have been discussed earlier (Section 2.1.3b).

With Rb6Bnr, however, significant differences in the frequencies of marked young were found at the highest dose in both of the reciprocal crosses and this would appear to indicate that X-ray exposure can increase the frequency of nondisjunction in Rb heterozygotes. Detailed analysis of the data demonstrated that the higher yield of marked young in the 200-rad group was unlikely to have been due to an age effect. It should be noted, however, that the scale of these experiments was very small and the possible influence of interanimal variation upon the results cannot really be evaluated.

3.2.1c. Assessment. The high rates of nondisjunction associated with wild mouse-derived Rbs have suggested that such Rbs could be useful tools for nondisjunction screening following mutagen/carcinogen exposure. The rationale for the Rb complementation test is that the segregational impairment found in Rb heterozygotes might be more vulnerable to further disturbance than that which may occur with normal autosomal bivalents. Higher degrees of response to nondisjunction-inducing agents might therefore be found. The limited data so far obtained are equivocal on this point; the experiment using Rb(9.14)6Bnr suggested that high rates of nondisjunction involving chromosome 14 were induced by X-irradiation, but with Rb(4.6)2Bnr no indication of any increase over the background nondisjunction levels for chromosome 4 were found.

The method is clearly an effective one for detecting any nondisjunction associated with Rb heterozygosity, but the numerous factors found to modify these events and the uncertainty as to whether Rb-associated nondisjunction is a valid model for the process in normal chromosomes renders the method suspect for screening. In view of the encouraging results with one Rb, further investigation is warranted, however.

Because of the wide range of marker genes available with most chromosomes, the problems encountered with the markers needed for the NSA method can be avoided.

3.2.2. The Rb (Tester) Method

3.2.2a. Methods. This test is a variation of the Rb complementation test method, which differs in one important respect. Nondisjunction is assessed in chromosomally normal (acrocentric chromosome) mice and therefore the uncertainties concerning the causes of Rb-associated nondisjunction are avoided. However, detection of nondisjunction products in viable, chromosomally balanced young is accomplished as before with the use of Rbs, which are introduced into the crosses in tester mice to provide high levels of chromosome gains and losses, which can complement any losses or gains in the chromosomally normal mice being investigated.[21] All basic procedures (crosses, markers, etc.) may be the same as in the Rb complementation test method, but only two rather than four stocks need to be maintained for any one type of test, i.e., for gains or losses. Nondisjunction frequencies and frequencies of gametes with specific additional or missing chromosomes may be calculated from the frequencies of marked young and the known frequencies of nondisjunction in the tester parents.

3.2.2b. Results. To date the Rb method has only been used in preliminary experiments to screen for X-ray-induced nondisjunction in female mice.[21] The tester parents were heterozygous for only a single

TABLE 16. Chromosome 1 Losses and Gains Detected with the Rb Method following X-Irradiation of Females[a]

Event	Dose cGy	Progeny: Number marked	Progeny: Total	Progeny: Percent marked	Calculated frequency of abnormal gametes, %
Loss	0	0	4385	0.0	0.0
	100	1	2168	0.05 ± 0.05	0.71
	200	3	2038	0.15 ± 0.09	2.12
	400	7	2706	0.26 ± 0.10[b]	3.50
Gain	400	1	3353	0.03 ± 0.03[b]	0.46

[a] From Cattanach *et al.*[21]
[b] Significantly different (p = 0.026).

Rb, Rb(1.3)1Bnr, and events involving only one chromosome, chromosome 1, were detected with *ln* or both *fz* and *ln* as markers. Two types of test were conducted (Table 16); one screened for chromosome 1 loss due to nondisjunction or other causes following the irradiation of chromosomally normal female mice with three different X-ray doses; the other screened for chromosome 1 gains following female irradiation with one high X-ray dose (Table 16). Losses were detected as *ln* young appearing among the wild-type progeny of irradiated wild-type (+ + +/+ + +) females mated with Rb + *ln*/+ *fz ln* tester males; gains were detected as *fz ln* young appearing among the wild-type progeny of irradiated + *fz ln*/ + *fz ln* females mated with Rb + +/+ + + tester males. The females were left with tester males in trio matings for only the first week postirradiation. Hence all their progeny derived from treated mature oocytes. Procedures were adopted in both crosses to reduce the effects of any interanimal variation in the Rb-associated nondisjunction among the tester males.

The data obtained (Table 16) clearly show that chromosome 1 losses increase significantly with dose (p = 0.00045) and this increase closely approximates linearity (p = 0.92 for goodness of fit to a straight line; slope $6.44 \pm 1.94 \times 10^{-6}$). Large numbers of animals were produced in order to demonstrate this response, but it may be noted that the estimated frequencies of gametes lacking a chromosome were remarkably high at the higher doses. Comparison with appropriate NSA method data indicates that the recovery frequency of marked young is similar at equivalent doses.[21] However, the recovery rate of XO females is much lower as a proportion of total progeny reared (including males) and, if comparison is made on the basis of frequency of gametes lacking a chromosome, maternal chromosome 1 loss following 400-cGy X-rays is about 16 times higher than sex-chromosome loss.

Only one chromosome 1 gain was detected with the Rb method (Table 16), and therefore it could have been of spontaneous origin rather than radiation-induced. The frequency of occurrence based on this single case was significantly lower ($p = 0.026$) than the frequency of chromosome 1 loss at the same dose. This suggests that most chromosome 1 losses, like sex-chromosome losses, derive from events other than nondisjunction. The single *fz ln* animal recovered is unique, however; it represents the first established example of autosomal nondisjunction in chromosomally normal mice, although the occurrence of *d se* double mutations in certain specific locus experiments has been attributed to this cause.[104]

3.2.2c. Assessment. It is evident from comparison of appropriate NSA and Rb data[21,101] that the Rb method compares very favorably with the NSA method for the detection of X-ray-induced chromosome loss from mature oocytes of female mice. However, like the NSA method, it does not seem to be very effective in its present form for detecting chromosomal gains attributable to nondisjunction, although this presupposes that the X-irradiation is in fact causing nondisjunction, as the cytogenetic methods suggest (Sections 2.1.3 and 2.3.1).

It must be recognized, however, that the Rb tester method has potential for further development. The data were obtained using only one Rb with only one of its chromosome arms marked, thus allowing the detection of gains or losses in only one chromosome. If both arms of one Rb are marked, which is perfectly feasible, the effectiveness of the method can be doubled. If two doubly marked Rbs are used and do not seriously impair fertility, then the effectiveness of the test can be further enhanced. Furthermore, as suggested for the Rb complementation test, semidominant markers could be used with certain chromosomes to allow the detection of both gains and losses in single crosses. With the wide range of Rbs available, every chromosome in the complement could be subjected to test.

3.2.3. The Monobrachial Homology (MBH) Method

Like the Rb method, this method screens for nondisjunction and/or chromosome loss in chromosomally normal mice and uses tester animals carrying Rbs. However, instead of carrying one or more independent Rbs in the heterozygous condition, the tester animals are doubly heterozygous for two Rbs which share a common arm[21] (monobrachial homology). This chromosome constitution forces a high frequency of nondisjunction for the common chromosome arm, and for this reason has been used as a means of producing specific trisomies for developmental studies.[47] Exact figures for the nondisjunction produced are difficult to deduce, however: metaphase II chromosome counts tend to underestimate the frequency because doubly aneusomic cells cannot easily be distinguished from normal cells and the trisomic fetus frequencies may also be underestimated due to loss of some zygotes. Nevertheless, estimates of nondisjunction range from 28 to 42%.[47]

3.2.3a. Methods. All basic procedures (crosses, markers) for the MBH method may be the same as with the preceding Rb methods and, given that various parameters are known, the frequencies of gametes with specific additional or missing chromosomes can be calculated from the frequency of marked young.

3.2.3b. Results. The method has only been used in one set of preliminary experiments, which screened for nondisjunction following X-irradiation of males.[18] The two Rbs used were Rb(1.3)1Bnr and Rb(1.15)2Ct, which have monobrachial homology for chromosome 1. The markers were *ln* or both *fz* and *ln*. Two types of test were conducted: one screened for chromosome 1 loss detected as *ln* young in crosses of X-irradiated (300 cGy) wild-type (+ +/+ +) males mated with Rb1 *ln*/Rb2 *ln* tester females in trio mating; the other screened for chromosome 1 gains detected as *fz ln* young in crosses of X-irradiated + *fz ln*/+ *fz ln* males similarly mated with Rb1 + +/Rb2 + + tester females. The germ cells tested were spermatozoa and spermatids (weeks 1–3 postirradiation) in cross 1 and spermatocytes (weeks 4–6 postirradiation) in cross 2. Since the losses were recovered from postmeiotic germ cells, they cannot derive from nondisjunction. However, the data on these germ cell stages are presented here only to demonstrate the method.

The results (Table 17) clearly show that the MBH method is quite effective in detecting chromosome loss, although, as with the other genetic methods, large numbers of animals are required. Data on the Rb1/Rb2 compound that would allow estimates of the frequency of gametes lacking chromosome 1 to be calculated are not yet available. However, using the nondisjunction frequencies obtained by Gropp *et al.*[47] with other Rb compounds as guidelines and ignoring nondisjunction other than that forced by the monobrachial homology, it has been calculated[21] that the MBH method is about two to three times more effective than the Rb tester method for detecting chromosome 1 losses and this will also be true for chromosome 1 gains due to nondisjunction. Using the data of Gropp *et al.*[47] as a guideline for the chromosome 1 nondisjunction frequencies in the tester animals, it has

TABLE 17. Chromosome 1 Losses and Gains Detected with the Rb Method following X-Irradiation of Males[a]

Event	Cell stage sampled	Dose cGy	Progeny: Number marked	Progeny: Total	Progeny: Percent marked	Calculated frequency of abnormal gametes, %
Loss	—	0	2	4150	0.05 ± 0.03	0.20
	Spermatozoa + spermatids	300	9	2586	0.35 ± 0.12	1.36
Gain	Spermatocytes	75	0	1477	0.0	0.0
		100	0	943	0.0	0.0

[a] From Cattanach.[18]

further been concluded that chromosome 1 loss from X-irradiated postmeiotic germ cell stages in the male is very similar to that for the sex chromosomes as detected by the NSA method.[21]

No evidence of nondisjunction was obtained following spermatocyte irradiation. The data are limited, however. This was partly due to the low fertility of the cross, deriving both from the high rates of spermatocyte killing by the X-irradiation and from the high rates of zygotic loss in the tester females attributable to nondisjunction.

3.2.3c. Assessment. The range of Rbs available is now so large that compounds can be made that have monobrachial homologies for any selected chromosome in the mouse chromosome complement. In principle, therefore, every chromosome could be investigated in nondisjunction screening experiments using the MBH method. Since the method appears to be more efficient than the Rb method and since the autosomes may be more sensitive to loss than the sex chromosomes, the MBH method might appear to be the genetic method of choice for nondisjunction screening. It suffers two major limitations, however: first, Rb compounds tend to be sterile in the male.[31,46] Therefore, the method can only be applied for investigating nondisjunction/chromosome loss in males. Second, although it is theoretically possible that female mice that are doubly compound for two sets of Rbs having different monobrachial homologies may be fertile, the method may in practice only allow one chromosome to be investigated at a time. The method therefore has less potential than the Rb method. A further problem is that some Rb-associated nondisjunction other than that forced by the monobrachial homology does occur in Rb compounds and, since the frequency of this event can only be assessed with considerable difficulty, any nondisjunction induced in the test parents of the cross can only be roughly calculated.

3.2.4. A Triple Compound Method—Under Development

A disadvantage of the Rb and MBH methods is that even with the high rates of nondisjunction in the tester parents, only a small proportion of the nullisomic and disomic gametes that contribute toward the formation of zygotes are detected and for this reason large numbers of progeny from the test crosses have to be produced in order to detect any increases in the nondisjunction frequency. In principle, a more efficient tester system might be one in which three Rbs are combined in a single compound with tribrachial homology. Searle and Beechey[111] are attempting to produce such a stock by combining Rb(4.6)2Bnr, Rb(4.15)Rma, and Rb(6.15)1Ald. Such a compound would generate only aneuploid gametes, nullosomic or disomic for chromosome 4, 6, or 15. The chances of any nullisomic or disomic gamete from the chromosomally normal parent of the test cross uniting with a complementary disomic or nullisomic gamete from the tester parent will therefore be higher than with any other test system. However, first, it has to be established that such a tester stock can be produced and be fertile, and, second, it will have to be shown that the normal sterility of the test cross, which results from the absence of any chromosomally balanced gametes from the tester parents, will allow the survival and subsequent detection of those few balanced zygotes that may appear as a consequence of complementary nondisjunction/chromosome loss from the normal parents of the cross.

4. Summary and Conclusions

The foregoing sections indicate the diversity of approaches available or potentially available for investigating meiotic chromosome nondisjunction. As is clear, some of these have yet to be fully developed, while others require validation. Of those that remain, the metaphase II gametocyte test, the first cleavage test, and tests upon pre- and postimplantation embryos have immediate value for nondisjunction testing.

A number of criteria are immediately obvious as influencing the choice of a nondisjunction test. These are (1) the quality of data generated, (2) the reliability of the test, (3) the time and effort to complete a test, and (4) the overall cost. The particular stress laid on any one of the criteria will necessarily depend on the purpose of the proposed experiment. For example, routine screening of potential nondisjunction-inducing agents requires methods that are both rapid and relatively inexpensive. Investigation of the nondisjunction process itself or of some of the consequences may, on the other hand, require methods that are more time-consuming or expensive.

For routine testing, clearly the metaphase II gametocyte and first cleavage division tests must be favored. The former is particularly valuable in males where very many scorable cells may be obtained from individual animals. Costs in terms of animal care and number of animals required are therefore minimal. The cytological preparations require little time and slide analysis is relatively easy. A disadvantage of the method is that chromosome identification, other than the X and Y chromosome, is impossible. In females, the metaphase II test is somewhat more laborious in that even with superovulation relatively few oocytes will be obtained from individual animals. Otherwise the method remains quick and reliable. No chromosome identification is possible in mouse metaphase II oocytes. The use of other laboratory mammals, such as the Chinese hamster, with a more varied karyotype than the mouse can permit a degree of chromosome identification at metaphase II.[121,122] An important restriction of the metaphase II test is that tertiary trisomies resulting from mutagen-induced chromosome rearrangements cannot be distinguished from trisomies deriving from primary nondisjunction.

The first cleavage division test has many of the same advantages as the metaphase II gametocyte method in terms of speed, labor requirements, and reliability. The time taken to reach the stage of slide preparation is slightly longer than for metaphase II oocytes, taking 5 days from PMS injection to cell fixation. However, chromosome analysis is somewhat easier, since the chromosomes do not have the irregular, convoluted appearance of metaphase II chromosomes. Moreover, both parental sets of chromosomes can be assessed separately (more reliably if a chromosome marker is used). Additionally, identification of chromosomes is theoretically possible if G or Q banding is used. Both the metaphase II and first cleavage division methods are affected by the problem of artifactual chromosome loss. Hence, only hyperhaploid and hyperdiploid cells can be considered for providing information on nondisjunction.

The remaining pre- and the postimplantation methods for detecting nondisjunction are laborious and therefore not to be recommended for routine screening. Thus, Chandley and Speed[23] estimated that only about 40 midterm fetuses could be processed to the slide-making stage in a single working day. For their experiment[23] on X-irradiation of spermatogenic stages in which over 2000 embryos were examined, a commitment of approximately 11 working weeks was required, and this was for slide preparation alone. Nevertheless, analysis of postimplantation embryos provides a reliable and clear method of demonstrating the occurrence of nondisjunction. These embryos have the additional advantages of providing sufficient material to identify unequivocally artifactual chromosome loss or gain and to karyotype aneuploidies by chromosome banding techniques.

The genetic tests, for the moment, are of little practical application. So far none have been developed to allow more than one chromosome to be studied at a time, that is, only 1/20 of the haploid genome. Hence they are indisputably inefficient. Moreover, they have as yet only proved effective in detecting chromosome loss, which may, of course, derive from events other than nondisjunction. However, it is conceivable that the Rb method may establish itself as a nondisjunction test. Its potential in this regard lies in the fact that it could move the study of nondisjunction from the skilled expertise of the cytology laboratory to the routine animal care of the toxicology laboratory, where the requirement for appropriate sized animal facilities is less of a limiting factor. The low spontaneous levels of nondisjunction suggested may prove to be an advantage to this method because, as Russell[102] pointed out for the NSA test, detection of a very few marked progeny provides clear evidence of nondisjunction. Although costs for the Rb test are obviously high in terms of animal care, this factor is offset, in part at least, by the costs of cytogenetic approaches, which require skilled, and therefore expensive, staff.

For the future, automated systems may offer the best prospects for routine nondisjunction screening. At present, only DNA content estimations of spermatids/spermatozoa have been attempted using the fully automated system of flow cytometry. However, increasingly sophisticated systems are being developed with the potential to scan slides automatically, record the positions of metaphase plates, attempt chromosome counts, and karyotype cells. It seems likely that in the near future much of the tedium inherent in many of the cytogenetic tests will be reduced by the introduction of at least partial automation.

Acknowledgments

We thank Dr. M. F. Lyon for her critical comments on the manuscript. We are also indebted to numerous colleagues who provided us with both preprints and unpublished material, and to P. Crocker for her patience and secretarial skills.

5. References

1. K. Adachi and T. H. Ingalls, Ovum aging and pH imbalance as a cause of chromosomal anomalies in the hamster, *Science 194*, 946–948 (1976).
2. I.-D. Adler, Male germ cell cytogenetics, in: *Cytogenetic Assays of Environmental Mutagens* (T. C. Hsu, ed.), pp. 249–276, Allanheld, Osmun (1982).
3. I.-D. Adler, A. Schmaltz, R. Rathenberg, D. Müller, F. F. Strasser, and R. Perret, Cytogenetic study in spermatocytes of mice and Chinese hamsters after treatment with Isoniazid (INH), *Hum. Genet. 42*, 50–54 (1978).
4. E. Alberman, P. E. Polani, J. A. Fraser-Roberts, C. C. Spicer, M. Elliot and E. Armstrong, Parental exposure to X-irradiation and Down's syndrome, *Ann. Hum. Genet. 36*, 195–208 (1972).
5. V. S. Baranov and N. A. Chebotar, Mutagenic effects of preovulatory administration of pyrimethamine on preimplantation rat embryos, *Cytol Genet. (USSR) 14*, 17–21 (1980).
6. P. Barlow and C. G. Vosa, The Y chromosome in human spermatozoa, *Nature 226*, 961–962 (1970).
7. A. Basler, A. Brucklacher, F. Nobis and G. Röhrborn, Comparative investigations with trypaflavin in metaphase II oocytes and in dominant lethal assay, *Hum. Genet. 40*, 87–92 (1977).
8. R. A. Beatty, F-bodies as Y chromosome markers in mature human sperm heads: A quantitative approach, *Cytogenet. Cell Genet. 18*, 33–49 (1977).
9. K. Becker and J. Schöneich, Expression of genetic damage induced by alkylating agents in germ cells of female mice, *Mutat. Res. 92*, 447–464 (1982).
10. C. V. Beechey, X–Y chromosome dissociation and sterility in the mouse, *Cytogenet. Cell Genet. 12*, 60–67 (1973).
11. D. J. Bond and A. C. Chandley, *Aneuploidy*, (Oxford Monographs on Medical Genetics No. 11), Oxford University Press, Oxford, (1983).
12. G. Breckon, A modified hypotonic treatment for increasing the frequency and quality of meiotic metaphases from spermatocytes of the Syrian hamster, *Stain Technol. 57*, 349–354 (1982).
13. J. D. Brook, The effect of 4 CBM on germ cells of the mouse, *Mutat. Res. 100*, 305–308 (1982).
14. B. M. Cattanach, XO mice, *Genet. Res. (Camb.) 3*, 487–490 (1962).
15. B. M. Cattanach, Induction of paternal sex-chromosome losses and deletions and of autosomal gene mutations by the treatment of mouse post-meiotic germ cells with triethylenemelamine, *Mutat. Res. 4*, 73–82 (1967).
16. B. M. Cattanach, Genetic disorders of sex determination in mice and other mammals, in: *Birth Defects* (A. G. Motulsky and W. Lenz, eds.), pp. 129–141, Exerpta Medica, Amsterdam (1974).
17. B. M. Cattanach, Non-disjunction tests with Robertsonian translocations, *Mouse News Letter 66*, 62–63 (1982).
18. B. M. Cattanach, Non-disjunction tests with Robertsonian translocations using the gene marker method, *Mouse News Letter 68*, 70 (1984).
19. B. M. Cattanach and H. Moseley, Non-disjunction and reduced fertility caused by

the tobacco mouse metacentric chromosomes, *Cytogenet. Cell Genet. 12*, 264–287 (1973).
20. B. M. Cattanach, I. Murray, and J. M. Tracey, Non-disjunction in heterozygotes for Robertsonian translocations, *Mouse News Letter 54*, 37–38 (1976).
21. B. M. Cattanach, D. Papworth and M. Kirk, Genetic tests for autosomal non-disjunction and chromosome loss in mice, *Mutat. Res. 126*, 189–204 (1984).
22. A. H. Cawood and G. Breckon, Synaptonemal complexes as indicators of induced structural change in chromosomes after irradiation of spermatogonia, *Mutat. Res. 122*, 149–154 (1983).
23. A. C. Chandley and R. M. Speed, Testing for non-disjunction in the mouse, *Environ. Health Perspect. 31*, 123–129 (1979).
24. N. A. Chebotar, Cytogenetic and morphogenetic changes in oogenesis and embryogenesis of albino rats induced by chloridin and 2,4,5-trichlorophenoxyacetic acid in the pre-ovulatory phase of meiosis, *Sov. Genet. 16*, 781–787 (1980).
25. A. H. C. Choi, Three dimensional reconstruction of quadrivalents and mapping of translocation breakpoints of the mouse translocations T(2;8)26H and T(9;17)138Ca, *Can. J. Genet. Cytol. 22*, 261–270 (1980).
26. P. de Boer and M. E. A. B. van Beek, Meiosis of T70H translocation trisomic male mice. I. Meiotic configurations and segregation, *Chromosoma 87*, 303–313 (1982).
27. P. de Boer and F. A. van der Hoeven, The use of translocation-derived "marker-bivalents" for studying the origin of meiotic instability in female mice, *Cytogenet. Cell Genet. 26*, 49–58 (1980).
28. R. P. Donahue, Cytogenetic analysis of the first cleavage division in mouse embryos, *Proc. Natl. Acad. Sci. USA 69*, 74–77 (1972).
29. R. G. Edwards, Observations on meiosis in normal males and females, in: *Human Population Cytogenetics, Pfizer Medical Monographs 5* (P. A. Jacobs, W. H. Price, and P. Law, eds.), pp. 9–21, Edinburgh University Press, Edinburgh (1970).
30. R. G. Edwards and A. H. Gates, Timing of the stages of the maturation divisions, ovulation, fertilization and the first cleavage of eggs of adult mice treated with gonadotrophins, *J. Endocrinol. 18*, 292–304 (1959).
31. E. P. Evans, Male sterility and double heterozygosity for Robertsonian translocations in mouse, in: *Chromosomes Today* (P. L. Pearson and K. R. Lewis, eds.), pp. 75–81, Wiley, New York (1976).
32. E. P. Evans, G. Breckon, and C. E, Ford, An air-drying method for meiotic preparations from mammalian testes, *Cytogenetics 3*, 289–294 (1964).
33. E. P. Evans, M. D. Burtenshaw, and C. E. Ford, Chromosomes of mouse embryos and newborn young: Preparations from membranes and tail tips, *Stain Technol. 47*, 229–234 (1972).
34. E. P. Evans, M. Burtenshaw, and B. M. Cattanach, Cytological evidence for meiotic crossing over between the X and Y chromosomes in male mice carrying the sex-reversing (*Sxr*) factor, *Nature 300*, 443–445 (1982).
35. C. E. Ford, The time in development at which gross genome unbalance is expressed, in: *The Early Development of Mammals* (M. Balls and A. E. Wild, eds.), pp. 285–304, Cambridge University Press, Cambridge (1975).
36. C. E. Ford, Non-disjunction, in: *Trisomy 21* (G. R. Burgio, M. Fraccaro, L. Tiepolo, and U. Wolf, eds.), pp. 103–143, Springer, Berlin (1981).
37. C. E. Ford and E. P. Evans, Non-expression of genome unbalance in haplophase and early diplophase of the mouse and incidence of karyotypic abnormality in post-implantation embryos, in: *Chromosomal Errors in Relation to Reproductive Failure* (A. Bouc and C. Thibault, eds.), pp. 271–285, Institut de la Santé et de la Recherche Médicale-Paris (1973).

38. L. R. Fraser and I. Maudlin, Analysis of aneuploidy in first-cleavage mouse embryos fertilized *in vitro* and *in vivo*, *Environ. Health Perspect. 31*, 141–149 (1979).
39. L. Fraser and I. Maudlin, Incidence of polyspermy in mouse eggs fertilized *in vivo* and *in vitro* after administration of progesterone and oestradiol, *J. Reprod. Fertil. 55*, 407–410 (1979).
40. A. H. Gates, C. H. Donaldson and D. Melissa. Oocyte aneuploidy screening using superovulating prepubertal mice: Effect of methotrexate, *Teratology 24*, 321–327 (1981).
41. C. B. Gillies, Synaptonemal complex and chromosome structure, *Annu. Rev. Genet. 9*, 91–109 (1975).
42. M. S. Golbus, The influence of strain, maternal age, and method of maturation on mouse oocyte aneuploidy, *Cytogenet. Cell Genet. 31*, 84–90 (1981).
43. R. G. Gosden, Chromosomal anomalies of preimplantation mouse embryos in relation to maternal age, *J. Reprod. Fertil. 35*, 351–354 (1973).
44. R. G. Gosden and D. E. Walters, Effects of low-dose X-irradiation on chromosomal non-disjunction in aged mice, *Nature 248*, 54–56 (1974).
45. A. Gropp, Value of an animal model for trisomy, *Virchows Arch. Pathol. Anat. 395*, 117–131 (1982).
46. A. Gropp and H. Winking, Robertsonian translocations: Cytology, meiosis, segregation patterns and biological consequences of heterozygosity, in: *Biology of the House Mouse* (R. J. Berry, ed.), pp. 141–181, Academic Press, London (1981).
47. A. Gropp, V. Kolbus, and D. Giers, Systematic approach to the study of trisomy in the mouse, *Cytogenet. Cell Genet. 14*, 42–62 (1975).
48. I. Hansmann, Induced chromosomal aberrations in pronuclei, 2-cell stages and morulae of mice, *Mutat. Res. 20*, 353–367 (1973).
49. I. Hansmann, Chromosome aberrations in metaphase II oocytes. Stage sensitivity in the mouse oogenesis to amethopterin and cyclophosphamide, *Mutat. Res. 22*, 175–191 (1974).
50. I. Hansmann and E. El Nahass, Incidence of non-disjunction in mouse oocytes, *Cytogenet. Cell Genet. 24*, 115–121 (1979).
51. I. Hansmann and H. D. Probeck, Chromosomal imbalance in ovulated oocytes from Syrian hamsters (*Mesocricetus auratus*) and Chinese hamsters (*Cricetulus griseus*), *Cytogenet. Cell Genet. 23*, 70–76 (1979).
52. I. Hansmann and H. D. Probeck, The induction of non-disjunction by irradiation in mammalian oogenesis and spermatogenesis, *Mutat. Res. 61*, 69–76 (1979).
53. I. Hansmann and G. Röhrborn, Chromosome aberrations in preimplantation stages of mice after treatment with triazoquinone, *Humangenetik, 18* 101–109 (1973).
54. I. Hansmann, J. Neher, and G. Röhrborn, Chromosome aberrations in metaphase II oocytes of Chinese hamster (*Cricetulus griseus*). I. The sensitivity of the preovulatory phase to triaziquone, *Mutat. Res. 25*, 347–359 (1974).
55. I. Hansmann, R. Zmarsly, H. D. Probeck, J. Schäfer, and J. Jenderney, Aneuploidy in mouse fetuses after paternal exposure to X-rays, *Nature 280*, 228–229 (1979).
56. I. Hansmann, J. Jenderney, and H. D. Probeck, Non-disjunction and chromosome breakage in mouse oocytes after various X-ray doses, *Hum. Genet. 61*, 190–192 (1982).
57. J. H. Heller, Human chromosome abnormalities as related to physical and mental dysfunction, *J. Hered. 60*, 239–248 (1969).
58. S. A. Henderson and R. G. Edwards, Chiasma frequency and maternal age in mammals, *Nature 218*, 22–28 (1968).
59. P. B. Holm and S. W. Rasmussen, Human meiosis I. The human pachytene

karyotype analysed by three dimensional reconstruction of the synaptonemal complex, *Carlsberg Res. Commun. 42*, 283–323 (1977).
60. P. B. Holm, S. W. Rasmussen, and D. von Wettstein, The possible contribution of electron microscopy to the understanding of the mechanism of non-disjunction in man, *Mutat. Res. 61*, 115–119 (1979).
61. E. B. Hook and P. K. Cross, Paternal age and Down's syndrome genotypes diagnosed prenatally: No association in New York State data, *Hum. Genet. 62*, 167–174 (1982).
62. T. H. Ingalls and M. Yamamoto, Hypoxia as a chromosomal mutagen, *Arch. Environ. Health 24*, 305–315 (1972).
63. T. H. Ingalls, T. Shimada, and M. Yamamoto, Hypoxia and induced mutations in Syrian (golden) hamsters, *Arch. Environ. Health 31*, 153–159 (1976).
64. Y. Kamiguchi, K. Funaki, and K. Mikamo, A new technique for chromosome study of murine oocytes, *Proc. Jpn. Acad. 52*, 316–319 (1976).
65. R. W. Kapp, Jr., Detection of aneuploidy in human sperm, *Environ. Health Perspect. 31*, 27–72 (1979).
66. R. W. Kapp, Jr., and C. B. Jacobson, Analysis of human spermatozoa for chromosomal non-disjunction, *Teratogen. Cancerogen. Mutagen. 1*, 193–211 (1980).
67. R. W. Kapp, Jr., D. J. Picciano, and C. B. Jacobson, Y-Chromosomal non-disjunction in dibromochloropropane-exposed workmen, *Mutat. Res. 64*, 47–51 (1979).
68. M. H. Kaufman, The chromosome complement of single-pronuclear haploid mouse embryos following activation by ethanol treatment, *J. Embryol. Exp. Morphol. 71*, 139–154 (1982).
69. M. H. Kaufman, Ethanol-induced chromosomal abnormalities at conception, *Nature 302*, 258–260 (1983).
70. A Léonard and J. H. Schröder, Incidence of XO mice after X-irradiation of spermatogonia, *Mol. Gen. Genet. 101*, 116–119 (1968).
71. M. F. Lyon, The use of Robertsonian translocations for studies on non-disjunction, in: *Chromosome Damage in Man* (T. Isihara, ed.), pp. 329–346, Liss, New York (1983).
72. M. F. Lyon, H. C. Ward, and G. M. Simpson, A genetic method for measuring non-disjunction in mice with Robertsonian translocations, *Genet. Res. (Camb.) 26*, 283–295 (1976).
73. M. F. Lyon, B. M. Cattanach, and H. M. Charlton, Genes affecting sex-differentiation in mammals, in: *Mechanisms of Sex Differentiation in Animals and Man* (C. R. Austin and R. J. Edwards, eds.), pp. 329–386, Academic Press, New York (1981).
74. J. B. Mailhes, Methylmercury effects on Syrian hamster metaphase II oocyte chromosomes, *Environ. Mutagen. 5*, 679–686 (1983).
75. R. H. Martin, W. Balkan, and K. Burns, Cytogenetic analysis of Q-banded pronuclear chromosomes in fertilized Syrian hamster eggs, *Cytogenet. Cell Genet. 35*, 41–45 (1983).
76. P. A. Martin-De-Leon and M. L. Boice, Spontaneous heteroploidy in one-cell mouse embryos, *Cytogenet. Cell Genet. 35*, 57–63 (1983).
77. I. Maudlin and L. R. Fraser, Maternal age and the incidence of aneuploidy in first cleavage mouse embryos, *J. Reprod. Fertil. 54*, 423–426 (1978).
78. C. Max, Cytological investigation of embryos in low-dose X-irradiated young and old female inbred mice, *Hereditas 85*, 199–206 (1977).
79. M. L. Meistrich, S. Lake, L. L. Steinmetz, and B. L. Gledhill, Increased nuclear DNA of testis cells and spermatozoa from mice with irregular meiotic segregation, *Mutat. Res. 49*, 397–405 (1978).
80. M. L. Meistrich, W. Göhde, R. A. White, and J. L. Longtin, "Cytogenetic" studies of spermatids of mice carrying Cattanach's translocation by flow cytometry, *Chromosoma 74*, 141–151 (1979).

81. K. Mikamo and Y. Kamiguchi, The primary incidences of spontaneous chromosomal anomalies and their origins and causal mechanisms in the Chinese hamster, *Mutat. Res. 108*, 265–278 (1983).
82. W. D. Morrison, V. Huff, L. G. Littlefield, and R. J. DuFrain, Chromosome preparations from rabbit preovulatory oocytes, *Mutat. Res. 119*, 169–175 (1983).
83. M. J. Moses, New cytogenetic studies on mammalian meiosis, in: *Animal Models in Human Reproduction* (M. Serio and L. Martini, eds.), pp. 169–190, Raven Press, New York (1980).
84. J. Moutschen, Mutagenesis with methyl methanesulfonate in the mouse, *Mutat. Res. 8*, 581–589 (1969).
85. M. N. Nesbitt and R. P. Donahue, Chromosome banding patterns in preimplantation mouse embryos, *Science 197*, 805–806 (1972).
86. J. H. Nijhoff and P. de Boer, Radiation induced meiotic autosomal non-disjunction in male mice. The effects of low doses of fission neutrons and X-rays in meiosis I and II of a Robertsonian translocation heterozygote, *Mutat. Res. 72*, 431–446.
87. F. Pacchierotti, D. Bellincampi, and D. Civitareale, Cytogenetic observations in mouse secondary spermatocytes on numerical and structural chromosome aberrations induced by cyclophosphamide in various stages of spermatogenesis. *Mutat. Res. 119*, 177–183 (1983).
88. H. S. Payne and K. P. Jones, Technique for mass isolation and culture of mouse ova for cytogenetic analysis of the first cleavage mitosis, *Mutat. Res. 33*, 239–250 (1975).
89. P. L. Pearson and M. Bobrow, Fluorescent staining of the Y chromosome in meiotic stages of the human male, *J. Reprod. Fertil, 22*, 177–179 (1970).
90. P. L. Pearson, M. Bobrow, and C. G. Vosa, Technique for identifying Y chromosomes in human interphase nuclei, *Nature 226*, 78–80 (1970).
91. D. Pinkel, B. L. Gledhill, M. A. van Dilla, S. Lake, and A. J. Wyrobeck, Radiation-induced DNA content variability in mouse sperm, *Radiat. Res. 95*, 550–565 (1983).
92. P. E. Polani and G. M. Jagiello, Chiasmata, meiotic univalents, and age in relation to aneuploid imbalance in mice, *Cytogenet. Cell Genet. 16*, 505–529 (1976).
93. W. I. Reichert, I. Hansmann, and G. Röhrborn, Chromosome anomalies in mouse oocytes after irradiation, *Hum. Genet. 28*, 25–38 (1975).
94. G. W. Rieck, H. Höhn, and A. Herzog, X-trisomie beim Rind mit Anzeichen familiärer Disposition für Meiosestörungen, *Cytogenetics 9*, 401–409 (1970).
95. G. Röhrborn and I. Hansmann, Induced chromosome aberrations in unfertilized oocytes of mice, *Hum Genet. 13*, 184–198 (1971).
96. G. Röhrborn and I. Hansmann, Oral contraceptives and chromosome segregation in oocytes of mice, *Mutat. Res. 26*, 535–544 (1974).
97. G. Röhrborn, O. Kühn, I. Hansmann, and K. Thon, Induced chromosome aberrations in early embryogenesis of mice, *Humangenetik 11*, 316–322 (1971).
98. G. Röhrborn, J. Sezer, I.-D. Adler, A. Schmaltz, and I. H. Pawlowitzki, Effects of Isoniazid (INH) on the oogenesis of mice, *Hum. Genet. 42*, 55–58 (1978).
99. L. B. Russell, Chromosome aberrations in experimental mammals, *Prog. Med. Genet. 2*, 230–294 (1962).
100. L. B. Russell, The use of sex chromosome anomalies for measuring radiation effects in different germ-cell stages of the mouse, in: *Effects of Radiation on Meiotic Systems*, pp. 27–41, International Atomic Energy Agency, Vienna (1968).
101. L. B. Russell, Numerical sex chromosome anomalies in mammals: Their spontaneous occurrence and use in mutagenesis studies, in: *Chemical Mutagens* (A. Hollaender, ed.), Vol. 4, pp. 55–91, Plenum Press, New York (1976).

102. L. B. Russell, Meiotic non-disjunction in the mouse: Methodology for genetic testing and comparison with other methods, *Environ. Health Perspect. 31*, 113–128 (1979).
103. L. B. Russell and C. S. Montgomery, The incidence of sex-chromosome anomalies following irradiation of mouse spermatogonia with single or fractionated doses of X-rays, *Mutat. Res. 25*, 367–376 (1974).
104. L. B. Russell and W. L. Russell, Genetic analysis of induced deletions and of spontaneous non-disjunction involving chromosome 2 of the mouse, *J. Cell Comp. Physiol. 56*(Suppl. 1), 169–188 (1960).
105. A. Russo, F. Pacchierotti, and P. Metalli, Meiotic nondisjunction induced by fission neutrons relative to X-rays in mouse secondary spermatocytes. I. The response of different cell stages to a single radiation dose, *Mutat. Res. 108*, 359–372 (1983).
106. K. Sankaranarayanan, The role of non-disjunction in aneuploidy in man, An overview, *Mutat. Res. 61*, 1–28 (1979).
107. K. Sankaranarayanan, *Genetic Effects of Ionizing Radiation in Multicellular Eukaryotes and the Assessment of Genetic Radiation Hazards in Man*, Elsevier Biomedical Press, Amsterdam (1982).
108. A. G. Searle, Nature and consequences of induced chromosome damage in mammals, *Genetics 78*, 173–186 (1974).
109. A. G. Searle, Radiation-induced chromosome damage and assessment of genetic risk, in: *Modern Trends in Human Genetics* (A. E. H. Emery, ed.), Vol. 2, pp. 83–110, Butterworths, London (1975).
110. A. G. Searle, The genetics of sterility in the mouse, in: *Genetic Control of Gamete Production and Function* (P. G. Crosignani and B. L. Rubin, eds.), pp. 93–114, Academic Press, London (1982).
111. A. G. Searle and C. V. Beechey, The use of Robertsonian translocations in the mouse for studies on non-disjunction, *Cytogenet. Cell Genet. 33*, 81–87 (1982).
112. T. Shimada. T. Watanabe, and A. Endo, Potential mutagenicity of cadmium in mammalian oocytes, *Mutat. Res. 40*, 389–396 (1976).
113. T. Shimada, G. Watanabe, and T. H. Ingalls, Trisomies and triploidies in hamster embryos: Induction by low-pressure hypoxia and pH imbalances, *Arch. Environ. Health 35*, 101–105 (1980).
114. R. M. Speed, The effects of ageing on the meiotic chromosomes of male and female mice, *Chromosoma 64*, 241–254 (1977).
115. R. M. Speed, Meiosis in the foetal mouse ovary. I. An analysis at the light microscope level using surface-spreading, *Chromosoma 85*, 427–437 (1982).
116. R. M. Speed and A. C. Chandley, The response of germ cells of the mouse to the induction of non-disjunction by X-rays, *Mutat. Res. 84*, 409–418 (1981).
117. R. M. Speed and A. C. Chandley, Meiosis in the foetal mouse ovary. II. Oocyte development and age-related aneuploidy. Does a production line exist? *Chromosoma 88*, 184–189 (1983).
118. J. Stene, E. Stene, S. Stengel-Rutkowski, and J.-D. Murken, Paternal age and Down's syndrome. Data from prenatal diagnoses (DFG), *Hum. Genet. 59*, 119–124 (1981).
119. R. Stolla and A. Gropp, Variation of the DNA content of morphologically normal and abnormal spermatozoa in mice susceptible to irregular meiotic segregation, *J. Reprod. Fertil. 38*, 335–346 (1974).
120. R. Strausmanis, I.-B. Henrikson, M. Holberg, and C. Röhnnbäck, Lack of effect on the chromosomal non-disjunction in aged female mice after low dose X-irradiation, *Mutat. Res. 49*, 269–274 (1978).
121. S. Suguwara and K. Mikamo, An experimental approach to the analysis of mechanisms of meiotic non-disjunction and anaphase lagging in primary oocytes, *Cytogenet. Cell Genet. 28*, 251–264 (1980).

122. S. Sugawara and K. Mikamo. Absence of correlation between changed chiasma configurations and meiotic non-disjunction in the aged female Chinese hamsters, *Cytogenet. Cell Genet. 35*, 34–40 (1983).
123. A. T. Sumner and J. A. Robinson, A difference in dry mass between heads of X- and Y-bearing human spermatozoa, *J. Reprod. Fertil. 48*, 9–15 (1976).
124. A. T. Sumner, J. A. Robinson, and H. J. Evans, Distinguishing between X, Y and YY-bearing human spermatozoa by fluorescence and DNA content, *Nature New Biol. 229*, 231–233 (1971).
125. G. Szemere and A. C. Chandley, Trisomy and triploidy induced by X-irradiation of mouse spermatocytes, *Mutat. Res. 33*, 229–238 (1975).
126. G. Szemere and I. Marczinovits, EMS and MMS-induced meiotic non-disjunction in the male mouse, in: *Progress in Genetic Toxicology* (D. Scott, B. A. Bridges, and F. H. Sobels, eds.), pp. 215–222, Elsevier/North-Holland Biomedical Press, Amsterdam (1977).
127. N. Tanaka, Studies on chemical induction of chromosomal aberrations in postcopulation germ cells and zygotes of female mice. I. Comparative studies on the frequency of first-cleavage chromosomal aberrations and dominant lethal mutations, *Jpn. J. Genet. 56*, 117–129 (1981).
128. N. Tanaka, M. Katoh, and S. Iwahara, Formation of chromosome-type aberrations at the first cleavage after MMS treatment in late spermatids of mice, *Cytogenet. Cell Genet. 31*, 145–152 (1981).
129. A. K. Tarkowski, An air-drying method for chromosome preparation from mouse eggs, *Cytogenetics 5*, 394–400 (1966).
130. A. D. Tates, *Microtus oeconomus* (Rodentia), a useful mammal for studying the induction of sex-chromosome non-disjunction and diploid gametes in male germ cells, *Environ. Health Prespect. 31*, 151–159 (1979).
131. A. D. Tates and N. de Vogel, Further studies on effects of X-irradiation of prespermatid stages of the Northern vole *Microtus oeconomus*. Low induction of sex-chromosomal non-disjunction and very high induction of diploid spermatocytes, *Mutat. Res. 82*, 323–330 (1981).
132. A. D. Tates, P. L. Pearson, M. V. D. Ploeg, and N. de Vogel, The induction of sex chromosomal non-disjunction and diploid spermatids following X-irradiation of prespermatid stages in the northern vole, *Microtus oeconomus*, *Mutat. Res. 61*, 87–101 (1979).
133. C. Tease, Similar dose-related chromosome non-disjunction in young and old female mice after X-irradiation, *Mutat. Res. 95*, 287–296 (1982).
134. C. Tease, Radiation induced chromosome non-disjunction in oocytes stimulated by different doses of superovulating hormones, *Mutat. Res. 105*, 95–100 (1982).
135. U. Tetteborn and A. Gropp, Meiotic non-disjunction in mice and mouse hybrids, *Cytogenetics 9*, 272–283 (1970).
136. I. A. Uchida and C. P. V. Freeman, Radiation-induced non-disjunction in oocytes of aged mice, *Nature 265*, 186–187 (1977).
137. I. A. Uchida and C. P. V. Lee, Radiation-induced non-disjunction in mouse oocytes, *Nature 250*, 601–602 (1974).
138. T. Watanabe and A. Endo, Chromosome analysis of preimplantation embryos after cadmium treatment of oocytes at meiosis I, *Environ. Mutagen. 4*, 563–567 (1982).
139. T. Watanabe, T. Shimada, and A. Endo, Mutagenic effects of cadmium on mammalian oocyte chromosomes, *Mutat. Res. 67*, 349–356 (1979).
140. T. Watanabe, T. Shimada, and A. Endo, Effects of mercury compounds on ovulation and meiotic and mitotic chromosomes in female golden hamsters, *Teratology 25*, 381–384 (1982).

141. P. J. J. Wauben-Penris and S. C. van Buul-Offers, Meiotic non-disjunction in male Snell dwarf mice, *J. Hered. 73*, 365–369 (1982).
142. W. J. Welshons and L. B. Russell, The Y-chromosome as the bearer of male determining factors in the mouse, *Proc. Natl. Acad. Sci USA 45*, 560–566 (1959).
143. M. Westergaard and D. von Wettstein, The synaptinemal complex, *Annu. Rev. Genet. 6*, 71–110 (1972).
144. J. Wróblewska, Chromosome preparation from mouse embryos during early organogenesis: Dissociation after fixation, followed by air drying, *Stain Technol. 44*, 147–50 (1969).
145. A. J. Wyrobek, G. Watchmaker, L. Gordon, K. Wong, D. Moore II, and D. Whorton, Sperm shape abnormalities in carbaryl-exposed employees, *Environ. Health Perspect. 40*, 255–265 (1981).
146. M. Yamamoto and T. H. Ingalls, Delayed fertilization and chromosomal anomalies in the hamster embryo, *Science 176*, 518–521 (1972).
147. M. Yamamoto and G. Watanabe, Effects of maternal aging on developmental defects in mice, *Jpn. J. Hyg. 25*, 558–562 (1975).
148. M. Yamamoto, A. Endo, and G. Watanabe, Maternal age dependence on chromosomal anomalies, *Nature New Biol. 241*, 141–142 (1973).
149. M. Yamamoto, T. Shimada, A. Endo, and G. Watanabe, Effects of low-dose X-irradiation on the chromosomal non-disjunction in aged mice, *Nature New Biol. 244*, 206–208 (1973).
150. C. Tease, Dose-related chromosome non-disjunction in female mice after X-irradiation of dictyate oocytes, *Mutat. Res. 151*, 104–119 (1985).

CHAPTER 7

Using Repair-Deficient Chinese Hamster Ovary Cells to Study Mutagenesis

Larry H. Thompson and Cynthia A. Hoy

1. Introduction

The experimental induction of single-gene mutations in cultured mammalian cells was first reported by two laboratories in 1968. Kao and Puck[58] showed that alkylating agents such as ethyl methanesulfonate would induce genetic changes causing auxotrophy for glycine or other nutrients in the Chinese hamster ovary (CHO) cell line. They described a selection procedure employing bromodeoxyuridine (BrdUrd) incorporation followed by exposure to fluorescent light to enrich for these mutant cells. Chu and Malling,[17] using the V79 line of hamster cells, reported the direct selection of chemically induced mutations that confer resistance to the purine analogue 8-azaguanine due to changes in the gene coding for the salvage enzyme hypoxanthine-phosphoribosyl transferase (*hprt*). At the time, concerns were voiced about the genetic basis of phenotypic changes in cultured cells.[42,71,104] Nevertheless, the

Larry H. Thompson and Cynthia A. Hoy • Biomedical Sciences Division, Lawrence Livermore National Laboratory, University of California, Livermore, California 94550. *Present address of Dr. Hoy*: Department of Biological Sciences, Stanford University, Stanford, California 94305-2493.

induced variants were presumed to originate, at least in some cases, from alterations in the structural genes coding for the target enzymes.

During the next 15 years there followed a rapid expansion of studies directed at isolating many different kinds of mutants that could be used to address basic genetic and biochemical issues.[38,82,84] Studies of the mechanisms of mutation production and the quantitative detection of mutagens also became numerous. With time, the technical difficulties associated with using purine analogue selections were overcome, and the highly reproducible measurement of induced mutation was possible.[74,50] Mammalian cells now occupy an important place beside microbial systems in the short-term testing and evaluation of environmental mutagens and carcinogens.[7,22,51,102] Moreover, spontaneous and induced mutants for genes such as *hprt* that have recently been cloned can now be analyzed for the precise base-sequence changes by using powerful recombinant DNA techniques.[35,39,69]

In spite of these methodological advances, the molecular basis of mutagenesis in mammalian cells is largely unresolved. A detailed understanding of the enzymes involved in DNA replication, repair, and recombination in both untreated cells and cells exposed to mutagens must be developed to relate observed genetic alterations to initial chemical changes in genomic DNA. For example, the proteins that mediate sister chromatid exchange, a widely used indicator of genotoxicity, need to be indentified to elucidate this intriguing phenomenon. Through the study of mutant cells defective in specific components of DNA metabolism, especially repair, knowledge about mutagenesis at the gene and chromosomal levels can be acquired.

Toward this goal, our laboratory began to isolate and characterize such mutants in the CHO cell line about 5 years ago.[103,107] This chapter discusses what is known about these mutants and describes their use in a rapid assay for detecting DNA-damaging agents that is based on the endpoint of differential cytotoxicity between normal and repair-deficient cells. Applying this assay to several classes of mutagens has helped to characterize the repair mutants as well as to provide a simple way to detect several major classes of mutagens. The DNA repair mutants of CHO and other rodent cells, developed in our laboratory and elsewhere, were recently summarized.[103]

2. Genetic and Biochemical Properties of Repair-Deficient Mutants

2.1. Complementation Groups of UV-Sensitive Mutants

The CHO cell line was chosen for mutant isolation because of its ease of experimental manipulation and proven ability to provide a wide variety of mutants.[38] In many organisms mutations that reduce DNA repair capacity confer increased sensitivity to cell killing for a class of DNA lesions. Thus, the general approach to isolate CHO mutants was to induce mutations by exposure to potent mutagens and recover mutant clones that were hypersensitive to killing by germicidal UV radiation. First, manual procedures[11,107,123] and then automated ones[10,11] were devised to identify sensitive colonies and recover viable cells from them. Among more than 200 mutant clones, most of which were obtained by Busch,[10] ~70% showed high sensitivity, i.e., about sevenfold (see examples in Figure 1a, discussed further in Section 3.1).

More than 140 of the highly sensitive clones were analyzed by tests for genetic complementation to determine whether the UV-sensitive phenotype arose from several different genotypes. A rapid assay for complementation in hybrid cells was devised. The method consisted in chemically fusing pairs of clones, allowing for an expression of ~24 hr, challenging the cells with a low dose of UV (5 J/m^2), and measuring the frequency of surviving colonies.[105] In mutant self-crosses the frequency was about 10^{-5}, but when complementation occurred the frequency increased 10- to 100-fold.

The complementation analysis revealed that genetic diversity underlay the common phenotype of high UV sensitivity. Five genetic groups were identified.[105,111] However, the distribution of mutants among the groups was highly skewed. Most of the mutants fell into the first two groups, and only one clone was obtained for the fifth group.[111] The two common groups most likely arose through mutations in regions of the CHO genome that are functionally hemizygous because of partial monosomy.[40,94,95,124] The fact that only one or a few representatives were obtained for three groups suggests that additional rare groups might be identified if additional clones were screened.

In cells obtained from xeroderma pigmentosum (XP) individuals, eight complementation groups have been recognized.[5,6,59,73] Whether all of these groups represent different gene loci essential for repair is unclear, but nucleotide excision repair is obviously genetically complex. Four complementation groups of UV-sensitive mutants have also been obtained in the mouse L5178Y line,[90,92] which has been widely used for mutagenesis testing.[22] Although this mouse line appears to have a lower level of UV excision repair than CHO cells,[91] the mutants from it provide further evidence for a nucleotide excision repair pathway that is common to mammalian species and multigenic in nature, as in simpler eukaryotes such as yeast.[33]

Because the CHO mutants were isolated after mutagenesis with ICR-170, ethyl methanesulfonate, UV radiation, or other agents, one should keep in mind that these mutants are not isogenic with the wild-type line. In fact, the frequency of random point mutations is likely of the order of 10^{-6} per base pair of DNA, since the probability of a mutation in the coding sequence of a typical gene ~1 kb in length (such as *hprt*) is ~10^{-3}. Thus, a given mutant line may show altered properties with respect to growth rate, DNA replication, and other processes that are not a consequence of the repair deficiency. Such adventitious changes in the mutants may complicate their analysis. Studying more than one isolate from a given complementation group will help to circumvent this problem.

We have characterized CHO mutants from each UV complementation group, namely lines UV5, UV20 (or UV4), UV24, UV41, and UV135 for groups 1–5, respectively. The growth properties of these mutants (as well as the mutant line EM9 introduced in Section 2.4) are favorable, but the mutants tend to grow less vigorously than the normal cells. Plating efficiencies of the mutants are in the range 0.6–0.9 versus 0.9 for the parental line (clone AA8). All the lines are typically grown in suspension culture in our laboratory. Under these conditions the AA8 has a doubling time of 12 hr, and the mutants are in the range 13–16 hr. The modal chromosome number has been determined for mutants UV5, UV20, UV41, and EM9 and shown to be 21, like that of the parental line.

2.2. Defective Excision Repair in UV-Sensitive Mutants

Initial studies with mutants from the first three complementation groups, lines UV5, UV20, and UV24, showed that UV-induced repair replication (measured by alkaline CsCl/$CsSO_4$ isopycnic gradients) was essentially undetectable.[11,107] This deficiency in repair synthesis was found to be associated with an inability of the mutants in each of the five complementation groups to perform the incision or nicking step in response to UV lesions.[109] Each mutant studied had <10% of the incision activity of the normal cells as determined by alkaline elution measurements of DNA size. Moreover, the five mutants showed no significant differences from each other. Thus, they are biochemically altered either in an endonuclease that is responsible for incising at the lesions or in some other factor(s) that precedes or assists the action of the endonuclease. The same type of defect has been recognized in XP cells.[6,31,44,126] Moreover, the defect in XP cells could be corrected by introducing a foreign pyrimidine-dimer endonuclease derived from bacteriophage T4.[101] This result shows that the steps in repair subsequent to incision are functional in XP cells.

In XP cells, the UV repair defects also impair the ability of the cells to remove bulky convalent adducts.[45,53,68,125] This aspect of repair was studied in the CHO mutants using the direct-acting polyaromatic hydrocarbon 7-bromomethylbenz[*a*]anthracene (7-BrMeBA). This compound was found to kill the UV mutants more efficiently than the normal CHO cells (Figure 1b), a result similar to that seen with XP cells.[68] As shown in Figure 2, normal CHO cells were able to remove 7-BrMeBA adducts from their DNA, with more than 50% of the adducts disappearing by 24 hr after exposure. None of the mutant lines from the five complementation groups was able to remove a measurable percentage of the adducts. (Mutant EM9, introduced in the next section, had normal removal.) From these results we can conclude that the UV mutants are repair-defective by the criterion of excision of lesions as well as by the incision assay. By all properties discussed so far, the defects in nucleotide excision repair of the CHO mutants seem to closely parallel the phenotype of classic XP cells.

An issue that has been raised about the nature of the CHO mutants concerns their degree of sensitivity to killing by UV relative to the level of repair present in the normal CHO cells. Historically, rodent cells have been considered weak in repair capacity compared with human cells.[43] Thus, the relatively strong phenotypes of the CHO mutants might not be expected if the parental cells had little repair. The rate of removal of pyrimidine dimers in normal CHO cells tends to be significantly lower.[70,78] than what is observed in normal human fibroblasts, although the results reported for human cells vary widely.[3,20,29,60] Whereas the UV-sensitive mutants of CHO are about sevenfold more sensitive to killing than normal CHO cells (based on D_{37} doses; see Figure 1a), the most repair-deficient XP strains show 10- to 20-fold hypersensitivity compared with normal human fibroblasts.[4,36,65] On the other hand, the mouse L5178Y line seems to have even less efficient dimer removal than CHO cells,[91] and the UV repair mutants derived from it have approximately fourfold hypersensitivity.[92] Thus, there is a correlation between the efficiency of repair of parental lines and the degree of hypersensitivity of the corresponding mutants.

However, the UV resistance of the normal CHO and mouse cell lines tends to be similar to that of normal human cells, suggesting that rodent cells may be intrinsically able to tolerate and survive higher levels of unrepaired damage. If this is the case, then one would expect to find for fully excision-deficient mutants that the CHO mutants are more resistant than the human mutants. Measurements performed under common conditions have shown such a difference. XP group A cells were found to be at least twofold more sensitive than CHO mutants UV4, UV5, and UV41 (L. Thompson, unpublished results).

2.3. Defective Strand-Break Repair in Mutant EM9

The CHO mutant cell line referred to as EM9 has a phenotype distinct from the UV-sensitive mutants. EM9 was identified and isolated by virtue of its tenfold hypersensitivity to killing by ethyl methanesulfonate (EMS).[107] In the initial characterization EM9 showed similar sensitivity to methyl methanesulfonate (MMS), but only twofold hypersensitivity to ionizing radiation.[107] The UV sensitivity of EM9 is near normal.[108] Subsequent analysis of a variety of agents that react with DNA has shown that the sensitivity profile is narrow and specific, even among methylating and ethylating agents (cf. Tables 2–4 and 6, to be discussed later). A common property of many of the agents to which EM9 shows sensitivity is their production of abundant DNA strand breaks (either directly or through a DNA repair response).

Analysis of the DNA in mutagen-treated EM9 cells suggested that they are defective in rejoining strand breaks.[108] EM9 cells showed greater numbers of breaks than normal cells immediately after short treatments (10 min) with MMS or EMS, and the breaks were more persistent in EM9 during 2 hr posttreatment incubation. After X rays, EM9 cells showed initially the same number of breaks as normal cells, but again the breaks disappeared more slowly in EM9 cells.[108] Measurement of repair replication using alkaline CsCl/$CsSO_4$ gradients showed that the amount of repair synthesis appeared normal in EM9 cells treated with EMS or MMS (J. Cleaver, personal communication). Together these observations suggest that the biochemical defect in EM9 might be at a late step in the sequence of repair events, i.e., at the stage of ligation of DNA strands following repair synthesis.

Attempts to determine the primary defect in EM9 have been unsuccessful. Many of the properties of EM9 resemble the properties of normal CHO cells treated with inhibitors of ADP-ribosyl transferase.[111] However, assay of the activity of this enzyme and associated parameters, such as the distribution and decay of the polymer product, and the intracellular NAD pool, did not reveal any alterations in EM9 cells.[52] Both of the major forms of DNA ligase appear to be intact in EM9 cells, as measured in both untreated cultures and cultures treated with MMS.[16] Two classes of apurinic/apyrimidinic endonucleases are also present at normal levels in EM9 cells.[61]

2.4. Elevated Sister Chromatid Exchange in Mutant EM9

The most remarkable property of EM9 is its greatly elevated frequency of sister chromatid exchange (SCE). Under standard conditions of 10 μM BrdUrd in the medium for one or two cycles of incorporation, the frequency is 10–12 times the value for normal CHO cells.[25,108] The enhanced SCEs in EM9 seem to arise when DNA replication occurs on a template containing BrdUrd.[25] Therefore, the enhanced SCEs in EM9 can be attributed mainly to the presence of BrdUrd in the DNA. This interpretation was confirmed by using a sensitive monoclonal antibody against BrdUrd-containing DNA and showing that the frequency of SCEs declined as the percentage of BrdUrd substitution decreased.[79] The magnitude of induced SCEs is also greatly altered in EM9 cells. In EMS-treated EM9 cells, the efficiency of SCE production is seven times greater than in normal CHO cells.[108]

The characteristic of elevated SCE in EM9 cells is shared by cells from individuals with Bloom syndrome (BS).[15,93] In BS cells, as in EM9, the incorporated BrdUrd appears to be responsible for the elevated frequency of SCEs.[93] The frequency of chromosomal aberrations was also greatly elevated in EM9 cells exposed to BrdUrd,[12,83] but in BS lymphoblasts this effect was not observed.[83] Moreover, BS fibroblasts were shown to repair normally the DNA strand breaks produced by ionizing radiation[119] or EMS.[111] Thus, the molecular defect(s) in BS cells may differ from that of EM9.

2.5. Complementation of CHO Mutations by Human Genes

An important question about the nature of the CHO repair-deficiency mutations is whether the genes involved correspond directly to genes in human cells. Tests for complementation of CHO mutants by fusion with human cells to form cell hybrids are used to address this issue. The experiments to date show that there are indeed human genes that can correct the hamster cell repair defects. To select for complementing hybrid cells, fusion is first performed between a CHO mutant and either human fibroblasts or peripheral lymphocytes. Hybrid cells are then isolated by selecting for cells that survive exposure to a DNA-damaging agent that is present in the culture medium at low concentration. For example, mutant UV20 was shown to be complemented by selecting hybrids in the presence of mitomycin C (MC).[116] Similarly, mutants UV5 and EM9 were complemented in lymphocyte-derived hybrids selected for resistance to decarbamoyl MC,[47] or CldUrd,[26] respectively (L. Thompson, unpublished results).

By analyzing the hybrids for their makeup of human chromosomes, the chromosomal locations of the complementing genes can be determined. The human gene that complements mutant UV20 was mapped to chromosome 19.[116] Mutant UV41, which is in a different complementation group but has the same phenotype as UV20 in terms of strong hypersensitivity to DNA cross-linking agents (see Section 4.4), appears to be complemented by human chromosome 16 (L. Thompson and M. Siciliano, unpublished results. These results illustrate that the complementation groups can be assigned to specific chromosomal locations, which can differ, as would be expected if each group represents a different genetic locus.

An alternative method of testing for complementation of the CHO-cell mutations by human genes involves the method of DNA-mediated gene transfer. Calcium phosphate precipitates of human DNA have been shown to correct the DNA repair defects in UV20[88] and EM9.[115] In each case the transfer of a functional human repair gene can be inferred by the fact that independent transformant clones exhibited a common restriction fragment when the cellular DNA was digested and probed for human repetitive sequences in a Southern blot. Such transformants provide a means of isolating the human repair genes by their linkage to repetitive interspersed human sequences or cotransferred marker sequences. The first molecular cloning of a human DNA repair gene was achieved using a cosmid library containing the DNA of a secondary transformant of a CHO mutant belonging to UV complementation group 2.[121] The availability of cloned repair genes will allow detailed analysis of gene structure and should accelerate efforts to identify the proteins encoded by these genes.

3. Hypersensitivity of Repair Mutants to DNA-Damaging Agents

3.1. Sensitivity Determined by Colony Formation

Characterization of representative UV-sensitive mutants initially involved analysis of their survival curves in response to germicidal UV radiation and several chemical mutagens that form large covalent DNA adducts. Figure 1a shows results for mutants UV5 and UV20, which belong to complementation groups 1 and 2, respectively, and represent the two phenotypic subgroups of UV-sensitive mutants (see below). The survival responses of the two mutants are identical. The normal cells show a D_{37} dose that is about seven times higher than that of the

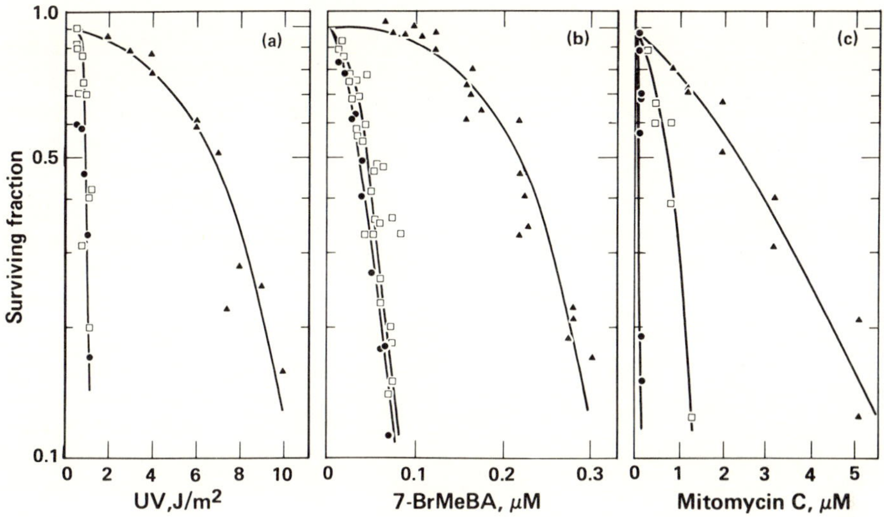

FIGURE 1. Survival curves of normal and UV-sensitive mutants in two genetic complementation groups. (a) Germicidal UV radiation (254 nm); (b) 7-BrMeBA, 60 min exposure; (c) mitomycin C, 90 min exposure. (▲) Normal AA8 cells; (□) UV5; (●) UV20. [Data taken from previous reports of Thompson and co-workers .[107,110]]

mutants. Mutants from the other three complementation groups have UV sensitivity similar to that of UV5 and UV20.[105,109]

The direct-acting polyaromatic hydrocarbon 7-BrMeBA is an example of a monofunctional agent whose bulky adducts are not excised normally by the mutants (cf. Figure 2). Mutants UV5 and UV20 had almost identical cell-killing responses to 7-BrMeBA and were about fivefold more sensitive than the normal CHO cells (Figure 1b). In a study using synchronized cells, the pattern of sensitivity to 7-BrMeBA through the cell cycle was compared in UV5 and parental AA8 cells and found to differ qualitatively.[114] AA8 cells were most sensitive in S phase, whereas UV5 cells were most sensitive in early G_1 and showed less dependence on phase of the cell cycle.

Mitomycin C (MC) was also examined with the mutants because of its well-known ability to cross-link cellular DNA. Figure 1c shows that this compound produces very different responses for UV5 and UV20. While UV5 is about threefold sensitive, UV20 shows 100-fold hypersensitivity![107] Analysis of other compounds that produce DNA cross-links has shown that this differential sensitivity is not restricted to MC, but rather is a general property of bifunctional compounds (Section 4.4). Other mutants in complementation groups 1 and 2 show the same

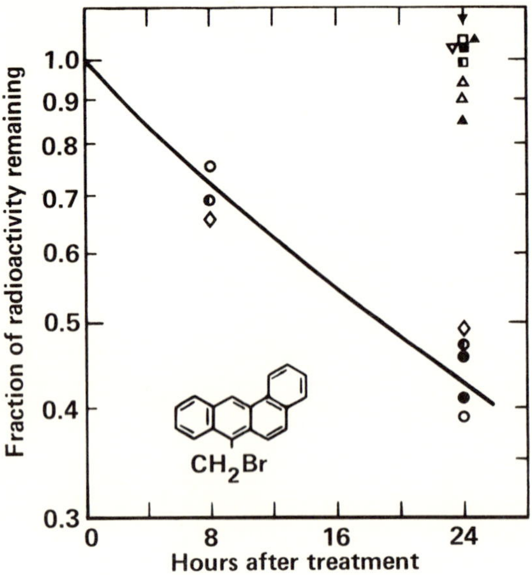

FIGURE 2. Removal of 7-BrMeBA adducts in normal and mutant cell lines during posttreatment incubation. The doses of tritium-labeled 7-BrMeBA were 0.1–.3 μM for the normal (AA8) cells and mutant EM9 and 0.05–0.10 μM for the five UV-sensitive mutants; exposure was 1 hr. Corrections for cell growth were made using a dual labeling procedure with ^{14}C-thymidine.[110] The symbols are: (○, ◑,●) AA8; (◇) EM9; (□, ◧ ■) UV5; (△) UV20; (▲) UV24; (▽) UV41; (▼) UV135. [Results taken from Thompson *et al.*[114] with permission from the publisher.]

differential sensitivity to MC, indicating that these properties are not peculiar to mutants UV5 and UV20.[48] Moreover, the MC sensitivity of mutants in groups 3 and 5 coincides with that of UV5, and the sensitivity of mutants in group 4 is like that of UV20.[48] On the basis of these results, the CHO UV mutants subdivide into two discrete phenotypic classes, and therby seem to differ from XP cells, which are not markedly sensitive to cross-linking agents.

These survival results with three very different agents point out that the mutants show a substantial hypersensitivity to various lesions that distort the DNA helix. Knowing whether there were similar degrees of enhancement in mutation induction by these and other agents was of interest and is discussed below.

3.2. Methodology for Measuring Mutations

In the studies described here, mutations were measured at two loci that code for nonessential enzymes in the salvage pathways of purine utilization. Mutations at the X-linked *hprt* locus were measured by resistance to 6-thioguanine (TG^r)[106] under conditions similar to those described by Hsie and co-workers.[49,50,74,75] Mutations could also be measured at the autosomal adenine phosphoribosyltransferase (*aprt*) locus because clone AA8, from which the repair mutants were derived, was isolated as being functionally heterozygous at this locus.[106] The *aprt* mutations are detected using 8-azaadenine (AA) as the selective agent.[106] Assay of AA^r mutations has the advantage of requiring an expression time of ~3 days,[13,106] as compared with a value of ~6 days[106] for TG^r (or longer in some CHO cell lines[50,74]).

The conditions used in our laboratory for mutation assay differ in several respects from those used by others.[13,50,75] with CHO cells. Usually the exposure to mutagen and growth during phenotypic expression are done in suspension culture using 200-ml spinner flasks. In suspension, cell growth after treatment can be readily monitored, and the requisite expression of 6 days is achieved with only one subculturing, 3 days after treatment. Keeping the cell concentration in the range of 3×10^4 to 8×10^5 cells per ml maintains a minimum of 6×10^6 cells, which is a good compromise between the sample size needed for accurate measurement of low mutant frequencies versus the expense of large cultures.

For mutation assay, cells were plated for viability (300 cells per 9-cm dish in replicate) and for resistance to TG or AA. Since mutant recovery was found not to be influenced by cell density up to 1×10^6 cells per dish,[106] we used 6×10^5 cells per dish (four to six replicates). The drug media are prepared with 10% dialyzed fetal bovine serum; removing purines results in a more reliable selection, particularly for AA^r mutants. These conditions for mutation assay have been used with the normal CHO cells and various mutant lines.[109]

3.3. Sensitivity to Mutation Induction

Mutation assays were done in parallel with survival curves for several agents over dose ranges that reduced the survival from 1.0 to as low as 0.1. Figure 3 shows mutation frequency induced by UV radiation in mutants UV5 and UV20 compared with the normal CHO cells. The mutants are at least tenfold hypermutable for both TG^r and AA^r mutations. The two mutant lines show similar if not identical behavior, and their dose–response curves are linear. The normal CHO cells have more complex curves, with decreasing efficiency of mutation induction at the higher doses. (This nonlinearity might be accounted for by differential sensitivity through the cell cycle.)

These curves for normal CHO cells differ from those of normal human fibroblasts, which exhibit a threshold of ~1 J/m^2 for UV-induced mutations.[65] This threshold can be attributed to repair, since XP cells do not show this property. The reasons for this difference

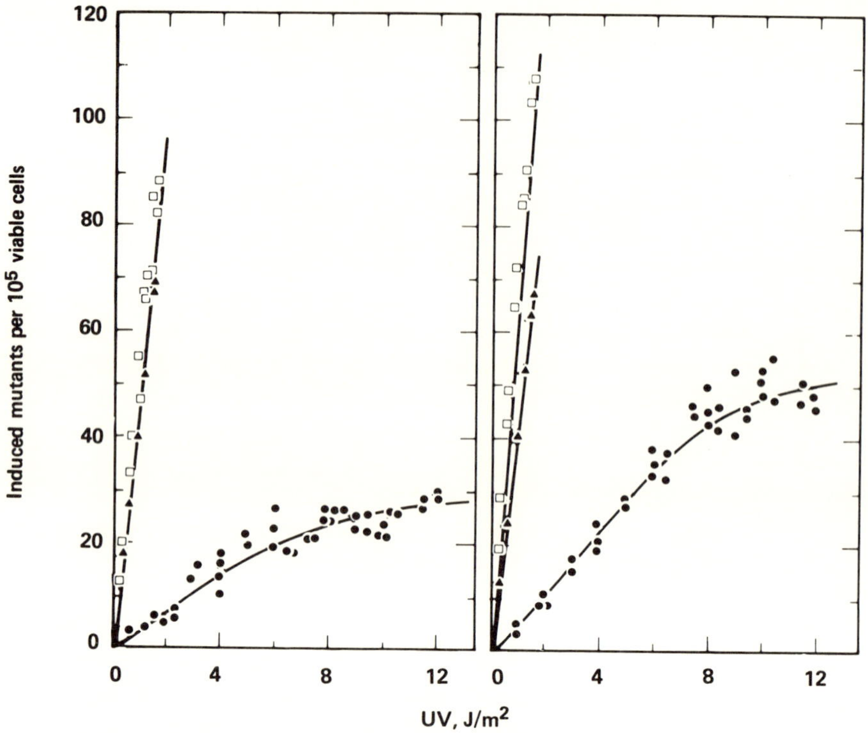

FIGURE 3. Dose response for mutations induced by UV radiation in normal and UV-sensitive mutants. Left: TG^r mutations; right: AA^r mutations. (●) Normal AA8 cells; (□) UV5; (▲) UV20. [Data taken from Thompson *et al.*[109]]

between CHO and human cells are not apparent, but could be due to differences in the cell cycle phases or other parameters.

With the mutagen 7-BrMeBA, UV5 and UV20 also showed hypersensitivity to mutation induction, although the differences were not as great as with UV radiation (Figure 4). Both mutant lines exhibited linear dose–response curves as with UV radiation. Mutation frequencies were slightly lower in UV20 than in UV5, but the significance of this difference is unclear, since both lines fail to excise 7-BrMeBA adducts (see Figure 2). The mutation induction curves for the normal CHO cells were distinctly nonlinear. The most interesting feature is the curvilinear behavior of both loci below 0.15 μM, where there is no killing of the normal cells. The efficiency of mutation induction increases with dose. This response may reflect saturation of the nucleotide excision repair system.

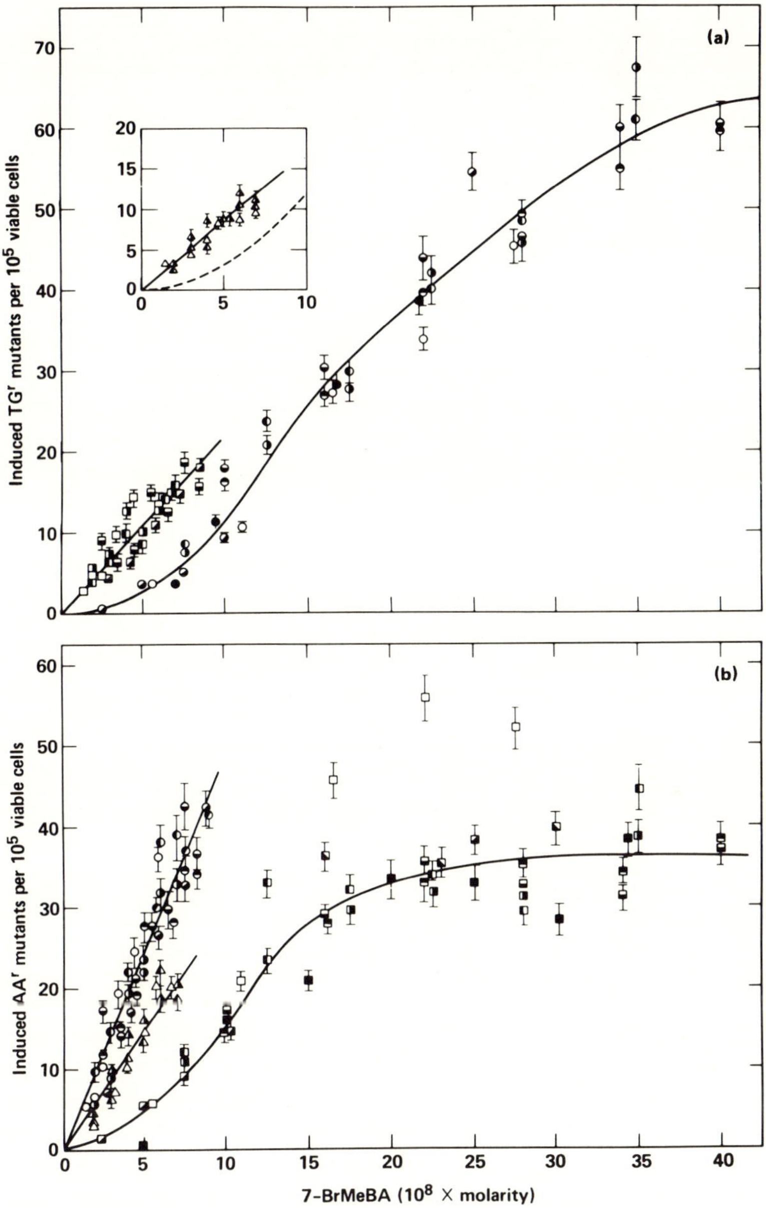

FIGURE 4. Dose response for mutations induced by 7-BrMeBA in normal and UV-sensitive mutants. (A) TG^r mutations: (circles) normal AA8 cells; (squares) UV5; (triangles) UV20. (B) AA^r mutations: (squares) normal AA8 cells; (circles) UV5; (triangles) UV20. [Data taken from Thompson *et al.*[110]]

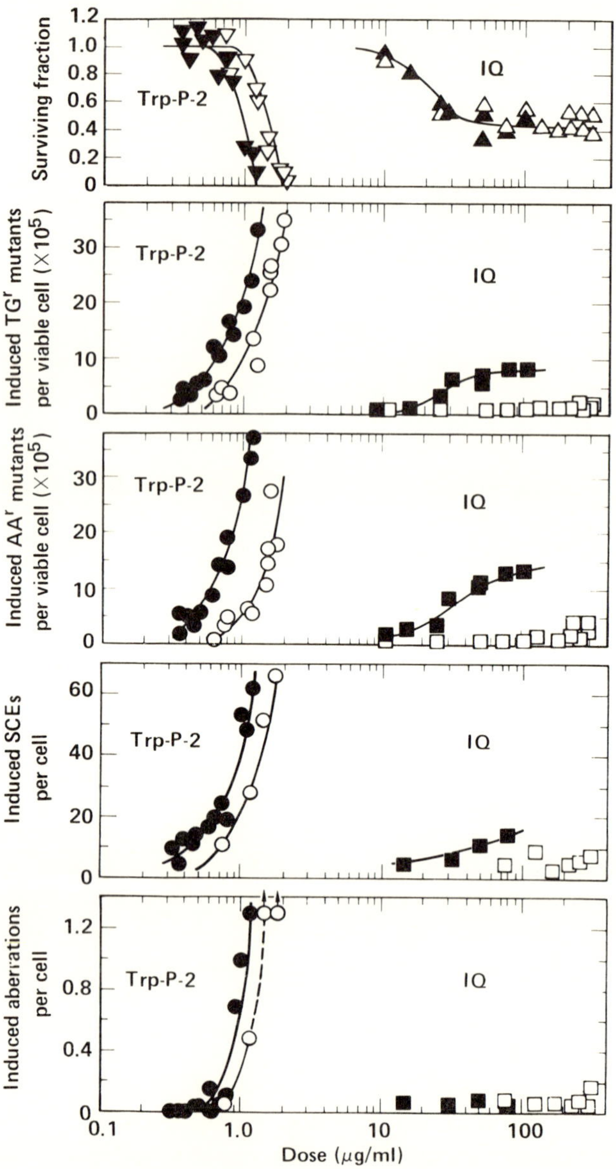

FIGURE 5. Dose response for cell killing, mutations, SCEs, and chromosomal aberrations produced by Trp-P-2 (3-amino-1-methyl-5*H*-pyrido[4,3-*b*]indole) and IQ (2-amino-3-methylimidazo[4,5-*f*]quinoline). Open symbols are for normal cells; closed symbols are for UV5. [Reproduced from Thompson *et al.*[112] with permission of the publisher.]

A third example of the enhanced sensitivity of the repair-deficient cells to mutation induction is illustrated with two compounds (Trp-P-2 and IQ) that are known to be produced by the cooking of proteinaceous foods (Figure 5). These compounds are of interest because of their carcinogenicity[67,100] and extremely high potency in the Ames *Salmonella* assay.[112] Both chemicals require metabolic activation to produce mutagenic activity. Trp-P-2 killed UV5 cells at twofold lower doses compared with normal cells. A similar difference in sensitivity was seen for mutation induction at both loci and for the cytogenetic endpoints of sister chromatid exchange and chromosomal aberrations. The behavior of IQ was more unusual. IQ was toxic only at much higher doses than Trp-P-2 and showed no difference in killing between UV5 and normal cells. However, UV5 had highly significant mutation induction at both loci with IQ, while normal cells did not. The failure to see differential killing with IQ suggests that a major component of the toxicity is probably not due to DNA damage. This interpretation is consistent with the absence of chromosomal aberrations in IQ-treated cells. These results provide a clear example of how the repair-deficient cells can be used to detect the weak mutagenic activity of an agent whose toxicity is predominantly nongenetic.

4. Rapid Detection of DNA-Damaging Agents by Differential Cytotoxicity (DC)

4.1. Concept and Methodology of DC Assay

Repair-deficient cells are noted for their hypersensitivity to killing by certain classes of DNA-damaging agents (Figure 1). Hypersensitivity to mutation induction is closely coupled with hypersensitivity to cell killing, as described in the preceding section and elsewhere[113] for the UV-sensitive strains of CHO. Thus, hypersensitivity of repair mutants to killing provides presumptive evidence for DNA damage, of which the unrepaired component can lead to mutation or lethality. Several bacterial assays have employed this concept as a means of detecting mutagens.[55,85] We sought to develop a simple assay of a similar nature using the repair-deficient CHO cells to detect DNA-damaging compounds by the endpoint of differential killing between the normal and mutant cells. In theory, by using different types of repair mutants in such an assay, the pattern of toxicity response seen among the mutants can also provide information about the type(s) of DNA lesions produced by the test agent. The utility of such an approach hinges on having a variety of mutants that are defective in the repair of different types of DNA lesions produced by chemicals and are reasonably well characterized. Although the mutants described in the preceding sections undoubtedly need more detailed biochemical analysis, an evaluation of this approach was undertaken. Besides determining the technical feasibility of the method, we expected that the results obtained would in turn improve our understanding of the mutants.

A method was devised to measure cell growth over a 72 hr period (or in some cases 96 hr) with the normal and mutant lines in parallel. The term *differential cytotoxicity* (DC) is used to refer to the amount of growth of the normal cells relative to each mutant line. The details of this procedure have been described.[47] Cells are inoculated into the wells of a 24-well cluster tray (Costar) under conditions that promote uniform attachment of the cells. After ~2 hr incubation, the cells are exposed to an agent and then further incubated until the cells of the control wells are confluent. After fixing and staining, relative growth as a function of dose is determined from the intensity of staining.

In order to have a reference scale by which to measure growth, a tray was prepared with graded inocula of each of the four cell lines chosen as the initial set. Figure 6 shows a tray in which the relative number of cells was varied from 100% on the left to 5% on right. The difference in cell inoculum between each vertical row of wells was a factor of 1.8. The tray was incubated 72 hr. An integer from 5 to 0 was assigned to the staining intensity of each vertical row of wells from left to right. For each experimental tray, numbers were similarly assigned to each well by using the tray in Figure 6 as a standard. The *lowest effective concentration* (LEC) of test compound is defined as the dose that produces a reduction in staining of 2 units compared to the control well. The LEC is determined for each cell line. A differential cytotoxicity (DC) ratio for each mutant line is then calculated by dividing the LEC of the normal line by the LEC of the mutant.

An example of the assay is given in Figure 7 for mitomycin C (MC), a bifunctional compound. The cell lines used were the normal AA8, mutants UV5 and UV4 from complementation groups 1 and 2, respectively (UV4 is an alternate line to UV20, which was described in Figure 1–4), and the EM9 mutant. The doses of MMC for AA8, UV5, and EM9 are shown above the tray, and the doses for UV4 are shown below. Immediately under each well of AA8 and UV4, a value of staining intensity is given. The LEC of AA8, which lies between 1.25 and 2.5 μg/ml, is estimated as 1.5 μg/ml. Similarly, the LEC of UV4 is about 0.019 μg/ml, resulting in a DC ratio of 79 for UV4. The DC values for UV5 and EM9 were 2 and 1, respectively.

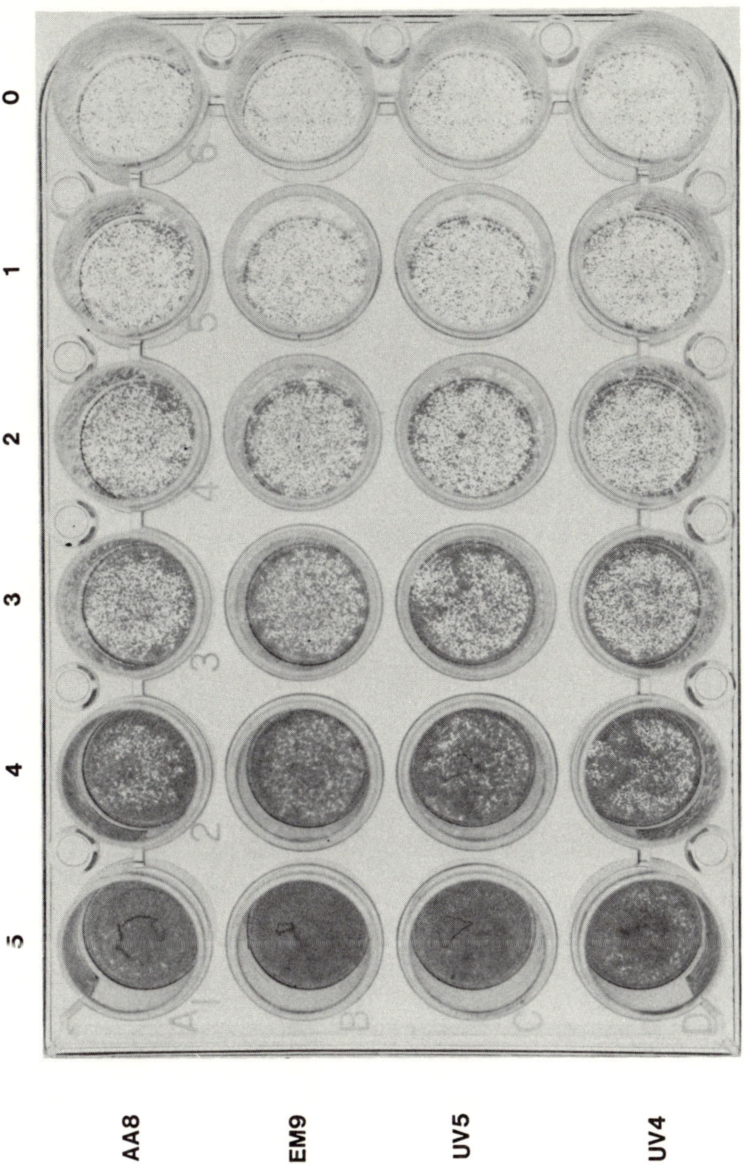

FIGURE 6. Reference tray for 24-well differential cytotoxicity assay. The column of wells on the extreme left was inoculated with 3×10^4 AA8 and UV5 or 4×10^4 EM9 and UV4 cells as described.[47] In each successive column to the right the number of cells was reduced by a factor of 1.8. After 72 hr incubation the wells were stained. The integers assigned to the top of each column provide standard levels of staining intensity against which experimental trays are compared.

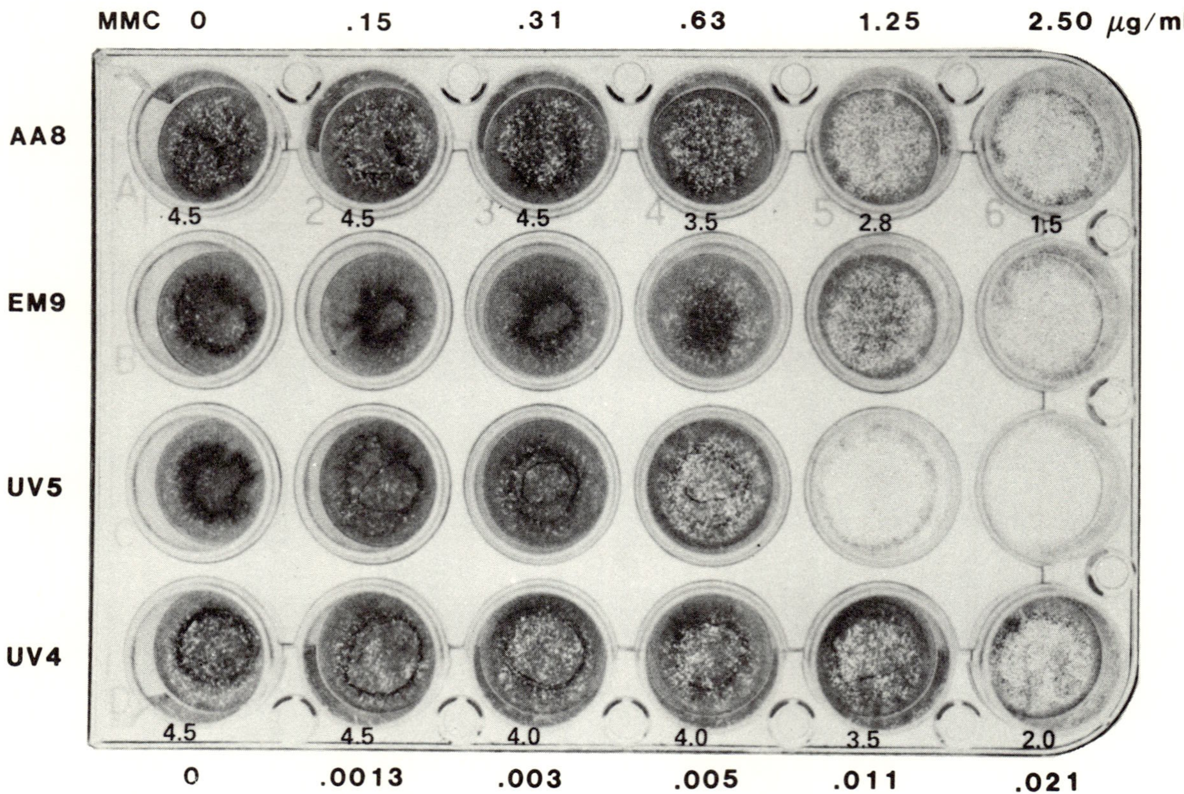

FIGURE 7. Dose response of mitomycin C in the DC assay with normal AA8 cells and mutants UV4, UV5, and EM9. Doses are shown across the top for AA8, UV5, and EM9; doses for UV4 are at the bottom. The LEC values for AA8 and UV4 are given below each well.

In validation experiments,[47] the DC values were shown to be highly reproducible for a given compound tested on different days. The assay was shown to be capable of reliably determining a DC value of ~2. For several compounds the DC values also compared favorably to relative sensitivity measured by survival curves.[47,48]

Subsequent sections present the results of DC assays for several series of compounds, first, ones that are not considered DNA-damaging agents, and then chemicals of several classes that are well-documented for their modification of DNA structure. The results of these tests serve to characterize and validate the DC assay. Initially, the cell lines shown in Figure 7 were used to assess mutagens forming bulky covalent adducts. For evaluating alkylating compounds that donate methyl, ethyl, or longer alkyl groups, and other compounds an additional repair mutant, CHO line UV-1, was obtained from Dr. C. Waldren[98] and added to the assay.

4.2. Response of Repair-Deficient Lines to Nonmutagens

It was important to substantiate that DC values significantly greater than 1.0 were specific for DNA damage and not associated with other modes of toxicity. To evaluate the likelihood of obtaining "false positives," the DC responses were measured for a variety of agents that are generally not thought to cause DNA damage as the primary mode of action. Table 1 summarizes the results of 13 compounds tested. None of the agents, which include several types of toxic inhibitors, resulted in a significant DC response in any of the mutants. Pyrene and γ-butyrolactone are non-DNA-binding structural analogues of the mutagens benzo(*a*)pyrene and β-propiolactone (these mutagens were positive in the assay; see Table 2). Caffeine, naladixic acid, and novobiocin are known to interfere with various aspects of DNA metabolism, including repair. Caffeine may inhibit postreplication recovery[9] and binds to single-stranded DNA.[28] Novobiocin and naladixic acid appear to inhibit excision repair[18,23] as well as normal DNA replication.[21] The normal responses of the mutants to these compounds suggest that the mutants are not deficient in the components of DNA metabolism affected by these compounds. The mutants were also not hypersensitive to the well-known metabolic inhibitors cycloheximide, hydroxyurea, and actinomycin D. The results in Table 1 lead to the inference that the mutants are not generally hypersensitive to toxic agents that act by nongenotoxic pathways.

TABLE 1. Response of DC Assay to 13 Presumed Non-DNA-Damaging Agents[a]

Compound	LEC,[b] μg/ml	Exposure time,[c] hr	DC ratio for UV4, UV5, EM9[d]
Pyrene	8.0	4	1
γ-Butyrolactone	4×10^3	24	1
Dimethyl formamide	1×10^4	72	1
Dimethysulfoxide	2.2×10^3	24	1
6-Thioguanine	0.16	24	1
Ouabain	360	48	1
Naladixic acid	62	4	1
Caffeine	800	72	1
Ascorbic acid	4×10^3	24	1
Novobiocin	500	4	1
Cycloheximide	0.2	72	1
Actinomycin D	1.6	72	1
Hydroxyurea	19	24	1

[a] Data are taken from Hoy *et al.*[47]
[b] LEC is the lowest effective concentration of chemical to produce a measurable loss of staining intensity (see text).
[c] Exposures greater than 4 hr were needed in some cases to obtain appreciable loss of staining intensity.
[d] DC is the differential cytotoxicity ratio defined as the LEC of the wild type AA8 divided by the LEC of the mutant.

4.3. Response of Repair-Deficient Lines to Monofunctional Bulky Mutagens

Compounds that react with DNA to form bulky, helix-distorting adducts represent a major portion of environmental mutagens. This broad class of DNA damage has generally been considered to be repaired by the nucleotide excision repair pathway[20] that is defective in classical XP cells[32] and the CHO mutants such as UV5 and UV4. To further develop the DC assay and assess the repairability of this class of damage in CHO cells, the 22 compounds shown in Table 2 were tested in the initial set of mutants. Each of these compounds has been shown to produce DNA adducts (references given in Table 2); some required activation with liver microsomal fractions. In every case, mutants UV4 (or UV20) and UV5 gave a positive response and had a similar DC ratio, ranging from 2 to 16.

The results with the simple compound formaldehyde were somewhat unexpected since it would not be considered a bulky mutagen. The adducts from this chemical are methylol derivatives,[46] relatively nonbulky groups. The fact that both UV4 and UV5 responded positively, although weakly, suggests that the excision repair pathway can act on these adducts at some level of efficiency.

Seven of the compounds in Table 2 also produced a DC response in EM9 cells. Since EM9 appears normal in excising bulky adducts (Figure 2), the results with these seven compounds suggest that they cause other types of DNA lesions, such as apurinic sites leading to excessive DNA strand breaks, to which the EM9 cells are responsive. The fact that the assay responded to all compounds shows that the initial set of mutants of the DC assay is reliable for detecting this important, general class of damage.

4.4. Response of Repair-Deficient Lines to Bifunctional Bulky Mutagens

As illustrated earlier, mutants UV4, UV20, and others belonging to complementation group 2 were extremely hypersensitive to killing by the cross-linking agent MC (Figures 1 and 7). Therefore a survey of bifunctional compounds was made to determine whether high sensitivity was correlated with the cross-linking property. Table 3 summarizes the results of testing 22 bifunctional compounds known to produce DNA interstrand cross-links (see table for references).

UV4 cells showed a dramatic (15- to 80-fold) DC response to all but two of the bifunctional compounds, giving a strong correlation between extreme hypersensitivity and cross-linking ability of the compounds. *trans*-Diamminedichloroplatinum and quinacrine mustard were exceptional in giving DC values less than 10. These lower values may be explained by the presence of monoadducts (which are normally repaired less efficiently than cross-links) or by other DNA lesions that are not repaired by the pathway that is defective in UV4 and that add to the overall cytotoxicity. For example, with quinacrine mustard, cross-links proved to be a minor lesion; protein-concealed breaks were the predominant lesion.[48] *Trans* Diamminedichloroplatinum is known to produce high levels of DNA–protein cross-links.[80,127]

UV5 cells consistently gave a much lower response than UV4 cells with the bifunctional compounds. With eight of the compounds, the DC ratio of UV5 was 2–4, and with all other compounds the value was 1.0. Thus, UV5 cells apparently repair cross-links with normal or nearly normal efficiency. The low, but significant, hypersensitivity that is seen with some of these compounds might be due simply to the inability of UV5 cells to repair monoadducts, which are generally present with the cross-links and are not expected to be as toxic as cross-links (on a per lesion basis).

TABLE 2. Response of DC Assay to 22 Monofunctional DNA-Damaging Agents That Form Covalent Adducts[a]

Compound	LEC,[b] μg/ml	DC ratio[c]	Ref.
Compounds that produce a similar DC ratio in both UV mutants			
Decarbamoyl mitomycin C	0.16	16, UV4; 8, UV5; 1, EM9	77
4-Nitropyridine-*N*-oxide	56	2, UV4; 2, UV5; 1, EM9	19
bis-Chloromethyl ether	30	2, UV4; 2, UV5; 1, EM9	37
4-Nitroquinoline-1-oxide	0.26	8, UV4; 8, UV5; 1, EM9	46
Formaldehyde	11	2, UV4; 2, UV5; 1, EM9	46
3-Amino-1-methyl-5*H*-pyrido(4,3-*b*)indole[d]	2.5	2, UV4; 2, UV5; 1, EM9	46
7,12-Dimethylbenz[*a*]anthracene[e]	8	4, UV4; 4, UV5; 1, EM9	46
Benzo[*a*]pyrene[e]	8	4, UV4; 4, UV5; 1, EM9	96
3-Methylcholanthrene[e]	9	6, UV4; 6, UV5; 1, EM9	72
Chrysene[e]	8	4, UV4; 4, UV5; 1, EM9	27
Benzyl chloride[f]	50	2, UV4; 2, UV5; 1, EM9	46
N-[4-(5-Nitro-2-furyl)thiazoyl]formamide[f]	25	2, UV4; 2, UV5; 1, EM9	19
Fluoranthene[e]	64	16, UV4; 16, UV5; 1, EM9	27, 102
7-Bromomethylbenz[*a*]anthracene	0.3	3, UV4; 3, UV5; 1, EM9	46
Aflatoxin B_1[d]	0.08	4, UV4; 4, UV5; 1, EM9	46

Compounds that produce a DC response in all three mutants			
Epichlorohydrin	375	3, UV4; 2, UV5; 3, EM9	46
β-Propiolactone	30	4, UV4; 2, UV5; 4, EM9	46
Propane sultone	140	4, UV4; 2, UV5; 4, EM9	46
Styrene oxide	300	2, UV4; 2, UV5; 2, EM9	46
Epoxybutane	600	2, UV4; 3, UV5; 2, EM9	46
N-Acetoxy-2-acetylaminofluorene	2.2	2, UV20; 2, UV5; 2, EM9	19
Propyleneimine	24	4, UV4; 2, UV5; 4, EM9	62

[a] Data are taken from Hoy *et al.*[47] The references listed give evidence for DNA adducts. Exposure to chemical was 4 hr except as noted otherwise.

[b] LEC is the lowest effective concentration of chemical to produce a measurable loss of staining (see text). The values given are for the wild-type cell line AA8.

[c] DC is the differential cytotoxicity ratio defined as the LEC of the wild type divided by the LEC of the mutant.

[d] Hamster S9 at 0.1–0.5 mg/ml was used.

[e] Rat microsomes at 0.5–0.7 mg/ml were used.

[f] Exposure to chemical, 24 hr.

These results with bifunctional chemicals point to an extraordinary difference between UV4 and UV5 in response to bifunctional agents in general and suggest that the repair pathways of UV lesions and cross-links have steps in common as well as steps that differ.

It is important to note that EM9 cells were not hypersensitive to any of the 22 bifunctional agents. This finding is consistent with what is known about the biochemical defect of this mutant. EM9 has reduced repair of DNA strand breaks, which are not produced to a substantial degree by most cross-linking agents. The fact that EM9 consistently gave a normal response with these compounds also strengthens the hypothesis that the positive responses seen with the UV mutants are due to their DNA repair defects.

In summary, the results given in Tables 2 and 3 provide evidence that a positive DC response is specifically associated with DNA damage. The results also illustrate that the patterns of mutant responses can be used to distinguish between certain general types of DNA damage produced by the test compound. By comparing the responses of the two UV mutants, one can discriminate between monofunctional versus bifunctional bulky chemicals: a similar positive response with mutants UV4 and UV5 indicates monoadduct damage, and a much larger response in UV4 and UV5 signals the presence of interstrand cross-links. A response by all three mutants in the initial set probably reflects the presence of strand breaks as well as monoadducts. These relationships are summarized later along with results for other classes of chemicals (see Table 8).

4.5. Response of Mutants EM9 and UV-1 to Methylating and Ethylating Agents

In addition to the two broad classes of DNA damage evaluated in the preceding section, it is important to consider whether the DC assay can detect other classes of damage. For the purpose of discussion, other DNA lesions are categorized as follows: (1) methylation or ethylation damage, as from simple alkylating agents such as *N*-methyl-*N*′-nitro-*N*-nitrosoguanidine (MMNG), MMS, or EMS, (2) longer alkyl groups, e.g., propyl or butyl, (3) intercalating agents (which usually do not bind covalently), and (4) strand breaks. For detecting the first class of damage, EM9 cells initially seemed promising, since they were markedly hypersensitive to MMS and EMS. However, certain other ethylating and methylating agents (e.g., MNNG) were not detected by EM9 cells in the DC assay. Thus, another CHO mutant, UV-1, which had been reported to be hypersensitive to MNNG,[120] was added to the assay to evaluate

TABLE 3. Response of DC Assay to 22 Bifunctional Agents[a]

Compounds	LEC,[b] μg/ml	DC ratio[c]	Ref.
Compounds that produce a differential response only in group 2 UV mutant			
bis-Chloroethylnitrosourea	12	15, UV4; 1, UV5; 1, EM9	81
Chlorozotocin	7.5	15, UV4; 1, UV5; 1, EM9	30
Chlorocyclohexylnitrosourea	5.0	20, UV4; 1, UV5; 1, EM9	81
Choroethylnitrosourea	4.5	30, UV4; 1, UV5; 1, EM9	30
p-Chloroethylnitrosourea	5.7	15, UV4; 1, UV5; 1, EM9	—
1,2,3,4-Diepoxybutane (D, L)	0.8	45, UV4; 1, UV5; 1, EM9	117
1,2,3,4-Diepoxyoctane (D, L)	7.5	15, UV4; 1, UV5; 1, EM9	62
Uracil mustard	3.0	15, UV4; 1, UV5; 1, EM9	81
Melphelan	2.5	20, UV4; 1, UV5; 1, EM9	81
Cyclophosphamide (S9-activated)[d]	200	20, UV20; 1, UV5; 1, EM9	81
Triethylene phosphoramide	13	16, UV4; 1, UV5; 1, EM9	81
Triethylene thiophosphoramide	13	16, UV4; 1, UV5; 1, EM9	81
Quinacrine mustard	4.5	6, UV4; 1, UV5; 1, EM9	—
trans-Diamminedichloroplatinum	72	8, UV4; 1, UV5; 1, EM9	127
Compounds that produce a strong differential response in group 2 mutant and weak response in group 1 mutant			
Triethylenemelamine	3.8	30, UV4; 2, UV5; 1, EM9	81
Methyl chloroethylnitrosourea	10	20, UV4; 2, UV5; 1, EM9	81
Mitomycin C	0.5	60, UV4; 2, UV5; 1, EM9	34
Busulfan (Myleran)	180	30, UV20; 2, UV5; 1, EM9	81
cis-Diamminedichloroplatinum	5.0	50, UV20; 3, UV5; 1, EM9	127
Nitrogen mustard	1.0	30, UV20; 2, UV5; 1, EM9	81
Chloroambucil	5.2	70, UV20; 2, UV5; 1, EM9	81
8-Methoxypsoralen + near-UV	1.2	32, UV4; 4, UV5; 1, EM9	97

[a] Data are taken from Hoy *et al.*[48] The references listed give evidence for DNA adducts.
[b] LEC is the lowest effective concentration to produce a measurable loss of staining intensity for wild-type AA8 cells (see text).
[c] DC is the differential cytotoxicity ratio defined as the LEC of the wild type divided by the LEC of the mutant.
[d] Hamster liver microsomal fraction was used at 1.0 mg/ml as described.[47]

this class of mutagens. UV-1 was isolated on the basis of its sensitivity to UV radiation, but its degree of hypersensitivity with this agent is only twofold.[98] Because the doubling time (~18 hr) of UV-1 is longer than that of the other mutants, 4 days incubation after exposure was usually allowed.

Table 4 summarizes the responses of mutants EM9 and UV-1 to a variety of alkylating agents. Because the UV-1 cells were derived from a different parental line of CHO (see table), the assays were performed using both parental lines. DC ratios for UV-1 were based on the response of its parental line. While EM9 did not respond to several of these compounds, UV-1 showed hypersensitivity in every instance. Particularly striking is the response of UV-1 to methylating agents. For these compounds (the first six entries in Table 4), the DC value was often 10 or more. For ethylating agents, the DC values for UV-1 were usually about 3. The fact that the sensitivity of UV-1 to methylation damage was similar to that of the AA8 parental cell line is especially noteworthy. This similarity suggests that AA8 cells may be deficient in the same component of methylation repair that is altered in UV-1.

4.6. Response of Mutant UV-1 to Various Other Agents

The mutant UV-1 was originally isolated on the basis of hypersensitivity to killing by UV radiation.[98] Therefore, its sensitivity to methylation damage may seem surprising, since the repair of UV damage and methylation damage are generally considered to proceed by different pathways.[20,41,63] UV-1 was also reported to be hypersensitive to MC.[120] The results in Table 5 confirm the sensitivity of UV-1 to MC in the DC assay and show pronounced hypersensitivity to three other cross-linking agents, *cis*-diamminedichloroplatinum, diepoxybutane, and chlorambucil. The hypersensitivity to UV radiation was only twofold, similar to that originally determined from survival curves.[98] UV-1 was also sensitive to the monofunctional mutagen decarbamoyl mitomycin C and to the intercalating agents adriamycin and daunomycin. The latter two compounds are associated with protein-concealed strand breaks in DNA.[86,87,128] ICR-170 and ICR-191 are intercalating agents that have reactive side chains[24] and are very potent mutagens in CHO cells[76,106]; only ICR-170 gave a positive response (see below). Of the two topoisomerase inhibitors novobiocin and naladixic acid, the former had a DC value greater than 1, and UV-1 was nonresponsive to caffeine.

The pattern of mutagen sensitivity of UV-1 is broad and obviously complex. The most pronounced sensitivity is to methylating agents and

TABLE 4. Response of Mutants EM9 and UV-1 to Simple Alkylating Agents[a]

Compound	LEC,[b] μg/ml				DC ratio[c]	
	Gly^-A	UV-1	AA8	EM9	LEC(Gly^-A)/ LEC(UV-1)	LEC(AA8)/ LEC(EM9)
Methylnitrosourea	375	15	13	9	25	1
Methylazoxy methanol acetate	450	10	20	20	45	1
N-Methyl-*N*′-nitro-*N*-nitrosoguanidine (MNNG)	1.0	0.05	0.05	0.05	20	1
Methyl methanesulfonate[d,e]	50	10	10	1.2	5	8
Dimethyl sulfate	18	4	12	0.6	4.5	20
Streptozotocin[e]	1000	32	62	62	30	1
Ethylnitrosourea	300	100	400	300	3	1.3
Ethyl methanesulfonate[d,e]	600	200	>400	40	3	10
N-Ethyl-*N*′-nitro-*N*-nitrosoguanidine	2	0.75	3	1.5	2.6	2
Diethyl sulfate	500	150	200	25	3.3	8

[a] Cells were treated for 4 hr and incubated for 96 hr after exposure unless otherwise specified. For Gly^-A and UV-1 the initial inoculum was 8×10^3 and 2×10^4, respectively. Gly^-A refers to strain 77256, the glycine auxotrophic clone from which UV-1 was isolated.

[b] LEC is the lowest effective concentration to produce a measurable killing of the cells.

[c] DC is the differential cytotoxicity ratio defined as the LEC of the wild type divided by the LEC of the mutant.

[d] Exposure 24 hr for AA8 and EM9.

[e] AA8 and EM9 were incubated for 72 hr after exposure.

TABLE 5. Response of Mutant UV-1 to Various Agents in the DC Assay[a]

Compound	LEC, μg/ml		DC ratio
	UV-1	Gly⁻A(77256)	
Mitomycin C[b]	0.3	3	10
cis-Diamminedichloroplatinum	0.2	2	10
Diepoxybutane	0.07	3	40
Chlorambucil	0.8	4.8	6
UV light (254 nm)	6 (J/m^2)	12 (J/m^2)	2
Decarbamoyl mitomycin C	0.063	0.25	4
Adriamycin	0.02	0.10	5
Daunomycin	0.1	0.2	2
ICR-170[c]	0.37	0.75	2
ICR-191[c]	8	8	1
Naladixic acid[d]	600	600	1
Novobiocin[e]	500	1000	2
Caffeine[e]	800	800	1

[a] Exposures are 4 hr unless otherwise specified. See text for definitions of LEC and DC.
[b] Exposure 90 min.
[c] Exposure 60 min.
[d] Exposure 24 hr.
[e] Continuous exposure.

cross-linking compounds. UV-1 also appears slightly hypersensitive to the bulky monoadduct class of damage. This broad sensitivity of UV-1 raises the question of whether the phenotype of this line can be accounted for by a single mutational change. A genetic analysis of the properties of revertants and cell hybrids derived from UV-1 has indicated that the phenotype likely results from more than one mutation (L. Thompson and C. Hoy, unpublished results). The sensitivity to methylating and ethylating agents is probably a phenotypic change that was acquired adventitiously during or after the mutation conferring UV sensitivity occurred. This conclusion is based on the observation that in cell hybridization experiments mutants UV-1 and UV4 complement to give resistance to MC but not resistance to the methylating agent MMS. The sensitivity of UV-1 to methylation damage is apparently not associated with diminished binding of the agents to DNA. UV-1 cells and the parental Gly-A cells were shown to bind the same amount of radiolabeled methylnitrosourea (C. Hoy and L. Thompson, unpublished results).

4.7. Response of DC Assay to Alkylation Congeners

In the previous sections we considered the responses of the DC assay to two broad classes of alkylating agents: compounds forming

TABLE 6. Response of DC Assay to Congeners of Simple Alkylating Agents[a]

Compound	LEC, μg/ml					
	AA8	Gly⁻A	EM9	UV5	UV4	UV-1
Isopropyl methanesulfonate[b]	400	400	300 (1.3)	400 (1)	200 (2)	200 (2)
N-Propyl-*N*′-nitro-*N*-nitrosoguanidine[c]	10	10	9 (1.1)	10 (1)	5 (2)	5 (2)
Butylnitrosourea[c]	800	600	500 (1.6)	500 (1.6)	400 (2)	300 (2)
N-Nitrosodibutylamine[d]	500	370	500 (1)	370 (1.3)	500 (2)	370 (1.3)

[a] Unless indicated otherwise, exposures were 4 hr and subsequent incubation was 96 hr; see text for definition of LEC. Differential cytotoxicity (DC) values are given in parentheses; mutants are compared with their respective parental lines.
[b] Incubation of AA8, EM9, UV4, and UV5 was 72 hr.
[c] Exposures were 1 hr.
[d] Exposure was done in the presence of 2 mg/ml hamster S9 fraction.

bulky adducts (often multiring structures) and compounds that add small groups (methyl or ethyl). In summary, mutants UV4 and UV5 respond to the first class, and mutants EM9 and UV-1 respond to the second class. The excision repair mutants UV4 and UV5 generally do not respond to the damage from these small alkyl groups. To understand better the specificity profile of the mutants in terms of the size and structure of the adduct group, compounds were tested that should produce adducts of an intermediate size. Congeners were chosen that were related to the compounds in Table 4 and have the ability to donate alkyl groups composed of three or four carbon atoms.[89]

The results for the four compounds tested are given in Table 6. Several patterns emerge from the data. First, none of the mutants showed high sensitivity, which suggests that the lesions produced by these compounds are either inefficiently repaired or are repaired mostly by pathways besides those defective in these mutants. Nongenetic toxicity, if present, could also result in a weak DC response. Second, the most consistent positive response was with UV4, followed by UV-1. The results with UV4 suggest that a component of nucleotide excision repair acts on the alkyl groups donated by these compounds. Third, EM9 tended to have a near normal response, which suggests that these compounds probably do not generate large numbers of DNA strand breaks. Fourth, UV-1 usually had a DC of 2, consistent with its sensitivity to a broad range of DNA-damaging agents. In summary, the DC assay

TABLE 7. DNA-Damaging Agents Not Detected in the DC Assay[a]

	Ref.
Intercalating agents	
ICR-191	76
Compounds with apparently inadequate activation	
4-Methylaminoazobenzene[b]	1, 14, 66
3-Amino-1*H*-1,2,4-triazole[b]	66
N-Nitrosodiethylamine	62
Compounds with nongenotoxic cytotoxicity	
4,4′-Methylene-*bis*(2-chloraniline)[b]	66
Hydrazine sulfate[b]	1, 66
Safrole[c]	54
Isopropyl-*N*-(3-chlorophenyl)carbamate[b]	1, 54
N-Hydroxy-2-naphthylamine[c]	56
2-Naphthylamine	57
2-Amino-3-methylimidazo[4,5-*f*]-quinoline (IQ)	112

[a] The references cited provide either direct evidence for DNA damage or suggestive evidence from mammalian cell assays of mutation and/or unscheduled DNA synthesis.
[b] Compound were not tested in UV-1.
[c] Compound tested only in mutant UV5.

is capable of detecting compounds that donate longer alkyl groups than ethyl, although the response is consistently weak.

4.8. DNA-Damaging Agents Not Detected by the DC Assay

A number of chemicals known to interact with DNA did not give a positive response with any of the four mutants in the DC assay. These compounds are listed in three groups in Table 7. One chemical that stood out was ICR-191, a heterocyclic intercalating agent having a reactive side chain.[24] ICR-170, which is structurally very similar to ICR-191, was weakly positive in mutants UV-1 and EM9. Although ICR-170 was reported to induce repair synthesis in human cells,[2] an excision repair response may not always occur with ICR compounds.[118] In experiments with ICR-170 no difference in induced mutation frequency was observed between UV5 and AA8 cells, suggesting that the mutagenic lesions produced are not repairable by the UV excision pathway (C. Hoy and L. Thompson, unpublished results). Other intercalating agents, such as adriamycin, are noted for producing protein-associated strand breaks,[86,87] and apparently do not give lesions that are processed by excision repair.[122] Consistent with this idea was our finding that adriamycin and daunomycin produced a response only in UV-1 (Table 5), which has normal excision repair.[99] More information about the changes in DNA produced by these classes of chemicals is needed for evaluating the extent to which repair processes are involved.

A second group of mutagenic agents that were negative in the assay were those having apparently inadequate metabolic activation. In these cases toxicity either did not occur or occurred abruptly at the highest concentration (e.g., change from 5 to 0 in staining intensity from one well to the next) independent of the presence of S9. Thus, a valid test for differential cytotoxicity was not possible. In these assays more than one concentration of S9 microsomal fraction was normally tested. An example of these inactive compounds is *N*-nitrosodiethylamine, which was not toxic at doses up to 10 mg/ml in the presence of 2 mg/ml S9 (4 hr incubation).

With a third group of compounds, both direct-acting and S9-dependent, cytotoxicity occurred, but did not appear to be genotoxic in nature. For example, direct-acting *N*-hydroxy-2-naphthylamine seems to produce toxicity through a mechanism other than DNA damage, since it has very little mutagenic activity in this system.[113] The S9-dependent compound IQ is also a weak mutagen (Figure 5) and produces a very low level of DNA adducts at the high doses required to produce toxicity.[8] Thus, in these instances compounds that are weak but positive mutagens probably are toxic by acting mainly on cellular components other than DNA. The DC assay will be inherently unable to detect this class of mutagens.

4.9. Merits and Limitations of the DC Assay

The results discussed in preceding sections show that the DC assay is methodologically simple and practical. The studies were done with four mutant lines and their two parental lines. For direct-acting compounds that react to form covalent bonds with DNA, the efficiency of detection is high, with a small percentage of "false negatives." The response patterns of the mutants vary, depending on whether the compounds form cross-links or monoadducts, and whether the monadducts are methyl, ethyl, longer alkyl groups, or bulkier structures. Certain response patterns are diagnostic of particular classes of DNA damage, as summarized in Table 8. For example, positive but different responses in mutants UV4 and UV5 point to cross-link damage. One class of mutagens, those that intercalate, did not always give a positive response in the assay (e.g., ICR-191). Little is known about how bound

TABLE 8. Summary of Mutant Responses in DC Assay[a]

Class of DNA lesion	Responsive line and DC ratio
Bulky monoadduct	UV4 and UV5, both 2–10
Cross-linking	UV4, >10; UV5, 2–10
Methylating	UV-1, ≳10; EM9, 1–10
Ethylating	UV-1, ~3; EM9, 1–10
Propylating, butylating	UV4, ~2
Intercalating	UV-1, 2–5
Strand-breaking (bleomycin)	EM9, ~3; UV4 and UV5, 2

[a] Values are given only for the mutant lines that were consistently positive.

intercalators interact with the enzymes in DNA metabolism and whether a DNA repair response is involved. Thus, it is unclear whether the assay can be broadened to respond to such agents.

One example of a compound known to produce DNA strand breaks directly, bleomycin,[(64)] was evaluated (Table 8). The positive response in EM9 is consistent with the deficiency in break rejoining mentioned in Section 2.3.

As discussed, some compounds that cause mutations or DNA damage in other mammalian systems are not detected in the DC assay apparently because they are not activated by the microsomal fractions, or because their major reactions occur with cellular components other than DNA (Table 7). Although additional repair mutations might partially alleviate this problem, the assay is not expected to perform well with compounds that interact weakly with DNA. With respect to "false positives" the assay has performed well. None of the 13 toxic compounds tested that do not bind covalently to DNA produced a positive response in the assay (Table 1).

Besides its usefulness in detecting DNA-damaging agents, the DC assay also provides a convenient means of rapidly evaluating many aspects of the interactions of mutagens with cells. The assay is much cheaper to perform than survival curves based on colony formation, and many conditions of dose, exposure time, and other parameters can be examined simultaneously.

5. Conclusions

The collection of DNA repair mutants of CHO cells that we have discussed provides an approach to studying the molecular details of repair processes in both hamster and human cells. The UV-sensitive mutants belonging to five complementation groups presumably represent genes encoding different proteins required for the incision step of this complex repair pathway. Our findings that human genes will complement many of the CHO mutants show that these proteins are likely to be very similar biochemically between the two species. This complementarity allows the mapping of repair genes in the human karyotype and also provides a means of isolating these genes.

A differential cytotoxicity assay was devised to test classes of DNA-damaging agents rapidly by comparing their responses in the normal and mutant cells. One of the important patterns that emerged is the extreme hypersensitivity of mutant UV4, but not UV5, to bifunctional agents that form DNA interstrand cross-links. Both mutants have similar sensitivity to far-UV radiation and monofunctional bulky chemicals. These relationships suggest that certain steps are common in monoadduct and cross-link repair. While mutant EM9, noted especially for its elevated sister chromatid exchange, is hypersensitive to only certain simple alkylating agents, mutant UV-1 (from a different parental line) has much broader sensitivity. However, the complex phenotype of UV-1 (notable high sensitivity to methylating and cross-linking agents) appears to stem from mutations in more than one gene. Using this mutant line significantly extends the sensitivity of the DC assay, but requires a second parental line.

The DC assay employing all four mutant cell lines responds to a wide range of chemicals that are known to react with DNA. Most toxic compounds that do not react covalently with DNA do not show hypersensitivity in the mutants. In addition to its value in detecting new mutagens, the assay has important applications as a efficient method for evaluating the toxic effects of mutagens.

Acknowledgments

This work was performed under the auspices of the U. S. Department of Energy by the Lawrence Livermore National Laboratory under contract W-7405-ENG-48 and a grant from the U.S. Environmental Protection Agency (No. R808641-10). We thank our co-workers, Kerry Brookman, Ed Salazar, Sheri Stewart, Carolyn Mooney, and Jason Minkler, for their excellent assistance during this study. We also wish to recognize Dr. A. V. Carrano's stimulating interaction.

6. References

1. C. Agrelo and H. Amos, DNA repair in human fibroblasts, in: *Progress in Mutation Research, Evaluation of Short-Term Tests for Carcinogens* (F. J. de Serres and J. Ashby, eds.) Vol. 1, pp. 528–532, Elsevier/North-Holland, New York (1981).
2. F. E. Ahmed and R. B. Setlow, DNA excision in repair proficient and deficient human cells treated with a combination of ultraviolet radiation and acridine mustard (ICR-170) or 4-nitroquinoline 1-oxide, *Chem. Biol. Interact. 29,* 31–42 (1980).
3. D. E. Amacher, J. A. Elliott, and M. W. Lieberman, Differences in removal of acetylaminofluorene and pyrimidine dimers from the DNA of cultured mammalian cells, *Proc. Natl. Acad. Sci. USA 74,* 1553–1557 (1977).
4. A. D. Andrews, S. F. Barrett, and J. H. Robbins, Xeroderma pigmentosum neurological abnormalities correlate with colony-forming ability after ultraviolet radiation, *Proc. Natl. Acad. Sci. USA 75,* 1984–1988 (1978).
5. S. Arase, T. Kozuka, K. Tanaka, M. Ikenaga, and H. Takebe, A sixth complementation group in xeroderma pigmentosum, *Mutation Res. 59,* 143–146 (1979).
6. D. Bootsma, Xeroderma pigmentosum, in: *DNA Repair Mechanisms* (P. C. Hanawalt, E. C. Friedberg, and C. F. Fox, eds.), pp. 589–601, Academic Press, New York (1978).
7. M. O. Bradley, B. Bhuyan, M. C. Francis, R. Langenbach, A. Peterson, and E. Huberman, Mutagenesis by chemical agents in V79 Chinese hamster cells: A review and analysis of the literature, A report of the Gene-Tox Program, *Mutat. Res. 87,* 81–142 (1981).
8. K. W. Brookman, E. P. Salazar, and L. H. Thompson, Comparative mutagenic efficiencies of the DNA adducts from the cooked-food-related mutagens Trp-P-2 and IQ in CHO cells, *Mutat. Res. 149,* 249–255 (1985).
9. S. N. Buhl and J. D. Regan, Effect of caffeine on post-replication repair in human cells, *Biophys. J. 14,* 519–527 (1974).
10. D. B. Busch, Large Scale Isolation of DNA Repair Mutants of Chinese Hamster Ovary Cells, Ph.D. Thesis, University of California, Berkeley, California (1980).
11. D. B. Busch, J. E. Cleaver, and D. A. Glaser, Large-scale isolation of UV-sensitive clones of CHO cells, *Somat. Cell Genet. 6,* 407–418 (1980).
12. A. V. Carrano, J. L. Minkler, L. E. Dillehay, and L. H. Thompson, Cytogenetic characteristics of the CHO cell mutant EM9 (abstract), *Environ. Mutagen. 6,* 377 (1984).
13. J. H. Carver, G. M. Adair, and D. L. Wandres, Mutagenicity testing in mammalian cells II. Validation of multiple drug-resistance markers having practical application for screening potential mutagens, *Mutat. Res. 72,* 207–230 (1980).
14. J. H. Carver, E. P. Salazar, M. G. Knize, and D. L. Wandress, Mutation induction at multiple loci in Chinese hamster ovary cells: The genetic activity of 15 coded carcinogens and noncarcinogens, in: *Progress in Mutation Research, Evaluation of Short-Term Tests for Carcinogens* (F. J. de Serres and J. Ashby, eds.) Vol. 1, pp. 594–601, Elsevier/North-Holland, New York (1981).
15. R. S. K. Chaganti, S. Schonberg, and J. German, A manyfold increase in sister chromatid exchanges in Bloom's syndrome lymphocytes, *Proc. Natl. Acad. Sci. USA 71,* 4508–4512 (1974).
16. J. Y. H. Chan, L. H. Thompson, and F. F. Becker, DNA ligase activities appear normal in the CHO mutant EM9, *Mutat. Res. 131,* 209–214 (1984).
17. E. H. Y. Chu and H. V. Malling, Mammalian cell genetics, II. Chemical induction

of specific locus mutations in Chinese hamster cells *in vitro, Proc. Natl. Acad. Sci. USA 61,* 1306–1312 (1968).
18. J. M. Clarkson and D. L. Mitchell, The effect of various inhibitors of DNA synthesis on the repair of DNA photoproducts, *Biochim. Biophys. Acta 740,* 355–361 (1983).
19. D. B. Clayson and R. C. Garner, Carcinogenic aromatic amines and other compounds, in: *Chemical Carcinogens* (C. E. Searle, ed.) ACS Monograph 173, pp. 366–461, American Chemical Society, Washington D.C. (1976).
20. J. E. Cleaver, DNA repair and its coupling to DNA replication in eukaryotic cells, *Biochim. Biophys. Acta 516,* 489–516 (1978).
21. J. E. Cleaver, Specificity and completeness of inhibition of DNA repair by novobiocin and aphidicolin, *Carcinogenesis 3,* 1171–1174 (1982).
22. D. Clive, R. McCuen, J. F. S. Spector, C. Piper, and K. H. Mavournin, Specific gene mutations in L5178Y cells in culture, A report of the U.S. Environmental Protection Agency Gene-Tox Program, *Mutat. Res. 115,* 225–251 (1983).
23. A. R. S. Collins and R. T. Johnson, Use of metabolic inhibitors in repair studies, in: *DNA Repair, A Laboratory Manual of Research Procedures* (P. C. Hanawalt and E. C. Friedberg, eds.), Vol. 1, Part B, pp. 341–360, Dekker, New York (1981).
24. H. J. Creech, R. K. Preston, R. M. Peck, A. P. O'Connell, and B. N. Ames, Antitumor and mutagenic properties of a variety of heterocyclic nitrogen and sulfur mustards, *J. Med. Chem. 15,* 739–746 (1972).
25. L. E. Dillehay, L. H. Thompson, J. L. Minkler, and A. V. Carrano, The relationship between sister-chromatid exchange and perturbations in DNA replication in mutant EM9 and normal CHO cells, *Mutat. Res. 109,* 283–296 (1983).
26. L. E. Dillehay, L. H. Thompson, and A. V. Carrano, DNA strand-breaks associated with halogenated pyrimidine incorporation, *Mutat. Res. 31,* 129–136 (1984).
27. A. Dipple, Polynuclear aromatic hydrocarbons, in: *Chemical Carcinogens* (ACS Monograph 173) (C. E. Searle, ed.), pp. 245–314, American Chemical Society, Washington, D.C. (1976).
28. R. Domon, B. Barton, A. Porte, and A. M. Rauth, The interaction of caffeine with ultraviolet-light-irradiated DNA, *Int. J. Radiat. Biol. 17,* 395–399 (1970).
29. U. K. Ehmann, K. H. Cook, and E. C. Friedberg, The kinetics of thymine dimer excision in ultraviolet irradiated human cells, *Biohys. J. 22,* 249–264 (1978).
30. L. C. Erickson, M. O. Bradley, and K. W. Kohn, Measurements of DNA damage on Chinese hamster cells treated with equitoxic and equimutagenic doses of nitrosoureas, *Cancer Res. 38,* 3379–3384 (1979).
31. A. J. Fornace, K. W. Kohn, and H. E. Kann, Jr., DNA single-strand breaks during repair of UV damage in human fibroblasts and abnormalities of repair in xeroderma pigmentosum, *Proc. Natl. Acad. Sci. USA 73,* 39–43 (1976).
32. E. C. Friedberg, U. K. Ehmann, and J. I. Williams, Human diseases associated with defective DNA repair, *Adv. Radiat. Biol. 8,* 85–174 (1979).
33. E. C. Friedberg, L. Naumovski, E. Yang, G. A. Pure, R. A. Schultz, W. Weiss, and J. D. Love, Approaching the biochemistry of excision repair in eukaryotic cells: The use of cloned genes from *Saccharomyces cerevisiae,* in: *Cellular Responses to DNA Damage* (UCLA Symposia on Molecular and Cellular Biology, New Series, Vol. 11) (E. C. Friedberg and B. R. Bridges, eds.), pp. 63–75, Liss, New York (1983).
34. Y. Fujiwara, M. Tatsumi, and M. S. Sasaki, Cross-link repair in human cells and its possible defect in Fanconi's anemia cells, *J. Mol. Biol. 113,* 635–649 (1977).
35. J. C. Fuscoe, R. G. Fenwick, Jr., D. H. Ledbetter, and C. T. Caskey, Deletion and amplification of the HGPRT locus in Chinese hamster cells, *Mol. Cell. Biol. 3,* 1086–1096 (1983).

36. T. W. Glover, C.-C. Chang, J. E. Trosko, and S. S.-L. Li, Ultraviolet light induction of diptheria toxin-resistant mutants in normal and xeroderma pigmentosum human fibroblasts, *Proc. Natl. Acad. Sci. USA 76,* 3982–3986 (1979).
37. B. M. Goldschmidt, B. L. Van Duuren, and K. Frenkel, The reaction of ^{14}C-labeled bis(chloromethyl) ether with DNA, *Proc. Am. Assoc. Cancer Res. 16,* 66 (abstract no. 163) (1975).
38. M. M. Gottesman (ed.), *Molecular Cell Genetics,* Wiley, New York (1985).
39. L. H. Graf, Jr. and L. A. Chasin, Direct demonstration of genetic alterations at the dihydrofolate reductase locus after gamma irradiation, *Mol. Cell. Biol. 2,* 93–96 (1982).
40. R. S. Gupta, D. Y. H. Chan, and L. Siminovitch, Evidence for functional hemizygosity at the Emt^r locus in CHO cells through segregation analysis, *Cell 14,* 1007–1013 (1978).
41. P. C. Hanawalt, P. K. Cooper, A. K. Ganesan, and C. A. Smith, DNA repair in bacteria and mammalian cells, *Annu. Rev. Biochem. 48,* 783–836 (1979).
42. M. Harris, Mutation rates in cells at different ploidy levels, *J. Cell. Physiol. 78,* 177–184 (1971).
43. R. W. Hart and R. B. Setlow, Correlation between deoxyribonucleic acid excision-repair and life-span in a number of mammalian species, *Proc. Natl. Acad. Sci. USA 71,* 2169–2173 (1974).
44. H. Hayakawa, K. Ishizaki, M. Inoue, T. Yagi, M. Sekiguchi, and H. Takebe, Repair of ultraviolet radiation damage in xeroderma pigmentosum cells belonging to complementation group F, *Mutat. Res. 80,* 381–388 (1981).
45. R. H. Heflich, R. M. Hazard, L. Lommel, J. D. Scribner, V. M. Maher, and J. J. McCormick, A comparison of the DNA binding, cytotoxicity, and repair synthesis induced in human fibroblasts by reactive derivatives of aromatic amide carcinogens, *Chem. Biol. Interact. 29,* 43–56 (1980).
46. K. Hemminki, Nucleic acid adducts of chemical carcinogens and mutagens, *Arch. Toxicol. 52,* 249–285 (1983).
47. C. A. Hoy, E. P. Salazar, and L. H. Thompson, Rapid detection of DNA damaging agents using repair-deficient CHO cells, *Mutat. Res. 130,* 321–332 (1984).
48. C. A. Hoy, L. H. Thompson, C. L. Mooney, and E. P. Salazar, Defective DNA cross-link removal in Chinese hamster cell mutants hypersensitive to bifunctional alkylating agents, *Cancer Res. 45,* 1737–1743 (1985).
49. A. W. Hsie, J. P. O'Neill, D. B. Couch, J. R. SanSebastian, P. A. Brimer, R. Machanoff, J. C. Fuscoe, J. C. Riddle, A. P. Li, N. L. Forbes, and M. H. Hsie, Quantitative analysis of radiation- and chemical-induced lethality and mutagenesis in Chinese hamster ovary cells, *Radiat. Res. 76,* 471–492 (1978).
50. A. W. Hsie, J. P. O'Neill, and V. K. McElheny (eds.), *Banbury Report 2, Mammalian Cell Mutagenesis: The Maturation of Test Systems,* Cold Spring Harbor Laboratory, Cold Spring Harbor, New York (1979).
51. A. W. Hsie, D. A. Casciano, D. B. Couch, D. F. Krahn, J. P. O'Neill, and B. L. Whitfield, The use of Chinese hamster ovary cells to quantify specific locus mutation and to determine mutagenicity of chemicals, A report of the Gene-Tox Program, *Mutat. Res. 86,* 193–214 (1981).
52. M. Ikejima, D. Bohannon, D. M. Gill, and L. H. Thompson, Poly(ADP-ribose) metabolism appears normal in EM9, a mutagen-sensitive mutant of CHO cells, *Mutat. Res. 128,* 213–220 (1984).
53. M. Ikenaga, H. Takebe, and Y. Ishii, Excision repair of DNA base damage in human cells treated with the chemical carcinogen 4-nitroquinoline-1-oxide, *Mutat. Res. 43,* 415–427 (1977).

54. M. M. Jotz and A. D. Mitchell, Effects of 20 coded chemicals on the forward mutation frequency at the thymidine kinase locus in L5178Y mouse lymphoma cells, in: *Progress in Mutation Research, Evaluation of Short-Term Tests for Carcinogens* (F. J. de Serres and J. Ashby, eds.) Vol 1, pp. 580–593, Elsevier/North-Holland, New York (1981).
55. T. Kada, K. Hirano, and Y. Shirasu, Screening of environmental chemical mutagens by the Rec-assay system with *Bacillus subtilis,* in: *Chemical Mutagens: Principles and Methods for Their Detection* (F. J. de Serres and A. Hollaender, eds.), Vol. 6, pp. 149–173, Plenum Press, New York (1980).
56. F. F. Kadlubar, L. E. Unruh, F. A. Beland, K. M. Straub, and F. E. Evans, *In vitro* reaction of the carcinogen, *N*-hydroxy-2-naphthylamine with DNA at the C-8 and N^2 atoms of guanine and the N^6 atom of adenine, *Carcinogenesis 1,* 139–150 (1980).
57. F. F. Kadlubar, J. F. Anson, K. L. Dooley, and F. A. Beland, Formation of urothelial and hepatic DNA adducts from the carcinogen 2-naphthylamine, *Carcinogenesis 2,* 467–470 (1981).
58. F.-T. Kao and T. T. Puck, Genetics of somatic mammalian cells, VII. Induction and isolation of nutritional mutants in Chinese hamster cells, *Proc. Natl. Acad. Sci. USA 60,* 1275–1281 (1968).
59. W. Keijzer, N. G. J. Jaspers, P. J. Abrahams, A. M. R. Taylor, C. F. Arlett, B. Zelle, H. Takebe, P. D. S. Kinmont, and D. Bootsma, A seventh complementation group in excision-deficient xeroderma pigmentosum, *Mutat. Res. 62,* 183–190 (1979).
60. B. Konze-Thomas, J. W. Levinson, V. M. Maher, and J. J. McCormick, Correlation among the rates of dimer excision, DNA repair replication, and recovery of human cells from potentially lethal damage induced by ultraviolet radiation, *Biophys. J. 28,* 315–325 (1979).
61. M. La Belle, S. Linn, and L. H. Thompson, Apurinic/apyrimidinic endonuclease activities appear normal in the CHO-cell ethyl methanesulfonate-sensitive mutant, EM9, *Mutat. Res. 141,* 41–44 (1984).
62. P. D. Lawley, Carcinogenesis by alkylating agents, in: *Chemical Carcinogens* (ACS Monograph 173) (C. E. Searle, ed.), pp. 83–244, American Chemical Society, Washington, D.C. (1976).
63. T. Lindahl, DNA repair enzymes, *Annu. Rev. Biochem. 51,* 61–87 (1982).
64. R. S. Lloyd, C. W. Haidle, and R. R. Hewitt, Bleomycin-induced alkaline-labile damage and direct strand breakage of PM2 DNA, *Cancer Res. 38,* 3191–3196 (1978).
65. V. M. Maher, D. J. Dorney, A. L. Mendrala, B. Konze-Thomas, and J. J. McCormick, DNA excision-repair processes in human cells can eliminate the cytotoxic and mutagenic consequences of ultraviolet irradiation, *Mutat. Res. 62,* 311–323 (1979).
66. G. N. Martin and A. G. McDermid, Testing of 42 compounds for their ability to induce unscheduled DNA repair synthesis in HeLa cells, in: *Progress in Mutation Research, Evaluation of Short-Term Tests for Carcinogens* (F. J. de Serres and J. Ashby, eds.) Vol 1, pp. 533–537, Elsevier/North-Holland, New York (1981).
67. N. Matsukura, T. Kawachi, K. Morino, H. Ohgaki, T. Sugimura, and S. Takayama, Carcinogenicity in mice of mutagenic compounds from tryptophan pyrolyzate, *Science 213,* 346–347 (1981).
68. B. A. McCaw, A. Dipple, S. Young, and J. J. Roberts, Excision of hydrocarbon–DNA adducts and consequent survival in normal and repair defective human cells, *Chem. Biol. Interact. 22,* 139–151 (1978).
69. M. Meuth and J. E. Arrand, Alterations of gene structure in ethyl methane sulfonate-induced mutants of mammalian cells, *Mol. Cell. Biol. 2,* 1459–1462 (1982).
70. R. E. Meyn, D. L. Vizard, R. R. Hewitt, and R. M. Humphrey, The fate of pyrimidine

dimers in the DNA of ultraviolet-irradiated Chinese hamster cells, *Photochem. Photobiol. 20,* 221–226 (1974).

71. L. Mezger-Freed, Effect of ploidy and mutagens on bromodeoxyuridine resistance in haploid and diploid frog cells, *Nature New Biol. 235,* 245–246 (1972).
72. E. C. Miller and J. A. Miller, The metabolism of chemical carcinogens to reactive electrophiles and their possible mechanisms of action in carcinogenesis, in: *Chemical Carcinogens* (C. E. Searle, ed.) ACS Monograph 173, pp. 737–764, American Chemical Society, Washington, D.C. (1976).
73. A. N. Moshell, M. B. Ganges, M. A. Lutzner, H. G. Coon, S. F. Barrett, J.-M. Dupuy, and J. H. Robbins, A new patient with both xeroderma pigmentosum and Cockayne syndrome comprises the new xeroderma pigmentosum complementation group H (abstract), *J. Cell. Biochem. Suppl. 7B,* 202 (1983).
74. J. P. O'Neill and A. W. Hsie, Chemical mutagenesis of mammalian cells can be quantified, *Nature 269,* 815–817 (1977).
75. J. P. O'Neill, P. A. Brimer, R. Machanoff, G. P. Hirsch, and A. W. Hsie, A quantitative assay of mutation induction at the hypoxanthine-guanine phosphoribosyl transferase locus in Chinese hamster ovary cells (CHO/HPRT system): Development and definition of the system, *Mutat. Res. 45,* 91–101 (1977).
76. J. P. O'Neill, J. C. Fuscoe, and A. W. Hsie, Mutagenicity of heterocyclic nitrogen mustards (ICR compounds) in cultured mammalian cells, *Cancer Res. 38,* 506–509 (1978).
77. N. Otsuji and I. Murayama, Deoxyribonucleic acid damage of monofunctional mitomycins and its repair in *Escherichia coli, J. Bacteriol. 105,* 475–483 (1972).
78. S. D. Park, K. H. Choi, S. W. Hong, and J. E. Cleaver, Inhibition of excision-repair of ultraviolet damage in human cells by exposure to methyl methanesulfonate, *Mutat. Res. 82,* 365–371 (1981).
79. D. Pinkel, L. H. Thompson, J. W. Gray, and M. Vanderlaan, Measurement of sister chromatid exchanges at very low BrdUrd substitution levels using a monoclonal antibody, *Cancer Res. 45,* in press (1985).
80. A. C. Plooy, M. van Dijk, and P. H. M. Lohman, Induction and repair of DNA cross-links in Chinese hamster ovary cells treated with various platinum coordination compounds in relation to platinum binding to DNA, cytotoxicity, mutagenicity, and antitumor activity, *Cancer Res. 44,* 2043–2051 (1984).
81. W. B. Pratt and R. W. Ruddon, The alkylating agents, in: *The Anti-Cancer Drugs,* pp. 64–97, Oxford University Press, Oxford (1979).
82. T. T. Puck and F.-T. Kao, Somatic cell genetics and its application to medicine, *Annu. Rev. Genet. 16,* 225–271 (1982).
83. J. H. Ray and J. German, Bloom's syndrome and EM9 cells in BrdU-containing medium exhibit similarly elevated frequencies of sister chromatid exchange but dissimilar amounts of cellular proliferation and chromosome disruption, *Chromosoma 90,* 383–388 (1984).
84. P. N. Ray and L. Siminovitch, Genetic variants of cultured animal cells, in: *Somatic Cell Genetics* (C. T. Caskey and D. C. Robbins, eds.), pp. 127–167, Plenum Press, New York (1982).
85. H. S. Rosenkranz and Z. Leifer, Determining the DNA-modifying activity of chemicals using DNA-polymerase-deficient *Escherichia Coli,* in: *Chemical Mutagens: Principles and Methods for Their Detection* (F. J. de Serres and A. Hollaender, eds.), Vol. 6, pp. 109–147, Plenum Press, New York (1980).
86. W. E. Ross and M. C. Smith, Repair of deoxyribonucleic acid lesions caused by adriamycin and ellipticine, *Biochem. Pharmacol. 31,* 1931–1935 (1982).

87. W. E. Ross, D. Glaubiger, and K. W. Kohn, Qualitative and quantitative aspects of intercalator-induced DNA strand breaks, *Biochim. Biophys. Acta 562,* 41–50 (1979).
88. J. S. Rubin, A. L. Joyner, A. Bernstein, and G. F. Whitmore, Molecular identification of a human DNA repair gene following DNA-mediated gene transfer, *Nature 306,* 206–208 (1983).
89. R. Saffhill, *In vitro* reaction of *N*-*n*-butyl-*N*-nitrosourea and *n*-butyl methanesulfonate with guanine and thymine bases of DNA, *Carcinogenesis 5,* 621–625 (1984).
90. K. Sato and N. Hieda, Isolation of a mammalian cell mutant sensitive to 4-nitroquinoline-1-oxide, *Int. J. Radiat. Biol. 35,* 83–87 (1979).
91. K. Sato and R. B. Setlow, DNA repair in a UV-sensitive mutant of a mouse cell line, *Mutat. Res. 84,* 443–455 (1981).
92. T. Shiomi, N. Hieda-Shiomi, and K. Sato, Isolation of UV-sensitive mutants of mouse L5178Y cells by a cell suspension spotting method, *Somat. Cell Genet. 8,* 329–345 (1982).
93. Y. Shiraishi, T. H. Yosida, and A. A. Sandberg, Analyses of bromodeoxyuridine-associated sister chromatid exchanges (SCEs) in Bloom syndrome based on cell fusion: Single and twin SCEs in endoreduplication, *Proc. Natl. Acad. Sci. USA 80,* 4369–4373 (1983).
94. M. J. Siciliano, J. Siciliano, and R. M. Humphrey, Electrophoretic shift mutants in Chinese hamster ovary cells: Evidence for genetic diploidy, *Proc. Natl. Acad. Sci. USA 75,* 1919–1923 (1978).
95. M. J. Siciliano, R. L. Stallings, G. M. Adair, R. M. Humphrey, and J. Siciliano, Provisional assignment of TPI, GPI, and PEPD to Chinese hamster autosomes 8 and 9: A cytogenetic basis for functional haploidy of an autosomal linkage group in CHO cells, *Cytogenet. Cell Genet. 35,* 15–20 (1983).
96. B. Singer and J. T. Kusmierek, Chemical mutagenesis, *Annu. Rev. Biochem. 52,* 655–693 (1982).
97. P. S. Song and K. J. Tapley, Jr., Photochemistry and photobiology of psoralens, *Photochem. Photobiol. 29,* 1177–1197 (1979).
98. T. D. Stamato and C. A. Waldren, Isolation of UV-sensitive variants of CHO-Kl by nylon cloth replica plating, *Somat. Cell Genet. 3,* 431–440, (1977).
99. T. D. Stamato, L. Hinkle, A. R. S. Collins, and C. A. Waldren, Chinese hamster ovary mutant UV-1 is hypomutable and defective in a postreplication recovery process, *Somat. Cell Genet. 7,* 307–320 (1981).
100. S. Takayama, Y. Nakatsuru, M. Masuda, H. Ohgaki, S. Sato, and T. Sugimura, Demonstration of carcinogenicity in F344 rats of 2-amino-3-methyl-imidazo[4,5-f]quinoline from broiled sardine, fried beef and beef extract, *Gann 75,* 467–470 (1984).
101. K. Tanaka, H. Hayakawa, M. Sekiguchi, and Y. Okada, Specific action of T4 endonuclease V on damaged DNA in xeroderma pigmentosum cells *in vivo, Proc. Natl. Acad. Sci. USA 74,* 2958–2962 (1977).
102. W. G. Thilly, J. G. DeLuca, E. E. Furth, H. Hoppe IV, D. A. Kaden, J. J. Krolewski, H. L. Liber, T. R. Skopek, S. A. Slapikoff, R. J. Tizard, and B. W. Penman, Gene-locus mutation assays in diploid human lymphoblast lines, in: *Chemical Mutagens: Principles and Methods for Their Detection* (F. J. de Serres and A. Hollaender, eds.), Vol. 6, pp. 331–363, Plenum Press, New York (1980).
103. L. H. Thompson, DNA repair mutants, in: *Molecular Cell Genetics* (M. M. Gottesman, ed.), pp. 641–667, Wiley, New York (1985).
104. L. H. Thompson and R. M. Baker, Isolation of mutants of cultured mammalian cells, in: *Methods in Cell Biology* (D. M. Prescott, ed.), Vol. VI, pp. 209–281, Academic Press, New York (1973).

105. L. H. Thompson, D. B. Busch, K. Brookman, C. L. Mooney, and D. A. Glaser, Genetic diversity of UV-sensitive DNA repair mutants of Chinese hamster ovary cells, *Proc. Natl. Acad. Sci. USA 78,* 3734–3737 (1980).
106. L. H. Thompson, S. Fong, and K. Brookman, Validation of conditions for efficient detection of HPRT and APRT mutations in suspension-cultured Chinese hamster ovary cells, *Mutat. Res. 74,* 21–36 (1980).
107. L. H. Thompson, J. S. Rubin, J. E. Cleaver, G. F. Whitmore, and K. Brookman, A screening method for isolating DNA repair-deficient mutants of CHO cells, *Somat. Cell Genet. 6,* 391–405 (1980).
108. L. H. Thompson, K. W. Brookman, L. D. Dillehay, A. V. Carrano, J. A. Mazrimas, C. L. Mooney, and J. L. Minkler, A CHO-cell strain having hypersensitivity to mutagens, a defect in DNA strand-break repair, and an extraordinary baseline frequency of sister chromatid exchange, *Mutat. Res. 95,* 427–440 (1982).
109. L. H. Thompson, K. W. Brookman, L. E. Dillehay, C. L. Mooney, and A. V. Carrano, Hypersensitivity to mutation and sister-chromatid exchange induction in CHO cell mutants defective in incising DNA containing UV lesions, *Somat. Cell Genet. 8,* 759–773 (1982).
110. L. H. Thompson, K. W. Brookman, A. V. Carrano, and L. E. Dillehay, Role of DNA repair in mutagenesis of Chinese hamster ovary cells by 7-bromomethylbenz[*a*]anthracene, *Proc. Natl. Acad. Sci. USA 79,* 534–538 (1982).
111. L. H. Thompson and A. V. Carrano, Analysis of mammalian cell mutagenesis and DNA repair using *in vitro* selected CHO cell mutants, in: *Cellular Responses to DNA Damage* (UCLA Symposia on Molecular and Cellular Biology, New Series, Vol. 11) (E. C. Friedberg and B. R. Bridges, eds.), pp. 125–143, Liss, New York (1983).
112. L. H. Thompson, A. V. Carrano, E. Salazar, J. S. Felton, and F. T. Hatch, Comparative genotoxic effects of the cooked-food-related mutagens Trp-P-2 and IQ in bacteria and cultured mammalian cells, *Mutat. Res. 117,* 243–257 (1983).
113. L. H. Thompson, E. P. Salazar, K. W. Brookman, and C. A. Hoy, Hypersensitivity to cell killing and mutation induction by chemical carcinogens in an excision repair-deficient mutant of CHO cells, *Mutat. Res. 112,* 329–344 (1983).
114. L. H. Thompson, K. W. Brookman, and C. L. Mooney, Repair of DNA adducts in asynchronous CHO cells and the role of repair in cell killing and mutation induction in synchronous cells treated with 7-bromomethylbenz[*a*]anthracene, *Somat. Cell Mol. Genet. 10,* 183–194 (1984).
115. L. H. Thompson, K. W. Brookman, J. L. Minkler, J. C. Fuscoe, K. A. Henning, and A. V. Carrano, DNA-mediated transfer of a human DNA repair gene that controls sister-chromatid exchange, *Mol. Cell. Biol. 5,* 881–884 (1985).
116. L. H. Thompson, C. L. Mooney, K. Burkhart-Schultz, A. V. Carrano, and M. J. Siciliano, Correction of a nucleotide-excision-repair mutation by human chromosome 19 in hamster–human hybrid cells, *Somat. Cell Mol. Genet. 11,* 87–92 (1985).
117. W. G. Verly, L. Brakier, and P. W. Felt, Inactivation of the T7 coliphage by diepoxybutane stereoisomers, *Biochim Biophys. Acta 228,* 400–406 (1971).
118. D. Viceps-Madore and L. Mezger-Freed, Studies on DNA repair in frog and human cells exposed to an acridine half-mustard (ICR191) and to MNNG, *Mutat. Res. 49,* 407–419 (1978).
119. R. A. Vincent, Jr., M. D. Hays, and R. C. Johnson, Single-strand DNA breakage and repair in Bloom's syndrome cells, in: *DNA Repair Mechanisms* (P. C. Hanawalt, E. C. Friedberg, and C. F. Fox, eds.), pp. 663–666, Academic Press, New York (1978).
120. C. Waldren, D. Snead, and T. Stamato, Restoration of normal resistance to killing

and of post-replication recovery (PRR) in CHO-UV-1 cells by transformation with hamster or human DNA, in: *Cellular Responses to DNA Damage* (UCLA Symposia on Molecular and Cellular Biology, New Series, Vol. 11) (E. C. Friedberg and B. R. Bridges, eds.), pp. 637–646, Liss, New York (1983).

121. A. Westerveld, J. H. J. Hoeijmakers, M. van Duin, J. de Wit, H. Odijk, A. Pastink, R. D. Wood, and D. Bootsma, Molecular cloning of a human DNA repair gene, *Nature 310,* 425–429 (1984).
122. R. J. Wilkins, Failure of the intercalating agent 4′-(9-acridylamino)-methanesulphon-*m*-anisidide to induce DNA-repair replication in cultured human cells, *Mutat. Res. 122,* 211–216 (1983).
123. R. D. Wood and H. J. Burki, Repair capability and the cellular age response for killing and mutation induction after UV, *Mutat. Res. 95,* 505–514 (1982).
124. R. G. Worton, C. C. Ho, and C. Duff, Chromosome stability in CHO cells, *Somat. Cell Genet. 3,* 27–45 (1977).
125. L. L. Yang, V. M. Maher, and J. J. McCormick, Error-free excision of the cytotoxic, mutagenic N^2-deoxyguanosine DNA adduct formed in human fibroblasts by (±)-7β,8α-dihyroxy-9α,10α-epoxy-7,8,9,10-tetrahydro-benzo[*a*]pyrene, *Proc. Natl. Acad. Sci. USA 77,* 5933–5937 (1980).
126. B. Zelle and P. H. M. Lohman, Repair of UV-endonuclease-susceptible sites in the 7 complementation groups of xeroderma pigmentosum A through G, *Mutat. Res. 62,* 363–368 (1979).
127. L. A. Zwelling, M. O. Bradley, N. A. Sharkey, T. Anderson, and K. W. Kohn, Mutagenicity, cytotoxicity and DNA crosslinking in V79 Chinese hamster cells treated with *cis*- and *trans*-Pt(II) diamminedichloride, *Mutat. Res. 67,* 271–280 (1979).
128. L. A. Zwelling, S. Michaels, L. C. Erickson, R. S. Ungerleider, N. Nichols, and K. W. Kohn, Protein-associated deoxyribonucleic acid strand breaks in L1210 cells treated with the deoxyribonucleic acid intercalating agents 4′-(acridinylamino)methanesulfon-*m*-anisidide and adriamycin, *Biochemistry 20,* 6553–6563 (1981).

CHAPTER 8

Transplacental Genotoxic Agents: Cytogenetic Methods for Their Detection

Leigh Henderson

1. Introduction

Transplacental exposure to agents that produce lesions in DNA may induce embryotoxicity, congenital malformations, tumors in the postnatal period, heritable mutations in the germ line, or a range of potentially adverse outcomes such as altered immunity[1] or altered metabolism.[2] The gestational age at the time of exposure is a major factor in determining which of these outcomes predominates. In rodents, embryotoxic effects are generally found in early pregnancy, followed by a sensitive period for congenital malformations in mid-pregnancy and tumor induction in the perinatal period. Since animal carcinogenicity and teratogenicity studies are time-consuming and expensive, the development of short-term assays for transplacentally active genotoxic chemicals is highly desirable. Cytogenetic assays for chromosome aberrations and sister chromatid exchanges (SCEs) have been developed in a variety of *in vitro* and *in vivo* test systems. The relationship of these endpoints to tumorigenesis and teratogenesis, and their appli-

Leigh Henderson • Centre for Medical Research, School of Biological Sciences, University of Sussex, Falmer, Brighton BN1 9QG, England.

cation to the prediction of fetal risk and to the elucidation of the biological basis of transplacental carcinogenesis and teratogenesis are discussed in this chapter.

It has been known for some time that specific chromosome changes are found in some tumors, particularly those of hematopoietic cells. However, only recently have major advances been reported in studies of the roles of oncogenes in carcinogenesis, which are leading to an understanding of the correlation between chromosome aberrations and cancer.[3] Several of the breakpoints involved in chromosome translocations and deletions in tumors occur at the locations of human cellular oncogenes. The mechanisms by which translocations or deletions of chromosomes containing cellular oncogenes are related to malignant transformations remains to be elucidated, but this observation suggests that chromosome aberrations may be among the causes of tumors and gives confidence in the predictive value of assays for chromosome aberrations in short-term tests for carcinogenicity.

The relationship of genetic lesions to teratogenesis is even less well understood. While it is known that many carcinogens are teratogens,[4,5] the mode of action of many teratogens does not seem to be by induction of DNA damage[6] and therefore any short-term tests for genotoxic agents will only be predictive for some teratogens. Several mechanisms may be involved in teratogenicity produced by chemicals that act on DNA. Cell death at critical developmental stages leads to depletion of cell numbers required for organogenesis. Delays in DNA synthesis and cell cycle progression, resulting from DNA damage, will have a similar effect. The role of mutations leading to an alteration in the programming of the developmental fate of the cell has only recently been investigated.[7] Some chemicals, e.g., bromodeoxyuridine and iododeoxyuridine, are mutagenic, teratogenic, and carcinogenic at concentrations that do not induce cytotoxicity, suggesting that mutagenicity rather than cell death may be important in the teratogenicity of these compounds.

The first attempts to develop cytogenetic tests that can be applied to transplacental exposure were made in the 1960s.[8,9] Currently, two cytogenetic endpoints are used to investigate chromosomal damage to the fetus. Chromosome aberrations are detected by analysis of metaphase chromosomes or indirectly by the induction of micronuclei. Chromosome breakage and subsequent loss of genetic material is usually a lethal event and may therefore serve as an indication of possible teratogenicity. As discussed above, chromosome breakage and rearrangements may be important in the development of neoplasia. The sister chromatid exchange assay has gained widespread acceptance as an indicator of DNA damage, despite an ignorance of its genetic significance or molecular basis. SCEs are induced by a variety of genotoxic agents and appear to indicate unrepaired damage in DNA, as evidence by their longer persistence in repair-deficient cell lines.[10] While most mutagens induce SCEs, the ratio of mutants to SCEs induced by different chemicals varies,[11] suggesting that the two events are not the consequence of the same lesion, although they may share some common pathways. Thus, while SCEs may indicate the interaction of a chemical with DNA, they may not be a quantitative indicator of mutagenesis. In addition to indicating the interaction of a chemical with DNA, the use of sister chromatid differentiation may be applicable to short-term testing for teratogens by indicating toxicity based on the inhibition of cell replication.[12,13]

2. Animal Studies

2.1. Metaphase Analysis

The rate of cell division in the developing fetus provides a rich source of cells for metaphase analysis. The mitotic index and the cell populations contributing to the analyzed metaphases will be affected by gestational age. The *in vivo* technique involves the administration of the agent to the pregnant dam, followed by an appropriate expression time to allow replication and then metaphase arrest *in vivo*. Most reports of metaphase analysis involve the use of whole-embryo preparations.[14–16] Alternatively, a protocol which involves *in vivo* treatment of the fetus followed by an *in vitro* culture period can be used. This latter method is desirable where tissue-specific effects are under investigation and the culture procedure produces an enrichment of the desired cell types. For example, Tsuda *et al.*[17,18] used this protocol to investigate the tissue-specific effects of diethylnitrosamine on tracheal epithelial cells in the Syrian hamster. Since tissue explantation and culture initiation will disrupt normal cell cycle progression, this technique requires investigation of onset of division to ensure chromosome aberrations are scored in first-division cells. The 24-hr *in vivo* exposure period followed by a 24-hr culture duration used by Inui *et al.*[19,20] in their studies of chromosome aberrations induced by furylfuramide (AF-2) and sodium nitrite on Syrian hamster fetuses may well result in later divisions than the first following exposure being scored. This could result in lower levels of scorable damage and possibly contribute to the lower sensitivity of the chromosome aberration assay than the analysis of 6-thioguanine and 8-azaguanine mutants reported by these authors. Different chemicals are effective at different stages of the cell cycle and vary in their rates of metabolism and uptake. In addition, they may differ in their abilities to induce chromosome damage at divisions subsequent to the first posttreatment division and therefore it is desirable to determine experimentally the optimum time interval between treatment and metaphase harvest. In a study of the induction of chromatid aberrations in whole embryos exposed *in utero,* Adler[16] reported optimal times ranging from 3 hr for bleomycin, which is effective in the G_2 phase of the cell cycle, to 18 hr for benzo(*a*)pyrene.

2.2. Micronuclei

An alternative to the traditional chromosome analysis technique for the detection of chromosome breakage is provided by the micronucleus test. Micronuclei result from chromosome breakage and spindle damage. The relative advantages and disadvantages of the micronucleus assay and metaphase analysis have been discussed extensively.[21,22] The assessment of micronuclei is considerably simpler than conventional metaphase analysis, more cells are available for examination, and the test does not require a highly trained observer. The background frequency of micronuclei in polychromatic erythrocytes is low, with little variation in control animals, allowing small elevations after treatment to be statistically significant. A shorter time is required to demonstrate a dose response when micronuclei are scored rather than chromosome aberrations.[23]

The micronucleus test detects compounds that are positive in conventional chromosome aberration analyses and also compounds that are spindle poisons. Few studies have considered the relative sensitivities of the two techniques. An indication that the micronucleus assay may be less sensitive than chromosome aberration analysis has been shown by Kliesch and co-workers. The lowest effective dose for both micronuclei and chromosome aberrations induced in adult bone marrow was found to be the same for mitomycin C (MC) and procarbazine[24] but higher for micronuclei induction by methyl methanesulfonate (MMS)[24] and benzo(*a*)pyrene.[25]

The original micronucleus assay developed by Schmid,[22] using bone marrow polychromatic erythrocytes, has been adapted to polychromatic erythrocytes in fetal mouse liver and blood.[26–28] Fetal erythropoiesis occurs in the liver, and immature polychromatic erythrocytes persist in the liver before entering the peripheral blood circulation. Both tissues can therefore be scored for micronuclei, although

TABLE 1. Effect of Sampling Time on Frequencies of Micronucleated Polychromatic Erythrocytes in Mice[a]

		Frequency for given exposure time[c]							
Treatment	Tissue[b]	Control	6 hr	12 hr	18 hr	24 hr	30 hr	36 hr	48 hr
γ-Irradiation	BM	0.3	0.3	1.3	2.2	2.6	2.4	2.0	—
(50 rad)	FL	0.3	0.6	2.0	1.8	1.6	1.2	1.0	—
MMS	BM	0.2	—	0.8	3.7	3.9	3.1	2.2	—
(50 mg/kg)	FL	0.3	—	4.6	7.1	3.6	1.2	0.8	—
Procarbazine	BM	0.2	—	—	0.7	3.4	3.5	—	—
(30 mg/kg)	FL	0.4	—	0.7	7.9	5.9	3.0	—	—
MC	BM	0.2	0.2	2.4	4.1	6.2	8.0	5.7	—
(2 mg/kg)	FL	0.4	0.4	4.0	3.5	2.0	—	0.9	—
Benzo(*a*)pyrene	BM	0.2	—	—	0.3	0.6	1.2	1.6	2.0
(300 mg/kg)	FL	0.3	—	—	0.3	0.5	0.8	0.8	0.6

[a] From Cole *et al.*[27]
[b] BM, Maternal bone marrow; FL, fetal liver.
[c] Frequencies are given as number of micronucleated polychromatic erythrocytes per 100 polychromatic erythrocytes. Peak frequencies are underlined.

peak frequencies will occur at later times in fetal blood and may be lower due to the dilution of exposed polychromatic erythrocytes with nonexposed cells in the blood. The cell cycle duration of erythroblasts in adult bone marrow is longer than in fetal liver and erythrocytes persist for longer in bone marrow than in fetal liver before entering the peripheral circulation. Consideration of cell cycle times and persistence times based on kinetic data determined by autoradiography led to predictions on the induction of micronuclei at various sampling times in bone marrow, fetal liver, and fetal blood.[26,27]

In practice, many factors, such as cell cycle delays, the persistence of slowly metabolized substances, delays in metabolic activation or compound uptake, the stages of the cell cycle where compounds are effective, and the persistence of lesions leading to chromosome breakage, will affect the sampling time at which peak micronuclei incidences are observed. For this reason it is important that time course data be obtained for each new chemical under test and the optimum expression time determined. Time course data should be obtained at doses that do not delay cell division or maturation. To demonstrate variation in peak sampling times, time course data for γ-irradiation, MMS, MC, benzo(*a*)pyrene, and procarbazine are shown in Table 1. Optimal sampling times vary in fetal liver from 12 hr for MC to 30–36 hr for benzo(*a*)pyrene and in bone marrow from 24 hr for γ-irradiation, MMS, and procarbazine to 48 hr for benzo(*a*)pyrene. While a single sampling time of 30 hr (as suggested by Yamamoto and Kikuchi[29]) would not produce a false negative value for any of the compounds for either tissue, it would very significantly underestimate the effectiveness in some cases. In addition, other authors have reported that some compounds may not be detected at a single sample time of 30 hr.[25]

Optimal regimes for dosing schedules in the micronucleus test have been the subject of some debate. The original protocol for adult bone marrow proposed by Schmid[22] involved two doses given 30 and 6 hr before cell harvest. The second injection occurs after the final cell division and, while possibly inhibiting maturation of polychromatic erythrocytes, will not increase the incidence of micronuclei. Multiple doses given at appropriate intervals spanning consecutive cell cycles may increase the yield of micronuclei compared with a single dose, although producing lower frequencies of micronuclei than the equivalent dosage given in the last cell cycle.[30] It was suggested that this could improve the sensitivity of the test, particularly where rapidly metabolized substances are being investigated or for marginally effective substances where high dosages are too toxic, but a split dose can be tolerated. As yet, no compound has been found to be detectable by the use of multiple doses and not by the use of a single dose. For the sake of simplicity and sensitivity in routine testing it is more important to use multiple sampling times following a single treatment.

Since fetal erythropoiesis occurs in the liver, erythroblast maturation occurs at sites where blood cells may be exposed to metabolites formed by fetal hepatocytes. Fetal liver has some activating enzymes not found in bone marrow cells and therefore fetal erythroblasts may be exposed to some chromosome-breaking agents that do not affect adult erythroblasts maturing in the bone marrow. The transplacental micronucleus test will be particularly sensitive to compounds activated by adult and fetal liver but not bone marrow that are metabolized to species with short half-lives. To date, lasiocarpine,[31] dimethylnitrosamine,[32] and diethylnitrosamine[33] have been found to be positive in the transplacental micronucleus test but negative in adult bone marrow. However, dimethylnitrosamine has been found to be active in adult bone marrow in another study.[34]

A variation on the transplacental micronucleus test involves the exposure of the fetus toward the end of pregnancy and the collection of blood from the tail vein at birth.[33] This method is considerably simpler than the preparation of erythrocytes from the fetus and allows the survival of the animal when correlation with longer term endpoints is desired.

It has been suggested that the scoring of normochromatic erythrocytes in peripheral blood provides a means of monitoring chronic exposure to chemicals.[35] If damaged polychromatic erythrocytes are not selectively removed in fetal mice, but enter the peripheral circulation, this method could be used for monitoring the effects of chronic exposure during pregnancy. This may have advantages over the SCE assay, which may detect acute doses but not chronic exposures (see Section 2.3).

While the micronucleus test has found widest application with polychromatic erythrocytes, in practice it is applicable to any cell type that has a sufficient cytoplasmic to nuclear ratio and also has proliferative potential. King and Wild[28] scored micronuclei in nucleated fetal blood cells in parallel with fetal blood erythrocytes, although this was found to be less sensitive than scoring erythrocytes. Inui *et al.*[20] found an increase in micronuclei in nucleated hamster embryo cells after *in vivo* treatment by sodium nitrite followed by a 30-hr culture period. The disadvantages of scoring micronuclei in nucleated cells are that the cohort of exposed cells cannot be defined as erythrocytes can and scoring is made more difficult by the presence of a nucleus. However, where organ-specific effects are under investigation, analysis of micronuclei in particular cell types may be desirable.

2.3. Sister Chromatid Exchange

Techniques for sister chromatid differentiation developed in adult systems have been extended to the detection of SCEs in fetal animals. Studies utilizing wholely *in utero* techniques depend on the release of bromodeoxyuridine-implanted tablets[36–40] or the administration of bromodeoxyuridine by infusion intravenously[12,41,42] or intraperitoneally.[43] Cultivation of the cells *in vitro* for 1–3 hr after dissection has been suggested to improve metaphase yield.[36] The two divisions required for sister chromatid differentiation can occur after explantation into culture medium following exposure of the fetal cells in their normal tissue environment. This *in vivo/in vitro* method allows the investigation of cell-specific effects, either by the use of specific cell growth regulators (e.g., erythropoietin to promote proliferation of fetal erythroblasts[44] or pregnant mouse uterus extract to stimulate granulocyte-macrophage progenitor cells[45]) or potentially by the use of cell-sorting by immunological or physical methods. Comparisons of the sensitivities of different systems for detecting SCEs *in utero* are rendered difficult by the fact that each protocol has been applied to different species, chemicals, cell types, or stages of development. Table 2 shows data compiled from

TABLE 2. Induction of SCEs by Cyclophosphamide in Different *in Utero* Systems[a]

Protocol	Species	GA, days	Cell type	Dose, mg/kg	SCEs/cell Background	SCEs/cell Induced	Ref.
Infusion	Mouse	13	Whole embryo	10	4	28	12
	(C57BL/6J)	13	Whole embryo	5	4	16	12
		17	Whole embryo	10	—	11	12
		13	Liver	10	4	36	12
		13	Lung	10	4	21	12
		11–17	BM	10	4	18–22	12
		11–17	BM	5	4	10	12
In vivo/ in vitro	Mouse (Porton)	13	FL-E	12.5	7	25.5	44
		17	FL-GM	12.5	5.6	13.7	45
		17	Lung	12.5	13.7	24.5	45
		17	BM-GM	12.5	5	16	46
Implantation	Rat	14	FL	5	5	27.9	36
		14	BM	5	4.7	25.2	36
		11	Yolk sac	5	3.8	8.0	36
	Syrian hamster	11–15	FL	10	6.2	12–15	47
	Chinese hamster	17	FL	5	4.1	8.8	39
			BM	5	3.3	8.8	39

[a] GA, Gestational age; BM, bone marrow; FL-E, fetal liver erythroblasts; FL-GM, fetal liver-granulocyte-macrophage progenitor cells; BM-GM, BM-granulocyte-macrophage progenitor cells.

reports of SCE induction by cyclophosphamide (CP), which has been tested in several different systems. The background levels of SCEs reported in the *in vivo* systems are slightly lower than in the *in vivo/in vitro* method. Species differences in the induction of chromosome breakage by CP have been reported.[(48)] ICR mice were more sensitive to CP-induced chromosome damage than Wistar rats, which in turn were more sensitive than Chinese hamsters. The lower sensitivity of hamsters reported is supported by the lower levels of SCEs seen in the two studies on hamsters, although there is no evidence from these SCE data that rats are less sensitive than mice. Strain differences in the response of mice to CP have also been found[(49)] and this might affect the levels of SCEs found in the two studies on mice. At comparable gestational ages, the levels of SCEs induced in the *in vivo* system are slightly higher than in the *in vivo/in vitro* system, suggesting that this might be a more sensitive technique to use when knowledge of the cell type being scored is not important. Cells containing high levels of SCE-inducing lesions may be selected against in the culture period, or a longer time may be available for repair of lesions before the first S phase in bromodeoxyuridine, resulting in lower levels of SCEs in the *in vivo/in vitro* protocol.

Current *in vivo* protocols involve the administration of test substances after bromodeoxyuridine administration, either during the first *in vivo* replication cycle[12] or in the second cycle,[36] so that lesions are produced in DNA substituted with bromodeoxyuridine. The incorporation of bromodeoxyuridine into chromatin before exposure to genotoxic agents may affect the number of SCEs induced. The frequency of SCEs induced by UV irradiation and MMS, but not nitrogen mustard, was reduced if bromodeoxyuridine-substituted CHO cells were exposed, compared to nonsubstituted cells.[50] Similarly, when bromodeoxyuridine incorporation preceded methyl nitronitrosoguanidine treatment, fewer SCEs were induced than if it succeeded it.[51] This suggests that the relative timing of administration of chemicals and the level of bromodeoxyuridine incorporation produced by the various protocols might affect the incidence of SCEs.

The SCE technique may not be particularly suitable for the investigation of chronic exposure to chemicals. An example of the insensitivity of SCEs in detecting damage induced before bromodeoxyuridine incorporation was given in a study of the effect of alcohol on fetal mice.[37] Chronic alcohol exposure in pregnant mice up to the time of bromodeoxyuridine administration did not lead to increased SCEs, whereas exposure to alcohol at the same time as bromodeoxyuridine produced a threefold increase. While it is possible that this is due to synergism between bromodeoxyuridine and alcohol, as suggested by the authors, it may also be explained by the induction of short-lived lesions by alcohol or its metabolites. It has been demonstrated that several chemicals (e.g., ethyl methanesulfonate, MMS; 4-nitroquinoline 1-oxide; and acetoxy acetylaminofluorene) produce lesions that induce SCEs until at least the second round of replication. However, acetoxy acetylaminofluorene did not induce SCEs in the third round of replication.[52]

2.4. Maternal–Fetal Comparisons

An important factor in consideration of the risk to the fetus of a chemical is the metabolic capacity of the fetus. If activation of a xenobiotic occurs only in the mother, the stability and physical properties of the metabolite determine whether it will cross the placenta and reach the fetus in significant quantities. Prenatal animals, including humans, have a low biotransformation capacity compared to adults. The development of enzyme systems involved in metabolism of xenobiotics has been extensively reviewed.[53–55] Since active metabolites may reach the fetus as a result of maternal metabolism, the development of detoxification reactions is also important. Many of these develop postnatally in rodents.

One approach to the problem of detection of enzyme systems capable of activating xenobiotics that may occur at very low levels or whose identity is not known is to look at the consequences of activation as assayed by cytogenetic damage. This involves either the exposure of the whole embryo or isolated fetal livers to the chemical under study. The ability of mouse embryos at $8\frac{1}{2}$–$9\frac{1}{2}$ days gestation to metabolize benzo(*a*)pyrene was demonstrated by the induction of SCEs in embryonic cells following whole-embryo culture in media containing benzo(*a*)pyrene.[56] The ability to metabolize benzo(*a*)pyrene in these embryos was related to the inducibility of the cytochrome P-450 monooxygenases. Embryos from strains lacking the receptor that mediates the induction showed no SCE increases, whereas genetically responsive strains showed increases. This experiment illustrates the usefulness of the SCE technique in detecting development of metabolic competence in the embryo.

A protocol involving a short exposure of isolated 13-day mouse fetal livers to genotoxic chemicals, followed by a culture period for two cell cycles in erythropoietin to stimulate erythroblast proliferation, has been developed to investigate the role of the fetal liver in metabolic activation.[44] MMS, which is a direct-acting chemical, and MC, which is activated by a ubiquitously distributed enzyme, induced SCEs in isolated livers. A small but significant increase was found for both benzo(*a*)pyrene and CP, but the response to CP suggests that the spectrum of metabolites produced in the intact pregnant animal is different from that produced by the fetal liver, since greater cytotoxicity relative to SCE induction was observed in isolated livers. An inability to metabolize CP, as shown by a lack of induction of SCEs by this compound (although one of its metabolites, phosphoramide mustard, induced SCEs) was observed in the rat embryo yolk sac system developed by Allen *et al.*[36] In this system, 11-day rat embryos were labeled *in vivo* and then the yolk sacs were removed into culture media for exposure to the test chemical for their last replication cycle before harvest. The yield of mitotic cells from rat yolk sacs decreased from day 11 and at day 14 few cells were available for analysis. Similarly, in the *in vitro* isolated liver system described above, responsiveness to erythropoietin declines after day 14, so that both these systems as they exist are limited to the detection of fetal metabolism as midpregnancy, whereas many enzyme reactions develop later in pregnancy and the fetus is more susceptible to tumor induction at later gestational ages. However, granulocyte-macrophage precursor cells in fetal liver can still be stimulated to proliferate *in vitro* at day 17 of gestation in mice[45] and may serve as a suitable indicator of fetal liver activity later in gestation.

In the intact animal, several studies have compared the induction of genetic damage in the mother and fetus. Since the best source of proliferating cells in the adult is the bone marrow, all studies have compared genetic damage (either micronuclei or SCE) in this tissue, which is known to have a low metabolic capacity. The extent to which these results can be generalized to an estimate of relative fetal and maternal risks is therefore doubtful. Kram *et al.*[12] compared levels of SCEs in the whole embryo with maternal bone marrow cells at different ages of gestation following CP treatment and found that similar levels were induced in both tissues at day 11, but the level in the fetus declined with age, so that by day 15 it was half the level in the adult bone marrow. A similar decline in frequency was noted in fetal cells following treatment with MC. It is not clear to what extent differences in cellular composition in the developing embryo at different gestational ages influence these observations. Allen *et al.*[36] reported similar frequencies of CP-induced SCEs in maternal rat bone marrow and 14-day fetal liver cells. This is in agreement with the similar SCE frequencies induced by CP in granulocyte-macrophage precursor cells in maternal mouse bone marrow and fetal liver at day 17 of gestation.[46] We have also compared SCEs induced by MC and procarbazine in this system and found similar levels of procarbazine-induced SCEs in maternal and fetal cells, but fewer MC-induced SCEs in fetal liver than in maternal bone marrow.

Comparisons of maternal and fetal sensitivity to genotoxic chemicals can also be made with the micronucleus test.[27,28,33] Normally the adult cells analyzed are bone marrow erythrocytes, but it has been suggested that the pregnant mouse spleen may provide a source of proliferating erythroid cells.[57] Interestingly, it seems that risk estimates ascertained by the micronucleus method may differ from those obtained using the SCE technique.[46] Relative levels of damage assayed by the two techniques for three chemicals are shown in Table 3. Much higher levels of micronucleated polychromatic erythrocytes were induced in 17-day fetal livers by CP than in maternal bone marrow. Similarly, 15-day fetal livers show higher levels of micronucleated polychromatic erythrocytes induced by procarbazine than bone marrow. These values are similar to those reported by King and Wild[28] in mouse fetal blood and bone marrow after CP, MC, and procarbazine. One difficulty in comparing these two predictors of fetal risk is that micronuclei and SCEs are scored

TABLE 3. Ratio of Levels of Damage Assayed As Micronuclei or Sister Chromatid Exchange in Mouse Liver and Bone Marrow Cells[a]

	Fetal liver:bone marrow	
	SCE	Micronuclei
Cyclophosphamide (25 mg/kg)	0.90[b]	6.5[b]
Mitomycin C (2 mg/kg)	0.80[b]	0.5[c]
Procarbazine (100 mg/kg)	0.91[b]	3.6[c]

[a] From Henderson *et al.*(46)
[b] Seventeen days gestation.
[c] Fifteen days gestation.

in different cell types. However, comparison of levels of SCEs induced by CP and procarbazine in fetal erythroblasts and granulocyte-macrophage cells(45) suggests that differential sensitivity of cell types is an inadequate explanation of higher levels of fetal damage as indicated by the micronucleus technique. Discrepancies between relative levels of damage induced in the fetus and mother as assayed by these two genetic endpoints suggest that caution is required in extrapolating these findings to estimates of maternal and fetal risk.

2.5. Cell and Tissue Specificity

Many chemicals are highly specific in their ability to transform particular cell types to a neoplastic state. Critical variables in this tissue specificity are the capacity of target cells to metabolize a carcinogen to its ultimate reactive form, the extent of reaction of ultimate metabolites with target DNA, the ability of the cell to repair lesions in its DNA, the state of differentiation and proliferative potential, and the ability of a transformed cell to escape host control mechanisms that limit further proliferation.(58) Because short-term cytogenetic tests can only reflect the initial steps of the multistage transformation process, it is unclear to what extent they will be predictive of differential risk for different cell types. Several fetal cell types can be examined for either SCEs or chromosome aberrations. Fetal liver, lung, and gut from 13-day gestation mice were analyzed for SCEs following CP, MC, and daunomycin.[12] None of these fetal organs displayed differential sensitivity to either daunomycin or MC, but fetal liver was more sensitive to CP than were gut and lung. Cell culture following genotoxic exposure allows differentiation of particular cell types within an organ. The culture of fetal liver erythroblasts, granulocyte-macrophage progenitors, and lung and brain cells following exposure to CP and procarbazine indicated differential responses to these agents.[45] The cells most sensitive to SCE induction by both CP and procarbazine were fetal liver erythroblasts.

The ability of short-term cytogenetic assays to predict tissue- or cell-specific activity of genotoxic chemicals requires critical assessment by administration of chemicals of known species and cell specificities. Most studies reported have used model mutagens of unknown transplacental tissue specificity. Tsuda *et al.*,[17,18] in a study of a tissue-specific agent, reported the induction of chromosomal aberrations after transplacental diethylnitrosamine administration to Syrian hamsters in its target cells, tracheal epithelial cells. They demonstrated an increase in chromosome aberrations in these cells but no increase of chromosome damage in nontarget mesenchymal cells.[18]

A simultaneous assay for chromosome aberrations, mutation, and neoplastic transformation in Syrian golden hamster brain, lung, liver, and kidney cells has also been reported.[59] Benzo(*a*)pyrene, methylnitrosourea, and diethylnitrosamine, which are carcinogenic *in utero* in rats and mice, induced different levels of chromosome aberrations in this *in vivo*/*in vitro* test. However, interpretation of the ability of this chromosome aberration assay to predict organ specificity is made difficult by lack of control values and because variations in cell cycle times that could lead to different times for peak frequencies for each tissue were not considered. The higher chromosome aberration frequency in total body cells than in individual organ cultures suggests that the most sensitive cell type(s) was not included in the organs cultured individually.

2.6. Detection of Transplacental Carcinogens and Teratogens

A summary of all the chemicals tested in a variety of *in utero* cytogenetic assays is presented in Table 4, together with information on their transplacental activity where this is available. The data on transplacental activity have been collated from several species and these do not always coincide with the species in which short-term tests were employed. Only a few chemicals have been tested primarily because they were known to be transplacental carcinogens. Some have been

TABLE 4. Summary of Results of Agents Tested following *in Utero* Exposure for Induction of Chromosome Aberrations (CA), Micronuclei (MN), and SCEs

Treatment	Transplacental effect	CA	MN	SCE	Ref.
Adriamycin	Not teratogenic[41]			+	41
Alcohol	Teratogenic[60]			+	37, 61
Benzo(*a*)pyrene	Embryotoxic,[62]	+			16, 59
	carcinogenic,[63]			+	56, 47
	teratogenic[62]		+		27
Bleomycin	Teratogenic[6]	+			16
Busulfan	Teratogenic[6]	+			64
Chlorambucil	Teratogenic[4]	+			14
Cyclophosphamide	Teratogenic[4]	+			64
			+		28, 44
				+	39, 12, 36, 45, 47
Daunomycin	Teratogenic[6]			+	41, 12
(Rubidomycin)		+			66
Diesel particulate exhaust	—			+	67
Diethylnitrosamine	Carcinogenic[63]	+			17, 18
			+		33
				+	44
Diethylstilbestrol	Carcinogenic[88]		−	−	117
7,12-Dimethylbenz(*a*) anthracene	Carcinogenic,[68] teratogenic[69]			+	43
Dimethylnitrosamine	Carcinogenic,[63]	+			59, 20
	not teratogenic[70]		+		32
Ethylnitrosourea	Carcinogenic,[71] teratogenic[71]	+			72
Furylfuramide	Carcinogenic,[63] teratogenic[73]	+			19
Heliotrine	Teratogenic[4]		+		31
Hydroxyurea	Teratogenic[4]	−			14
γ-Irradiation	Teratogenic,[4]		+		27
	carcinogenic[74]	+			9
Isoniazid	Teratogenic[6]	+			15
Lasiocarpine	—		+		31
Methyl	Carcinogenic[63]		+		27, 30
methanesulfonate				+	40
Methylnitrosourea	Carcinogenic,[63] teratogenic[4]	+			59
Mitomycin C	Teratogenic[4]			+	12, 44, 45
			+		27, 28
		+			16

TABLE 4. (*continued*)

Treatment	Transplacental effect	CA	MN	SCE	Ref.
Nitrogen mustard	Teratogenic,[4] not carcinogenic[5]	+			65, 14
Phenobarbitone	Teratogenic[4]		−	−	[a]
Procarbazine	Teratogenic,[4] carcinogenic[63]		+		27, 28
				+	44, 45
		+			16
Sodium nitrite	—	−	+		20
Streptonigrin	Teratogenic[4]	+			14
Thalidomide	Teratogenic[4]	−			14
		+			66
Trenimon	—		+		28
Triethylenemelamine	Teratogenic,[4] not carcinogenic[5]	+			16
Trypan blue	Teratogenic,[4] not carcinogenic[5]	+			66
Ultrasound	Embryotoxic,[38] teratogenic[38]			−	38

[a] L. Henderson and H. Cole, unpublished results.

tested because fetal cytogenetic assays were potentially more sensitive than adult assays (e.g., heliotrine, lasiocarpine), because they were compounds to which a large number of people are exposed and were being investigated in a variety of tests (e.g., diesel exhaust, sodium nitrite), or because they were especially applicable to the prenatal period (e.g., ultrasound).

The SCE-inducing activity of 57 compounds in a variety of *in vitro* and *in vivo* systems has previously been compared with their teratogenicity.(13) The present survey is limited to transplacental studies in order to take into account the ability of active metabolites to reach fetal tissues and their consequent action *in utero*. Considering the data reported here, 20/24 teratogens produced a positive result in at least one cytogenetic assay. Three did not (hydroxyurea, phenobarbitone, and ultrasound). Thalidomide was positive in one chromosome aberration study and negative in another. Two chemicals claimed not to be teratogenic, adriamycin and dimethlynitrosamine, induced fetal chromosome damage. Caution should be used regarding a chemical reported to be nonteratogenic, because of the number of confounding variables (species, age, strain, sex, age at treatment, method of administration, etc.) inherent in a teratogenicity assay. It is likely that the high correlation between teratogenicity and positive short-term cytogenetic assays found to date is due mainly to the choice of compounds with known genotoxicity. While short-term assays might detect teratogens that act by producing lesions in DNA, they are unlikely to detect those that act by other mechanisms.

All ten known transplacental carcinogens tested to date have been found to be detectable by short-term cytogenetic tests. The only discrepancies found between different short-term tests were for sodium nitrite and thalidomide. Sodium nitrite was negative in the chromosome aberration assay but positive in the micronucleus test and this may possibly be explained by the use of suboptimal timing. The transplacental activity of this food additive has not been tested in animals, although it has been suggested to be a risk factor in brain tumors in children.(75) The transplacental clastogenicity of thalidomide is controversial, with one positive and one negative chromosome aberration assay.

The human transplacental carcinogen diethylstilbestrol induces aneuploidy in mouse embryos.(76,117) Diethylstilbestrol does not induce SCEs or micronuclei in mouse fetal liver cells exposed in utero,(117) although it induces chromosome aberrations but not SCEs or aneuploidy in adult mouse bone marrow.(77)

The relative sensitivity of the different short-term tests in detecting damage requires further investigation. In a study of SCE and micronucleus induction in fetal liver erythroblasts, the lowest detectable doses for MC, MMS, CP, and benzo(*a*)pyrene were similar for both endpoints, but for procarbazine the micronucleus assay was more sensitive at lower doses.(44)

With the exception of ultrasound, there have been no reports of the failure of transplacental cytogenetic assays to detect agents shown to be genotoxic in other systems. Contradictory findings have been reported on genetic effects of ultrasound on cells in culture, which need to be resolved with further studies.(38)

The ability of short-term tests to detect transplacental carcinogens at doses equal to or lower than the doses that induce tumors is demonstrated in Table 5. Generally dose–response studies have not been made for tumor induction and lower doses than those presented may be carcinogenic, since high levels of tumors were produced by some treatments, e.g., 42% of transplacentally exposed hamsters have tumors after a dose of 20–25 mg/kg diethylnitrosamine. However, the data give some confidence that short-term assays are in the range of sensitivity required to detect carcinogenic doses.

TABLE 5. Comparison of Carcinogenic Doses and Doses Required to Produce a Positive Cytogenetic Response

	Carcinogenicity			Cytogenetic assays		
Compound	Dose, mg/kg	Species	Ref.	Dose, mg/kg	Species	Ref.
Benzo(*a*)pyrene	80	Mouse	2	25	Mouse	16
Diethylnitrosamine	20	Hamster	78	20	Hamster	17
	50	Mouse	79	25	Mouse	33
Dimethylnitrosamine	~7 mg/kg per day for 7 days	Rat	80	5	Rat	32
Furylfuramide	50 mg/kg per day for 3 days	Mouse	81	20	Hamster	19
Procarbazine	5–10 mg/kg per day for 7 days	Rat	4	10	Mouse	27

2.7. General Comments on Animal Studies

Several short-term cytogenetic assays have been developed for the detection of transplacental genotoxic agents. The extent to which these can determine relative risks to the fetus and mother and whether they can indicate tissue specificity are still under investigation. Many confounding factors, such as effect of gestational age, route of administration, chronic versus acute exposure, and species and strain differences in sensitivity, are vitally important in transplacental studies. Within a species, the genetic background may also affect levels of damage.[56] In investigations to determine whether a chemical is genotoxic *in utero,* it is debatable whether whole-embryo cell preparations or individual organs should be investigated. When a compound of unknown genotoxicity is being investigated, its organ specificity will not be known. Therefore, methods of analyzing cytogenetic damage in different cell types, while possibly enhancing understanding of the basis of tissue specificity of some carcinogens, will have limited application in the identification of transplacental genotoxic agents. The use of the whole embryo to some extent averages the overall fetal effects,[16] but the use of the most sensitive tissue should maximize the sensitivity of the assay.

Further comparisons of the response of different tissues to a variety of chemicals are required to determine whether a single tissue, possibly liver, would be the most sensitive tissue for routine studies.

The feasibility of a standard protocol has been discussed briefly in the context of time course studies (see Sections 2.1 and 2.2). While investigation of sampling times, particularly for chromosome aberrations and micronuclei, is essential if sensitivity is to be achieved, this will conflict with simplicity, which may be an important consideration if transplacental effects are to be included in routine carcinogenicity bioassays.

3. Human Studies

3.1. Risk Factors in Childhood Cancers and Congenital Malformations

Extrapolating from animal studies to humans, it seems reasonable that environmental factors will be important in the etiology of childhood cancer, congenital malformations, and fetal loss. Many factors, such as differences in metabolism, placenta structure and function, and routes of exposure, complicate direct comparisons of fetal risk in animals and humans. In fact, solving the problem of how to extrapolate from animal studies to human risk remains a major priority in genetic toxicology.

Many human teratogens have been detected epidemiologically.[6] Although several risk factors for childhood cancer have been suggested (Table 6), the only validated human carcinogens effective *in utero* are X-irradiation and diethylstilbestrol. It is probable that diphenylhydantoin (dilantin) also induces brain tumors in children following *in utero* exposure. Childhood cancer, while the second largest cause of death in children, is still a comparatively rare event, occurring at incidences ranging from 10 per 10^5 children per year in Manchester, England,[89] to 11.3 per 10^5 per year in Queensland, Australia,[90] to 11.9 per 10^5 per year in the greater Delaware Valley, U.S.,[91] to 17.6–22.3 per 10^5 per year in Finland.[92] The rarity of childhood cancers means few registries are large enough to carry out independent epidemiologic studies that will show statistically significant correlations with possible causative agents.[90] Since the endpoint in these studies is diagnosis of malignancy, sometimes several years after exposure, any carcinogen detected by this means may have widespread effects before preventative measures can be taken. More indirect methods of assessing transplacental human genotoxins, such as epidemiologic investigations of abor-

TABLE 6. Possible Risk Factors for Childhood Cancer

Agent	Ref.
Barbiturates	82
Parental exposure (maternal occupation involving chemical exposure; paternal occupation involving solvents; employment of father in aircraft industry)	83
Drugs	84
Smoking	85
Illness (epilepsy and tuberculosis)	86
Alcohol, benzene, mephalan, vinyl chloride	2
Diphenylhydantoin (dilantin)	87
X-Irradiation	74
Diethylstilbestrol	88
N-Nitroso compounds	75

tions and congenital malformations, which occur at much higher frequencies than tumors, have many confounding factors (see discussion by Hook.[93]) Progress is being made on several alternative methods for monitoring exposure to possible genotoxic agents, including the analysis of hemoglobin variants[94] and somatic mutations at the hypoxanthine-phosphoribosyl transferase locus,[95,96] and these may be applicable to tests of transplacental exposure.

3.2. Choice of Cell Types and Assay Systems

The measurement of chromosome breakage and SCE induction in human cord blood cells may provide a sensitive indication of genetic damage induced *in utero*. Important factors to be considered in the application of short-term cytogenetic assays to monitor damage are the longevity of lesions giving rise to these endpoints, the life span of the cell, and the extent to which it is exposed to genotoxic metabolites. Polychromatic erythrocytes occur in human cord blood at frequencies sufficient to make their scoring for micronuclei feasible. However, fetal erythropoiesis in humans changes from hepatic to myeloid in the second trimester,[97] whereas in rodents this switch from liver to bone marrow occurs perinatally. Consequently, erythrocytes will lack sensitivity as an indicator of genetic damage due to the low metabolic capacity of fetal bone marrow. Lymphocytes, being circulating nucleated cells, may be a more sensitive cell type for monitoring genetic lesions induced prenatally. The detection of cells with unstable chromosome aberrations provides information on the life span of lymphocytes.[98] *In utero*-exposed survivors of the atomic bombings of Hiroshima and Nagasaki provide a unique group for the investigation of the life span of lymphocytes formed in fetal life. The lymphocytes of these individuals had higher levels of chromosome aberrations than controls 20 years after exposure.[99] The initial damage could have been induced in stem cells or committed lymphocytes, but the presence of some unstable chromosome aberrations (rings, dicentrics, and fragments) suggests that some of the cells were in their first postexposure division, indicating that some lymphocytes formed in early fetal life may have the capacity for long-term survival. The authors suggested that the presence of lymphocytes with unstable chromosome aberrations in three first-trimester-exposed individuals is due to damaged lymphocyte stem cells with the damage maintained through many cell divisions. Alternatively, they could arise from committed lymphocytes present in the fetus from the first trimester. This explanation is supported by the observation that E-rosette-forming cells (T lymphocytes) are detectable in the human fetus from 12–14 weeks of gestation.[100] Assays of chromosome damage at birth should therefore be capable of detecting damage in early pregnancy either to stem cells or committed lymphocytes.

Micronuclei are essentially manifestations of unstable aberrations, since they are mainly due to acentric fragments, but *in vitro* studies on micronuclei induced by X-irradiation and bleomycin have shown that they are scorable in human lymphocyte cultures which contain cells that have undergone several divisions.[101,106] Their persistence may be due either to inclusion and exclusion of micronuclei from the main nucleus during cell divisions as visualized by premature chromosome condensation,[103] generation of micronuclei in divisions subsequent to the first, accumulation of micronucleated lymphocytes without further division, or persistence of micronuclei in the cytoplasm for more than one cycle. Therefore the scoring of micronuclei in human cord blood lymphocyte cultures is potentially capable of detecting lesions leading to chromosome breakage for some time after induction. The micronucleus technique has been adapted for human adult lymphocytes.[101,104] We have modified this technique using cord blood lymphocytes and investigated its sensitivity by exposing cord blood lymphocytes to bleomycin and irradiation.[102,105,106] The method is sensitive to low doses of bleomycin (1.25 μg/ml) and irradiation (10 rad) and therefore is applicable to population monitoring.

Experimental studies in animals and in cell culture systems have shown that some chemicals produce lesions that give rise to SCEs only in the first post-treatment division, whereas others give rise to SCEs also in second and subsequent divisions (see Section 2.3). Animal studies have indicated that SCE frequencies may stay elevated only briefly after an acute exposure *in vivo*. For example, rabbits injected with a cytotoxic dose of CP have transiently elevated SCE frequencies in their lymphocytes, which return to control levels 10–11 days after exposure.[107] It was suggested that the decline in SCE levels with time post-treatment is a function of lymphocyte turnover. These studies indicate that SCEs may not be useful in monitoring stem cell damage, which will be removed in the divisions leading to the mature lymphocyte, and the ability of SCEs to indicate transplacental damage will be dependent on turnover of lymphocytes in the prenatal period and the time interval between exposure and birth.

3.3. Results of Population Monitoring

A prerequisite of cytogenetic surveys of exposed populations is the determination of baseline frequencies and criteria for a significant deviation. Methods of analysis of SCEs in lymphocytes of exposed populations have been thoroughly investigated by Carrano and Moore,[108] who compared efficiencies of analyses based on individuals with excessive "high-frequency cells" (statistically determined "outlier" cells with high numbers of SCEs) with analyses based on deviations of the mean SCE value. The methods can be applied to comparisons of two population groups or to detect deviation of a single individual from a control group. The incidence of SCEs is lower in cord bloods than in adult samples and shows a smaller range of mean SCEs between individuals. We analyzed cord blood samples from 60 babies for SCE frequencies (L. Henderson and Z. Aghamohammadi, unpublished results). The mean frequency was 4.4, with a range from 2 to 8.25. Calculations based on this population indicate that an individual baby must have three cells with more than nine SCEs in a sample of 20 cells to deviate from the population. Larger numbers of cells scored per individual would reduce the relative number of outlier cells required for a significant result.

The number of studies reported on cytogenetic damage in newborn babies is very limited (Table 7). One study, which linked higher levels

TABLE 7. Cytogenetic Analyses of Chromosome Damage in Somatic Cells of Newborn Babies

Group studied	Endpoint[a]	Result	Ref.
Children of smoking/ nonsmoking mothers	SCE	–	109, 110
Low-birth-weight babies	CA	–	111
	SCE	–	112
Children of smokers, heavy alcohol users, epileptics, patients with malignancy	SCE	–	113
Babies exposed to ultrasound	SCE	–	116
	MN	–	116
Babies with neural tube defects	SCE	+	113

[a] SCE, Sister chromatid exchange; CA, chromosome aberrations; MN, micronuclei.

of chromosome aberrations and SCEs in children of workers in chemical laboratories and the printing industry, is not included in the table, since it covered children aged 4 days to 11 years, which introduces the complication of postnatal environmental effects.[114] Only one group of babies, those with neural tube defects, had an elevated SCE level. Numbers of babies in each group were small, with the exception of the low-birth-weight babies. These studies need to be extended before general conclusions can be made on the usefulness of cytogenetic assays for monitoring lesions in newborn babies. Caution must be used when conditions exist (e.g., certain disease states) that may alter the proportion of B and T lymphocytes or subsets of T lymphocytes, since it is not yet clear whether different lymphocyte subsets may be characterized by different SCE values.[115] As pointed out by Carrano and Moore,[108] studies on occupational exposure have shown small SCE increases from 10 to 60% higher than controls. The likely magnitude of SCE increases implies that an adequate control data base should be obtained and confounding variables identified and taken into consideration before the SCE technique can be used as a sensitive method of monitoring humans. The inability of this assay to detect short-lived lesions and its insensitivity to certain types of chemicals means that it should be used in conjunction with an assay for chromosome breakage for monitoring newborn humans. The lymphocyte micronucleus technique offers a sensitive and simple alternative to metaphase analysis for this dual monitoring system.

ACKNOWLEDGMENTS

The author is grateful to Prof. R. Cole and Dr. J. Cole for helpful discussions and to Helen Cole for assistance in the preparation of this manuscript.

4. References

1. P. Urso and N. Gengozian, Alterations in the humoral immune response and tumor frequencies in mice exposed to benzo(*a*)pyrene and X-rays before and after birth, *J. Toxicol Environ. Health 10,* 817–836 (1982).
2. L. F. Soyka, Hepatic drug metabolizing enzyme activity and tumorigenesis in mice following perinatal exposure to benzo(*a*)pyrene, *Pediatric Pharmacol. 1,* 85–96 (1980).
3. J. D. Rowley, Human oncogene locations and chromosome aberrations, *Nature 301,* 290–291 (1983).
4. T. H. Shepard, *Catalog of Teratogenic Agents,* John Hopkins University Press, Maryland (1973).
5. R. P. Bolande, Teratogenesis and oncogenesis, in: *Handbook of Teratology* (J. G. Wilson and F. C. Fraser, eds.) Vol. 2, pp. 293–325, Plenum Press, New York (1977).
6. H. Nishimura and T. Tanimura, *Clinical Aspects of the Teratogenicity of Drugs,* Excerpta Medica, Amsterdam (1976).
7. J. D. Manson, Developmental toxicity of alkylating agents: mechanism of action, *The Biochemical Basis of Chemical Teratogenesis* (M. R. Juchau, ed.), pp. 95–135, Elsevier/North-Holland, New York (1981).
8. T. H. Ingalls, E. F. Ingenito, and F. J. Curley, Acquired chromosomal anomalies induced in mice by injection of a teratogen in pregnancy, *Science 141,* 810–812 (1963).
9. S. W. Soukup, E. Takacs, and J. Warkany, Chromosome changes in rat embryos following X-irradiation, *Cytogenetics 4,* 130–144 (1965).
10. S. Wolff, Chromosome aberrations, sister chromatid exchanges and the lesions that produce them, in: *Sister Chromatid Exchange* (S. Wolff, ed.), pp. 41–57, Wiley, New York (1982).
11. R. H. Heflich, D. T. Beranek, R. L. Kodell, and S. M. Morris, Induction of mutations and sister-chromatid exchanges in Chinese hamster ovary cells by ethylating agents. Relationship to specific DNA adducts, *Mutat. Res. 106,* 147–161 (1982).
12. D. Kram, G. D. Bynum, C. C. Senula, C. K. Bickings, and E. L. Schneider, *In utero* analysis of sister chromatid exchange: Alterations in susceptibility to mutagenic damage as a function of fetal cell type and gestational age, *Proc. Natl. Acad. Sci. USA 77,* 4784–4787 (1980).
13. D. Kram, In utero analysis of sister chromatid exchanges, in: *Sister Chromatid Exchange* (A. A. Sandberg, ed.), pp. 205–216, Liss, New York (1982).
14. S. Soukup, E. Takacs, and J. Warkany, Chromosome changes in embryos treated with various teratogens, *J. Embryol. Exp. Morphol. 18,* 215–226 (1967).
15. J. Franz, Mutagenicity of isoniazid: Testing for somatic chromosome mutations in mouse embryos, *Hum. Genet. 42,* 28–30 (1978).
16. I. D. Adler, New approaches to mutagenicity studies in animals for carcinogenic and mutagenic agents. II. Clastogenic effects determined in transplacentally treated mouse embryos, *Teratogen. Carcinogen. Mutagen. 3,* 321–334 (1983).

17. H. Tsuda, M. Emura, H. Richter-Reichhelm, W. Boning, and U. Mohr, Chromosomal aberrations in tracheal epithelial cells of the fetal Syrian golden hamster after transplacental DEN administration, *Toxicology 25,* 61–66 (1982).
18. H. Tsuda, M. Emura, H. Richter-Reichhelm, W. Boning, and U. Mohr, Transplacental effect of *N*-diethylnitrosamine on chromosomal aberrations in fetal trachael cells of the Syrian hamster: Comparison between epithelial and mesenchymal cells, *Carcinogenesis 4,* 1385–1388 (1983).
19. N. Inui, Y. Nishi, and M. Taketomi, Mutagenic effect of orally given AF-2 on embryonic cells in pregnant Syrian hamsters, *Mutat. Res. 57,* 69–75 (1978).
20. N. Inui, Y. Nishi, M. Taketomi, and M. Mori, Transplacental action of sodium nitrite on embryonic cells of Syrian golden hamster, *Mutat. Res. 66,* 149–158 (1979).
21. B. E. Matter and T. Tsuchimoto, Mutagenicity test systems for the detection of chromosome aberrations *in vivo, Arch. Toxicol. 46,* 89–98 (1980).
22. W. Schmid, The micronucleus test, *Mutat. Res. 31,* 9–15 (1975).
23. D. W. Frank, R. J. Trzos, and P. I. Good, A comparison of two methods for evaluating drug-induced chromosome alterations, *Mutat. Res. 56,* 311–317 (1978).
24. U. Kliesch, N. Danford, and I.-D. Adler, Micronucleus test and bone-marrow chromosome analysis. A comparison of 2 methods *in vivo* for evaluating chemically induced chromosomal alterations, *Mutat. Res. 80,* 321–332 (1981).
25. U. Kliesch, I. Rapova, and I.-D. Adler, Induction of chromosome damage in mouse bone marrow by benzo(*a*)pyrene, *Mutat. Res. 102,* 265–273 (1982).
26. R. J. Cole, N. A. Taylor, J. Cole, and C. F. Arlett, Transplacental effects of chemical mutagens detected by the micronucleus test, *Nature 277,* 317–318 (1979).
27. R. J. Cole, N. A. Taylor, J. Cole, and C. F. Arlett, Short term tests for transplacentally active carcinogens 1. Micronucleus formation in fetal and maternal mouse erythroblasts, *Mutat. Res. 80,* 141–157 (1981).
28. M. T. King, and D. Wild, Transplacental mutagenesis: The micronucleus test on fetal mouse blood, *Hum. Genet. 51,* 183–194 (1979).
29. K. I. Yamamoto and Y. Kikuchi, Studies on micronuclei time response and on the effects of multiple treatments of mutagens on induction of micronuclei, *Mutat. Res. 90,* 163–173 (1981).
30. R. J. Cole, Z. Aghamahammadi, J. Cole, and L. Henderson, Short-term tests for transplacentally active carcinogens. Multiple-dose regimes in the transplacental micronucleus test, *Mutat. Res. 105,* 115–120 (1982).
31. C. J. Stoyel and A. M. Clark, The transplacental micronucleus test, *Mutat. Res. 74,* 393–398 (1980).
32. S. Fumero, S. Peano, J. Meriggi, G. Berruto, S. Silvestri, and A. Mondino, Micronucleus induction by dimethylnitrosamine in rat fetal blood, *Mutat. Res. 91,* 71–73 (1981).
33. R. J. Cole, N. Taylor, J. Cole, L. Henderson, and C. F. Arlett, Short-term tests for transplacentally active carcinogens. Sensitivity of the transplacental micronucleus test to diethylnitrosamine, *Mutat. Res. 104,* 165–171 (1982).
34. D. Wild, Cytogenetic effects in the mouse of 17 chemical mutagens and carcinogens evaluated by the micronucleus test, *Mutat. Res. 56,* 319–327 (1978).
35. R. Schlegel and J. T. MacGregor, The persistence of micronuclei in peripheral blood erythrocytes: Detection of chronic chromosome breakage in mice, *Mutat. Res. 104,* 367–369 (1982).
36. J. W. Allen, E. El-Nahass, M. K. Sanyal, R. L. Dunn, B. Gladen, and R. L. Dixon, Sister chromatid exchange analyses in rodent maternal, embryonic and extra-embryonic tissues. Transplacental and direct mutagen exposures, *Mutat. Res. 80,* 297–311 (1981).

37. M. R. Alvarez, L. E. Cimino, and T. J. Pusateri, Induction of sister chromatid exchanges in mouse fetuses resulting from maternal alcohol consumption during pregnancy, *Cytogenet. Cell Genet. 28,* 173–180 (1980).
38. W. W. Au, N. Obergoemer, K. L. Goldenthal, P. M. Corry, and W. Willingham, Sister chromatid exchanges in mouse embryos after exposure to ultrasound *in utero, Mutat. Res. 103,* 315–320 (1982).
39. A. Basler, Sister chromatid exchanges *in vivo* in Chinese hamster embryonic liver cells exposed transplacentally to BrdU, *Cytogenet. Cell Genet. 24,* 193–196 (1979).
40. A. Basler, Sister chromatid exchanges in Chinese hamster embryonic liver cells exposed transplacentally to bromodeoxyuridine (BrdU) and chemical mutagens, *Arch. Toxicol. Suppl. 4,* 22–24 (1980).
41. D. Kram, G. D. Bynum, G. C. Senula, and E. L. Schneider, *In utero* sister chromatid exchange analysis for detection of transplacental mutagens, *Nature 279,* 531 (1979).
42. S. Knuutila, A. Harkki, L. Rossi, T. Westermarch, L. Lappalainen, and P. Rantanen, *In vivo* method for sister chromatid exchanges in Chinese hamster foetal and bone marrow cells, *Hereditas 91,* 23–26 (1979).
43. N. P. Singh, A. Turturro, M. J. W. Chang, and R. W. Hart, The measurement of sister chromatid exchanges induced *in utero* utilizing intraperitoneal infusion of BrdU: A novel technique, *Cytogenet. Cell Genet. 35,* 81–86 (1983).
44. R. J. Cole, J. Cole, L. Henderson, N. A. Taylor, C. F. Arlett, and T. Regan, Short-term tests for transplacentally active carcinogens. A comparison of sister-chromatid exchange and the micronucleus test in mouse foetal liver erythroblasts, *Mutat. Res. 113,* 61–75 (1983).
45. L. M. Henderson, R. J. Cole, J. Whittaker, T. Regan, and H. Cole, Short-term tests for transplacentally active carcinogens. Induction of sister chromatid exchanges in foetal brain, lung and blood-forming cells by procarbazine and cyclophosphamide, *Mutat. Res. 113,* 489–498 (1983).
46. L. Henderson, R. Cole, J. Cole, H. Cole, Z. Aghamohammadi, and T. Regan, Sister chromatid exchange and micronucleus induction as indicators of genetic damage in maternal and foetal cells, *Mutation Res. 126,* 47–52 (1983).
47. M. A. Pereira, L. McMillan, P. Kaur, D. K. Gulati, and P. S. Sabharwal, Effect of benzo(*a*)pyrene on sister chromatid exchange in fetal hamster liver exposed *in utero, Mutat. Res. 105,* 343–374 (1982).
48. P. Goetz, R. J. Sram, and J. Dohnalova, Relationship between experimental results in mammals and man 1. Cytogenetic analysis of bone marrow injury induced by a single dose of cyclophosphamide, *Mutat. Res. 31,* 247–254 (1975).
49. T. A. Dragani, A. Zunino, and G. Sozzi, Differences in sister chromatid exchange (SCE)-induction *in vivo* by cyclophosphamide in murine strains, *Carcinogenesis 2,* 210 222 (1081).
50. C. H. Ockey, Methyl methane-sulphonate (MMS) induced SCEs are reduced by the BrdU used to visualize them, *Chromosoma 84,* 243–256 (1981).
51. N. C. Popescu, S. A. Amsbaugh, and J. A. DiPaolo, Reduced *N*-methyl-*N*-nitro-*N*-nitroguanidine sister chromatid exchange induction in Chinese hamster V79 cells pre-exposed to 5-bromodeoxyuridine, *Chromosoma 76,* 329–338 (1980).
52. S. Wolff, Induced chromosome variation, in: *Chromosomes Today* (M. D. Bennett, M. Bobrow, and G. M. Hewitt, eds.), Vol. 7, pp. 226–241, Allen and Unwin, London (1981).
53. C. R. Short, D. A. Kinden, and R. Stith, Fetal and neonatal development of the microsomal monooxygenase system, *Drug Metab. Rev. 5,* 1–42 (1976).
54. G. W. Lucier, E. M. K. Lui, and C. A. Lemartinierc, Metabolic activation/deactivation reactions during prenatal development, *Environ. Health Perspect. 29,* 7–16 (1979).

55. W. Klinger, D. Muller, U. Kleeberg, and A. Barth, Peri- and postnatal development of phase I reactions, in: *Developmental Toxicology* (C. A. Kimmel and J. Buelkesam, eds.), pp. 83–99, Raven Press, New York (1981).
56. S. M. Galloway, P. E. Perry, J. Meneses, D. W. Nebert, and R. A. Petersen, Cultured mouse embryos metabolize benzo(*a*)pyrene during early gestation: Genetic differences detectable by sister chromatid exchange, *Proc. Natl. Acad. Sci. USA 77,* 3524–3528 (1980).
57. M. T. King and D. Wild, Micronucleus test in fetal mouse blood and in spleen from pregnant mice (abstract), *Mutat. Res. 85,* 301 (1981).
58. M. F. Rajewsky, Possible determinants for the differential susceptibility of mammalian cells and tissues to chemical carcinogens, *Arch. Toxicol. Suppl. 3,* 229–236 (1980).
59. N. Inui and Y. Nishi, A short-term test to detect organotropic specificity of carcinogens, *Toxicol. Lett. 2,* 277–283 (1978).
60. J. W. Hanson, K. Jones, and D. W. Smith, Fetal alcohol syndrome: Experience with 41 patients, *J. Am. Med. Assoc. 235,* 1458–1460 (1976).
61. M. R. Czajka, S. M. Tucci, and G. I. Kaye, Sister chromatid exchange frequency in mouse embryo chromosomes after *in utero* ethanol exposure, *Toxicol. Lett. 6,* 257–261 (1980).
62. S. Shum, N. M. Jensen, and D. W. Nebert, The murine *Ah* locus: *In utero* toxicity and teratogenesis associated with genetic differences in benzo(*a*)pyrene metabolism, *Teratology 20*, 365–376 (1979).
63. S. Ivankovic, Teratogenic and carcinogenic effects of some chemicals during prenatal life in rats, syrian golden hamsters and mini pigs, in: *Perinatal Carcinogenesis* (J. M. Rice, ed.), pp. 103–115, U. S. Dept. of Health, Education and Welfare, Bethesda, Maryland (1979).
64. A. Basler, I. Theiss, and G. Rohrborn, Cytogenetic effects of busulfan *in vivo* on bone marrow cells and oocytes of adult mice and liver cells of transplacentally exposed embryos, *Environ. Mutagen. 1,* 233–238 (1979).
65. J. Meyne and M. S. Legator, Clastogenic effects of transplacental exposure of mouse embryos to nitrogen mustard or cyclophosphamide, *Teratogen. Carcinogen. Mutagen. 3,* 281–287 (1983).
66. C. Roux, I. Emerit, and J. L. Tallemite, Chromosomal breakage and teratogenesis, *Teratology 4,* 303–316 (1971).
67. M. A. Pereira, L. McMillan, P. Kaur, D. K. Gulati, and P. S. Sabharwal, Effect of diesel exhaust emissions, particulates and extract on sister chromatid exchange in transplacentally exposed fetal hamster liver, *Environ. Mutagen. 4,* 215–220 (1982).
68. N. P. Napalkov and V. A. Alexandrov, Neurotropic effects of 7,12-dimethylbenz(*a*)anthracene in transplacental carcinogenesis, *J. Natl. Cancer Inst. 52,* 1365–1366 (1974).
69. V. A. Alexandrov, Embryotoxic and teratogonic effects of chemical carcinogens, in: *Transplacental Carcinogenesis* (L. Tomatis and U. Mohr, eds.), pp. 112–126, International Agency for Research on Cancer, Lyon (1973).
70. T. Nomura, Similarity of the mechanism of chemical carcinogen-initiated teratogenesis and carcinogenesis in mice, *Cancer Res. 37,* 969–973 (1977).
71. W. Wechsler, Carcinogenic and teratogenic effects of ethylnitrosourea and methylnitrosourea during pregnancy in experimental rats, in: *Transplacental Carcinogenesis* (L. Tomatis and U. Mohr, eds.), pp. 127–142, International Agency for Research on Cancer, Lyon (1973).
72. I. Theiss, A. Basler, and G. Rohrborn, Transplacental and direct exposure of mouse

and marmoset to ethylnitrosourea: Analysis of chromosome aberrations, *Teratogen. Carcinogen. Mutagen. 3,* 219–230 (1983).
73. T. Nomura and S. Konda, The enhancement effect of phenobarbital on toxicity of furylfuramide in mouse embryos, *Mutat. Res. 35,* 167–171 (1976).
74. G. W. Kneale and A. M. Stewart, Mantel–Haenszel analyses of Oxford data. I. Independent effects of several birth factors including fetal irradiation, *J. Natl. Cancer. Inst. 56,* 879–883 (1976).
75. S. Preston-Martin, M. C. Yu, B. Benton, and B. E. Henderson, *N*-nitroso compounds and childhood brain tumours: A case–control study, *Cancer Res. 42,* 5240–5245 (1982).
76. C. L. Chrisman, Aneuploidy in mouse embryos induced by diethylstilbestrol diphosphate, *Teratology 9,* 229–232 (1974).
77. J. L. Ivett and R. R. Tice, Diethylstilbestrol-diphosphate induces chromosomal aberrations but not sister-chromatid exchanges in murine bone marrow cells *in vivo, Environ. Mutagen. 3,* 445–452 (1981).
78. U. Mohr, J. Althoff, and A. Authaler, Diaplacental effect of the carcinogen diethylnitrosamine in the golden hamster, *Cancer Res. 26,* 2349–2352 (1966).
79. B. A. Diwan and H. Meier, Transplacental carcinogenic effects of diethylnitrosamine in mice, *Naturwissenschaften 63,* 487–488 (1976).
80. V. A. Alexandrov, Blastomogenic effect of dimethylnitrosamine on pregnant rats and their offspring, *Nature 218,* 280–281 (1968).
81. T. Nomura, Carcinogenicity of the food additive furylfuramide in foetal and young mice, *Nature 258,* 610–611 (1975).
82. E. Gold, L. Gordis, J. Tonascia, and M. Szklo, Increased risk of brain tumours in children exposed to barbiturates, *J. Natl. Cancer Inst. 61,* 1031–1034 (1978).
83. J. M. Peters, S. Preston-Martin, and M. C. Yu, Brain tumours in children and occupational exposure of parents, *Science 213,* 235–237 (1981).
84. L. M. Kinnier-Wilson, G. W. Kneale, and A. M. Stewart, Childhood cancer and pregnancy drugs, *Lancet 2,* 314–315 (1981).
85. R. B. Everson, Individuals transplacentaly exposed to maternal smoking may be at increased cancer risk in adult life, *Lancet 2,* 123–127 (1980).
86. B. M. Sanders and G. J. Draper, Childhood cancer and drugs in pregnancy, *Br. Med. J. 1,* 717–718 (1979).
87. R. A. Seeler, J. N. Israel, J. E. Royal, C. I. Kaye, S. Rao, and M. Abulaban, Ganglioneuroblastoma and fetal-hydantoin-alcohol syndromes, *Pediatrics 63,* 524–527 (1979).
88. A. L. Herbst, H. Ulfelder, and D. C. Poskanzer, Adenocarcinoma of the vagina: Association of maternal stilboestrol therapy with tumor appearance in young women, *N. Engl. J. Med. 284,* 878–881 (1971).
89. J. M. Birch, H. B. Marsden, and R. Swindell, Incidence of malignant disease in childhood: A 24-year review of the Manchester Children's Tumour Registry data, *Br. J. Cancer 42,* 215–223 (1980).
90. W. R. McWhirter and J. E. Brown, Incidence of childhood tumours in Queensland, *Br. J. Cancer 44,* 637–642 (1981).
91. S. Kramer, A. T. Meadows, P. Jarrett, and A. E. Evans, Incidence of childhood cancer: Experience of a decade in a population-based registry, *J. Natl. Cancer Inst. 70,* 49–55 (1983).
92. K. Hemminki, I. Saloniemi, K. Luona, T. Salanen, T. Partanen, H. Vanio, and E. Hemminki, Transplacental carcinogens and mutagens: Childhood cancer, malformations and abortions as risk indicators, *J. Toxicol. Environ. Health 6,* 1115–1126 (1980).

93. E. B. Hook, The value and limitations of clinical observations in assessing chemically-induced genetic damage in humans, in: *Banbury Report 13, Indicators of Genotoxic Exposure* (B. A. Bridges, B. E. Butterworth, and I. B. Weinstein, ed.), pp. 19–32, Cold Spring Harbor Laboratory, Cold Spring Harbor, New York (1982).
94. G. Stamatoyannopoulos and P. E. Nute, Screening of human erythrocytes for products of somatic mutation: an approach and a critique, in: *Population and Biological Aspects of Human Mutation* (E. B. Hook and I. H. Porter, eds.), pp. 265–273, Academic Press, New York (1981).
95. R. J. Albertini, K. L. Castle, and W. R. Borcherding, T-Cell cloning to detect the mutant 6-thioguanine resistant lymphocytes present in human peripheral blood, *Proc. Natl. Acad. Sci. USA 79,* 6617–6621 (1982).
96. A. A. Morley, K. J. Trainor, R. Seshadri, and R. G. Ryall, Measurements of *in vivo* mutations in human lymphocytes, *Nature 302,* 155–156 (1983).
97. F. Stohlman, Fetal erythropoiesis, in: *Regulation of Hematopoiesis* (A. S. Gordon, ed.), pp. 471–485, Appleton-Century-Crofts, New York (1970).
98. K. E. Buckton, P. G. Smith, and W. M. Court Brown, The estimation of lymphocyte lifespan from studies on males treated with X-rays for ankylosing spondylitis, in: *Human Radiation Cytogenetics* (H. J. Evans, W. M. Court Brown, and A. S. McLean, eds.), pp. 106–114, North-Holland, Amsterdam (1967).
99. A. D. Bloom, S. Neriishi, and P. G. Archer, Cytogenetics of the *in utero* exposed survivors of Hiroshima and Nagasaki, *Lancet 2,* 10–12 (1968).
100. L. S. Rayfield, L. Brent, and C. H. Roderick, Development of cell-mediated lympholysis in human foetal blood lymphocytes, *Clin. Exp. Immunol. 42,* 561–570 (1980).
101. P. I. Countryman and J. A. Heddle, The production of micronuclei from chromosome aberrations in irradiated cultures of human lymphocytes, *Mutat. Res. 41,* 321–322 (1976).
102. R. J. Cole and L. M. Henderson, Measuring prenatal genotoxic effects in mice and men, in: *Banbury Report 13: Indicators of Genotoxic Exposure* (B. A. Bridges, B. E. Butterworth, and I. B. Weinstein, ed.), pp. 355–365, Cold Spring Harbor Laboratory, Cold Spring Harbor, New York (1982).
103. G. Obe and B. Beek, Premature chromosome condensation in micronuclei, in: *Premature Chromosome Condensation. Application in Basic, Clinical and Mutation Research* (P. N. Rao, R. T. Johnson, and K. Sperling, eds.), pp. 113–130, Academic Press, New York (1982).
104. O. Iskandar, An improved method for the detection of micronuclei in human lymphocytes, *Stain Technol. 54,* 221–223 (1979).
105. R. J. Cole, Z. Aghamohammadi, and L. Henderson, Frequencies of genetic lesions in lymphocytes from newborn humans, in: *Mutagens in Our Environment* (M. Sorsa and H. Vanio, eds.), pp. 375–386, Liss, New York (1982).
106. S. Z. Aghamohammadi, L. Henderson, and R. J. Cole, The human lymphocyte micronucleus assay. Response of cord blood lymphocytes to γ-irradiation and bleomycin, *Mutat. Res. 130,* 395–401 (1984).
107. V. Huff, R. J. DuFrain, and L. G. Littlefield, SCE frequencies in rabbit lymphocytes as a function of time after an acute dose of cyclophosphamide, *Mutat. Res. 94,* 349–357 (1982).
108. A. V. Carrano and D. H. Moore, The rationale and methodology for quantifying sister chromatid exchange in humans, in: *Mutagenicity: New Horizons in Genetic Toxicology* (J. Heddle, ed.), pp. 267–304, Academic Press, New York (1982).
109. G. Ardito, L. Lamberti, E. Ansaldi, and P. Ponzetto, Sister chromatid exchanges in

cigarette-smoking human females and their newborns, *Mutat. Res. 78*, 209–212 (1980).

110. M. S. Lundberg and G. K. Livingston, Sister-chromatid exchange frequency in lymphocytes of smoking and nonsmoking mothers and their newborn infants, *Mutat. Res. 121*, 241–246 (1983).
111. N. H. Hatcher and E. B. Hook, Somatic chromosome breakage in low birth weight newborns, *Mutat. Res. 83,* 291–299 (1981).
112. N. H. Hatcher and E. B. Hook, Sister chromatid exchanges in newborns, *Hum. Genet. 59,* 389–391 (1981).
113. R. Seshadri, E. Baker, and G. R. Sutherland, Sister chromatid exchange (SCE) analysis in mothers exposed to DNA-damaging agents and their newborn infants, *Mutat. Res. 97*, 139–146 (1982).
114. F. Funes-Cravioto, C. Zapata-Gayon, B. Kohlmodin-Hedman, B. Lambert, J. Lindsten, E. Norberg, M. Nordenskjold, and R. Olin, Chromosome aberrations and sister chromatid exchange in workers in chemical laboratories and a rotoprinting factory and in children of women laboratory workers, *Lancet 2,* 322–325 (1977).
115. B. Lambert, A. Lindblad, K. Holmberg, and D. Francesconi, The use of sister chromatid exchange to monitor human populations for exposure to toxicologically harmful agents, in: *Sister Chromatid Exchange* (S. Wolff, ed.), pp. 149–182, Wiley, New York (1982).
116. L. Henderson, S. Z. Aghamahammadi, C. F. Arlett, and R. J. Cole, Lack of discernible effect of diagnostic ultrasound on the chromosomes of cord blood lymphocytes exposed *in utero, British J. Radiology* accepted for publication.
117. L. Henderson and T. Regan, Effects of diethylstilbestrol-diproprionate on SCES, micronuclei, cytotoxicity, aneuploidy, and cell proliferation in maternal and foetal mouse cells treated *in vivo, Mutat. Res. 144,* 27–31 (1985).

CHAPTER 9

Computer Automation of Metaphase Finding, Sister Chromatid Exchange, and Chromosome Damage Analysis

David A. Shafer, Kenneth I. Mandelberg, and Arthur Falek

1. Introduction

1.1. Value of Short-Term Cytogenetic Tests for Mutagenesis Testing

Current methodology provides few means for assessing the risk of birth defects, cancer, or other pathology arising from exposure to potential genotoxic agents. Exposures to mutagens, carcinogens, or teratogens can be estimated as to probable dosage and potential harm based on experimental or epidemiologic study of such agents, but such extrapolation is largely theoretical.[89] In any case, relatively few agents have been tested, and individual variability in the susceptibility to different genotoxic agents is largely undefined,[11,12,16] aside from the sensitivities associated with a number of unique genetic disorders.[75] In addition, individuals in various industrial, laboratory, or even residential settings

David A. Shafer and Arthur Falek • Human and Behavioral Genetics Research Laboratory, Georgia Mental Health Institute, and Department of Psychiatry, Emory University School of Medicine, Altanta, Georgia 30306. **Kenneth I. Mandelberg** • Department of Mathematics and Computer Science, Emory University, Atlanta, Georgia 30322.

come in daily contact with thousands of chemical agents that have never been and may never be tested with currently available genotoxic methodologies. While cytogenetic assays have proved useful for directly evaluating a small number of the estimated 50,000 chemical agents that may pose genotoxic risk,[61] current procedures are slow and labor intensive. There is thus an immediate need for more efficient cytogenetic testing to evaluate genotoxic damage that may have occurred in environmentally exposed individuals or populations, and to allow routine surveillance of workers in industrial or laboratory settings where there is a reasonable risk of accidental or chronic exposure to genotoxic agents.[37] Short-term *in vivo* mutation assays have been proposed[91,92] as an alternative test, but no such procedure has been put into routine use. Thus far, the most useful routine procedures for evaluating *in vivo* genotoxic effects in humans have been the chromosome damage (CD) and the sister chromatid exchange (SCE) assays.

Chromosome damage is well established as an index of genetic alterations causally associated with mutations and cancer.[13,21,60] While observed CD, such as breaks, gaps, and chromosome interchanges, is unstable and results in cell death,[14] such damage is believed to reflect the parallel occurrence of stable, undetectable genetic alterations which can pass through cell division. However, with the development of bromodeoxyuridine-mediated staining procedures,[42] SCEs have become more extensively studied, since they occur more frequently, are more benign to cell survival, and are much more sensitive to mutagen or carcinogen induction.[36,43,67,71] Comparative studies indicate that SCE and CD effects are linearly correlated with increased mutagenesis and correlated with carcinogenesis,[7,13,21,22,70,74] but that they are produced by independent mechanisms which reveal different types of DNA damage.[22,34,49,72,81,82,98] Therefore, both assays provide important means for assessing genetic risk. Moreover, these assays are best coupled with additional CD and SCE cultures that are challenged *in vitro* with different mutagenic agents,[81,85] since individuals differ in DNA repair capacity and other metabolic factors important to CD or SCE responses.[12,53,54,59]

1.2. Need for High-Speed Automated Cytogenetic Analysis

For statistical reliability, a CD and SCE test battery using two or three different mutagens could reasonably require a minimum of three to six separate peripheral blood cultures per subject and the consequent analysis of at least 150 cells up to a conservative maximum of about 920 cells (150 cells = simultaneous SCE and CD analysis on same cultures with 50 cells/culture for three cultures using one for the base level and two for mutagen induction; 920 cells = separate SCE and CD cultures with 30 cells/culture for SCE analysis and 200 cells/culture for CD analysis using one each for the base levels and three each for three mutagens). However, for low-level exposure effects *in vivo*, evaluating each CD base level per subject could require up to 1000 cells analyzed. The only feasible way so many cells could be evaluated on a routine basis for large groups of people would be with a high-speed computerized cytogenetic system.

The principal rate-limiting factors impeding wide-scale use of CD and SCE assays are the very tedious and repetitive aspects of manual microscopic evaluation. A technician must do considerable searching at low power (10–20×) interrupted by frequent alternation to high power (100×), which requires oil immersion and refocusing in order to find and select useful metaphase cells. Typical SCE evaluations of 20–30 cells per test involves counting the chromosomes and then scoring the SCE events for each acceptable metaphase found. Both of these tasks require considerable concentration, since the technician must retain a mental picture relevant to each chromosome counted or scored. Due to interruptions or other distractions during this process, recounting or rescoring is often necessary. Typical CD evaluations of 100–200 metaphase complements also require considerable scanning and a similar mental intensity for chromosome counting and aberration scoring. Each SCE or CD test per subject or experiment typically requires 1–4 hr to complete an analysis, depending on the type of test, the mitotic index, and the quality of the preparation. For CD analysis, which requires far less time per cell evaluation, a scoring rate of 30–60 cells/hr can be achieved. However, because of the tedium of the process, a technician cannot be expected to remain at the microscope and score much more than about 200–300 CD cells per day.[56] Due to these difficulties, it is not surprising that few chemicals have been evaluated and the cytogenetic studies of human populations have been highly restricted in scope.

Other factors have also augmented the need for automated cytogenetics. This laboratory and others have demonstrated that there is significant interindividual variation in both resident CD and SCE levels and in the subsequent CD and SCE responses to different chemical and physical agents in normal human populations.[11,12,16,33,41,59,85] Further clarification of the degree and frequency of such variability in demographically defined human populations is therefore needed in order to establish benchmarks for subsequent evaluations of the human risk for mutagenesis or carcinogenesis consequent to various environmental or industrial exposures. Furthermore, the public has become acutely sensitized to the potential genotoxic hazards of accidental or chronic exposure to chemicals or radiation either via the workplace, by various industrial products, or by environmental contamination. Industrial management is likewise becoming sensitized to their possible liability for harm to their customers or employees and a number of major industries have initiated their own cytogenetic testing and monitoring. Consequently, such testing must not only be able to evaluate the relative risk for exposed versus unexposed populations, but will also have to clearly discriminate between individual exposure levels and the actual or expected levels of genotoxic effect for specific individuals. These conditions thus additionally require improvement in the reliability as well as the speed of cytogenetic monitoring. Since CD and SCE scoring varies consistently from technician to technician,[14,84] an arbitrary computer scoring system could considerably increase the validity and comparability of CD and SCE scores conducted in different laboratories and at different times. In order to overcome the slow pace and inconsistency of manual microscopic analysis, a number of automated cytogenetic systems have been developed. This chapter will briefly review previous approaches taken in other laboratories and will describe our recent advances in developing a small-scale, integrated system for automating the processes of finding metaphase chromosome spreads and for scoring SCE and CD effects.

2. Historical Trends in Computer Automation

Computerized analysis of chromosomes was first proposed about 20 years ago by Ledley[44] (Washington). Since that time there has been considerable progress in a number of pioneering laboratories. However, most of these efforts have been directed toward clinical karyotyping.[8,10,20,23,26,27,38,40,45,46,51,52,55,57,58,65,79,94,97,99] Consequently, such systems have provided little software development in common with the functions required for testing mutagenic alterations, except for procedures to locate metaphase chromosome complements.

Automatic karyotyping requires complex pattern recognition and sorting routines. Consequently, images must be fully digitized, stored, manipulated, and outputted with sufficient resolution for human recognition and verification. Moreover, karyotyping requires essentially 100% accuracy in decision-making in order to complete analysis. Each inaccurate automatic decision burdens the operator to correct it. For mutagenesis testing, it is sufficient to record scored aberrations or SCEs in association with cell coordinates. The latter procedures can thus utilize high-speed, real-time processing of images and only need to extract and retain very limited information. Second, automation of cytogenetic mutagenicity testing would easily allow 90–95% scoring accuracy for statistically acceptable reliability. Therefore, the hardware configurations, though broadly similar, can be significantly scaled down for mutagenicity testing as compared to karyotyping. Similarly, software development is a manageable task and need not remain endlessly on the verge of practical utility, as has repeatedly happened in automatic karyotyping projects. However, for automatic SCE scoring and particularly for CD scoring, speed is a very critical and primary requirement.

2.1. Metaphase Finding and Karyotyping

Automated metaphase finding was first proposed by Wald and Preston(35,63,95) (Pittsburgh) and more than a half dozen different types of metaphase-finding systems have been built since then.(3–5,18,19,29–31,35,39,47,48,63,64,66,73,77,78,84,87,88,95) Highly divergent hardware and software systems have been utilized to detect presumed chromosomes and to interpret the presence and location of suitable clusters. Imaging has been achieved with Vidicon or other television scanners, flying spot scanners, lasers, and linear photodiode arrays. Computing systems have ranged from large mainframes to small minicomputers, with some specific functions served by dedicated hardware devices such as analog/digital converters, shading correctors, and thresholding logic units. Most of the major prototypes were developed 10–15 years ago using what was then considered large and expensive computer hardware in order to provide the extensive memory and processing requirements needed for automatic karyotyping.(4,8,35,38,51,55,62,79,96) There were several attempts, however, to build smaller, single-function metaphase-finding systems based on the belief that greater speed could be achieved for this particular function, since high image resolution is not required to detect chromosome clusters accurately.

In association with Wald's laboratory, Norgen(64) used a laser scanning system to locate chromosomes based on the Weiner spectrum of light scattering. This approach was then improved by Herron *et al.*(35) by capturing the scattered light in annular rings at different angles and by the addition of other devices to approximate the size of the features. In another extension of this optical approach, Paton(66) used Fourier analysis to determine the periodicity of light scattering by the chromosomes. Ledley *et al.*(45) designed a different device using white light to scan for features within a preset length range that meet certain distribution criteria. In general, these early optical systems have not been adapted for routine use or been extensively evaluated, nor have any of them apparently emerged from the prototype stage.

The second major approach toward dedicated metaphase finders was the use of arrays of small photodiodes (0.5–1.5 μm) coupled with logic devices. A matrix or line of such devices provides multiple simultaneous inputs and can thus arithmetically increase speed by adding diodes. LeGo[47] (France) first developed this approach with a nine-photodiode array and eventually built a system with a 512-diode array.[48] Farrow *et al.*,[18,19] at Edinburgh, used a 256-diode array and took the approach of detecting anticoincidence points, which are in effect the ends of features. Neurath's group at Tufts (Boston) designed a very high-speed system using a linear array of 1024 photodiodes, which could scan 2.5 cm^2/min.[5,63] The slide was continuously swept over the array with the diode sensors detecting the endpoints of dark features passing over it. The relative distribution of those endpoints was then used to interpret the location of metaphase clusters. The speed of these systems, which could scan a complete slide in 2–5 min, was obtained, however, at the expense of accuracy, since they all suffered from high false-positive rates of 50% or greater.[77,78] Recently, a modified version of the Edinburgh system[88] has emerged as the CYTOSCAN commercial metaphase finder manufactured by Shandon Instruments in Great Britain.

The third major approach to dedicated metaphase finders has been to utilize television scanners to simply detect features that exceed a certain darkness threshold, match a size range, and exceed a preset feature count in a limited area.[10,29,30,39] These early versions, often with hardware logic designed for speed, also suffered from accuracy problems. However, at the same time television scanners were also being used in commercial image analyzers not initially directed at cytogenetic applications and in more elaborate systems in attempts to achieve automated karyotyping. In the latter systems, speed was secondary to the objective of obtaining the high image resolution needed for precise feature identification and measurement. Most image processing was software-based after fully digitized image information was piped to a separate, general-purpose computer. Castleman *et al.*,[10] at the Jet Propulsion Laboratory in California, built the first such automated karyotyping system, which was considered portable,[50] since it used largely commercial detection equipment and minicomputer hardware. However, like other systems built during this time, it ignored the banding information and based chromosome assignments on comparisons of the centromere position and relative chromosome arm lengths. This system was improved by Graham and Taylor[26] in Manchester and has emerged as the commercial MAGISCAN chromosome analysis system manufactured by Joyce Loebl Ltd. Although this system allows easy manipulation and output of the karyotype, automatic chromosome identification is still limited to structural criteria. Moreover, the extensive interaction and time required to set up and run metaphase finding might be considered burdensome. Indeed, most other commercial systems that have emerged in recent years seem to suffer from related difficulties when applied in a similar manner to routine clinical cytogenetic tasks.

The principal commerical systems based on Plumbicon or Chalnicon television scanners include the QUANTIMET from Cambridge Instruments-Imanco (England), the TAS from the Leitz Corp. (West Germany), the OMNICON from Bausch and Lomb (Rochester, New York), and the IBAS from Zeiss (West Germany). In Leiden, Pearson and Ploem in association with Serra[80,94] have developed the LEYTAS system, a refinement of the TAS system for chromosome analysis. The QUANTIMET, TAS, and OMNICON systems gain efficiency by only capturing binary data from a single gray level or a gray level range for each image scan, and they have been offered with metaphase-finding software based on a dilation/erosion agglomeration technique or related procedures to find clusters of chromosome-sized features. In comparison to dedicated hardware-based metaphase finders, these commercial metaphase-finding programs provided fairly slow scanners, ranging from 0.02 (TAS) to 0.24 cm^2/min (OMNICON), but they nonetheless reduce both the false-positive and the false-negative rate to about 10–30%.[78] These systems are also fairly flexible because their measurement and analysis procedures are largely software-based. This chapter will focus extensively on the development of new application software for accurate metaphase finding and SCE and CD analysis using an OMNICON system. There has been considerable development of fully digitized image analysis systems for banding analysis and karyotyping,[6,10,20,21,24,26,27,40,46,52,55,79,97,99] but they will not be reviewed here. Although some of the latter sophisticated systems can automatically straighten chromosomes and precisely measure and interpret banding or replication patterns, they are generally slow, research prototypes lacking the efficiency required for mutagenesis testing.

2.2. Automated Chromosome Aberration and Sister Chromatid Exchange Scoring

There have been relatively few computerized cytogenetics programs directed at automatic or semiautomatic CD scoring for mutagenicity testing.[1,2,9,19,56,68,84,87,96,100,101] Wald was among the first to attempt automatic CD analysis by using a separate minicomputer system to process fully digitized image information captured by a flying spot scanner.[96] Image detail made it possible to extract centromere locations, shape characteristics, and the chromosomal skeleton[1] in order to detect three types of chromosome damage—dicentrics, rings, and acentric fragments. Since these algorithms were prone to error with touching or overlapping chromosomes, each decision had to be reviewed and/or corrected by a trained technician. When Wald simply used the stand-alone metaphase-finding system to present TV images of found metaphases to a technician for manual recognition and scoring of CD, a three- to five-fold increase in completing the analysis was reported.[96] However, use of the automatic CD scoring system did not appreciably increase the analysis rate,[56,96] and we are unaware of subsequent refinements of that system. Also during the mid-1970s Rutovitz's group in Edinburgh proposed that high-speed CD detection could be achieved by a refinement of their linear array scanner,[19] and Castleman *et al.*[9] similarly suggested that their system provided a useful prototype, since abnormal chromosomes were automatically rejected from the karyotype. To date, no system has been reported from either group that has been adapted for routine CD analysis. While all of the television-based commercial image analyzers described above should also be capable of simulating the aberration detection routines used by Wald, no such applications have been reported. If such algorithms are basically inefficient,[56] then either image data must be input and processed faster than the rate characteristic of most serial computers or aberration detection decisions must be made by some simpler criteria. Recently, Bille *et al.*[2] have reported initial computer development work toward automatic detection of aberrations with a parallel processor (interconnected multiple processors), although analysis results are not yet forthcoming. Such a parallel or a pipeline architecture could provide the speed required for truly automatic CD analysis.[2,25,69,76,90] It should be noted here that a number of laboratories have followed a very different approach by attempting to score CD by flow cytometry rather than microscopy.[17,28] This approach provides very rapid scoring by precisely measuring the DNA content of the chromosomes as they are forced in a narrow stream past a laser beam. With a flow system, Fantes *et al.*[17] achieved effective CD scoring of cells irradiated with ⩾100 rad, although lower levels of damage cannot be detected by this method.

Besides our laboratory,[83,84,87] only Zack *et al.*[100,101] has published attempts at automatic or semiautomatic SCE analysis. Their system used a Vidicon scanner, was minicomputer based, and had no metaphase-finding capability, and all SCE detection software was formulated *de novo*. While they reportedly were able to count up to 30 SCEs/cell, development and use of this system was discontinued. Recently, Lubs and Ledley announced that they expect to start building several prototypes for a commercial version of their automatic karyotyping system[46,51] and SCE analysis is listed as a future function. Otherwise, we are unaware of any active system development for automated SCE analysis.

Based on this short overview, it seems there has been a surprising lack of advancement in automated cytogenetic mutagenicity testing in the last decade despite the increased scientific and public awareness of the growing need for such testing.

3. Development of a System for Cytogenetic Mutagenicity Testing

3.1. Early Approaches

Our laboratory has long been engaged in cytogenetic studies of human populations exposed to chemicals, drugs, and radiation. Since Falek had been an early collaborator with Neurath,[15,62] and Shafer had developed a more recent collaboration for the measurement of chromosome replication and banding patterns,[86] computer automation offered a tangible hope of overcoming the slow pace of manual microscope analysis that had long impeded such cytogenetic studies. Rather than attempting to build another prototype, we pursued the use of commerically available image analysis hardware and the development of software systems for metaphase finding and SCE scoring.[83,84,87]

After a brief collaboration with Peach and McCarthy of Bausch and Lomb to develop preliminary metaphase-finding software,[84] our laboratory purchased an OMNICON 3500. We first redesigned and operationalized an efficient metaphase-finding program. We then developed new programs to relocate automatically the found metaphases at a high power, to count the chromosomes, and to scan the chromosomes for SCE scoring.

Shafer had first proposed that SCEs could be automatically scored if we could count the chromosomes, count the number of dark features in the SCE stained chromosomes, and then subtract the number of chromosomes. Previous exploration with the QUANTIMET and TAS systems and interaction with both Swenson and Oney at Cambridge Instruments[83] and Golden at Leitz Corp. had suggested that relative-

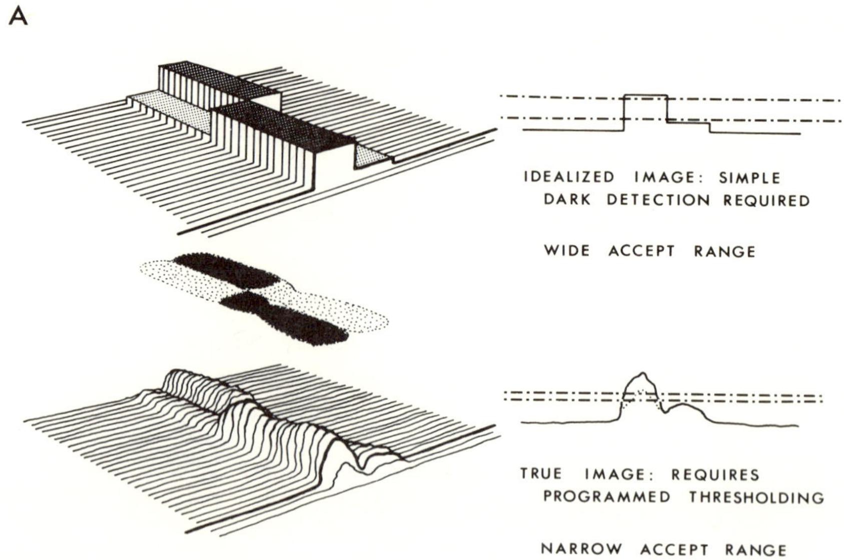

FIGURE 1. Thresholding for automatic SCE detection. (A) High-speed automatic SCE analysis using binary feature detection requires critical gray level thresholding in order to select that level above which the dark SCE-stained segments are fully but separately detected. If that level is too low, some of the light SCE-stained segments will be detected. If that level is too high, the dark SCE segments will be broken into smaller segments where there are valleys in the dark stain intensity. Such errors will artificially increase the SCE count. The deepest stain valley usually occurs at the centromere, and this problem

area criteria could be used to select the gray level that corresponded to optimal binary SCE feature detection (the level where the dark chromatid regions are separately and fully detected). We therefore developed routines for the OMNICON to optimize this selection process in order to score SCEs. Repeated trials, however, indicated that SCE scoring accuracy was highly variable when a single global threshold was used to score all the chromosomes in a spread. Apparently, there was often sufficient chromosome to chromosome variation in stain incorporation to compromise the accuracy of global thresholding (Figure 1). Development of an automatic SCE analysis system that could rapidly process each chromosome separately was recognized as a long-range project. To bring the OMNICON into routine use, we therefore pursued program development for a semiautomatic SCE scoring system. In the course of subsequent development we also designed a new metaphase-finding program with an entirely different clustering algorithm. The

B

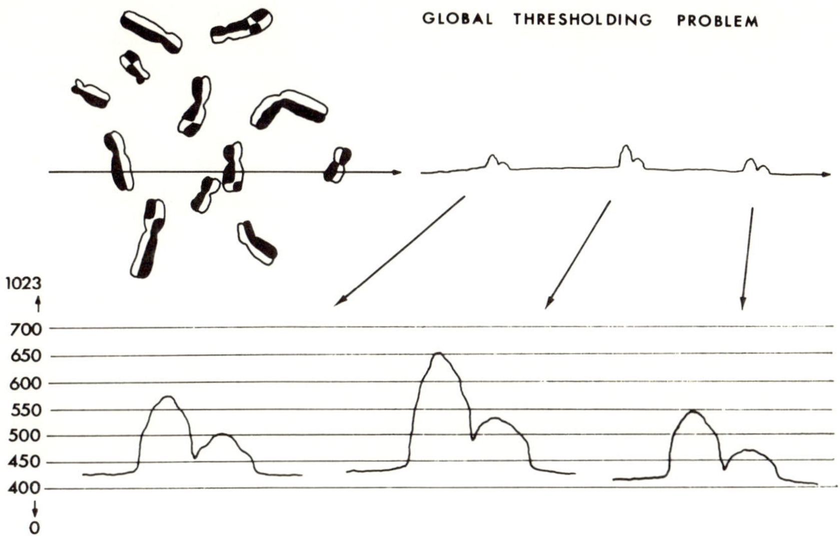

is common enough to require special subroutines to overcome it routinely. (B) The thresholding problem is accentuated when there are stain variations between chromosomes in a spread. The correct threshold for SCE detection with some chromosomes may give incorrect SCE scoring for other chromosomes if a single threshold is applied globally. These subtle variations, which are not usually discernible by manual observation, typically involve a reduced stain intensity in the chromosomes at the periphery of a metaphase cell.

new program, METAFIND, rapidly scans microscope slides at low magnification and detects and records metaphase locations. The semiautomatic scoring program, SCESCAN, relocates these sites at high magnification, automatically counts chromosomes, and allows lightpen SCE scoring. A subroutine to the last program is also being added to allow simultaneous scoring of CD.

3.2. Image Analysis Configuration

The METAFIND and SCESCAN programs were developed with an OMNICON 3500 Image Analysis System (Bausch and Lomb), which uses a Data General ECLIPSE S/140 minicomputer with 256 kilobytes of memory and a Data General model 6098 12.5-megabyte Winchester disk with a 1.2-megabyte floppy diskette subsystem. The standard stage on a Zeiss photomicroscope 1 was replaced with a computer-driven stage mechanism. This device used two 5 μm stepping motors that allow smooth, continuous, and rapid *X* and *Y* motion that is precise and repeatable. A third motor device provides autofocus control of the fine focus mechanism. The microscope image is detected via a Chalnicon scanner mounted on the microscope, which transmits the image to a television monitor. The analogue TV signal is secondarily processed by several modular devices, which can modify the image viewed on the monitor. One device is a shade correction module, which detects and corrects for variations in light intensity within a field of view caused by the physical characteristics of the specific condenser and objective in use. A second module consists of the basic image analysis hardware and software for detecting and extracting gray-level-specific features from the total scanned image. While thresholding can potentially discriminate from a range of 1024 gray levels, only one or two levels can be selected. Depending on the selected thresholding criteria, this device either captures all gray level data that are higher or lower than one specific threshold or it captures and compresses a slice of gray level data between two specific threshold levels. This process produces a binary image of the detected features in the computer memory as well as allows this binary image to be superimposed upon the normal image on the monitor as obvious bright or dark features. A third module is a lightpen-controlled image editor, which allows the binary detected image to be directly modified. With this device, binary features to be analyzed can be cut or connected, included or excluded, and created or deleted (Figure 2).

This characteristic of extracting and storing binary images from a single gray level or a gray level range is shared by the OMNICON, QUANTIMET, and TAS systems and provides the means for high-speed, real-time image processing. With these systems, image storage is limited strictly to feature boundary information for a specific gray level or a gray level range, while data from other gray levels must be input as separate image acquisitions (1/30 sec per acquisition). This approach contrasts with other systems discussed previously that store and process a fully digitized image with a quarter million to a half million pixels at 256 gray levels of discrimination. Each approach has distinct benefits and limitations for different applications.

METAFIND and SCESCAN were developed as separate Fortran IV programs. The subroutines utilize a number of OMNICON-designed general functions which are accessible by coded call names. These functions facilitate stage control and feature location and measurement.

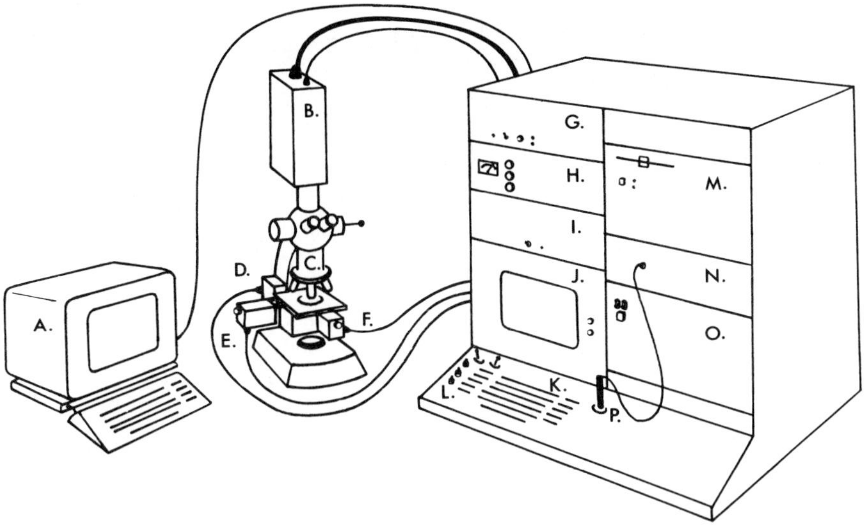

FIGURE 2. Cytogenetic image analysis system. Schematic drawing of the principal features of our current hardware configuration: (A) computer control consule, (B) Chalnicon camera, (C) Zeiss photomicroscope, (D) autofocus control, (E) *X*-motion control, (F) *Y*-motion control, (G) Chalnicon scanner module, (H) shade correction module, (I) OMNICON basic modules to control imaging, thresholding, measurement, and alphanumeric display, (J) television monitor to display real and/or binary images and/or alphanumeric characters, (K) OMNICON interactive run-time keyboard, (L) OMNICON manual threshold controls and frame joysticks, (M) Winchester and floppy disk mass memory units, (N) lightpen control module, (O) Eclipse S/140 computer, (P) lightpen. (A printer and an auxiliary terminal are not shown.)

3.3. Program Operation

3.3.1. Metaphase Scanning

The operator initiates the system via a CRT terminal and then control is switched to a keyboard below the monitor when the programs are running. The METAFIND program first requests slide identification and scanning information. The operator resets the shade corrector, selects the origin (the location for beginning the scan), and requests the number of fields to be scanned in the *X* and *Y* directions. The origin is recorded with reference to hardstop positions on the stage so that found metaphase cells can be consistently relocated again and again.

METAFIND scans slides at low power (16× objective). Each field image is stored and processed in memory while the stage is moving to the next field. In an earlier stage of development, METAFIND scanned at a single gray level selected by the operator. The clustering algorithm determines metaphases by first linking chromosome-sized features that are closely distributed and by then selecting those clusters that exceed a minimal numerical criterion. For each selected cluster of features, the feature distribution is resolved into a single mean centralized location, which is recorded along with the count. Coupled with this clustering scheme, single-level scanning proceeds essentially as fast as the stage can be moved from field to field. However, the single scan setting was sensitive to the operator's judgment of the correct level after viewing a few fields. We have further modified this approach to scan each field over a small series of gray levels to ensure metaphase detection of cells with variable stain intensity and to eliminate operator bias or uncertain judgment. The same clustering algorithm is still used, but each cluster might be detected on one to seven different gray levels. This multilevel scanning system operates in two modes. In the initial prescanning mode, multiple images in each field are acquired and processed at a wide range of gray levels, from the background staining level to the darkest dark level in nuclei and chromosomes. When a preset number of clusters have been found, a brief statistical process selects the optimal narrow range in which the most and the best clusters are found, and subsequent scanning only continues in this reduced gray-level range. At each field, the multi level cluster information is sorted to determine the optimal cluster count for each location within that field. For each of these selected clusters, the mean central coordinates and the count total are then recorded in an external data file on the disk while METAFIND is moving to the next field to be scanned. When scanning is completed, the slide is returned to the origin to either proceed with high-power analysis or to initiate scanning of another area or another slide.

3.3.2. Chromosome Analysis

SCESCAN begins by initially requesting a list of metaphase locations from the appropriate METAFIND datafile and by then moving to the first location. Since SCESCAN operates solely at high magnification (100 × objective) and since such magnification causes many typical metaphase distributions to fill the monitor frame, fine correction of the centering of each metaphase is necessary. SCESCAN initially centers each spread automatically by using algorithms that detect chromosomes in the field and then move their median location to the center. Alternatively, the operator can fine position the spread by touching the apparent center with the lightpen and that location is automatically moved to the center of the monitor. As each presumed metaphase spread is automatically moved into view, the operator is presented with a cell selection menu superimposed on the monitor. By pressing the approriate key, the operator automatically classifies and tabulates the cell type (SCE, third-cycle SCE, non-SCE-differentiated), notes any false positives, and initiates analysis on appropriate cells. When the analysis mode is selected, the chromosomes are automatically counted, each is marked with a bright tag at its center, and the count is displayed in the corner of the monitor. The operator then has the option of adjusting the count by adding or deleting tags with the lightpen. When the count is accepted, the lightpen shifts into an SCE scoring mode in which the operator tags each SCE site by touching its location. Again the count is maintained in the corner of the monitor and the operator can correct any errors or missed sites before accepting the count. The tags greatly facilitate SCE scoring over manual microscope analysis where the operator must maintain a mental picture of the chromosomes scored. Semiautomatic aberration scoring will utilize the same tagging system, although we are redesigning the on-screen menu for automatically recording of the different aberration types. After all previously recorded metaphase locations have been reviewed, the data are automatically compiled, the ratios are calculated, and these results are displayed on the computer monitor. For SCE analysis, these data currently include the number and proportion of cell types observed, the number of cells with SCE analysis, and the mean SCE ratio in SCEs/chromosome. All such raw and calculated data are recorded along with the cell coordinates in a separate datafile indexed to the subject or experiment number and the specific microscope slide.

3.3.3. Ergonomic Advantages

The system we have developed provides a high-speed but accurate means for finding and recording cytogenetic alterations such as SCEs. The nature of this system also transforms the general approach to conducting such analyses. One no longer switches back and forth between low-power scanning for appropriate metaphases and high-power analysis of each appropriate cell. The repetitive adjustments for parfocal differences between objectives, focusing, and refocusing and for oil immersion are thus eliminated. Rather, the automatic system remains in low-power mode throughout METAFIND scanning and requires minimal attendance. In SCESCAN, the operator's time and attention are only engaged for relatively brief periods for classifying cells, correcting automatic chromosome counts, and scoring SCEs. Oil immersion is usually only required at the beginning of high-power analysis and fine focusing need only be adjusted occasionally and to a small degree. The operator's time is concentrated on intelligent decision-making and is not absorbed by repetitive, tedious manual adjustments.

3.3.4. Performance Rates

In early trials with human lymphocyte preparations, using METAFIND with the new clustering algorithm and a single gray level scan, typical runs of 100 fields are scanned in about 1.8 min, yielding 15–30 metaphase locations. This rate does not include the initial setup time for loading the slide and encoding the filename and scan specifications, but it does include about 30 sec for periodic autofocusing every three to five fields in the scan range. When a variety of stained slides were scanned, covering some 2400 fields, about 500 metaphases were located, yielding 58% normal SCE differentiated cells, 2.5% third-cycle SCE cells, and 36.8% non-SCE cells. The false-positive and the false-negative rates were about 10% each. With the multilevel version of METAFIND, scan time for 100 fields is increased to about 3 or 4 min when three to four acquisitions are made per field. The clustering algorithm requires about $\frac{1}{4}$–$\frac{1}{3}$ sec per acquisition, while each field motion requires about $\frac{1}{2}$ sec. However, movement to the next field proceeds while the last acquisition per field is being processed. The yield of found cells is not significantly increased by multilevel scanning, but both the false-positive and the false-negative rates drop to about 5% and there is improvement in the overall quality of found cells and in the centering accuracy. When the semiautomatic SCESCAN program is subsequently run, considerably more time is required for the screening and analysis of found cells because of the operator interactive requirements. Besides the brief screening time for tallying nonanalyzed cells (1–3 sec/cell), the scoring of selected cells requires about 1 min/cell (about 2 sec/cell for automatic chromosome counting, 2–8 sec/cell for count corrections, 10–35 sec/cell for SCE tagging; CD tagging, which has not been tested, should require 5–20 sec/cell).

While repeated timed trials of the relative overall speed of the automated metaphase finding and SCE analysis system as compared to manual scanning and analysis have not yet been conducted, it is readily obvious that METAFIND is about fivefold faster than manual scanning and that semiautomatic SCESCAN is about threefold faster than manual SCE scoring. Moreover, the present testing versions of these programs are still encumbered with a variety of options and checks for performance validation that will subsequently be removed.

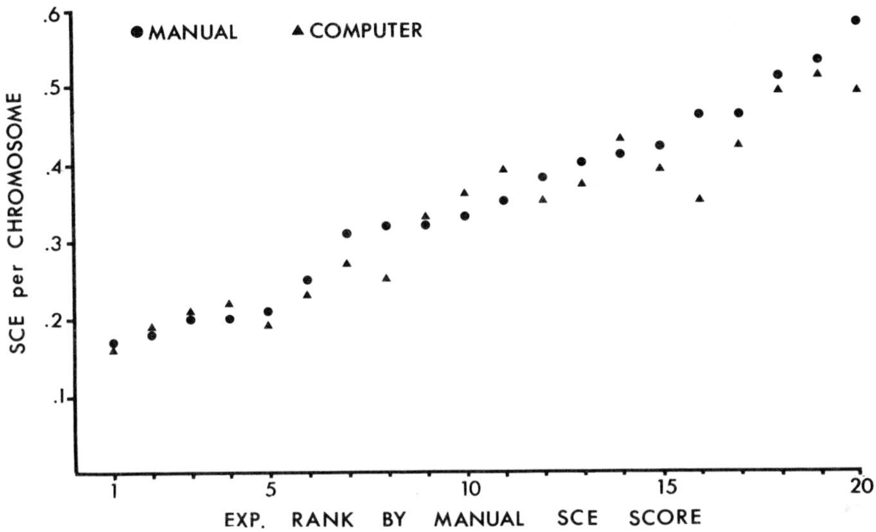

FIGURE 3. Correlation of computer versus manual SCE scoring. Independent manual and computer scoring of the same slides from 20 different experiments. Each data point reflects the mean SCE ratio from scoring 25 cells (SCEs/chromosomes counted). For visual clarity they have been arbitrarily ranked by increasing manual SCE score, although relative computer score could have served just as well as the comparative reference.

3.3.5. Comparative Reliability

An important issue in cytogenetic scoring is the reliability between different observers in analyzing the same data. In manual microscope analysis of chromosome damage, well-trained scorers show significant differences in their results, although they may reach similar general conclusions.[93] Scoring on an image analysis system apparently helps overcome such difficulties, although the microscope image is a little sharper than the television image. When preliminary trials of semiautomatic SCE analysis were conducted using earlier versions of the current programs, two trained technicians independently scored the same SCE cells both manually on the microscope and semiautomatically on the television monitor. In these blind comparisons, the two experienced operators were found to be consistently and significantly different from one another when scoring manually on the microscope. On the other hand, their separate semiautomatic SCE scores on the OMNICON monitor were highly correlated (Pearson correlation, $r = 0.94$).

We have also tested the current integrated METAFIND and

SCESCAN systems (in semiautomatic SCE mode) against manual operator scoring of the same microscope slides from a variety of experiments exhibiting diverse SCE frequencies. These comparisons indicated that computer and manual scoring produced highly correlated results (r = 0.95) (Figure 3), although the same technician consistently scored about one SCE point lower using the computer than when scoring manually. This difference may be partially caused by a tendency to proceed with SCESCAN processing while some peripheral chromosomes are off the monitor; however, operator judgment may also be consistently different on the computer. In any case, the presently demonstrated statistical consistency validates the reporting of such computer-scored SCEs on the same footing with manual analysis.

4. Future Prospects for Full Automation

While METAFIND and SCESCAN provide a useful and efficient integrated system for automated cytogenetic scoring, an ideal system would be fully automated and considerably faster. With our current system we have been able to develop initial algorithms for chromosome-specific automatic SCE detection, although semiautomatic tagging still remains the method of choice for routine analysis. Accurate automation awaits implementation of an algorithm to correct false counts caused by staining gaps at the centromere, and we have not yet achieved our initial speed goal of less than 12 sec/cell for this AUTOSCE program. While we expect to overcome these difficulties reasonably in the near future simply by software improvements, very rapid and accurate automatic SCE analysis will eventually depend on the incorporation of new imaging and processing technology currently under development. The same conditions are especially true for automatic CD analysis, which requires considerable artificial intelligence for pattern recognition of specific chromosome aberrations.

The relatively inexpensive microcomputers manufactured today are rapidly exceeding the processing capacity of yesterday's minicomputers, and these emerging devices hold the promise of desktop image analysis systems in the immediate future. New imaging technology, however, offers greater prospect for the development of highly accurate and fast, fully automatic systems. General image processing hardware is being developed that can input, transform, and enhance fully digitized images at processing rates fast enough for real-time analysis and display of video images. Other devices under development provide geometric arrays of arithmetic logic units that will essentially offer individual and simultaneous processing capacity for each pixel site in an image.[32] The immense processing power of such devices suggests that image analysis may become a commonplace capability and that the automation of cytogenetics may become a relatively easy task.

ACKNOWLEDGMENTS

Supported in part by grants from the National Institute on Drug Abuse (DA-0145) and the March of Dimes Birth Defects Foundation (15-45).

5. References

1. R. K. Aggarwal and K. S. Fu, Automatic recognition of irradiated chromosomes, *J. Histochem. Cytochem. 22*, 561 (1974).
2. J. Bille, H. Scharfenberg, and R. Manner, Biological dosimetry by chromosome aberration scoring with parallel image processing with the Heidleberg POLYP polyprocessor system, *Comput. Biol. Med. 13*, 49–79 (1983).
3. R. P. Bishop, The Automated Detection and Classification of Human Chromosome Spreads, M.S. Thesis, Department of Electrical Engineering and Computer Science, *Massachusetts Institute of Technology*, Cambridge (1976).
4. R. P. Bishop and I. T. Young. The automated classification of mitotic phase for human chromosome spreads, *J. Histochem. Cytochem. 25*, 730–740 (1977).
5. J. F. Brenner, B. Dew, J. B. Horton, T. King, P. W. Neurath, and W. D. Selles, An automated microscope for cytological research: A preliminary evaluation, *J. Histochem. Cytochem. 24*, 100–111 (1976).
6. C. Bruschi, F. Tedeschi, P. P. Puglisi, and N. Marmiroli, Computer-assisted karyotyping system of banded chromosomes, *Cytogenet. Cell Genet. 29*, 1–8 (1981).
7. A. V. Carrano and L. H. Thompson, Sister chromatid exchange and the single gene mutation, in: *Sister Chromatid Exchange* (S. Wolff, ed.), pp. 59–86, Wiley, New York (1982).
8. K. R. Castleman and R. J. Wall, Automatic systems for chromosome identification, in: *Nobel Symposium 23, Chromosome Identification* (T. Caspersson and L. Zech, eds.), Nobel Foundation and Academic Press (1973).
9. K. R. Castleman, J. Melnyk, H. J. Frieden, G. W. Persinger, and R. J. Wall, Karyotype analysis by computer and its application to mutagenicity testing of environmental chemicals, *Mutat. Res. 41*, 153–162 (1976).
10. K. R. Castleman, J. Melnyk, H. J. Frieden, G. W. Persinger, and R. J. Wall, A minicomputer based karyotyping system, in: *Automation of Cytogenetics* (M. L. Mendelsohn, ed.), pp. 46–50, National Technical Information Service, U. S. Department of Commerce (1976).
11. P. E. Crossen, Variation in the sensitivity of human lymphocytes to DNA damaging agents measured by SCE frequency, *Hum. Genet. 60*, 19–23 (1982).
12. F. J. De Serres and R. Pero (eds.), *Individual Susceptibility to Genotoxic Agents in the Human Population*, Plenum Press, New York (1984).
13. H. J. Evans and Vijayalaxmi, Induction of 8-azaguanine resistance and sister

chromatid exchange in human lymphocytes exposed to mitomycin C and x-rays *in vitro*, *Nature 292*, 601–605 (1981).

14. A. Falek, Chemical mutagens and drugs of abuse in man, *Int. J. Addictions 9*, 171–183 (1974).
15. A. Falek and P. W. Neurath, Human chromosome analysis by a semi-automatic system, *Lancet 2*, 1277 (1963).
16. A. Falek, J. J. Madden, D. A. Shafer, and R. Donahoe, Variation in metabolic sensitivity and genetic response to mutagens, presented at *American Association for the Advancement of Science Annual Meeting*, Toronto, 1981.
17. J. A. Fantes, D. K. Green, J. K. Elder, and P. Malloy, Detecting radiation damage to human chromosomes by flow cytometry, *Mutat. Res. 119*, 161–168 (1983).
18. A. S. J. Farrow and J. H. Tucker, A real time TV to computer image input system using run length coding, *J. Histochem, Cytochem. 24*, 112–121, (1976).
19. A. S. J. Farrow, D. K. Green, and D. Rutovitz, A cytogeneticists' microscope and a proposed system for aberration scoring, in: *Automation of Cytogenetics* (M. L. Mendlesohn, ed.). pp. 68–71, National Technical Information Service, U. S. Department of Commerce (1976).
20. A. J. Fornace, H. Nagasawa, and J. B. Little, Relation of DNA repair to chromosome aberrations, SCEs and survival during liquid-holding recovery in x-irradiated mammalian cells, *Mutal. Res. 70*, 323–336 (1980).
21. A. Forabosco, G. Fontana, R. Fabiani, E. Bolzani, and E. Sarti, Computer analysis of reverse banded chromosomes in human metaphases, in: *Biomedical Computing*, Pitmans (1976).
22. S. M. Galloway and S. Wolff, The relation between chemically induced SCEs and chromatid breakage, *Mutat. Res. 61*, 297–307 (1979).
23. G. Gallus and P. W. Neurath, Improved computer chromosome analysis incorporating preprocessing and boundary analysis, *Phys. Med. Biol. 15*, 435–445 (1970).
24. G. Gallus and G. Regoliosi, A decisional model of recognition applied to the chromosome boundaries, *J. Histochem. Cytochem 22*, 546–553 (1974).
25. B. K. Gilbert, T. M. Kinter, and L. M. Krueger, Advances in processor architecture, device technology, and computer-aided design for biomedical image processing, in: *Multicomputers and Image Processing*, pp. 385–405, Academic Press, New York (1982).
26. J. Graham and C. J. Taylor, Automated chromosome analysis using the Magiscan image analyser, *Anal. Quant. Cytol 2*, 237–242 (1980).
27. G. Granlund, The use of distribution functions to describe integrated density profiles of human chromosomes, *J. Theor. Biol. 40*, 573–589 (1973).
28. J. W. Gray, A. V. Carrano, L. L. Steinmetz, M. A. Van Dilla, D. H. Moore, B. H. Mayall, and M. L. Mendelsohn, Chromosome measuring and sorting by flow systems, *Proc. Natl. Acad. Sci. USA 72*, 1231–1234 (1975).
29. D. K. Green and J. Cameron, Metaphase cell finding by machine, *Cytogenetics 11*, 475–487 (1972).
30. D. K. Green and P. W. Neurath, The design, operation and evaluation of a high speed automatic metaphase finder, *J. Histochem. Cytochem. 22*, 531–535 (1974).
31. D. K. Greene, R. Bayley, and D. Rutovitz, A cytogeneticists' microscope, *Micro. Acta 79*, 237–245 (1977).
32. W. Hannaway, G. Shea, and W. R. Bishop, Handling real-time images comes naturally to systolic array chip, *Electronic Design 32*, 289–300 (1984).
33. I. L. Hansteen, SCE as a monitor of industrial and environmental toxins, in: *Progress and Topics in Cytogenetics, Sister Chromatid Exchange* (A. Sandberg, ed.), Vol. 2, pp. 675–697, Liss, New York (1982).

34. K. Hedner, B. Hogstedt, A. M. Kolnig, E. Mark-Vendel, B. Strombeck, and F. Mitelman, Relationship between SCE and structural chromosome aberrations in lymphocytes of 100 individuals, *Hereditas 97*, 237–245 (1982).
35. J. Herron, R. Ranshaw, J. G. Castle, and N. Wald, Automatic microscopy for mitotic cell location, *Comput. Biol. Med. 2*, 129–135 (1972).
36. A. Hollaender and R. Tice (eds.), *Sister Chromatid Exchange: 25 Years of Experimental Research*, Plenum Press, New York (1984).
37. E. Hook, Importance of epidemiology and population monitoring in genetic risk assessment, in: *Individual Susceptibility to Genotoxic Agents in the Human Population* (F. J. de Serres and R. Pero, eds.), Plenum Press, New York (1984).
38. P. S. Ing, R. S. Ledley, and H. A. Lubs, Chromosome analysis at the National Biomedical Research Foundation, in: *Automation of Cytogenetics* (M. L. Mendelsohn, ed), pp. 27–38, National Technical Information Service, U. S. Department of Commerce (1976).
39. E. T. Johnson and L. J. Goforth, Metaphase spread detection and focus using closed circuit television, *J. Histochem. Cytochem. 2*, 536–545 (1974).
40. V. Kliment, R. Tuscany, J. Dvorak, and L. Tomasek, Computer-assisted analysis of human G-banded chromosomes, *Comput. Biomed. Res. 16*, 20–28 (1983).
41. B. Lambert, S. Lindblad, K. Holmberg, and D. Francesconi, The use of sister chromatid exchange to monitor human populations for exposure to toxicologically harmful agents, in: *Sister Chromatid Exchange* (S. Wolff, ed.), pp. 149–182, Wiley, New York (1982).
42. S. A. Latt, Localization of sister chromatid exchanges in human chromosomes, *Science 185*, 74–76 (1974).
43. S. A. Latt and R. R. Schreck, Sister chromatid exchange analysis, *Am. J. Hum. Genet. 32*, 297–313 (1980).
44. R. S. Ledley, High speed automatic analysis of biomedical pictures, *Science 146*, 216–232 (1964).
45. R. S. Ledley, H. A. Lubs, and F. H. Ruddle, Introduction to chromosome analysis, *Comput. Biol. Med. 2*, 107–128 (1972).
46. R. S. Ledley, P. S. Ing, and H. A. Lubs, Human chromosome classification using discriminant analysis and Bayesian probability, *Comput. Biol. Med. 10*, 209–218 (1980).
47. R. LeGo, Un systeme de selection automatique des metaphases, etude au moyen d'un programme de stimulation, *C. R. Acad. Sci. Paris Ser. D 108*, 108–111 (1972).
48. R. LeGo, Image-processing automation for chromosome analysis, in: *Mutagen-Induced Chromosome Damage in Man* (H. J. Evans and D. C. Lloyd, eds.), pp. 322–325, Yale University Press, New Haven, Connecticut (1978).
49. M. S. Lin and W. Wertelecki, Evidence that sister chromatid exchanges and chromatid breaks are two independent events, *Chromosoma 85*, 413–419 (1982).
50. H. A. Lubs, The current state of the art, in: *Automation of Cytogenetics* (M. L. Mendlesohn, ed.), pp. 174–177, National Technical Information Service. U. S. Department of Commerce (1976).
51. H. A. Lubs and R. S. Ledley, Automated analysis of differentially stained human chromosomes, in: *Nobel Symposium 23, Chromosome Identification*, pp. 61–76, Nobel Foundation and Academic Press (1973).
52. C. Lundsteen, J. Philip, and E. Granum, Quantitative analysis of 6985 digitized tryspin G-banded human metaphase chromosomes, *Clin. Genet. 18*, 355–370 (1980).
53. J. J. Madden, A. Falek, D. A. Shafer, and J. H. Glick, Effects of opiates and demographic factors on DNA repair synthesis in human leucocytes, *Proc. Natl. Acad. Sci. USA 76*, 5769–5773 (1979).

54. J. J. Madden, A. Falek, D. A. Shafer, R. M. Donahue, D. C. Eltzroth, F. Hollingsworth, and P. J. Bokos, Influence of demographic factors on unscheduled DNA synthesis as measured in human peripheral leukocytes and fibroblasts, in: *Individual Susceptibility to Genotoxic Agents in the Human Populations* (F. J. de Serres and R. Pero, eds.), pp. 359–371, Plenum Press, New York (1984).
55. K. M. Marimuthu, S. D. Selles, and P. W. Neurath, Computer analysis of Giemsa banding patterns and automatic classification of human chromosomes, *Am. J. Hum. Genet. 26*, 369–377 (1974).
56. D. Mason and D. Rutovitz, The economics of automatic aberration scoring, in: *Mutagen-Induced Chromosome Damage in Man* (H. J. Evans and D. C. Lloyd, eds.), pp. 339–345, Yale University Press, New Haven, Connecticut (1978).
57. J. Melnyk, G. W. Persinger, B. Mount, and K. R. Castleman, A semi-automated specimen preparation system for cytogenetics, *J. Reprod. Med. 17*, 59–67 (1976).
58. M. L. Mendelsohn, B. H. Mayall, and J. M. S. Prewitt, Approaches to the automation of chromosome analysis, in: *Image Processing in the Biological Sciences*, pp. 119–136, University of California Press, Berkeley, California (1969).
59. W. F. Morgan and P. E. Crossen, Factors influencing SCE rate in cultured human lumphocytes, *Mutat. Res. 81*, 395–402 (1981).
60. A. T. Natarajan, B. A. Kihlman, and G. Obe, Use of the 5-bromodeoxyuridine labelling technique for exploring the mechanisms involved in the formation of chromosomal aberrations, *Mutat. Res. 73*, 307–317 (1980).
61. National Toxicology Program, *Second Annual Report on Carcinogens*, U. S. Department of Health and Human Services (1981).
62. P. W. Neurath, B. L. Bablouzian, T. M. Warms, R. C. Serbagi, and A. Falek, Human chromosome analysis by computer, *Ann. N.Y. Acad. Sci. 128*, 1013–1028 (1966).
63. P. W. Neurath, J. F. Brenner, and W. D. Selles, Automatic Metaphase Cell Location System Development, *NIH grant number GM 22965-03* (1977).
64. P. E. Norgren, High speed automatic laser beam detection of chromosome spreads, *Ann. N. Y. Acad. Sci. 157*, 514–524 (1969).
65. A. Oosterlinck, F. Dom, J. VanDaele, and H. Van den Berghe, Computer assisted karyotyping with human interaction, *J. Histochem. Cytochem. 15*, 754–762 (1977).
66. K. Paton, Properties for automatic metaphase selection, *Ann. Hum. Genet. 35*, 67–77 (1971).
67. P. Perry and H. J. Evans, Cytological detection of mutagen–carcinogen exposure by sister chromatid exchange, *Nature 258*, 121–125 (1975).
68. D. Picciano and D. J. Killian, A semi-automated cytogenetic analysis system, *Exp. Cell. Res. 107*, 431–434 (1977).
69. J. Piper, D. Mason, D. Rutovitz, H. Ruttledge, and L. Smith, Efficient interaction for automated chromosome analysis using asynchronous parallel processing, *J. Histochem. Cytochem. 27*, 560–563 (1979).
70. N. C. Popescu and J. A. Dipaolo, The relevance of sister chromatid exchange to the induction of neoplastic transformation, in: *Progress and Topics in Cytogenetics, Sister Chromatid Exchange* (A. Sandberg, ed.), Vol. 2, pp. 425–460, Liss, New York (1982).
71. N. C. Popescu, D. Turnbull, and J. A. Dipaolo, SCE and chromosome aberration analysis with the use of several carcinogens and non-carcinogens, *J. Natl. Cancer Inst. 58*, 289–293 (1977).
72. R. J. Preston and P. C. Gooch, The induction of chromosome type aberrations in G_1 by MMS and 4-nitroquinoline-*n*-oxide, and the non-requirement of an S-phase for their production, *Mutat. Res. 83*, 395–402 (1981).
73. R. J. Purrott and B. D. Stephenson, An Assessment of the Edinburgh Metaphase Finder, NRPB Memorandum M-37 (1978).

74. T. Raposa, SCE induction by cytostatics and its relation to iatrogenic leukemogenesis, in: *Sister Chromatid Exchange: 25 Years of Experimental Research* (R. Tice and A. Hollaender, eds.), pp. 859–884, Plenum Press, New York (1984).
75. J. H. Ray and J. German, Sister chromatid exchange in the chromosome breakage syndromes, in: *Progress and Topics in Cytogenetics, Sister Chromatid Exchange* (A. Sandberg, ed.), Vol. 2, pp. 553–577, Liss, New York (1982).
76. D. Rutovitz and J. Piper, The balance of special and conventional computer architecture requirements in an image processing application, in: *Multicomputers and Image Processing*, pp. 161–177, Academic Press, New York (1982).
77. M. S. Rzeszotarski, Pattern Recognition Techniques Applied to Metaphase Finding and the Evaluation of Quality in Human Chromosome Preparations, Ph.D. Thesis, *Case Western Reserve University*, Cleveland, Ohio (1982).
78. H. W. Sabrin, A. O. Martin, and M. S. Shaunnessey, Current status of metaphase locating devices: Preliminary clinical evaluation of one system, in: *Proceedings of the 5th Annual Symposium on Computer Applications in Medical Care*, pp. 1108–1112, IEEE (1981).
79. J. Serra, *Image Analysis and Mathematical Morphology*, Academic Press, London (1982).
80. W. D. Selles, P. W. Neurath, J. Conklin, and J. B. Horton, Progress and problems in cytophotometric anlaysis of G-banded chromosomes, in: *Automation of Cytogenetics* (M. L. Mendelsohn, ed.), pp. 3–16, National Technical Information Service, U. S. Department of Commerce (1976).
81. D. A. Shafer, Alternate replication bypass mechanisms for sister chromatid exchange formation, in: *Progress and Topics in Cytogenetics, Sister Chromatid Exchange* (A. Sandberg, ed.), Vol. 2, pp. 67–98, Liss, New York (1982).
82. D. A. Shafer, Replication bypass SCE mechanisms and the induction of SCE by single strand adducts or lesions of DNA, in: *Sister Chromatid Exchange: 25 Years of Experimental Research* (A. Hollaender and R. Tice, eds.), Plenum Press, New York (1984).
83. D. A. Shafer, A. Falek, R. Swenson, D. Oney, and J. J. Madden, Computer image analysis of sister chromatid exchanges, *Am. J. Hum. Genet, 32*, 88A (1980).
84. D. A. Shafer, A. Falek, C. McCarthy, J. Peach, J. J. Madden, and R. Donahoe, Computerized analysis of cytogenetic tests for mutagenesis, presented at *Environmental Mutagen Society Annual Meeting*, San Diego, 1981.
85. D. A. Shafer, A. Falek, J. J. Madden, F. Tadayon, M. Pline, P. Bokos, J. C. Kuehnle, and J. Mendelson, Parallel increases in sister chromatid exchanges at base level with UV treatment in human opiate users, *Mutat. Res. 109*, 73–82 (1981).
86. D. A. Shafer, W. D. Selles, and J. F. Brenner, Computer image analysis of variation between human chromosome replication sequences and G-bands, *Am. J. Hum. Genet. 34*, 307–321 (1982).
87. D. A. Shafer, K. I. Mandelberg, and A. Falek, Computer-automated metaphase finding, chromosome damage, and sister chromatid exchange analysis, in: *Proceedings International Symposium on Recent Trends in Medical Genetics* (K. M. Marimuthu, ed.), (1984).
88. G. Shippey, R. J. H. Bayley, A. S. J. Farrow, D. Rutovitz, and J. H. Tucker, A fast interval processor, *Pattern Recognition 14*, 345–365 (1981).
89. F. H. Sobels, Some problems associated with testing for environmental mutagens and a perspective for studies in comparative mutagenesis, *Mutat. Res. 46*, 245–260 (1977).
90. S. R. Sternberg, Pipeline architectures for image processing, in: *Multicomputers and Image Processing*, pp. 291–305, Academic Press, New York (1982).

91. G. H. Strauss and R. J. Albertini, Enumeration of 6-thioguanine resistant peripheral blood lymphocytes in man as a potential test for somatic cell mutations arising *in vivo*, *Mutat. Res. 61*, 353–379 (1979).
92. T. R. Skopek, P.-M. Leong, and W. G. Thilly, Discrimination between spontaneous and induced mutations in human cell populations by use of mutational spectra, in: *Individual Susceptibility to Genotoxic Agents in the Human Populations* (F. J. de Serres and R. Pero, eds.), pp. 281–291, Plenum Press, New York (1984).
93. H. T. Van den Berg, H. F. deFrance, J. D. F. Habbema, and J. W. Raatgever, Automated selection of metaphase cells by quality, *Cytometry 1*, 363–368 (1981).
94. J. Vrolijk, P. L. Pearson, and J. S. Ploem, LEYTAS, A system for the processing of microscopic images, *Anal. Quant. Cytol. 2*, 41–48 (1980).
95. N. Wald and P. Preston, Automatic screening of metaphase spreads for chromosome analysis, in: *Proceedings Conference on Image Processing in Biological Sciences* (UCLA Forum Medical Science, No. 9), (D. M. Ramsey, ed.), pp. 9–34, University of California Press, Berkeley, California (1968).
96. N. Wald, C. C. Li, J. M. Herron, L. Davis, and S. R. Fatora, Automated analysis of chromosome damage, in: *Automation of Cytogenetics* (M. L. Mendelsohn, ed.), pp. 39–45, National Technical Information Service, U. S. Department of Commerce (1976).
97. A. W. Wayne and J. C. Sharp, The use of high resolution microscope photometry in the discrimination of chromosome bands, *J. Microsc. 124*, 163–167 (1981).
98. S. Wolff, Relation between DNA repair, chromosome aberrations, and sister chromatid exchanges, in: *DNA Repair Mechanisms* (P. C. Hanawalt, E. C. Friedberg, and C. F. Fox, eds.), pp. 751–760, Academic Press, New York (1978).
99. H. C. Wulf, Operator-assisted semi-automatic karyotyping of banded metaphases, *Cytogenet. Cell Genet. 19*, 146–158 (1977).
100. G. W. Zack, J. A. Spriet, S. A. Latt, G. H. Granlund, and I. T. Young, Automated detection and localization of sister chromatid exchanges, *J. Histochem. Cytochem. 24*, 168 (1976).
101. G. W. Zack, W. E. Rogers, and S. A. Latt, Automatic measurement of sister chromatid exchange frequency, *J. Histochem. Cytochem. 25*, 741–753 (1977).

CHAPTER 10

Mutagen-Sensitive Mutants and Chemical Mutagenesis in *Drosophila*

Friedrich E. Würgler, Hansjörg Frei, and Ulrich Graf

1. Introduction

Drosophila melanogaster has a long history of use in the study of basic genetics (since 1910), radiation genetics (since 1928), and chemical mutagenesis (since 1946). Reviews on the present status of the classical *Drosophila* mutagenicity assays have recently been published.[69,120]

At present, the well-validated sex-linked recessive lethal assay is recommended for routine mutagenicity screening. A disadvantage of

Abbreviations used in this chapter: AAF, *N*-acetyl-2-aminofluorene; BP, benzo(*a*)pyrene; BrdUrd, 5-bromodeoxyuridine; DAS, 4-dimethylamino-*trans*-stilbene; DEN, diethylnitrosamine; DMBA, 7,12-dimethylbenz(*a*)anthracene; DMN, dimethylnitrosamine; EDTA, ethylenediaminetetraacetic acid; EMS, ethyl methanesulfonate; ENNG, *N*-ethyl-*N'*-nitro-*N*-nitrosoguanidine; ENU, ethylnitrosourea; 3-MC, 3-methylcholanthrene; HN2, nitrogen mustard; iPMS, isopropyl methanesulfonate; MMS, methyl methanesulfonate; MNNG, *N*-methyl-*N'*-nitro-*N*-nitrosoguanidine; MNU, methylnitrosourea; 4 NQO, 4-nitroquinoline-1-oxide; PR, photoreactivation; RD, repair deficient; RP, repair proficient; SCE, sister chromatid exchange; UDS, unscheduled DNA synthesis; UV, ultraviolet radiation (predominantly 254 nm).

Friedrich E. Würgler, Hansjörg Frei, and Ulrich Graf • Institute of Toxicology, Swiss Federal Institute of Technology and University of Zurich, CH-8603 Schwerzenbach/Zurich, Switzerland.

this test is that it takes several weeks to obtain results for a given chemical. Therefore, various studies are under way to improve and optimize the detection capacity of this and other *Drosophila* assays. One aim is to reduce as much as possible the time needed for testing.

It might be possible to increase the detection capacity of *Drosophila* assays by the use of mutagen-sensitive strains. Defects in DNA repair mechanisms are a well-known basis of increased mutagen sensitivity.[12] The theoretical approach with eukaryotes is the same as with microorganisms (e.g., in the Ames *Salmonella*/mammalian microsome test): error-free pathways of DNA repair are blocked so that as many premutational lesions as possible are channelled through error-prone repair pathways and/or DNA replication. This increases the probability that a premutational lesion will lead to a fixed heritable mutation.

The use of repair-defective mutants represents but one of several ways to improve the *Drosophila* testing routine. Two further approaches use strains with improved xenobiotic metabolism or strains resistant to toxic effects. Among *Drosophila* strains there is variation in the capacity to transform metabolically certain types of compounds into their respective mutagenic derivatives.[42,43,56,57,141] Optimally, tester strains should possess a broad spectrum of metabolically activating pathways. In addition, strains with an increased resistance to the toxic effects of particular compounds represent a further way of improving *Drosophila* test systems. With such resistant strains Vogel[121,123] succeeded in testing insecticides and promutagens for their mutagenic effects. We feel that the latter two possibilities have actually been underestimated with respect to their potential value, and may offer additional means to render test systems more sensitive.

In this chapter we review the different approaches undertaken to study the effects of defective DNA repair on chemical mutagenesis in *Drosophila*.

2. Isolation and Localization of Mutants

Smith[106] found the first MMS-sensitive mutant in *Drosophila*. The mutant was later shown to be DNA repair-defective.[19] Most mutagen-sensitive mutants available today were isolated on the basis of an increased sensitivity of the larvae to MMS. About 30 genes have been identified by classic genetic analyses.[26] Table 1 gives an overview of the genes and alleles identified in *Drosophila*.

Based on complementation tests it has been estimated that approximately 60 loci in the *Drosophila* genome control sensitivity to MMS.[76,109]

Some of the mutants were obtained from studies aimed at the identification of genes active in female meiosis[10] and were only later found to be mutagen-sensitive.[12] The mutants at the *mei-9* and *mei-41* loci are the most prominent ones derived from this approach. Mutants at these loci conferring mutagen sensitivity were found to increase the frequency of spontaneous somatic recombination and deletion events.[13] A number of mutants at these and other loci isolated on the basis of their hypersensitivity to MMS are also hypersensitive to a number of other mutagens, such as UV, HN2, or X rays.[109,110] Because of this cross-sensitivity, some of the mutants isolated as hypersensitive to ionizing radiation turned out to represent alleles at loci identified by MMS-sensitive mutations.[72,82,102]

On the other hand, mutants at loci such as *mus(1)103* or *mus(3)308*[22,24] seem to be more specific in that they respond particularly to one kind of mutagen. Genes of this type are only detected in direct searches using the particular mutagen. Most probably, the majority of genes responsible for a broad spectrum of hypersensitivity in *Drosophila* have now been identified. The same may not be true for loci that control specific sensitivity to particular mutagens (e.g., UV light).

3. Biochemical Characteristics

3.1. Introduction

Extensive knowledge of *Drosophila* genetics has made this organism a logical bridge between work on microorganisms and that on multicellular eukaryotes. Accordingly, it appeared interesting to investigate DNA repair in *Drosophila* by an analysis of mutants that are sensitive to mutagens. A summary of the biochemical characteristics of the various mutants is given in Table 2.

3.2. Photoreactivation

Several pieces of evidence have been presented for the existence of a system for photorepair in *Drosophila melanogaster*.[18,21,23,26,118] A highly stable photoreactivating enzyme was purified from tissue culture cells.[15] Yegorova *et al.*[140] and Boyd *et al.*[26] identified strains that are deficient in the photoreversal of UV-induced pyrimidine dimers. The gene responsible for photoreactivation ability was tentatively assigned to chromosome 2.[39,40]

TABLE 1. Mutagen-Sensitive Mutants in *Drosophila melanogaster*

Gene	Allele	Location	Reference	
			Original	Location
X Chromosome				
mei-9	*a, b*[a]	1-5	10	33
	L1 (formerly *mut-159*)	1-4.8	122	51
	AT1 [formerly *mus(1)110*^{A1}]	1-6	107	109
	AT2, AT3	—	109	—
	D1, . . . , D4	—	76	—
mei-41	—	1-53.8	10	107
	A1 (formerly *mut-s*)	—	106	—
	A2, . . . , A17	—	107	—
	AM	—	76	—
	D1	—	82	—
	D2, . . . , D5	—	22	—
	D6, . . . , D11	—	76	—
	195 (formerly *mei-195*)	—	10	—
	12-1007 (formerly *fs(1)*$^{M3\text{-}12\text{-}1007}$	—	80	—
mus(1)101	*A1*	1-44.8	109	108
	D1, D2	—	22	—
mus(1)102	*A1, . . . , A8*	1-0.5	107	107
	D1, D2	1-0.4	22	22
	D3	—	76	—
mus(1)103	*D1, D2*	1-(36–44)	22	108
[*mus(1)104*]	(*D1*) allele of *mei-41*	1-53	22	108
	(*D2*) allele of *mei-41*	—	76	101
mus(1)105	*A1*	1-14.8	107	108
	D1	1-(5.5–20)	22	22
mus(1)106	*D1*	1-(36–44)	22	108
[*mus(1)107*]	(*D1*) now *mus(1)109*^{D1}	1-(33–36)	82	109
	(*D2*) now *mus(1)109*^{D2}	—	—	46

mus(1)108	—	1-10.8	108	108
mus(1)109	*A1*	1-30.2	107	108
	D1	—	82	—
	D2	—	76	—
[*mus(1)110*]	(*A1*) now *mei-9*^{AT1}	—	107	—
mus(1)111	*D1*	1-27	76	109
"Unnamed" on chromosome carrying *fs(1)97*		—	76	—
Second chromosome				
mus(2)201	*A1, D1*	2-	113	—
mus(2)202	*A1*	2-	113	—
mus(2)203	*A1*	2-	113	—
mus(2)204	*A1*	2-	113	—
mus(2)205	*A1*	2-	113	—
mus(2)206	*A1*	2-	113	—
mus(2)207	*A1*	2-	113	—
Third chromosome				
mus(3)301	*D1*	3-23	19	24
	D2, . . . , D5	—	24	—
mus(3)302	*D1*	3-45	19	24
mus(3)303	*D1* (lost mutant)	3-	19	—
mus(3)304	*D1, D2, D3*	3-46	24	24
mus(3)305	*D1, D2, D3*	3-44	24	24
mus(3)306	*D1*	3-56	24	24
mus(3)307	*D1*[b]	3-59	24	24
mus(3)308	*D1, D2, D3, D4,*[b] *D5,*[b] *D6*[b]	3-55	24	24
mus(3)309	*D1, D2, D3*[b]	3-	24	—
mus(3)310	*D1*	3-47	24	24
mus(3)311	*D1, D2, D3*	3-47	24	24
mus(3)312	*D1, D2*	3-(16–18)	24	24

[a] Allele *mei-9*^{a} originally was designated as *mei-9* only; *mei-9*^{b} was found on a double mutant chromosome in combination with disjunction-defective *mei-254*^{a}.
[b] Associated with *mutator-2*.[(77)]

TABLE 2. Mutagen-Sensitive *Drosophila* Mutants[a] and Associated Defects in the Repair of UV-Induced Damage[b]

Mutant	Mutagen sensitivity				Excision repair capability	Postreplication repair capability	Recovery of	
	MMS	X rays	UV	HN2			Thymidine incorporation	Synthesis of long DNA
X Chromosome								
Wild type	–	–	–	–	Normal	Normal	+	+
mei-9	+	+	+	+	Defective	Normal	–	+
mei-41	+	+	+	+	Normal	Defective	–	–
mus(1)101	+	+	+	+	Normal	Intermediate	+/–	+/–
mus(1)102	+	+	–	–	Normal	Normal	+	+
mus(1)103	+	–	–	–	Normal	Normal	+/–	+
mus(1)105	+	+	–	–	Normal	Normal	+/–	–
mus(1)106	+	+	–	–				
mus(1)108	+							
mus(1)109	+	+	+	+				
Second chromosome								
mus(2)201	+	–	+	+	Defective	Normal	–	+

mus(2)202	+	+	+	+				
mus(2)203	+	+	+	+				
mus(2)204	+	−	+	−				
mus(2)205	+	−	+	−	Intermediate	Defective	−	−
mus(2)206	+	−	+	+				
mus(2)207	+	−	+	+	Normal	Normal	−	+/−
Third chromosome								
mus(3)301	+			+				
mus(3)302	+			+	Intermediate	Defective	−	+
mus(3)304	+			−	Intermediate	Intermediate	+	−
mus(3)305	+			+				
mus(3)306	+			−	Intermediate	Normal	−	−
mus(3)307	+			+	Normal	Normal	+/−	+
mus(3)308	−			+	Intermediate	Intermediate	−	+
mus(3)309	+			−				
mus(3)310	+			−	Normal	Defective	−	+
mus(3)311	+			−				
mus(3)312	+			−				

[a] From Smith *et al.*[110]: +, Hypersensitiv ty; −, equal sensitivity to wild type; no entry, the mutagen was not tested with this mutant.
[b] Adapted from Brown and Boyd[28,29]: +, Recovery equivalent to that of control cells; +/−, extent of recovery reduced; −, no recovery.

3.3. Excision Repair

Observations of unscheduled DNA synthesis in *Drosophila* polytene chromosomes following X-ray treatment suggested the existence of excision repair in this organism.[119] Subsequent analyses of the removal kinetics of pyrimidine dimers and repair replication firmly established its presence.[18,118] The *mei-9* mutants, which have been the object of extensive genetic and biochemical analyses,[16,21,30,33,59] are deficient in the ability to excise pyrimidine dimers. Characteristically, they are sensitive to a broad spectrum of mutagens (Table 2). Meiotic recombination is reduced by 90% in females homozygous for *mei-9* mutants, whereas the frequencies of recombination and chromosome breakage are increased in somatic cells of both sexes.[10,12] In meiosis, *mei-9* does not alter the frequency, distribution, or morphology of spherical recombination nodules, although it drastically decreases the frequency of exchanges.[31] On the other hand, the frequency of gene conversion at the *rosy* locus is not decreased. This led to the hypothesis[32] that the functioning of the *mei-9*$^{+}$ product in the correction of mismatch is integral to the transition of an event from one that can resolve only as gene conversion (i.e., without recombination of flanking markers) to one that can resolve as a crossover (isomerization step in the Meselson–Radding model).[78] The *mei-9* mutants are deficient in repair replication following both UV and X-ray exposure,[81] whereas they have normal capacity for photorepair, postreplication repair, and repair of X-ray-induced DNA single-strand and double-strand breaks.[21,35]

On chromosome 2 a second locus, *mus(2)201*, was identified whose normal function is also essential for excision repair. Unlike the *mei-9* mutants, mutants at this locus do not exhibit extreme hypersensitivity to X rays and do not influence meiocyte function.[25]

At present the total number of genes known to take part in excision repair in *Drosophila*is seven. In primary cell cultures, five mutagen-sensitive mutants [*mus(2)205, mus(3)302, mus(3)304, mus(3)306*, and *mus(3)308*] display a significantly reduced excision repair capacity relative to control cultures.[17] In addition, in two of the five cultured mutants [*mus(3)306* and *mus(3)308*] the accumulation of DNA single-strand breaks normally seen after UV irradiation is reduced or absent.

Smith and co-workers are presently studying the nature of induced DNA lesions, the mechanisms by which these lesions are repaired, and the mutational consequences that occur when alkylation repair is impaired.[37,110] It appears that in primary cell cultures the inhibition of DNA synthesis during the first 2 hr of MMS treatment is dose dependent and does not differ between excision repair-defective and control cells. The authors conclude that the introduction of alkylations and their immediate effect on semiconservative replication is identical in both cell types.

Unscheduled DNA synthesis (UDS) was examined in primary cell cultures after UV, MMS, and MNU treatment.[37] Excision repair-defective cells [*mei-9, mus(2)201,* or the double mutant *mei-9 mus(2)201*] were incapable of performing UDS. These studies provide the first direct biochemical evidence that the *mei-9* and *mus(2)201* loci are defective in the repair of alkylation-induced damage. The *mei-9* and *mus(2)201* mutants also show reduced apurinic endonuclease activity in ganglia extracts.[90]

Smith *et al.*[111] propose a function for the *mei-9*$^+$ gene of *Drosophila* corresponding to the normal function of the *uvrD* gene of *E. coli.* They base their hypothesis on the striking similarities between the effects brought about by the mutant alleles of the respective genes in *E. coli* and *Drosophila* (for details and references see Smith *et al.*[111]). Mutant alleles of both genes (1) exhibit a similar alkylation hypermutability pattern (modestly more mutable by EMS and significantly more mutable by MMS), (2) confer sensitivity to UV light, ionizing radiation, and MMS, (3) increase spontaneous mutation, (4) affect recombinational processes, (5) are defective in the correction of base mismatches in heteroduplex DNA, and (6) are defective in the excision repair of UV-induced thymine dimers.

The *uvrD* gene product has been identified as a polypeptide possessing DNA-dependent ATPase activity.[88] It has been suggested that it is identical to DNA helicase II.[94] Due to the similarity of effects of mutant alleles of *uvrD* and *mei-9*, it seems quite reasonable to assume that the *mei-9* locus encodes a DNA-unwinding protein. The pleiotropic phenotypic effects of the *mei-9* alleles would also be in agreement with the proposed gene function.

3.4. Postreplication Repair

Initially four mutant genes had been identified as defective in postreplication repair.[19] The early results suggested the existence of two postreplication repair pathways in *Drosophila.* That idea has recently been rejected[26] because of the reassignment of the *mus(1)104* mutants to the *mei-41* locus (Table 1).[101] In addition, caffeine, which played an important role in the original hypothesis, has been shown to have a minimal influence on postreplication repair in the *Drosophila* mutants,[20] quite in contrast to the situation in man.[26]

Primary cell cultures derived from repair-proficient embryos of *Drosophila melanogaster* respond to UV light within 1 hr after exposure with a decline in thymidine incorporation and by synthesizing nascent DNA in shorter segments than those observed in unirradiated cultures. Cells from all five postreplication repair-defective mutants studied respond to UV by synthesizing nascent DNA in even shorter segments (see Table 2).[28] UV radiation affects thymidine incorporation in cells of only two postreplication repair-defective mutants. Incorporation is abnormally low in irradiated cultures from the *mus(3)302*and *mus(3)310* mutants, but is normal in the mutants *mei-41*, *mus(1)101*, and *mus(2)205*. Therefore, only two of five postreplication repair functions govern the ability of irradiated cells to synthesize normal amounts of DNA, whereas all five functions are essential for the synthesis of normal-length DNA. The finding suggests that the defects in postreplication repair exhibited by the *mus(3)302* and *mus(3)310* mutants are basically different from those of the *mei-41*, *mus(1)101*, and *mus(2)205* mutants.

The results of Brown and Boyd[29] are consistent with the gapped-synthesis model[70] if it is assumed that cells from the *mus(3)302* and *mus(3)310* mutants are defective in functions that allow normal replication forks to proceed past dimers, and if it is assumed that cells from the *mei-41*, *mus(1)101*, and *mus(2)205* mutants are defective in functions that allow normal cells to seal gaps opposite dimers.

In addition, primary cell cultures of the mutants *mei-41*, *mus(1)101*, *mus(3)304*, and *mus(3)308* were irradiated with 10 kR of X rays, but none of the mutants was significantly deficient in DNA double-strand break repair.[35]

There is circumstantial evidence that the postreplication repair-defective mutant *mus(1)101* is defective in the repair of cross-links and other double-strand damage,[44,52,53] but the processing of DNA treated with cross-linking agents in normal and in defective cells has not yet been studied biochemically. Gatti *et al.*[48] have analyzed cytologically neuroblasts of larvae carrying the temperature-sensitive allele $mus(1)101^{ts1}$. They observed defective condensation of heterochromatic but not of euchromatic portions of the chromosomes. It appears that in the case of mutations at the *mus(1)101* locus a disruption of the chromosome organization is the cause for the increased mutagen sensitivity and the decreased repair capacity.

4. Larval Exposure

Considering that most of the mutants have been isolated because they render larvae hypersensitive to killing by MMS, it seems a reasonable first step to test exposed larval cells for mutation induction. Various approaches were undertaken to investigate mutagenesis in somatic cells as well as in germ cells of untreated and mutagen-treated larvae.

4.1. Spontaneous Mutations

4.1.1. Chromosome Aberrations

Recombination-defective meiotic mutants at five loci (*mei-218*, *mei-251*, *mei-352*, *mei-41*, and *mei-9*) and mutagen-sensitive mutants at six loci [*mus(1)101*, *mus(1)102*, *mus(1)103*, *mus(1)105*, *mus(1)106*, and *mus(1)109*] on the X chromosome have been examined cytologically in neuroblast metaphases for their effects on frequencies and types of spontaneous chromosome aberrations.[14,46] Mutants at five loci increase the frequency of chromosome aberrations. They differ markedly both in the types of aberrations produced and in the local effects along the chromosome. According to these criteria, the mutants can be assigned to four different classes: (1) mutants producing almost exclusively chromatid breaks in both euchromatin and heterochromatin (*mei-9*); (2) mutants producing chromatid and isochromatid breaks in both euchromatin and heterochromatin [*mei-41* and *mus(1)102*]; (3) mutants producing chromatid and isochromatid breaks localized preferentially in euchromatin [*mus(1)105*]; and (4) mutants producing chromatid and isochromatid breaks clustered in the heterochromatin [*mus(1)109*]. In addition, it appeared that the first three classes of mutants lead to a higher frequency of aberrations in females than in males, whereas the fourth does not.

Similar to *mei-9* and *mei-41*, *mus(3)302* also has an increased frequency of spontaneous breaks (see Table 3).[98–100] Moreover, it appeared that the double mutant *mei-9 mei-41* is particularly prone to spontaneous chromosome damage, a finding paralleled by the observation that these two mutants apparently affecting different repair pathways also show interaction with respect to mutagen sensitivity.[108]

4.1.2. Somatic Mosaic Spots

In flies heterozygous for appropriate recessive somatic cell markers (mostly abdominal cuticle and bristle markers) somatic spots expressing the markers can arise spontaneously from mitotic recombination, mutation, or deletion.[117] The effects of 13 mutagen-sensitive mutants (representing six loci) on mitotic chromosome stability in nonmutagenized cells have been examined in this way.[11,14] Mutants at two loci

TABLE 3. Neuroblasts of DNA Repair-Defective Mutants as Sensitive Indicators of Chromosome Aberrations Induced by Ionizing Radiation[a]

Mutant	Spontaneous frequency breaks/cell	Ratio of breaks/cell in mutant versus wild type following exposure of 200 rad X rays
Wild type	0.0067	—
mus(1)101	0.0060	1.1
mei-9	0.0381	5.9
mei-41	0.0472	7.2
mei-9 mei-41	0.0312	9.4
mus(3)302	0.0599	6.4

[a] After Schweizer.(100)

[*mus(1)103* and *mus(1)106*] do not affect the frequency of somatic spots, whereas *mus(1)101* and *mus(1)102*, *mus(1)105*, and *mus(1)109* alleles increase the frequency of clones. An examination of the frequency of spontaneous somatic spots in *mei-218*, *mei-9*, and *mei-41* shows that both *mei-41* and *mei-9* substantially increase the frequency of somatic spots, whereas *mei-218* (not a mutagen-sensitive mutant) does not differ from the control in either the frequency or the type of somatic spots produced(12–14) The particular pattern of effects caused by a mutant is characteristic of its locus, as shown by the similar effects observed among alleles of each of these loci, by the different spot frequencies observed among alleles of each locus and by the different effects of mutants of different complementation groups. In *mei-41*, the frequency of spontaneous spots on the abdomen is increased nearly 20-fold. The increase appears to be attributable to the production of deficiencies that delete the dominant wild-type allele in the cells of marker heterozygotes. In *mei-9*, the frequencies of both single and twin spots are elevated. A fraction of the single spots in *mei-9* may result from chromosome breakage, because approximately 20% of them include morphologically abnormal bristles. The increased frequency of twin spots in *mei-9* suggests that this mutant increases mitotic recombination as well as chromosome breakage.

4.1.3. Sister Chromatid Exchanges

The *mei-9* and *mei-41* mutants, which exhibit an about 30-fold increase in the frequency of spontaneous chromosome aberrations, did not show a correspondingly large increase in the frequency of SCEs, although a relatively high concentration of BrdUrd (9 μg/ml) was used in the study. The level of SCEs in *mei-41* mutants did not differ from that found in nonmutant controls. In *mei-9* mutants, the frequency of SCEs was about 70% of that in the control.[14,47] Lack of parallelism between effects on induction of chromosome aberrations and SCE formation has also been found in a number of other circumstances (for reviews see Wolff[132] and Kato[64]). Altogether the data indicate that the chromosomal lesions responsible for the formation of SCEs are generally different from those that cause chromosome aberrations. Additional support for this conclusion is provided by the observation that mutants at other loci [*mus(1)102*, *mus(1)105*, and *mus(1)109*] that produce elevated frequencies of spontaneous chromosome aberrations in somatic cells [46] do not produce a concomitant increase in the frequency of SCEs.[47]

4.2. Direct-Acting Mutagens

Although a systematic study of direct-acting mutagens has not been undertaken, a number of interesting key experiments have been conducted with different chemicals, stocks, and genetic endpoints. These studies represent starting points for more extensive systematic work.

Both *mei-9* and *mei-41* larvae are hypersensitive to X-ray induction of chromosome aberrations.[47] Hypersensitivity to chromosome break induction by X rays was observed in *mei-9*, *mei-41*, and *mus(3)302*, as well as in the double mutant *mei-9 mei-41*.[100] Table 3 shows the sensitization factors for the different hypersensitive mutants and for *mus(1)101*, which has wild-type sensitivity with respect to X-ray induction of chromosome aberrations.

The study of the oxygen effect in irradiated excision repair-deficient embryos led to a novel hypothesis. The *mei-9* embryos are up to four times more radiation sensitive than are repair-proficient embryos. The lack of an oxygen effect in repair-deficient 4-hr embryos and a reduced oxygen effect in 1.75-hr embryos suggest that the oxygen effect modifies repair ability.[45]

The germ cells of *mus(1)102* males showed a more than tenfold increase in mutation frequency when larvae were grown on formaldehyde-containing food,[3] whereas *mei-9* larvae exhibited wild-type sensitivity with respect to sex-linked recessive lethals induced by formaldehyde. The double mutant *mus(1)102 mei-41* showed a four- to fivefold increase, but no data could be obtained for *mei-41* alone because this type of larvae did not survive the treatment, due to the high toxicity of the standard formaldehyde exposure used in this approach (0.02 M formaldehyde).

When testing chemical mutagens in somatic cells with the *mwh* +/+ *flr* wing spot test (refs. 54, 55, 137, and 138; and this volume, Chapter 1) two different situations may be encountered with respect to toxic effects (induced lethality) (U. Graf, unpublished results): the excision repair-defective (*mei-9*) larvae can be more sensitive than the repair-proficient larvae or both can be equally sensitive. In the first case, for example with MMS, the mutagenicity tests have to be carried out at different exposure levels, depending on the genotype studied. With the monofunctional alkylating agent MMS as well as with the intercalating agent acridine orange the frequencies of spots were higher in the *mei-9* than in the repair-proficient larvae. While MMS produced significantly increased spot frequencies in both types of larvae, acridine orange did so only in the repair-defective type.

Equal lethality of the two types of larvae was observed after treatments with formaldehyde and with parachloroaniline. In the mutagenicity studies with formaldehyde, only the *mei-9* larvae gave a statistically significant frequency of induced spots.

4.3. Promutagens

The induction of sex-linked recessive lethals in germ cells of male larvae was first studied by Nguyen *et al.*[83] using BP. A second set of data included the effects of AAF, 3-MC, and DMBA.[129] Those data allowing for a comparison of repair-proficient and deficient larvae are summarized in Table 4.. The data for BP seem to signal an increased detection capacity, but the spontaneous frequency in the *mei-9 mei-41* control is about twice the frequency in the repair-proficient *y* strain. If one compares the data for both types of larvae (0.50% − 0.16% = 0.34% versus 0.91% − 0.37% = 0.54%), it appears that only small induced frequencies were obtained. Due to the high toxicity of BP in *mei-9 mei-41* larvae, only a low concentration, 0.025 mM, could be tested.

It was shown by Nguyen *et al.*[83] that *mei-9* as well as *mei-9 mei-41* larvae are very sensitive to the toxic effects of AAF, 4NQO, BP, and MNNG. Because of the high toxicity of AAF, 3-MC, and DMBA (Table 4) in the repair-deficient strains,[129] the exposure had to be reduced to 1/4–1/20, of the exposure used with the repair-proficient larvae. Under these circumstances no increase in the detection capacity was obtained with the *mei-9* or *mei-9 mei-41* larvae.

TABLE 4. Mutation Induction in Germ Cells Measured by the Frequency of Sex-Linked Recessive Lethals Induced in DNA Repair-Proficient (RP) and DNA Repair-Deficient (RD) Male Larvae[a]

Compound	Genotype treated	Concentration, mM	Sex-linked recessive lethals: Number	Sex-linked recessive lethals: Percent	Ref.
BP	RP: *y*	0	6/3683	0.16	83
		0.025	12/2384	0.50	
	RD: *y mei-9 mei-41*	0	11/2986	0.37	
		0.025	26/2839	0.91	
AAF	RP: *Berlin K*	0	8/6093	0.13	129
		1.0	30/7754	0.39	
	RD: *y mei-9 mei-41*	0	15/4864	0.31	
		0.1	29/4237	0.68	
3-MC	RP: *Berlin K*	0	17/6622	0.26	129
		4.0	45/8335	0.54	
	RD: *mei-9*	0	1/605	0.17	
		1.0	5/1392	0.36	
DMBA	RP: *Berlin K*	0	20/8395	0.24	129
		A 1.0	5/622	0.80	
		B 2.0	3/622	0.48	
		A+B	8/1244	0.64	
	RD: *mei-9*	0	2/1236	0.16	
		A 0.05	6/835	0.72	
		B 0.10	4/818	0.49	
		A+B	10/1653	0.60	

[a] Only data of the first brood of 0–2 or 0–3 days are shown; for BP, the data are from a 0- to 7-day brood.

5. Adult Exposure

The type of experiment currently used in *Drosophila* mutagenicity testing involves the exposure of adult males rather than that of larvae. Although a number of exposure techniques have been developed (injection, inhalation, topical application, etc.), the route of application most often used is feeding.[139] At the end of the feeding period, treated males are mated to virgin females; 2–3 days later, the males are transferred to new vials and the females are replaced. In this way a brood analysis is carried out. In the first brood, chromosomes exposed in mature sperm are analyzed; in later broods the chromosomes tested are those treated in spermatids, spermatocytes, or even in spermatogonia. With respect to the possible effects of repair deficiencies it is expected that if effects are seen, they should be most prominent in the cells exposed prior to or during the metabolically most active period of germ cell development (spermatogonia to spermatids), and virtually absent or nearly so in cells exposed as maturing sperm or mature spermatozoa.

5.1. Spontaneous Mutations

The discovery of the first mutagen-sensitive mutant (*mei-41*^{A1}) also provided evidence that increased spontaneous mutation frequencies may be observed in such mutants.[106] This suggested that mutagen-sensitive mutants could have mutator gene activity. An early study of *mei-9*^{L1} showed a similar effect in that spontaneous sex-linked recessive lethals were recovered at higher frequencies from mutant males than from wild-type males.[122]

Meanwhile, most mutagen-sensitive gene loci on the X chromosome have been examined for possible mutator activities by means of the sex-linked recessive lethal test. Although the effort was considerable, and in many cases the investigations included the study of several alleles or were based on repetitions with the same mutant allele, the general picture is far from clear (see Table 5 and references quoted therein). In most cases, the mutagen-sensitive mutants did not increase significantly the frequency of spontaneous lethals. Nevertheless, some frequencies clearly exceeded the control values, but repetitions with the same or with other alleles of the same complementation group did not provide a satisfactory confirmation. Elevated frequencies of spontaneous lethals may occasionally result from gonial clusters, be caused by mutator activity of other genes eventually contained in the strains (effects of the genetic background), or be due to random sampling variation. On the

TABLE 5. Spontaneous Mutation Frequencies in Male Germ Cells of Mutagen-Sensitive Mutants: Frequencies of Sex-Linked Recessive Lethals from Control and Mutant Males Crossed to *Basc* Females

Control						
Tests	Percent lethals	Mutant locus	Allele tested	Tests	Percent lethals[a]	Ref.
9,419	0.13	*mei-9*	*a*	3437	0.64*	74
	0.13		*AT1*	3320	0.09	74
5,082	0.24		*AT1*	5644	0.16	111
1,690	0.18		*L1*	3430	0.82*	122
9,662	0.13		*L1*	9696	0.24	129
	0.13		*L1*	4964	0.56*	129
2,460	0.12	*mei-41*	*A1*	2458	1.22*	106
11,710	0.15		*D3*	3807	0.11	74
	0.15		*D5*	4621	0.15	74
	0.15	*mus(1)104*	*D1*	4999	0.10	74
	0.15	*mus(1)101*	*D1*	5748	0.05	74
	0.15			3972	0.05	74
	0.15		*D2*	3107	0.03	74
	0.15			3268	0.06	74
	0.15	*mus(1)102*	*D1*	3213	0.12	74
	0.15		*D2*	2642	0.42*	74
	0.15	*mus(1)103*	*D1*	3492	0.11	74
	0.15		*D2*	5923	0.24	74
	0.15	*mus(1)105*	*D1*	4615	0.09	74
	0.15	*mus(1)106*	*D1*	3436	0.06	74
	0.13	*mus(1)109*	*A1*	4012	0.10	74
3,918	0.10	*mus(1)111*	*D1*	3144	0.19	76
	0.10	"Unnamed"	[*fs(1)97*]	3898	0.13	76
3,683	0.16	*mei-9*^{a} *mei-41*^{D5}		2986	0.37*	83
	0.13	*mei-9*^{a} *mei-41*^{D5}		7339	0.50*	129
	0.13	*mei-9*^{a} *mei-41*^{D5}		4864	0.31*	129

[a] Asterisks mark items that differ significantly from the pooled control value of 70/47,624 = 0.15% (Fisher's exact test, $p < 0.05$). The control values are combined from parallel controls obtained with nonmutant males published in refs. 74, 76, 83, 106, 111, 122, and 129.

other hand, it cannot be excluded that among mutants of the same complementation group, specific alleles exist that increase the spontaneous mutation frequency more than others. All three repetitions with a specific allele double mutant combination, *mei-9*^{a} *mei-41*^{D5}, gave spontaneous frequencies out of the control range, reaching a level roughly the double the pooled control average from nonmutant males (Table 5).

Particularly low spontaneous mutation frequencies are found with *mus(1)101* males.[74] A consistent reduction by a factor of about two or three was obtained with two different alleles, each tested twice. Although the statistical treatment of the data does not allow one to be firm on the point, the result nevertheless is suggestive. It may be assumed that in *mus(1)101*, spontaneous premutational lesions, which in repair-proficient males would be recovered as recessive lethals, are in part unrecoverable in the mutant due to the repair deficiency.

5.2. Direct-Acting Mutagens

Work with radiation has indicated that premutational damage induced in immature male germ cells is repaired by enzymatic systems in the germ cells themselves, so that error-prone repair pathways might lead to endpoints such as recessive lethals or chromosome aberrations. The processing of premutational lesions induced in mature sperm, however, takes place after fertilization and depends on the repair systems of the oocyte. Apparently, mature sperm cells, in contrast to immature stages, do not dispose of the whole machinery required to cope with the lesions caused by mutagenic treatment.

Graf *et al.*[52] obtained the first indication that the differential DNA repair capacities and the differential abilities of mutation fixation exhibited by the different stages of spermatogenesis depend on the activity state of repair genes. They showed that the oocytes of *mus(1)101* females lacked the ability to transform cross-links into recessive lethals,[53] and that recessive lethals could only be recovered from HN2-treated males crossed to *mus(1)101* females in the fourth and fifth 2-day broods. It was suggested that these lethals resulted from premutational lesions that were subject to mutation fixation under the influence of the nonmutant repair-proficient paternal DNA metabolism active in the male germ cells. The absence of lethals in the first three broods was explained as due to the normal regression of paternal metabolic activity toward the end of spermatogenesis combined with a failure of the mutant maternal system in the fertilized egg to transform cross-link damage into recessive lethals.

TABLE 6. Sex-Linked Recessive Lethals Induced in Germ Cells of Adult Males from Repair-Proficient (Wild-Type) and Repair-Defective (*mei-9*) Strains

Compound	Frequency of lethals from *mei-9* compared to wild-type males	Ref.
Direct-acting mono functional alkylating agents		
MMS	Enhanced (spermatids, spermatocytes)	128
	Enchanced (premeiotic)	
	Equal (postmeiotic)	111
	Equal	111
		86
EMS	Enhanced (spermatids, spermatocytes)	128
	Enhanced (premeiotic)	111
	Equal (postmeiotic)	111
	Equal	84
MNU	Enhanced (premeiotic)	111
	Equal (postmeiotic)	111
	Equal (spermatids)	128
	Reduced	86
ENU	Equal	128
Promutagens		
AAF	Negative[a]	129
DAS	Enhanced (postmeiotic)	129
DEN	Equal	128
	Equal	84
DMN	Reduced	84
		86
Physical agent		
X rays	Reduced	79

[a] Neither *mei-9* nor wild-type males gave induced mutations after feeding. Injection into wild-type males resulted in a low frequency of induced lethals.

Smith *et al.*[110,111] have recently undertaken a more direct approach to the problem. They exposed repair-proficient and excision repair-defective (*mei-9*) males to monofunctional alkylating agents and by means of the brooding technique determined the frequency of sex-linked recessive lethals induced in the different stages of spermatogenesis. For MNU, MMS, and EMS the postmeiotic broods showed no significant difference between the two types of males. In contrast, a highly significant difference between the two types of males was found for the premeiotic broods: with the *mei-9* males a much higher mutation frequency was obtained in comparison to repair-proficient males. This finding suggests that the low mutability of gonial cells normally observed after treatment with monofunctional alkylating agents is due to removal of damage by excision repair processes that in the male germ cells are effective up to the meiotic stage.

It appears that the proper analysis of repair in the germ line of repair-defective strains depends on a subtle distinction of the germ cell stages involved.[125,130] The seemingly controversial results of earlier studies with *mei-9*[84,86,124,128] may be explicable in part on these grounds (see Table 6).

5.3. Promutagens

Only a few promutagens have been tested with repair-deficient adult males. The results are included in Table 6. All reservations of the previous section maintained, it is obvious that with this group of chemicals the use of *mei-9* males did not result in an increased detection capacity. On the contrary, for DMN a reduced mutation frequency has been reported.[84,86] AAF induced low frequencies of mutations in repair-proficient male germ cells only.[129]

6. Maternal Effects

Maternal effects on mutation recovery that are due to mutagen-sensitive mutants are of special interest for the analysis of the repair mechanisms. The term "maternal effect" is defined as "any non-lasting effect of the maternal genotype or phenotype on the immediate offspring."[95] The genotype of the mother is expressed in the phenotype of the progeny because of the presence of products deposited into the oocytes during oogenesis. The case of maroonlike (*mal*) in *Drosophila* is well known: genotypically mutant *mal* progeny from genotypically and phenotypically mal^+ mothers are phenotypically mal^+ due to the deposition of a still unidentified "mal^+ factor" in the eggs.[49,87] Conversely, in the case of females homozygous for a DNA repair mutation, particular products normally present in the oocyte are not synthesized. The zygote, although genotypically wild type, is not transcriptionally active early during development. Thus, it depends on maternally derived DNA repair systems to process premutational DNA damage present in the sperm pronucleus. The absence of a gene product required to process premutational lesions may affect mutagenesis when an egg is fertilized by a sperm cell carrying such lesions. Evidently mature sperm are not capable of performing enzymatic DNA repair. However, changes in the state of the premutational lesions are possible by way of spontaneous reactions in the mutagenized mature sperm. These are evident after sperm storage in the female genital tract (see below).

Immediately after fertilization the sperm head nucleus starts to swell, as the extremely condensed chromatin gradually takes on interphase structure. During this process of swelling important components of the enzymatic systems responsible for DNA metabolism may enter the sperm head that is becoming the male pronucleus. The uptake of such components by the nucleus may have an impact on processes such as excision repair of DNA lesions, DNA replication, or postreplication DNA repair. These processes may be involved in the processing of DNA lesions and contribute to mutagenesis. Thus it becomes clear how the maternal genotype may exert an influence on the frequency of mutations that primarily are induced by the treatment of mature sperm.

The first indications of the existence of such maternal effects on mutagenized sperm were obtained by chance in experiments in which, for other reasons, identically mutagen-treated males were crossed to different types of females.[62,136] With the various DNA repair mutants isolated in recent years, a systematic approach to the question of maternally influenced repair has become possible.

6.1. Spontaneous Mutations

Spontaneous mutation frequencies can give an indication of the stability of chromosomes replicating for the first time in a repair-deficient cell. In a systematic study[135] X chromosomes were introduced, via sperm, into oocytes produced by wild-type, *mei-9*, *mei-41*, and *mus(1)101* females. Sex-linked recessive lethals as well as losses of ring-X chromosomes were consistently increased in frequency with oocytes from *mei-9* females. With *mei-41* and *mus(1)101* oocytes there was no effect on the recovery of spontaneous lethals; there was a tendency for a reduced frequency of ring-X loss, even though not all experimental series gave statistically significant decreases. The "mutator effect" of *mei-9* oocytes was also observed by Frei.[44] After crossing untreated *Basc* males to wild-type or *mus(1)101* females about 0.3% sex-linked recessive lethals were found, whereas in the cross to *mei-9* females the spontaneous frequency was increased to 1.46%.

For *mei-9*, the control data from the study of 21 mutagens carried out by Zimmering[151] can be used to illustrate the maternal effect on

TABLE 7. Maternal Effects on Sperm Exposed to Ionizing Radiation in DNA Repair-Proficient Males Crossed to Different Types of Females

Treatment	Females	Genetic endpoint	Relative effect compared with cross to repair-proficient females	Ref.
X rays	*mei-9*	Lethals	Increased	38, 44, 52
		Translocations	Equal	44
		Ring-X loss	Increased	1, 34, 50
	mei-41	Lethals	Equal/increased	38, 50
		Translocations	Equal	38
		Ring-X loss	Equal	1, 50
	mus(1)101	Lethals	Equal	44, 52
		Translocations	Reduced	44
		Ring-X loss	Increased	1
Neutrons	*mei-9*	Lethals	Reduced	44
		Translocations	Reduced	44
	mus(1)101	Lethals	Reduced	44
		Translocations	Reduced	44

the frequency of loss of paternally derived chromosomes from untreated repair-proficient males. Chromosome loss as a spontaneous event was studied in crosses of ring-X males to either repair-proficient females or to excision repair-deficient *mei-9* females. Complete loss, concerning mostly the ring-X chromosome and rarely the Y chromosome, can be distinguished from partial loss representing loss of a part of the genetically marked Y chromosome. The pooled control data indicate that in crosses to *mei-9* females, complete loss is roughly 1.5 times as frequent as in crosses to repair-proficient females, i.e., 1.41% on the average instead of 0.89%. Partial loss was increased about four times, with an average frequency of 0.062% in the crosses with *mei-9* females as compared to 0.015% in the crosses with repair-proficient females.

6.2. Direct-Acting Mutagens

Ionizing radiation is a classical mutagen that has been used in several studies dealing with maternal effects on mutagenized sperm introduced into oocytes with different repair capacities (Table 7 and references quoted therein).

X-ray induction led to increased frequencies of induced recessive lethals in the assays with *mei-9* and *mei-41*. With *mus(1)101*, no effect was detected on the recovery of induced lethals. With respect to the induction of translocations, no maternal effects were found with *mei-9* or *mei-41*, whereas with *mus(1)101* the yield of such chromosome rearrangements was strongly reduced. It has been concluded that the *mus(1)101*$^+$ gene specifies a function involved in the formation of chromosome rearrangements.[44] The rearrangements can be regarded as the product of repair, although in this case it is misrepair rather than error-free repair that leads to the resulting endpoint. The induction of ring-X loss was enhanced by excision repair-defective *mei-9* and by postreplication repair-defective *mus(1)101*, but not by postreplication repair-defective *mei-41*. Altogether it appears that with respect to the three endpoints considered each of the three mutant gene loci has its own pattern of effects.[38]

Irradiation with densely ionizing neutrons may lead to DNA damage requiring more complex processing than damage caused by sparsely ionizing X rays. This might explain the reduced frequencies of recessive lethals and translocations recovered from neutron-irradiated sperm in crosses with *mei-9* or *mus(1)101* females. Such reduced frequencies of recovery are interpreted as due to repair failure leading to dominant lethality, which reduces the possibility for misrepair produces recessive lethals and chromosome rearrangements under conditions of unimpaired processing of DNA damage.[44]

Experiments similar in design to those described for ionizing radiations were carried out with various chemicals. Table 8 gives an overview of the results obtained with repair-proficient females compared with those obtained with *mei-9* repair-deficient females. Eight potent direct-acting mutagens were assayed in the recessive lethal test. Mature sperm of mutagen-treated males was assayed in crosses to repair-proficient and *mei-9* females. The procedure is aimed at probing for differential recovery of induced lethals, an effect which can be ascribed to differences in the repair capacity of the fertilized oocytes. In comparing the results obtained with different compounds that interact in specific ways with DNA, additional information may be gained on the types of the DNA defects that are differentially repaired. With oocytes from *mei-9* females, increases in the recovery of induced lethals above the levels reached with repair-proficient oocytes were found following sperm treatment with EMS, MMS, MNU, and trimethylphosphate (see Table 8 and references quoted therein). Würgler and Graf[134] reported an increased recovery of lethals for ENU also, but could not confirm the effect in a later experiment with a larger number of tests,[50] and are inclined to agree with Vogel *et al.*[128] who did not obtain increased frequencies of recessive lethals with oocytes of

TABLE 8. Maternal Effects on Sex-linked Recessive Lethal Mutations and on Chromosome Loss in Sperm Mutagenized with Various Chemical Mutagens after Crossing of Treated Repair-Proficient Males with Repair-Proficient Females (RP) or with Repair-Deficient *mei-9* Females (RD)[a]

	Sex-linked lethals (SLRL)			Chromosome loss (LOSS)					Reference	
				Complete loss			Partial loss			
Compound	RP	RD	ME	RP	RD	ME	RP	RD	SLRL	LOSS
Direct-acting mutagens										
Acetin				–	+	↑	–	+		151
Allyl glycidyl ether				+ +	+	n.s.	–	+		151
5-Aminoacridine				–	–					b
Calcium chromate				–	+	n.s.	–	+		151
1,2-Dibromo-3-chloropropane				+	+ +	↑	–	+		151
2,3-Dibromo-1-propanol				–	+	↑	–	+		151
Dichloroacetonitrile				–	+	n.s.	–	+		151
Diepoxybutane				+ +	+	n.s.	+	+ +		151
Diethyl sulfate				–	+	n.s.	–	+		151
ENNG	+ +	+	n.s.						130	
Ethidium bromide				–	–					b
Ethyl methanesulfonate	+	+ +	↑	–	+	↑	–	+	52	50
									128	151
Ethylnitrosourea	+	+	n.s.	–	+	↑	+	+ +	128	50
									130	151
ICR-170				–	–					b
Isopropyl methanesulfonate	+	+	n.s.						130	
Methyl methanesulfonate	+	+ +	↑	–	+	↑	–	+	128	50
									130	151
Methylnitrosourea	+	+ +	↑	+	+ +	↑	–	+	128	50
									130	151

MNNG				+	+	n.s.				[b]
Nitrogen mustard	++	+	↓	++	+	↓	−	+	52	151
β-Propiolactone				−	+	↑	−	+		151
T-2 toxin				−	+	n.s.				[b]
Trenimon				+	−	n.s.				[b]
Trimethylphosphate	+	++	↑					125		
Promutagens										
N-Acetyl-2-aminofluorene	−	−		−	−				129	129
Benzo(*a*)pyrene	−	−		−	−				129	129,[b]
Cyclophosphamide				+	+	n.s.	−	(+)		151
				−	−					50
Diethylnitrosamine	+	+	n.s.	−	+	↑	−	+	125	151
									130	[b]
Dimethylnitrosamine	+	++	↑	+	++	↑	−	+	125	151
									130	
Hycanthone methanesulfonate				−	−		−	+		151
Hydralazine				−	−					[c]
Lasiocarpine				++	+	↓	−	+		151
Mitomycin C				+	++	n.s.				50
4-Nitroquinoline-*N*-oxide				−	−					[b]
N-Nitrosopiperidine				+	++	↑	−	+		151
Procarbazine	+	++	↑	−	+	↑	−	+	130	151,[b]
Senkirkine				+	+	n.s.				[b]
Intercalating Agents										
5-Aminoacridine				−	−					[b]
Ethidium bromide				−	−					[b]
ICR-170				−	−					[b]

(*continued*)

TABLE 8. *(continued)*

	Sex-linked lethals (SLRL)			Chromosome loss (LOSS)					Reference	
				Complete loss			Partial loss			
Compound	RP	RD	ME	RP	RD	ME	RP	RD	SLRL	LOSS
Cross-linking agents										
Diepoxybutane				++	+	n.s.	+	++		151
Lasiocarpine				++	+	↓	−	+		151
Mitomycin C				+	++	n.s.				50
Nitrogen mustard	++	+	↓	++	+	↓	−	+	52	151
Senkirkine				+	+	n.s.				[b]
Trenimon				+	−	n.s.				[b]

[a] Statistical significance of the maternal effect on complete loss was determined with a χ^2 approximation (see Section 6.5). RP, Repair-proficient females; RD, repair-deficient *mei-9* females; −, not significant; +, significant; ++, significant (frequency increased with respect to + in the pairwise comparison RP versus RD). ME, Maternal effect; ↑, frequency significantly increased by *mei-9*; ↓, frequency significantly decreased by *mei-9*; n.s., maternal effect of *mei-9* not significant.
[b] U. Graf and F. E. Würgler, unpublished results.
[c] H. Juon, unpublished results.

mei-9 females in a series of experiments with this particular compound. The absence of a *mei-9* effect with ENU might be related to the low Swain–Scott value ($s = 0.26$) and the high ratio of O-6/N-7 alkylation of guanine in DNA.[126] ENU as well as DEN may exert mutagenesis mainly by producing persistent lesions that lead directly to mispairing.[36] Based on the chemical similarities between ENU and DEN, one would expect that DEN induction of recessive lethals also should not be increased with *mei-9* females. That this is so has been recently confirmed.[130] In addition, iPMS and ENNG were also shown to yield the same frequency of sex-linked recessive lethals in crosses with excision repair-proficient and repair-deficient females. The order of hypermutability of monofunctional alkylating agents with *mei-9*^{L1} relative to *mei-9*$^{+}$ is MMS > MNU > DMN = EMS > iPMS = ENU = DEN = ENNG. This relative order compares very favorably to the relative proportion of 3-alkyladenine made by these agents both *in vivo* and *in vitro*.[104,130] It is known that 3-methyladenine in DNA is not well tolerated by cells and is released very rapidly both in bacteria and in mammalian cells by a DNA glycosylase as the initial step in base excision repair.[71] Alkylation of DNA at the N-3 position in *vitro* is also known to increase substantially spontaneous depurination *in vitro*.[27,68] Sherman and Loeb[103] have shown that both prokaryotic and eukaryotic DNA polymerases are able *in vitro* to polymerize past abasic sites in polynucleotide templates, and recent studies[96] have shown that apurinic sites induced in ϕX174 DNA are highly mutagenic when the altered DNA transfects SOS-induced *E. coli* cells. By analogy with these studies, Vogel *et al.*[130] hypothesized that the enhancement of mutability in the excision-defective *Drosophila* strains is caused by the formation of apurinic sites at 3-alkyladenine lesions, persistance of these apurinic sites following fertilization, and error-prone replication during the cleavage stages.

The cross linking agent HN2 did not lead to an increase in the frequency of recessive lethals with *mei-9* females, but rather to a decrease.[52] In this case, however, it is likely that DNA cross-links are not converted into recessive lethals at a normal rate under certain conditions of impaired repair. It may be recalled in this context that a very pronounced reduction in the recovery of recessive lethals was also observed in crosses of HN2-treated males to postreplication repair-defective *mus(1)101* females.[52,53] The effect has been interpreted as resulting from cross-links that rather than being converted into recessive lethals, act as dominant lethals if the normal *mus(1)101*$^{+}$ gene product is absent from the oocytes. This conclusion is further corroborated by a similar study of the maternal effects of *mus(1)101* on HN2-induced reciprocal translocations.[75]

Olsen and Green[89] crossed diepoxybutane-treated *Basc* males to females homozygous or heterozygous for the postreplication repair-defective mutation *mei-41* or for the excision repair-defective mutation *mus(2)201*. They found an increase in the frequency of sex-linked recessive lethals with *mei-41*, but a decrease with *mus(2)201*.

A number of chemical compounds have been studied for the induction of chromosome loss where mutagen-treated ring-X males were crossed to repair-proficient (RP) or *mei-9* repair-deficient (RD) females. Most data summarized in Table 8 are from Zimmering and co-workers.[151] Data refer to the endpoints of complete as well as of partial chromosome loss (see Section 6.1). Added are a number of published and unpublished results from our laboratory,[50,52] although only complete loss is considered, due to a technical difference between our mating scheme and that used by Zimmering. The general feature resulting from the observations listed is that most of the compounds tested show a significant increase in the chromosome loss frequencies if the tests are carried out with RD females, but not if they are carried out with RP females. The results may be interpreted as demonstrating that the use of *mei-9* females enhances the capacity for the detection of the mutagenic activity of the majority of the compounds tested. On the other hand, it might appear as rather unexpected that potent mutagens such as EMS or MMS should not provide a positive result with RP females. Most probably the negative outcome is circumstantial. Treatment of males with potent mutagens often results in marked mortality among the progeny from matings with RD females. Thus, to obtain reasonable numbers of offspring, concentrations or exposures have to be adjusted to levels at which in many cases a mutagenic effect is not detectable with RP females. Could the strength of the treatment be increased at will, the likely outcome with most of the substances would probably be an increased frequency of chromosome loss in the assays with RD females at exposure levels producing positive results in crosses with RP females. It can be seen in Table 8 that in some cases an outcome of this kind was realized. Most of the substances showing increased frequencies of loss or giving positive results only in the test with RD females are alkylating agents. For two of the substances, diepoxybutane and HN2, the situation was reversed for the endpoint of complete loss. Although both RD and RP females gave positive results, RP females gave the higher frequency of loss. The reversal may be characteristic for certain types of premutational DNA lesions produced specifically by these compounds, e.g., for cross-links. Note that the reversal was not observed for partial chromosome loss. In all cases in which this endpoint was studied, an enhancement with RD females was observed.

The three tested polycyclic aromatic compounds, which in microbial assays act as frameshift mutagens (listed together a second time in the third section of Table 8), could not be detected as mutagens in the chromosome loss test, neither with RP nor with RD females. Other *Drosophila* tests, such as the sex-linked recessive lethal assay, have shown poor performance for this group of mutagens. It seems as if for the frameshift mutagens the use of *mei-9* females does not bring the improvement one might have expected. However, only three polycyclic aromatic compounds have been studied so far and, due to the restricted sample size, only the endpoint of complete loss has been considered.

The only intercalating agent tested for the induction of partial loss, hycanthone, showed an increase with *mei-9*. A number of tests with other intercalating agents are needed to substantiate a possible maternal effect on partial loss induced by this class of mutagens.

6.3. Promutagens

In *Drosophila*, promutagens are biotransformed into their mutagenic metabolites in various tissues of the intestinal tract, in the fat body, and—important in the present context—to some extent also in the male gonadal tissues.(41,141) There is evidence that male germ cells during their metabolically most active phase (spermatocyte and spermatid stages) contain the enzymes involved in the activation of xenobiotics. It is likely that toward the end of sperm development this metabolic activity ceases, so that genetic damage caused by the ultimate mutagenic products in mature sperm depends on the possibility of intercellular transport or on persistence implying slow decay and/or incomplete detoxification.(128) Since there is evidence that DNA lesions produced in mature sperm persist in a premutational form until fertilization of the egg, maternal effects depending on the repair capacity of the egg can be seen, provided the promutagen is activated in the male and reaches the DNA contained in the sperm nucleus. Depending on the chemical nature of the compounds, their conversion follows different pathways of enzymatically catalyzed transformation steps. Apparently, *Drosophila* possesses the enzymes for various routes of biotransformation (refs. 58 and 92, and references cited therein).

In the chromosome loss test with RP and RD females, negative results were recorded with BP, hycanthone methanesulfonate, and hydralazine. For hycanthone methanesulfonate, however, this is only the case for the criterion of complete loss, partial loss having been increased in frequency in the assay with *mei-9* females.[151] On the other hand, Kramers[65] found this compound to be positive for complete loss in the chromosome loss test with RP females. At least for BP[129] and for hydralazine (H. Juon, unpublished results), *Drosophila* mutagenicity tests other than the chromosome loss assay have shown that these compounds can have mutagenic activity in this organism, although the effects are weak. In the case of BP it is likely that although the enzymes for metabolic biotransformation of the compound are present in *Drosophila*, the main routes of metabolism do not lead to the mutagenic products. In *Drosophila*, the production of BP dihydrodiols catalyzed by epoxyde hydrolase is generally low. Since these are the precursors of the reactive metabolites, it seems that the low mutagenic effect of the promutagen resides in this step of metabolism. However, since there was variability between the strains investigated for xenobiotic metabolism,[58,141] one cannot exclude the possibility that in addition to the use of RD females further improvements of the chromosome loss test are still possible by the choice or construction of metabolically more suitable ring-X strains.

For the majority of the other promutagens tested in the chromosome loss assay a significant increase was noted in the frequencies of loss, either with one or both types of females (RP and RD). For many of the compounds the maternal effect on loss was significant. In the majority of the cases an increase in the frequency of loss was observed in the crosses with *mei-9* females. However, there were also cases, usually with cross-linking agents, in which the use of *mei-9* females produced a reduced frequency of complete chromosome loss. To give an impression of the regularity with which this type of maternal effect is realized with the cross-linking compounds, they are listed together a second time in the fourth section of Table 8, irrespective of whether they are direct-acting agents or promutagens.

6.4. The Assay with *st mus(3)302* Females

In addition to the *mei-9* assay for complete and partial chromosome loss, Zimmering[149,150] developed an analogous assay with *st mus(3)302* females. The females contain the *mus(3)302* mutant, which is defective in postreplication repair and has an intermediate defect in excision repair. In addition, the stock contains the *mu-2* mutation,[77] which seems to facilitate the recovery of terminal deletions and partial chromosome losses.[77,154] The use of *st mus(3)302* females showed an enhanced detection capacity for complete and partial chromosome loss with a number of classic mutagens: MMS and DMN,[144,147] DEN and procarbazine,[145] β-propiolactone,[146] and nitrosopiperidine.[148] With the exception of nitrosopiperidine, the mutation frequencies with *st mus(3)302* females were higher than those obtained with the usually used *mei-9* females. Woodruff *et al.*[133] tested the assay for routine applications. They examined commercial preparations of fungicides (captan, maneb), herbicides (bromacil, paraquat, picloram, 2,4-D), and insecticides (carbaryl, chloropyrifos, dimethoate, DDT, diazinon, carbofuran, permethrin) for their ability to induce complete and partial chromsome losses. Only the positive controls DMN and MMS and the insecticide chlorpyrifos induced a significant frequency of ring-X chromosome loss. None of the pesticides induced partial losses.

6.5. Statistics

For the detection of mutagens with weak effects the statistical treatment of the results may be difficult, for two reasons.

1. Control data may be heterogeneous for the endpoint considered. If, as in the case of chromosome loss, this is demonstrably the case, it is strongly advised to carry out parallel controls for each experimental series. Heterogeneity may also be due to effects that cannot be corrected for in this way. Clustered occurrence of relatively rare segregants in control or experimental series may be particularly difficult to account for. Standard statistical tests, such as the χ^2 test, the test of Kastenbaum and Bowman, or Fisher's exact test, require that variation of the frequency estimates from individual samples around the true frequency be due solely to chance. Additional variation, often showing up as statistically significant heterogeneity among replicate samples, renders these tests overly optimistic because they give too many significant results. Unfortunately, the data structure is frequently such that variation due to mutagen treatment cannot be compared to the additional background variation from undetermined sources, which may be there in excess of random variation. In other words, it cannot be decided whether a given difference between control and experimental series is observed because background variation is important or because the effect of the treatment has been important. This is definitely the statistical crux of the application of the chromosome loss test and may be for other tests as well. This element of uncertainty may be responsible for disagreement in the interpretation of apparently similar experiments with seemingly contradictory results.

2. Another problem arises if one attempts to decide whether the use of a mutant defective in DNA repair improves the detection capacity of a particular mutagenicity test for a given mutagenic treatment. One

may carry out experiments comprising control series and treated series both with and without the mutant in question. If, in a particular assay, one obtains a significant result only in the experimental series in which the mutant was used, the conclusion would not be justified that there was no effect at all in the case in which no significant result was obtained. Therefore, such an outcome does not necessarily imply that the use of the mutant improves the test performance. Such a conclusion becomes more likely if the results are the same in a number of similar experiments or of replicate experiments. For a particular single treatment, however, the effect may be the same under both conditions whether the mutant was used or not. By chance, however, the experiments in comparison with their corresponding controls may have produced numerical results just above the level of statistical significance in one case and just below it in the opposite one. It is also clear that if significant results are obtained in both cases with and without the use of a repair mutant, possible inequalities in the induction frequencies would require a statistical assessment of the resulting frequency difference. As far as we are aware, appropriate statistical tests have not been developed for this purpose. One would particularly need tests accounting for unequal spontaneous frequencies. They should also be compatible with low absolute number of mutations in the individual samples used for testing. If the absolute numbers are not too small (say, >5), the following approximate χ^2 test may be used to verify whether the control corrected difference in induction frequency D is significantly different from zero:

$$D = (p_{t2} - p_{c2}) - (p_{t1} - p_{c1})$$

$$\chi^2_{\text{approx}} = \frac{D^2}{p_{t2}/N_{t2} + p_{c2}/N_{c2} + p_{t1}/N_{t1} + p_{c1}/N_{c1}}$$

with 1 degree of freedom

Here the indexed p values represent the (relative) mutation frequencies; the indices t and c designate treated and control series respectively; the indices 1 and 2 refer to the two conditions, i.e., with or without the repair mutant; and, according to the indexed p, the indexed N values correspond to the size of the respective samples.

7. Storage Effect

After mating, the spermatozoa transferred from the male to the female are stored in special organs, the receptacula seminis. In this way spermatozoa may inseminate eggs many days after mating. If mutagenized sperm are stored in females, the premutational lesions present in the condensed chromatin of the sperm head may undergo nonenzymatic changes, e.g., by hydrolysis of alkylated bases from DNA.[63] Such storage effects may result in an increased mutation frequency with stored compared to nonstored sperm. This was observed for chromosome aberrations after treatments with chromosome-breaking chemicals.[127] The effect has been interpreted as resulting from delayed opening and rejoining of initially latent breaks.[4] The storage effect is considered a more general characteristic of chemically produced breaks in *Drosophila* both with polyfunctional[61,93,97,105,112,131] as well as with monofunctional agents.[2,114–116] Typically, storage effects need days of storage to develop. Monofunctional alkylating agents with a known pattern of interaction with DNA are used to analyze the mechanism(s) involved. Storage of MMS-treated sperm for 9–12 days results in an increase in point mutations and deletions in certain regions of the X chromosome. Paradi *et al.*[91] consider apurinic sites as the likely primary lesion responsible for the storage effect. These sites result from bases alkylated at nitrogens. 7-Methylguanine, the major DNA adduct induced by MMS, is spontaneously lost by depurination with a half-life of 3–6 days at 37°C.[66,67] For other chemical mutagens corresponding hypotheses remain to be developed.

In the experiments concerning the *mei-9* maternal effect, possible storage effects can be studied in the ring-X loss test. The standard protocol[153] is to mate the mutagen-treated ring-X males for 3 days with the appropriate females (brood A), then to discard the males and allow the inseminated females to deposit eggs for an additional 3 days (brood B). Some early experiments seemed to indicate a substantially higher frequency of chromosome loss in brood B than in brood A.[153] Subsequent studies showed that, depending on the mutagen tested, brood B may yield higher, equal, or lower frequencies of chromosome loss in comparison to brood A.[50,143] For example, DEN and procarbazine[143] as well as HN2[50] did not show increased frequencies in brood B. Based on this experience, brood B is usually produced simply to increase the number of F_1 flies to be analyzed, but not as a means to improve the sensitivity of the assay. At present, the conclusions concerning the mutagenic activity and the presence or absence of a *mei-9* maternal effect are based on the pooled data from broods A and B.[50,151]

8. Conclusions

A large data base on chemical mutagenesis using mutagen-sensitive strains has been accumulated, but the studies are still in the exploratory phase. One reason for this is the need to study a variety of compounds, cell types, genetic endpoints, and repair deficiencies. The information has to be evaluated with regard to the potential applications of mutagen-sensitive mutants in routine testing as well as in basic research.

The term routine testing is used here in the sense that large numbers of compounds and complex mixtures are tested for mutagenicity using well-defined standard protocols. Routine testing demands validated test systems which have previously been characterized and optimized with respect to the two most important steps involved in chemical mutagenesis: (1) biotransformation of promutagens and procarcinogens, and (2) mutation fixation starting from premutational lesions.

The fact that a wide array of procarcinogens are potent in inducing sex-linked recessive lethals was the first circumstantial evidence for the presence of enzymes activating a great variety of indirectly acting chemicals in *Drosophila.* It is now well established that microsomal fractions from *Drosophila* adults and larvae contain cytochrome P-450 and monooxygenase activities similar to those found in mammalian tissues.[5–9,56,85] Whereas for several wild-type strains such as *Berlin K, Karsnäs,* and *Hikone R* many details of the relative activity of the xenobiotic metabolism are known, none of the mutagen-sensitive strains have been examined in biochemical analyses for their procarcinogen metabolism. This may be a critical point, because more and more strain differences in enzyme activity and inducibility of microsomal oxidative enzymes are being reported.[73,141] For a few promutagens it has been shown that they are active in certain mutagen-sensitive strains or in the ring-X males mutagenized for the study of maternal effects. Additional information on the spectrum of promutagens detectable in the assays using mutagen-sensitive mutants is indispensable before the assays can be considered for use in routine testing.

With respect to the problem of optimalization of mutation fixation from induced premutational lesions, most of the information stems from studies using *mei-9* alleles. It is characteristic, as, for example, in the ring-X loss assay with *mei-9* females, that certain types of lesions (e.g., those induced by most monofunctional alkylating agents) give increased mutation frequencies, whereas others (e.g., those induced by polyfunctional alkylating agents) give normal or reduced frequencies. Intercalating agents (see Table 8) may be completely negative (such as 5-aminoacridine and ethidium bromide). Thus, the use of *mei-9* mutants did not generally increase the detection capacity of the ring-X loss assay. On the other hand, partial Y loss, which can be studied with the same assay, was increased in *mei-9* with all compounds studied.[151] Results from further validation studies using chemicals from a wide spectrum of chemical classes will be of interest because certain chemicals, such as ENU or DEN, that are not potentiated in the sex-linked recessive lethal assay are potentiated with the partial chromosome loss test.

Work performed by a number of research groups has led to the appreciation of technical problems which might become critical in routine applications. The majority of the studies were performed with mutagen-sensitive mutants having meiotic effects (e.g. *mei-9*). Most of the alleles of meiotic mutants have low fertility and, in consequence, make the experiments more demanding with respect to stock breeding and collection of virgins. The substitution of *mei-9* by mutagen-sensitive mutants without meiotic effects is expected to improve the situation. This was examined in a study with *mus(2)201*.[155] It appeared that *mei-9*^{a} is a stronger potentiator than *mus(2)201* for loss induced by procarbazine and DEN.

Another aspect important in practice is the frequently observed strong reduction in the yield of surviving offspring if mutant mutagen-sensitive individuals are mutagenized or if mutagenized males are mated to mutant females.[129,155] Furthermore, special effort has to be invested in the periodic check of the mutant strains. Since no morphological criteria are available, the meiotic mutants have to be tested for the allele specific frequency of spontaneous nondisjunction. The other mutagen-sensitive strains require checks for the mutagen-sensitive phenotype of larvae by exposing them to MMS (or another agent suitable to identify the mutant).

It should be kept in mind that in *Drosophila*, probably in a similar way as in yeast,[60] a complex network of repair activities exists. Therefore, it may turn out that other mutant repair genes or combinations result in a more generalized sensitization of the mutagenicity assays in *Drosophila*. In addition to the *mei-9* studies, there is at present only limited information on the effects of other excision repair-defective mutants, such as *mus(2)201*,[155] postreplication repair-defective mutants such as *mei-41* and *mus(1)101*,[50,53,134,142] strains with multiple repair defects such as *st mus(3)302*,[147–149] or the *mei-9 mei-41* double mutant.[83,129,152] The mutant strain *st mus(3)302*, which is postreplication repair-deficient [19] and has a partially reduced excision repair activity as well,[17] is especially interesting because it sensitizes the chromosome loss assay with respect to partial Y chromosome losses.[77,154] These appear to originate from chromosome lesions ultimately resolved as chromosome breaks. Evaluation of additional mutagen-sensitive strains, preferably with multiple defects, seems to be promising with respect to a general improvement of the detection capacity as well as with respect to specific effects resulting from particular mutagenic events.

Although the assays with mutagen-sensitive mutants are not yet ready for general use in routine testing, they represent advantageous research tools with important potential for the future. They could be helpful, for example, in a detailed analysis of the role of different repair activities in mutagenesis in particular cell types. In turn this may allow the selection of appropriate mutants and appropriate assay systems with an optimized detection capacity for mutagens and promutagens.

ACKNOWLEDGMENTS

We thank Drs. J. B. Boyd, R. L. Dusenbery, J. M. Mason, P. D. Smith, and S. Zimmering for preprints. Work in our laboratory was supported by Swiss National Science Foundation grant 3.657–0.84.

9. References

1. C. S. Aaron and B. Leigh, X-ray induced sex-chromosome loss, when ring-X chromosome males are irradiated and mated to females carrying *mei-9, mei-41* or *mei-218, Mutat. Res. 82,* 381–388 (1981).
2. S. Abrahamson, W. C. Kiriazis, and E. M. Sabol, A storage effect of ethyl methanesulfonate (EMS) on the induction of translocations in *Drosophila* sperm, *Drosophila Inform. Serv. 44,* 110–111 (1969).
3. I. D. Alexandrov, S. E. Alexandrova, and M. A. Ankia, Formaldehyde-induced lethality and mutagenesis in *Drosophila* males defective in DNA metabolism, *Drosophila Inform. Serv. 58,* 12–14 (1982).
4. C. Auerbach, Past achievements and future tasks of research in chemical mutagenesis, in: *Genetics Today* (S. J. Geerts, ed.), Vol. 2, pp. 275–284, Pergamon, London (1965).
5. A. J. Baars, Biotransformation of xenobiotics in *Drosophila melanogaster* and its relevance for mutagenicity testing, *Drug Metabl. Rev. 11,* 191–221 (1980).
6. A. J. Baars, J. A. Zijlstra, E. Vogel, and D. D. Breimer, The occurrence of cytochrome P-450 and aryl hydrocarbon hydroxylase activity in *Drosophila melanogaster* microsomes, and the importance of this metabolizing capacity for the screening of carcinogenic and mutagenic properties of foreign compounds, *Mutat. Res. 44,* 257–268 (1977).
7. A. J. Baars, M. Jansen, and D. D. Breimer, Xenobiotica-metabolizing enzymes in *Drosophila melanogaster.* Activities of epoxide hydratase and glutathione *S*-transferase compared with similar activities in rat liver, *Mutat. Res. 62,* 279–291 (1979).
8. A. J. Baars, J. A. Zijlstra, M. Jansen, E. Vogel, and D. D. Breimer, Biotransformation and spectral interaction of xenobiotics with subcellular fractions from *Drosophila melanogaster, Arch. Toxicol. Suppl. 4,* 54–58 (1980).
9. A. J. Baars, W. G. H. Blijleven, G. R. Mohn, A. T. Natarajan, and D. D. Breimer, Preliminary studies on the ability of *Drosophila* microsomal preparations to activate mutagens and carcinogens, *Mutat. Res. 72,* 257–264 (1980).

10. B. S. Baker and A. T. C. Carpenter, Genetic analysis of sex chromosomal meiotic mutants in *Drosophila melanogaster, Genetics 71,* 255–286 (1972).
11. B. S. Baker and D. A. Smith, The effects of mutagen-sensitive mutants of *Drosophila melanogaster* in nonmutagenized cells, *Genetics 92,* 833–847 (1979).
12. B. S. Baker, J. B. Boyd, A. T. C. Carpenter, M. M. Green, T. D. Nguyen, P. Ripoll, and P. D. Smith, Genetic controls of meiotic recombination and somatic DNA metabolism in *Drosophila melanogaster, Proc. Natl. Acad. Sci. USA 73,* 4140–4144 (1976).
13. B. S. Baker, A. T. C. Carpenter, and P. Ripoll, The utilization during mitotic cell division of loci controlling meiotic recombination and disjunction in *Drosophila melanogaster, Genetics 90,* 531–578 (1978).
14. B. S. Baker, M. Gatti, A. T. C. Carpenter, S. Pimpinelli, and D. A. Smith, Effects of recombination-deficient and repair-deficient loci on meiotic and mitotic chromosome behavior in *Drosophila melanogaster,* in: *DNA Repair and Mutagenesis in Eukaryotes* (W. M. Generoso, M. D. Shelby, and F. J. de Serres, eds.), pp. 189–208, Plenum Press, New York (1980).
15. L. A. Beck and B. M. Sutherland, Purification of a photoreactivating enzyme from *Drosophila melanogaster, Am. Soc. Photobiol. 7,* 130 (1979).
16. J. B. Boyd, DNA repair in *Drosophila,* in: *DNA Repair Mechanisms* (P. C. Hanawalt, E. C. Friedberg, and C. F. Fox, eds.), pp. 449–452, Academic Press, New York (1978).
17. J. B. Boyd and P. V. Harris, Mutants partially defective in excision repair at five autosomal loci in *Drosophila melanogaster, Chromosoma 82,* 249–257 (1981).
18. J. B. Boyd and J. M. Presley, Repair replication and photorepair of DNA in larvae of *Drosophila melanogaster, Genetics 77,* 678–700 (1974).
19. J. B. Boyd and R. B. Setlow, Characterization of postreplication repair in mutagen-sensitive strains of *Drosophila melanogaster, Genetics 84,* 507–526 (1976).
20. J. B. Boyd and K. E. S. Shaw, Postreplication repair defects in mutants of *Drosophila melanogaster, Mol. Gen. Genet. 186,* 289–294 (1982).
21. J. B. Boyd, M. D. Golino, and R. B. Setlow, The *mei-9^{a}* mutant of *Drosophila melanogaster* increases mutagen sensitivity and decreases excision repair, *Genetics 84,* 527–544 (1976).
22. J. B. Boyd, M. D. Golino, T. D. Nguyen, and M. M. Green, Isolation and characterization of X-linked mutants of *Drosophila melanogaster* which are sensitive to mutagens, *Genetics 84,* 485–506 (1976).
23. J. B. Boyd, P. V. Harris, C. J. Osgood, and K. E. Smith, Biochemical characterization of repair-deficient mutants of *Drosophila,* in: *DNA Repair and Mutagenesis in Eukaryotes* (W. M. Generoso, M. D. Shelby, and F. J. de Serres, eds.), pp. 209–221, Plenum Press, New York (1980).
24. J. B. Boyd, M. D. Golino, K. E. S. Shaw, C. J. Osgood, and M. M. Green, Third-chromosome mutagen-sensitive mutants of *Drosophila melanogaster, Genetics 97,* 607–623 (1981).
25. J. B. Boyd, R. D. Snyder, P. V. Harris, J. M. Presley, S. F. Boyd, and P. D. Smith, Identification of a second locus in *Drosophila melanogaster* required for excision repair, *Genetics 100,* 239–257 (1982).
26. J. B. Boyd, P. V. Harris, J. M. Presley, and M. Narachi, *Drosophila melanogaster*: A model eukaryote for the study of DNA repair, in: *Cellular Responses to DNA Damage* (E. C. Friedberg and B. A. Bridges, eds.), pp. 107–123, Liss, New York (1983).
27. P. Brookes and P. D. Lawley, The reaction of mono- and di-functional alkylating agents with nucleic acids, *Biochem. J. 80,* 496–503 (1961).
28. T. C. Brown and J. B. Boyd, Postreplication repair defective mutants of *Drosophila melanogaster* fall into two classes, *Mol. Gen. Genet. 183,* 356–362 (1981).

29. T. C. Brown and J. B. Boyd, Abnormal recovery of DNA replication in ultraviolet-irradiated cell cultures of *Drosophila melanogaster* which are defective in DNA repair, *Mol. Gen. Genet. 183,* 363–368 (1981).
30. T. C. Brown, P. V. Harris, and J. B. Boyd, Effects of radiation on the survival of excision-defective cells from *Drosophila melanogaster, Somat. Cell Genet. 7,* 631–644 (1981).
31. A. T. C. Carpenter, Recombination nodules and synaptonemal complex in recombination-defective females of *Drosophila melanogaster, Chromosoma 75,* 259–292 (1979).
32. A. T. C. Carpenter, Mismatch repair, gene conversion, and crossing-over in two recombination-defective mutants of *Drosophila melanogaster, Proc. Natl. Acad. Sci. USA 79,* 5961–5965 (1982).
33. A. T. C. Carpenter and L. Sandler, On recombination-defective meiotic mutants in *Drosophila melanogaster, Genetics 76,* 453–475 (1974).
34. S. F. Cooper and S. Zimmering, A genetic study of the effects of the repair deficient *mei-9* mutation in *Drosophila* on spontaneous and X-ray-induced paternal sex chromosome loss, *Mutat. Res. 80,* 281–287 (1981).
35. W. Dezzani, P. V. Harris, and J. B. Boyd, Repair of double-strand DNA breaks in *Drosophila, Mutat. Res. 92,* 151–160 (1982).
36. J. W. Drake and R. H. Baltz, The biochemistry of mutagenesis, *Annu. Rev. Biochem. 45,* 11–37 (1976).
37. R. L. Dusenbery, S. C. McCormick, and P. D. Smith, *Drosophila* mutations at the *mei-9* and *mus(2)201* loci which block excision of thymine dimers also block induction of unscheduled DNA synthesis by methyl methanesulfonate, ethyl methanesulfonate, *N*-methyl-*N*-nitrosourea, UV light and X-rays, *Mutat. Res. 112,* 215–230 (1983).
38. J. C. J. Eeken, W. Ferro, H. Frei, B. Leigh, K. Sanakaranarayanan, and F. H. Sobels, Three repair deficient mutants of *Drosophila*: Their effect in oocytes on the recovery of X-ray-induced paternal damage, in: *DNA Repair, Chromosome Alterations and Chromatin Structure* (A. T. Natarajan, G. Obe, and H. Altman, eds.), pp. 99–102, Elsevier Biomedical Press, Amsterdam (1982).
39. W. Ferro, Biochemical studies on a *Drosophila* strain with impaired photorepair and increased sensitivity to UV and X-rays, presented at 8th European Drosophila Research Conference Cambridge, England, 1983.
40. W. Ferro, K. Sankaranarayanan, J. C. J. Eeken, and Vijayalaxmi, Studies on mutagen-sensitive strains of *Drosophila melanogaster*. I. The enhanced X-ray sensitivity of a mutant strain (*ebony*) which is UV-sensitive and also deficient in photorepair, *Mutat. Res. 106,* 33–53 (1982).
41. R. E. Foerster and F. E. Würgler, *In vitro* studies on the metabolism of aflatoxin B1 and aldrin in testes of genetically different strains of *Drosophila melanogaster, Arch. Toxicol. 56,* 12–17 (1984).
42. C. Forbes, Sex-linked lethal mutations induced in *Drosophila melanogaster* by 7,12-dimethylbenz(*a*)anthracene, *Mutat. Res. 79,* 231–237 (1980).
43. C. Forbes, Sex-linked lethal frequencies produced by 7,12-dimethylbenz(*a*)anthracene in five *Drosophila melanogaster* wild strains, *Mutat. Res. 90,* 255–260 (1981).
44. H. Frei, Genetic damage induced by X-rays or neutrons in spermatozoa of *Drosophila melanogaster*; differential processing in the oocytes carrying the DNA repair deficient mutants *mei-9* and *mus(1)101, Mutat. Res. 94,* 63–78 (1982).
45. H. Fritz-Niggli, C. Büchi, and P. Schweizer, Oxygen-effect as an inhibition of repair: Radiation studies on excision repair deficient *mei-9*^{L1} embryos of *Drosophila, Radiat. Environ. Biophys. 19,* 265–274 (1981).

46. M. Gatti, Genetic control of chromosome breakage and rejoining in *Drosophila melanogaster*: Spontaneous chromosome aberrations in X-linked mutants defective in DNA metabolism, *Proc. Natl. Acad. Sci. USA 76,* 1377–1381 (1979).
47. M. Gatti, S. Pimpinelli, and B. S. Baker, Relationships among chromatid interchanges, sister chromatid exchanges, and meiotic recombination in *Drosophila melanogaster, Proc. Natl. Acad. Sci. USA 77,* 1575–1579 (1980).
48. M. Gatti, D. A. Smith, and B. S. Baker, A gene controlling condensation of heterochromatin in *Drosophila melanogaster, Science 221,* 83–85 (1983).
49. E. Glassman and J. McLean, Maternal effect of *mal-1*$^+$ on xanthine dehydrogenase in *Drosophila melanogaster.* II. Xanthine dehydrogenase activity during development, *Proc. Natl. Acad. Sci. USA 48,* 1712–1718 (1962).
50. U. Graf and F. E. Würgler, DNA repair dependent mutagenesis in *Drosophila melanogaster,* in: *Advances in Genetics, Development, and Evolution of Drosophila* (S. Lakovaara, ed.), pp. 85–99, Plenum Press, New York (1982).
51. U. Graf, E. Vogel, U. P. Biber, and F. E. Würgler, Genetic control of mutagen sensitivity in *Drosophila melanogaster.* A new allele at the *mei-9* locus on the X-chromosome, *Mutat. Res. 59,* 129–133 (1979).
52. U. Graf, M. M. Green, and F. E. Würgler, Mutagen-sensitive mutants in *Drosophila melanogaster.* Effects on premutational damage, *Mutat. Res. 63,* 101–112 (1979).
53. U. Graf, A. Kägi, and F. E. Würgler, Mutagenesis in spermatozoa of *Drosophila melanogaster* by cross-linking agents depends on the *mus(1)101*$^+$ gene product in the oocyte, *Mutat. Res. 95,* 237–249 (1982).
54. U. Graf, H. Juon, A. J. Katz, H. J. Frei, and F. E. Würgler, A pilot study on a new *Drosophila* spot test, *Mutat. Res. 120,* 233–239 (1983).
55. U. Graf, F. E. Würgler, A. J. Katz, H. Frei, H. Juon, C. B. Hall, and P. G. Kale, Somatic mutation and recombination test in *Drosophila melanogaster, Environ. Mutagen. 6,* 153–188 (1984).
56. I. Hällström and R. Grafström, The metabolism of drugs and carcinogens in isolated subcellular fractions of *Drosophila melanogaster.* II. Enzyme induction and metabolism of benzo(*a*)pyrene, *Chem. Biol. Interact. 34,* 145–159 (1981).
57. I. Hällström, A. Sundvall, U. Rannug, R. Grafström, and C. Ramel, The metabolism of drugs and carcinogens in isolated subcellular fractions of *Drosophila melanogaster.* I. Activation of vinylchloride, 2-aminoanthracene and benzo(*a*)pyrene as measured by mutagenic effects in *Salmonella typhimurium, Chem. Biol. Interact. 34,* 129–143 (1981).
58. I. Hällström, A. Blanck, and S. Atuma, Genetic variation in cytochrome P-450 and xenobiotic metabolism in *Drosophila melanogaster, Biochem. Pharmacol. 33,* 13–20 (1084).
59. P. V. Harris and J. B. Boyd, Excision repair in *Drosophila*: Analysis of strand breaks appearing in DNA of *mei-9* mutants following mutagen treatment, *Biochim. Biophys. Acta 610,* 116–129 (1980).
60. R. H. Haynes and B. A. Kunz, DNA repair and mutagenesis in yeast, in: *The Molecular Biology of the Yeast Saccharomyces: Life Cycle and Inheritance* (J. N. Strathern, E. W. Jones, and J. R. Broach, eds.), pp. 371–414, Cold Spring Harbor Laboratory, Cold Spring Harbor, New York (1981).
61. I. H. Herskowitz, Mutagenesis in mature *Drosophila* spermatozoa by "triazine" applied in vaginal douches, *Genetics 41,* 605–609 (1956).
62. P. E. Hildreth and G. L. Carson, Influence of the type of inseminated females on the lethal frequency in the X-chromosome from the male, *Proc. Natl. Acad. Sci. USA 43,* 157–183 (1957).

63. F. C. Janca, W. R. Lee, and C. S. Aaron, Stability of induced DNA alkylations in sperm and embryos of *Drosophila melanogaster, Mutat. Res. 53,* 103–104 (1978).
64. H. Kato, Spontaneous and induced sister chromatid exchanges as revealed by the BUdR-labelling method, *Int. Rev. Cytol. 49,* 55–97 (1977).
65. P. G. N. Kramers, Mutagenicity of hycanthone in *Drosophila*: Additional results and a comparison with some analogs, *Mutat. Res. 82,* 111–123 (1981).
66. P. D. Lawley, Some chemical aspects of dose–response relationships in alkylation mutagenesis, *Mutat. Res. 23,* 283–295 (1974).
67. P. D. Lawley, Alkylation of nucleic acids and mutagens, in: *Molecular and Environmental Aspects of Mutagenesis* (L. Prakash, F. Sherman, M. W. Miller, C. W. Lawrence, and H. W. Taber, eds.), pp. 17–31, Thomas, Springfield, Illinois (1974).
68. P. D. Lawley and R. Brookes, Further studies on the alkylation of nucleic acids and their constituent nucleotides, *Biochem. J. 89,* 127–138 (1963).
69. W. R. Lee, S. Abrahamson, R. Valencia, E. S. Von Halle, F. E. Würgler, and S. Zimmering, The sex-linked recessive lethal test for mutagenesis in *Drosophila melanogaster*: A report of the U. S. Environmental Protection Agency Gene-Tox Program, *Mutat. Res. 123,* 183–279 (1983).
70. A. R. Lehmann, S. Kirk-Bell, C. F. Arlett, M. C. Paterson, P. H. M. Lohman, A. E. de Weerd-Kastelein, and D. Bootsma, Xeroderma pigmentosum cells with normal levels of excision repair have a defect in DNA synthesis after UV-irradiation, *Proc. Natl. Acad. Sci. USA 72,* 219–223 (1975).
71. T. Lindahl, DNA repair enzymes, *Annu. Rev. Biochem. 51,* 61–87 (1982).
72. L. A. Luchkina, Y. M. Khromykh, and W. I. Sharigin, Studies on *Drosophila* radiosensitive strains. V. Sensitivity of the mutant *mus(2)201-G1* to MMS, UV light and deficiencies of DNA repair in UV-irradiated cells, *Genetika 18,* 625–633 (1982).
73. J. Magnusson, I. Hällström, and C. Ramel, Studies on metabolic activation of vinylchloride in *Drosophila melanogaster* after pretreatment with phenobarbital and polychlorinated biphenyls, *Chem. Biol. Interact. 24,* 287–298 (1979).
74. J. M. Mason, Spontaneous mutation frequencies in mutagen-sensitive mutants of *Drosophila melanogaster, Mutat. Res. 72,* 323–326 (1980).
75. J. M. Mason, Nitrogen mustard induced translocations in mutagen-sensitive mutants, *Drosophila Inform. Serv. 59,* 81–82 (1983).
76. J. M. Mason, M. M. Green, K. E. S. Shaw, and J. B. Boyd, Genetic analysis of X-linked mutagen sensitive mutants of *Drosophila melanogaster, Mutat. Res. 81,* 329–343 (1981).
77. J. M. Mason, E. Strobel, and M. M. Green, *mu-2*: Mutator gene in *Drosophila* that potentiates the induction of terminal deficiencies, *Proc. Natl. Acad. Sci. USA 81,* 6090–6094 (1984).
78. M. S. Meselson and C. M. Radding, A general model for genetic recombination, *Proc. Natl. Acad. Sci. USA 72,* 358–361 (1975).
79. S. Mittler, Response of mutagen-sensitive *Drosophila melanogaster* to hyperthermia and radiation, *Mutat. Res. 72,* 101–114 (1980).
80. J. D. Mohler, Developmental genetics of the *Drosophila* egg. I. Identification of 59 sex-linked cistrons with maternal effects on embryonic development, *Genetics 85,* 259–272 (1977).
81. T. D. Nguyen and J. B. Boyd, The *meiotic-9* (*mei-9*) mutants of *Drosophila melanogaster* are deficient in repair replication of DNA, *Mol. Gen. Genet. 158,* 141–147 (1977).
82. T. D. Nguyen, M. M. Green, and J. B. Boyd, Isolation of two X-linked mutants in *Drosophila melanogaster* which are sensitive to gamma-rays, *Mutat. Res. 49,* 139–143 (1978).

83. T. D. Nguyen, J. B. Boyd, and M. M. Green, Sensitivity of *Drosophila* mutants to chemical carcinogens, *Mutat. Res. 63,* 67–77 (1979).
84. C. E. Nix, C. McKinley, and J. L. Epler, Examination of the effect of *mei-9* and *mei-41* alleles on mutation induction in *Drosophila* males, *Genetics 94,* s77 (1980).
85. C. E. Nix, B. Brewen, and J. L. Epler, Microsomal activation of selected polycyclic aromatic hydrocarbons and aromatic amines in *Drosophila melanogaster, Mutat. Res. 88,* 291–299 (1981).
86. C. E. Nix, J. L. Epler, N. L. Forbes, L. M. Kyte, C. M. McKinley, and R. D. Wilkerson, Effect of genotype on the induction of X-linked recessive lethals by alkylating agents in *Drosophila* males, *Environ. Mutagen. 3,* 381 (1981).
87. S. J. O'Brien and R. J. MacIntyre, Genetics and biochemistry of enzymes and specific proteins of *Drosophila,* in: *The Genetics and Biology of Drosophila* (M. Ashburner and T. R. F. Wright, eds.), Vol. 2a, pp. 395–551, Academic Press, New York (1978).
88. K. Oeda, T. Horiuchi, and M. Sekiguchi, The *uvr D* gene of *E. coli* encodes a DNA-dependent ATPase, *Nature 298,* 98–100 (1982).
89. O. A. Olsen and M. M. Green, The mutagenic effects of diepoxybutane in wild-type and mutagen-sensitive mutants of *Drosophila melanogaster, Mutat. Res. 92,* 107–115 (1982).
90. C. J. Osgood and J. B. Boyd, Apurinic endonuclease from *Drosophila melanogaster*: Reduced enzymatic activity in excision-deficient mutants of the *mei-9* and *mus(2)201* loci, *Mol. Gen. Genet. 186,* 235–239 (1982).
91. E. Paradi, E. W. Vogel, and E. Szilagyi, Effect of storage and dose on MMS-induced deletions. Complementation analysis of X-chromosomal recessive lethals in the *zeste–white* and *maroon-like* regions of *Drosophila melanogaster, Mutat. Res. 111,* 145–159 (1983).
92. I. Porsch-Hällström, Cytochrome P-450 in *Drosophila melanogaster*, Activity, Genetic Variation and Regulation, Thesis, University of Stockholm, Graphic Systems, Göteborg (1984).
93. W. E. Ratnayake, Effects of storage on dominant lethals induced by alkylating agents (triethylene melamine and ethylenimine), *Mutat. Res. 5,* 271–278 (1968).
94. E. Richet, R. Kern, M. Kobiyama, and M. Hirota, Isolation of DNA-dependent ATPase I mutants in *E. coli,* in: *Mechanistic Studies of DNA Replication and Genetic Recombination* (B. Alberts, ed.), pp. 606–608, Academic Press, New York (1980).
95. R. Rieger, A. Michaelis, and M. M. Green, *Glossary of Genetics and Cytogenetics,* 4th ed., Springer, Berlin (1976).
96. R. M. Schaaper, B. W. Glickman, and L. A. Loeb, Mutagenesis resulting from depurination is an SOS process, *Mutat. Res. 106,* 1–9 (1982).
97. A. Schalet, The relationship between the frequency of nitrogen mustard induced translocations in mature sperm of *Drosophila* and utilization of sperm by females, *Genetics 40,* 543 (1955).
98. P. M. Schweizer, Chromatid breaks induced by small doses of X-rays in somatic cells of *Drosophila, Experientia 35,* 979 (1979).
99. P. M. Schweizer, Genetic control of chromosome aberration formation in *Drosophila, Experientia 37,* 661 (1981).
100. P. M. Schweizer, Neuroblasten von DNA-Reparatur defekten Mutanten von *Drosophila* als empfindliche Indikatoren strahleninduzierter Chromosomenschäden (Neuroblasts of DNA-repair defective mutants of *Drosophila* as sensitive indicators of radiation-induced chromosome damage), in: *Proceedings of the "Journée Scientifique de la Société Suisse de Radiobiologie et Radiophysique",* pp. 30–37, Epalinges, Switzerland (1982).

101. N. N. Scobie and J. M. Mason, Genetic characterization of the *mei-41* locus of *Drosophila melanogaster*, *Genetics 97*, s94 (1981).
102. E. R. Sechkina, Isolation of radiation-sensitive strains of *Drosophila melanogaster* from natural populations, *Genetika 9*, 160–162 (1973).
103. C. W. Sherman and L. A. Loeb, Effects of depurination on the fidelity of DNA synthesis, *J. Mol. Biol. 128*, 197–218 (1979).
104. B. Singer, The chemical effects of nucleic acid alkylation and their relation to mutagenesis and carcinogenesis, in: *Progress in Nucleic Acid Research and Molecular Biology* (E. W. Cohn, ed.), Vol. XV, pp. 219–284, Academic Press, New York (1975).
105. H. Slizinska, The progressive approximation, with storage, of the spectrum of TEM-induced chromosomal changes in *Drosophila* sperm to that found after irradiation, *Mutat. Res. 8*, 165–175 (1969).
106. P. D. Smith, Mutagen sensitivity of *Drosophila melanogaster*. I. Isolation and preliminary characterization of a methyl methanesulfonate-sensitive strain, *Mutat. Res. 20*, 215–220 (1973).
107. P. D. Smith, Mutagen sensitivity of *Drosophila melanogaster*. III. X-linked loci governing sensitivity to methyl methanesulfonate, *Mol. Gen. Genet. 149*, 73–85 (1976).
108. P. D. Smith, Mutagen sensitivity of *Drosophila melanogaster*. IV. Interactions of X chromosome mutants, in: *DNA Repair Mechanisms* (P. C. Hanawalt, E. C. Friedberg, and C. F. Fox, eds.), pp. 453–456, Academic Press, New York (1978).
109. P. D. Smith, R. D. Snyder, and R. L. Dusenbery, Isolation and characterization of repair deficient mutants of *Drosophila melanogaster*, in: *DNA Repair and Mutagenesis in Eukaryotes* (W. M. Generoso, M. D. Shelby, and F. J. de Serres, eds.), pp. 175–188, Plenum Press, New York (1980).
110. P. D. Smith, R. L. Dusenbery, S. F. Cooper, and C. F. Baumen, Examining the mechanism of mutagenesis in DNA repair deficient strains of *Drosophila melanogaster*, in: *Environmental Mutagens and Carcinogens* (T. Sugimura, S. Kondo, and H. Takebe, eds.), pp. 147–155, University of Tokyo Press, Tokyo, and Liss, New York (1982).
111. P. D. Smith, C. F. Baumen, and R. L. Dusenbery, Mutagen sensitivity of *Drosophila melanogaster*. VI. Evidence from the excision-defective *mei-9*$^{\mathrm{AT1}}$ mutant for the timing of DNA-repair activity during spermatogenesis, *Mutat. Res. 108*, 175–184 (1983).
112. L. A. Snyder, Evidence of an essential difference between point mutations and chromosome breaks induced by triethylene melamine in *Drosophila* spermatozoa, *Z. Vererbungsl. 94*, 182–189 (1963).
113. R. D. Snyder and P. D. Smith, Mutagen sensitivity of *Drosophila melanogaster*. V. Identification of second chromosomal mutagen sensitive strains, *Mol. Gen. Genet. 188*, 249–255 (1982).
114. R. J. Sram, The effect of storage on the frequency of dominant lethals in *Drosophila melanogaster*, *Mol. Gen. Genet. 106*, 286–288 (1970).
115. R. J. Sram, The effect of storage on the frequency of translocations in *Drosophila melanogaster*, *Mutat. Res. 9*, 243–244 (1970).
116. R. J. Sram, The differences in the spectra of genetic changes in *Drosophila melanogaster* induced by chemosterilants TEPA and HEMPA, *Folia Biol.* (*Prague*) *18*, 139–148 (1972).
117. C. Stern, Somatic crossing-over and segregation in *Drosophila melanogaster*, *Genetics 21*, 625–730 (1936).
118. J. E. Trosko and K. Wilder, Repair of UV-induced pyrimidine dimers in *Drosophila melanogaster* cells *in vitro*, *Genetics 73*, 297–302 (1973).
119. J. I. Valencia and W. Plaut, X-Ray-induced DNA synthesis in polytene chromosomes (abstract), *J. Cell. Biol. 43*, 151a (1969).

120. R. Valencia, S. Abrahamson, W. R. Lee, E. S. Von Halle, R. C. Woodruff, F. E. Würgler, and S. Zimmering, Chromosome mutation tests for mutagenesis in *Drosophila melanogaster*: A report of the U. S. Environmental Protection Agency Gene-Tox Program, *Mutat. Res. 134,* 61–88 (1984).
121. E. Vogel, Mutagenic activity of the insecticide oxydemeton methyl in a resistant strain of *Drosophila melanogaster, Experientia 30,* 396–397 (1974).
122. E. Vogel, Mutagenicity of carcinogens in *Drosophila* as function of genotype-controlled metabolism, in: *In Vitro Metabolic Activation in Mutagenesis Testing* (F. J. de Serres, J. R. Fouts, J. R. Bend, and R. M. Philpot, eds.), pp. 63–79, Elsevier/North-Holland Biomedical Press, Amsterdam (1976).
123. E. Vogel, Genetical relationship between resistance to insecticides and procarcinogens in two *Drosophila* populations, *Arch. Toxicol. 43,* 201–211 (1980).
124. E. W. Vogel, Dependence of mutagenesis in *Drosophila* males on metabolism and germ cell stage, in: *Environmental Mutagens and Carcinogens* (T. Sugimura, S. Kondo, and H. Takebe, eds.), pp. 183–194, University of Tokyo Press, Tokyo, and Liss, New York (1982).
125. E. W. Vogel, Approaches to comparative mutagenesis in higher eukaryotes: Significance of DNA modifications with alkylating agents in *Drosophila melanogaster, Ann. N. Y. Acad. Sci. CVII,* 208–220 (1983).
126. E. Vogel and A. T. Natarajan, The relation between reaction kinetics and mutagenic action of alkylating agents in higher eukaryotic systems: Interspecies comparisons, in: *Chemical Mutagens: Principles and Methods for Their Detection* (F. J. de Serres and A. Hollaender, eds.), Vol. 7, pp. 295–336, Plenum Press, New York (1982).
127. E. Vogel and F. H. Sobels, The function of *Drosophila* in genetic toxicology testing, in: *Chemical Mutagens: Principles and Methods for Their Detection* (A. Hollaender, ed.), Vol. 4, pp. 93–142, Plenum Press, New York (1976).
128. E. W. Vogel, W. G. H. Blijleven, M. J. H. Kortselius, and J. A. Zijlstra, A search for some common characteristics of the effects of chemical mutagens in *Drosophila, Mutat. Res. 92,* 69–87 (1982).
129. E. W. Vogel, J. A. Zijlstra, and W. G. H. Blijleven, Mutagenic activity of selected aromatic amines and polycyclic hydrocarbons in *Drosophila melanogaster, Mutat. Res. 107,* 53–77 (1983).
130. E. W. Vogel, R. L. Dusenbery, and P. D. Smith, The relationship between reaction kinetics and mutagenic action of mono-functional alkylating agents in higher eukaryotic systems. IV. The effects of the excision-defective *mei-9*^{L1} and *mus(2)201*^{D1} mutants on alkylation-induced genetic damage in *Drosophila, Mutat. Res 149,* 193–207 (1985).
131. W. A. F. Watson, Further evidence of an essential difference between the genetical effects of mono- and bifunctional alkylating agents, *Mutat. Res. 3,* 455–457 (1966).
132. S. Wolff, Sister chromatid exchange, *Annu. Rev. Genet. 11,* 183–201 (1977).
133. R. C. Woodruff, J. P. Phillips, and D. Irwin, Pesticide-induced complete and partial chromosome loss in screens with repair-defective females of *Drosophila melanogaster, Environ. Mutagen. 5,* 835–846 (1983).
134. F. E. Würgler and U. Graf, Mutation induction in repair deficient strains of *Drosophila,* in: *DNA Repair and Mutagenesis in Eukaryotes* (W. M. Generoso, M. D. Shelby, and F. J. de Serres, eds.), pp. 223–240, Plenum Press, New York (1980).
135. F. E. Würgler and U. Graf, Stability of *Drosophila melanogaster* chromosomes replicating for the first time in a repair deficient cell, *Mutat. Res. 92,* 99–106 (1982).
136. F. E. Würgler and P. Maier, Genetic control of mutation induction in *Drosophila melanogaster.* I. Sex-chromosome loss in X-rayed mature sperm, *Mutat. Res. 15,* 41–53 (1972).

137. F. E. Würgler, U. Graf, H. Frei, and H. Juon, Genotoxic activity of the anti-cancer drug methotrexate, *Mutat. Res. 122,* 321–328 (1983).
138. F. E. Würgler, U. Graf, H. Frei, and H. Juon, Detection of point mutations, chromosome aberrations, and somatic recombination in *Drosophila melanogaster,* in: *Critical Evaluation of Mutagenicity Tests* (R. Bass, V. Glocklin, P. Grosdanoff, D. Henschler, B. Kilbey, D. Müller, and D. Neubert, eds.), pp. 185–223, Bundesgesundheitsamt, Berlin, and MMV Medizin Verlag München, Munich (1984).
139. F. E. Würgler, F. H. Sobels, and E. Vogel, *Drosophila* as an assay system for detecting genetic changes, in: *Handbook of Mutagenicity Test Procedures* (B. J. Kilbey, M. Legator, W. Nichols, and C. Ramel, eds.), 2nd ed., pp. 555–601, Elsevier, Amsterdam (1984).
140. L. A. Yegorova, V. L. Levin, and M. A. Kozlova, Heritability of UV-sensitivity and photoreactivation ability in *Drosophila* embryos, *Mutat. Res. 49,* 213–218 (1978).
141. J. A. Zijlstra, E. W. Vogel, and D. D. Breimer, Strain-differences and inducibility of microsomal oxidative enzymes in *Drosophila melanogaster, Chem. Biol. Interact. 48,* 317–338 (1984).
142. S. Zimmering, Chromosome loss induced by procarbazine and diethylnitrosamine in *Drosophila* from matings of treated males with repair deficient *mei-9*^{a}, *mei-41*^{D5}, *mus101* and *mus104* females, *Environ. Mutagen. 3,* 687–690 (1981).
143. S. Zimmering, Review of the current status of the *mei-9*^{a} test for chromosome loss in *Drosophila melanogaster*: An assay with radically improved detection capacity for chromosome lesions induced by methyl methanesulfonate (MMS), dimethylnitrosamine (DMN), and especially diethylnitrosamine (DEN) and procarbazine, *Mutat. Res. 83,* 69–80 (1981).
144. S. Zimmering, Induced chromosome loss following treatment of postmeiotic cells of the *Drosophila melanogaster* male with MMS and DMN and matings with repair proficient females and the repair deficient females *mei-9*^{a} and *st mus302, Mutat. Res. 94,* 79–86 (1982).
145. S. Zimmering, Preliminary data suggesting that females of the repair-deficient strain designated *st mus302* are strong potentiators of chromosome loss induced by procarbazine and diethylnitrosamine (DEN) in the male genome of *Drosophila, Mutat. Res. 103,* 141–144 (1982).
146. S. Zimmering, Note on the utility of the *st mus302* test for chromosome loss in a blind test of an environmental compound for chromosome breakage in *Drosophila, Mutat. Res. 104,* 117–119 (1982).
147. S. Zimmering, A note on the utility of repair-deficient *st mus302 Drosophila* females in detecting chromosome loss and sex-linked recessive lethals induced in the male genome by methyl methanesulfonate (MMS) and dimethylnitrosamine (DMN), *Environ. Mutagen. 4,* 291–293 (1982).
148. S. Zimmering, Induced chromosome loss with nitrosopiperidine in the male *Drosophila melanogaster, Environ. Mutagen. 4,* 521–524 (1982).
149. S. Zimmering, Utility of the *st mus302* test for chromosome loss in a blind test of an environmental compound for chromosome breakage in *Drosophila, Mutat. Res. 104,* 117–119 (1982).
150. S. Zimmering, Selective elimination of potential ring-X as opposed to rod-X in matings of males treated with diepoxybutane (DEB) to repair-deficient *st mus302* females of *Drosophila, Environ. Mutagen. 5,* 363–365 (1983).
151. S. Zimmering, The *mei-9*^{a} test for chromosome loss in *Drosophila*: A review of assays of 21 chemicals for chromosome breakage, *Environ. Mutagen. 5,* 907–921 (1983).
152. S. Zimmering and S. F. Cooper, A maternal effect in homozygous *mei-9*^{a} *mei-41*^{D5} repair deficient *Drosophila melanogaster* females influencing the recovery rate of progeny bearing a Y chromosome, *Environ. Mutagen. 2,* 543–545 (1980).

153. S. Zimmering and K. L. Kammermeyer, Potentiation of chromosome loss induced in the paternal genome by methyl methanesulfonate and procarbazine following matings with repair deficient *mei-9*^{a} females of *Drosophila melanogaster, Environ. Mutagen. 2,* 515–520 (1980).
154. S. Zimmering and K. L. Kammermeyer, On the nature of partial losses of the Y chromosome from treatment of ring-X/*B*^{S} *Y y*$^{+}$ males with diethylnitrosamine (DEN) or procarbazine and matings with repair-deficient *st mus302* females of *Drosophila, Mutat. Res. 104,* 121–123 (1982).
155. S. Zimmering and K. L. Kammermeyer, Comparison of excision repair deficient *mei9*^{a} and *mus201* females in the test for paternal sex chromosome loss in *Drosophila* with procarbazine and diethylnitrosamine (DEN), *Environ. Mutagen. 5,* 235–237 (1983).

CHAPTER 11

Chromosomal Causes for Fertility Reduction in Mammals

Peter de Boer

1. Introduction

It has long been realized that chromosome mutations, whatever their mode of induction, usually have consequences for the fertility of their carrier. In fact, this characteristic is used in mutagenicity tests, such as the heritable translocation assay, to establish the frequency of this genetic endpoint. There are two possible reasons for a reduction in fertility accompanying chromosome mutations. First, the production of gametes can be interrupted somewhere in the spermatogenesis or oogenesis process. A reduction of the number of stem cell spermatogonia or primary oocytes can be anticipated. This point has received little experimental attention, however. The second reason is that the chromosome mutation, through the course of the meiotic divisions, gives rise to genetically unbalanced gametes, which usually function properly but lead to embryonic and/or fetal death. It will be apparent in what follows that in situations where genetically unbalanced gametes are produced through chromosome mutation there can be an increased frequency of germ cell death throughout the various stages of spermatogenesis. In addition, the initial number of stem cell spermatogonia

Peter de Boer • Department of Genetics, Agricultural University, 6703 BM Wageningen, The Netherlands.

TABLE 1. Chromosomal Anomalies of Sex Chromosomes That Affect Fertility in Male Mice[a]

Type of anomaly	Nature of effect
1. XXY	No spermatogenic cells in adult
2. XX *Sxr*[b]	No spermatogenic cells in adult
3. XO *Sxr*[b]	Major spermatocyte death, very few abnormal spermatozoa
4. XYY	Variable degree of spermatogenic impairment, but some spermatids and spermatozoa usually formed
5. X,Y separation	No secondary spermatocytes as most severe consequence

[a] From de Boer and Searle.[39]
[b] Sex-reversed.

or primary oocytes may be reduced. In this chapter the effect of numerical chromosome aberrations of the sex chromosomes on gametogenesis and gamete quality are discussed first. Then the effects of structural chromosome abnormalities are treated. Breeding situations that yield phenotypes of reduced fertility with aspects of abnormal chromosome behavior during gametogenesis, such as sterile hybrids and XY females, are not treated in this chapter.

In the last section, an attempt is made to reconcile hypotheses with regard to the causes of chromosomal sterility, especially for the male. To this purpose, X-autosome translocations are introduced in this section. The mouse is the species mainly dealt with, because the bulk of mutagenesis work has been carried out in this mammal. For references to observations in man, the reader is referred to Searle, this volume, Chapter 3. For the mouse, no comprehensive treatment of all chromosome mutants has been attempted (see Searle[109] and the *Mouse News Letter*).

2. Numerical Chromosome Mutations of the Sex Chromosomes and Related Conditions

2.1. Sex-Chromosome Anomalies in the Male

A list of chromosomal mutations concerning the sex chromosomes and leading to a decrease in fertility in the male, excluding sex-chromosome autosome translocations, is given in Table 1 (see also Table I of de Boer and Searle[39]).

In mammals spermatogonia do not tolerate the presence of two X chromosomes.[20,78,117] Thus, adult XXY and XX *Sxr* (sex-reversed) males resemble each other and have seminiferous tubules that are devoid of germ cells.[20] A structural change within the Y chromosome must underlie the sex-reversed condition (probably a duplication in the telomeric region of male-determining factors of pericentric origin[56,118]). The factor(s) for male differentiation are transported by the process of crossing-over to the telomeric region of the X chromosome, thus mimicking the XXY chromosome constitution. In contrast, in XO *Sxr* mice, spermatogonial differentiation is achieved and spermatogenesis is mainly interrupted during the late primary spermatocyte stages.[20]

Recently, Eicher and co-workers[46,49] described a Y-chromosome structural mutation that leads to similar consequences as the *Sxr* structural change. Eicher[46] postulates that the X–Y pairing region within the Y chromosome has been transposed to a site adjacent to the centromere. Thus, X^Y and Y^X sex chromosomes originate by crossing-over. Almost all of the Y is translocated by this process to the telomeric region of the X chromosome, and vice versa, a distal segment of the X chromosome should adjoin the centromere of the Y chromosome. As expected, XX^Y males are sterile (and XY^X females fertile). However, X^YO males, although producing reduced numbers of sperm, are fertile.

The XYY condition has been described in the mouse and is said to be inducible by 6-mercaptopurine, which probably acts during the preleptotene stage of the resting primary spermatocytes. Evans *et al.*[54] give a detailed account of their fertility. They observed a gradual decrease in cell number over the successive spermatogenic stages with some additional cell death around diakinesis and metaphase I. The very low testis weights (14–31% of normal, four XYY males described) suggest that the actual number of stem cell spermatogonia forming the basis of spermatogenesis must already be reduced. It has been proposed[13] that to explain the spectrum of sex-chromosome constitutions of secondary spermatocytes in XYY males, those that arise from an XYY trivalent at meiotic metaphase I are most likely the ones to proceed beyond the first meiotic division. Thus, it has been suggested that in XYY male mice, spermatocytes with univalent X and/or Y chromosomes are the ones to die during the first meiotic division.

The condition of X–Y univalence during late diplotene, diakinesis, and metaphase I is also associated with interrupted spermatogenesis. In sterile males, up to 100% of the primary spermatocytes in the diakinesis–metaphase I stages can show univalent X and Y chromosomes.[5] When this condition is followed back into the pachytene stage of meiosis,[38] X–Y synapsis is already absent in 36.8% of the pachytene spermatocytes, whereas the length of the synapsed region is on average less in those cells where some pairing has been achieved. The DBA genome seems to be associated with X–Y univalence, since about 35% diakinesis–metaphase I cells with X–Y separation can be seen in this strain or in C57BL/6J × DBA/2J hybrids.[66,104] The males of concern were fertile, but no comment was made concerning testes development. When the *Mus musculus molossinus* genome is used to generate X–Y separation, very high levels (over 90%) of X–Y separation can be attained.[38,73,85,86] Testis weights in the F_1 were 40.1% of the BALB/c parent and 68.1% of the *Mus musculus molossinus* parent,[73] but no fertility reduction has been observed. In the study by de Boer and Nijhoff,[38] it was found that the effect of X–Y univalence (introduced through crosses with *Mus musculus molossinus*) on the sperm count could be greatly enhanced by heterozygosity for the T(1;13)70H translocation. This translocation has only a slight effect on sperm production. These authors also concluded that spermatocytes with univalent sex chromosomes do not contribute to the formation of secondary spermatocytes: the condition does not lead to secondary spermatocytes aneuploid for the sex chromosomes (see also ref. 5).

The *Sxr* structural chromosome mutation is very interesting with regard to X–Y meiotic pairing as well. During pachytene (all stages pooled) 24% of all sex vesicles contain unpaired sex chromosomes (controls 4%[55]). In diakinesis–metaphase I spermatocytes, sex-chromosome univalence can be as high as 75%.[26] A strong negative correlation is found between the frequency of X–Y univalence at this stage and testis weight. Sterile animals with a high incidence of morphologically abnormal spermatozoa are encountered more frequently than normal.[20] It may be that the material described in this paragraph is not homogeneous with respect to pachytene X–Y pairing. When pairing is affected more heavily and leads to the high univalent frequencies found later in first meiotic prophase and metaphase, primary spermatocyte death could follow. When meiotic pairing is not so heavily influenced, high metaphase I univalent frequencies could to some extent be tolerated and enough meiocytes survive to ensure the minimal 10% sperm production that is needed for normal fertility in the mouse.[41,111] It should be borne in mind that the air-drying technique used to display meiotic chromosomes will exaggerate the true sex-chromosome univalent rate more heavily when pachytene pairing is less disturbed.

There is positive evidence[17] that the X–Y separation condition leading to sterility can also arise in F_1 male offspring from mutagen-treated fathers (treatment with ethyl methanesulfonate or 200-R X rays during the spermatid and spermatozoal stages of spermatogenesis).

2.2. Sex-Chromosome Anomalies in the Female

The most common of these is the XO chromosome mutation, whereas the XXX condition known in man has never been found in the mouse. XO female mice have a reduced reproductive life span.[80] From day 12 of postnatal life onward the ovaries of XO females contain about half the number of oocytes of XX controls.[11] The number of oogonia entering meiosis is identical for the two karyotypes, as is the number of oocytes up to the pachytene stage.[12] The reduction in number seems to take place at the late pachytene and early diplotene stages (P. S. Burgoyne, personal communication). When meiotic stages are compared between 14.5 and 18.5 days post coitum, the meiotic rate is reduced in XO versus XX female fetuses.[12] The embryos of XO females culture less successfully *in vitro*.[14] The transmission of the trait usually is lower than the 33% expected on the basis of the production of an equal proportion of X and O secondary oocytes and under the justified assumption[15] of complete survival of XO embryos (incorporating the fact that the OY condition is the earliest genetic embryonic lethal known[83]). A transmission percentage of 16.4 ($n = 1534$) at weaning has been observed in a cross where XO females were *Ta* (tabby)/O and XX females *Ta*/+.[89] Kaufman[75] karyotyped secondary oocytes of XO females, mostly after superovulation, but also after *in vitro* culture of primary oocytes. Of the 99 oocytes counted, 34 had less than 20 chromosomes. Because the XO condition only produces hypomodal counts and some cell breakage always occurs when preparing secondary oocytes for chromosome inspection, the upper limit of the percentage of O oocytes is 34; under the assumption of no embryonic loss of the XO karyotype, this would amount to 20.5% transmission. However, in that case, given the normal number of corpora lutea counted in XO females,[89] the relative fertility of 63.8% at a prenatal age of 15 days is less than the 83% expected (100 − 17% OY embryos). Some increased mortality of XO, XX, and XY embryos seems likely. If prenatal death preferentially concerned the XO embryo class, then the upper limit of the expected frequency (20.5%) could not be so close to the actual frequency of 16.4% observed. It is known that the X chromosome inactivated in female embryos around the time of implantation is reactivated in the germ-line cells just before first meiotic prophase.[1] The lower dosage of the X chromosome and the unpaired status of this chromosome during meiotic prophase can affect the number of oocytes available for ovulation and can explain why *in vitro* culture of embryos from XO mothers is less successful and why there is reason to expect increased embryonic mortality in XO mothers.

3. Structural Chromosome Anomalies

3.1. Reciprocal Translocations between the Autosomes

Broadly speaking, two effects on fertility can result from the presence of a reciprocal translocation. The first acts through death *in utero* or, more rarely, thereafter of conceptuses originating from the fusion of a normal gamete with duplication/deficiency gametes that are characteristic of a reciprocal translocation heterozygote. The second effect originates in the fact that, especially in the male, gametogenesis can be influenced by translocation heterozygosity. We initially confine ourselves to the behavior of translocation heterozygosity at first and second meiotic anaphase, where the unbalanced gametes are produced. Then we treat the effects that translocation heterozygosity can have on the quantitative aspects of gametogenesis in both sexes.

3.1.1. The Production of Unbalanced Gametes by Reciprocal Translocations

For mouse reciprocal translocations between acrocentric chromosomes, Searle *et al.*[113] give the theoretical expectations with regard to the types of gametes produced. Four modes of segregation at anaphase I of the translocation multivalent derived at zygotene after homologous meiotic pairing can be distinguished (see also Figures 1 and 2);

1. *Alternate*: Of the four centromeres involved, the nonhomologous alternate ones move to the same pole.
2. *Adjacent-1*: Adjacently situated but nonhomologous centromeres move to the same pole.
3. *Adjacent-2*: Adjacently situated but homogous centromeres move to the same pole.
4. *Numerical nondisjunction*: The four chromosomes potentially involved in the multivalent configuration do not segregate two by two, but three by one.

Thus, reciprocal translocations can also produce aneuploid conceptuses. If at least one of the two interstitial segments (between the centromere and the translocation breakpoint; see Figures 1 and 2)

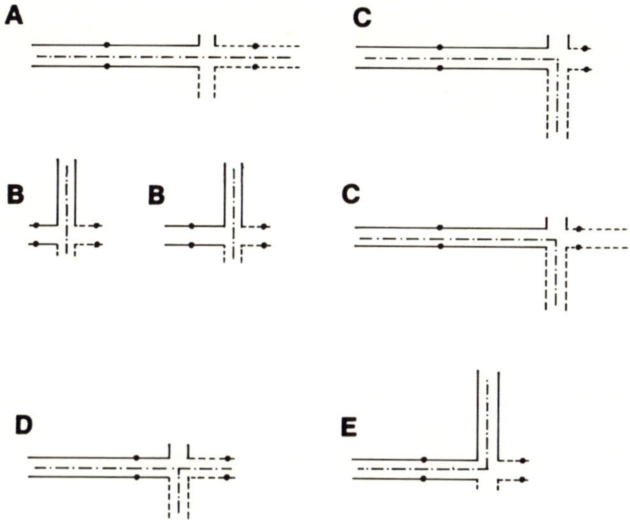

FIGURE 1. Reciprocal translocation types and their preferred manner of segregation as proposed by Jalbert *et al.*[74] (A) Adjacent-1; (B) adjacent-2; (C) "3:1" segregation leading to tertiary trisomic offspring; (D) a combination of (A) and (C); (E) "3:1" segregation leading to translocation trisomic progeny. Centromeres are marked by a dot.

carries a chiasma, alternate and adjacent-1 segregation produce the same spectrum of gamete types: the normal karyotype, the translocation karyotype, and two different duplication/deficiency karyotypes in the ratio 1:1:1:1. The combination of a duplication and a deficiency usually leads to embryonic death around and just after the time of implantation.[18,37] Only in cases where at least one breakpoint is near the telomeres, as in T(2;4)1Sn,[119] can duplication/deficiency gametes lead to the birth of viable offspring, because the deficiency and, less important in this respect, the duplication are small in size. However, in the case of T1Sn some extra genetic death of this type ensues after birth but before weaning. Thus, if the translocation complex with at least one interstitial segment bound by a chiasma exhibits only alternate/adjacent-1 segregation, the expected relative fertility* is 50%. The adjacent-2 and numerical nondisjunction types of segregation of the translocation multivalent give rise to neither normal nor translocation

* The relative fertility of a translocation heterozygote compares its fertility with chromosomally normal controls of the same sex in fertile matings with chromosomally normal individuals.

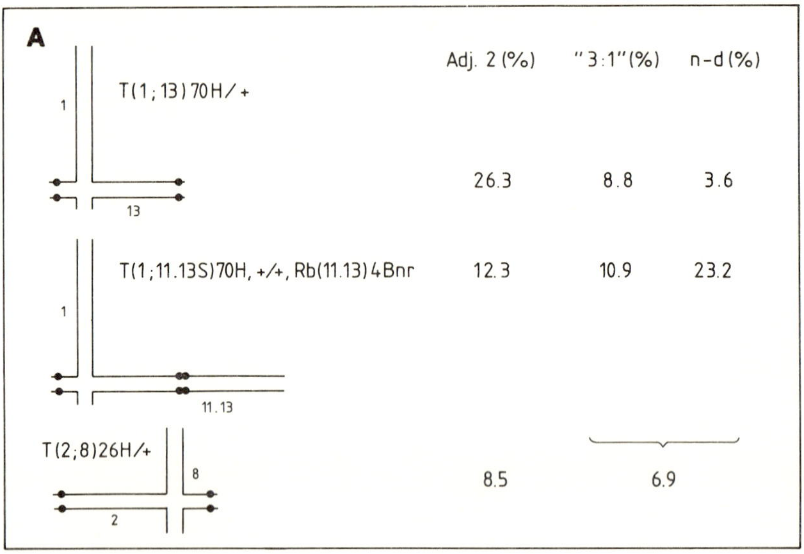

FIGURE 2. (A) Pachytene pairing diagrams (centromeres marked by a dot) of three mouse reciprocal translocations, together with some data concerning segregation behavior in the male.[33,35,95] (B1) The T70H/+ pairing diagram from an electron micrograph of the synaptonemal complex as produced after whole-mount spreading of pachytene spermatocytes on formvarcoated copper grids. (B2) As in (B1), but with failure of pairing in the proximal segment of chromosome 1. Note proximity of the XY bivalent. Adj. 2, Adjacent-2 segregation; "3:1", "3:1" segregation; n-d nondisjunction for normal bivalents.

heterozygote karyotypes. Provided the genetic imbalance of the gametic types produced is large enough and leads to prenatal death, the summed frequency p of these two can be estimated from the relative fertility (RF) score, where RF $= \frac{1}{2} - p$.[113] The only possible way for a reciprocal translocation to yield viable offspring after adjacent-2 segregation is for one interstitial segment to be so small or genetically insignificant that a deficiency for it is tolerated, whereas the other interstitial segment is long enough to be tolerated as a duplication. This situation is found when (1) both of the translocation breakpoints are very proximal,* or (2) at least one is very proximal and the other is not too distal.

1. In the first situation, since short interstitial segments are relatively devoid of chiasmata (whereas short translocated segments contain relatively many[33]), two genetically heteromorphic bivalents will be

* Proximal chromosome regions are near the centromere; distal chromosome regions are near the telomere.

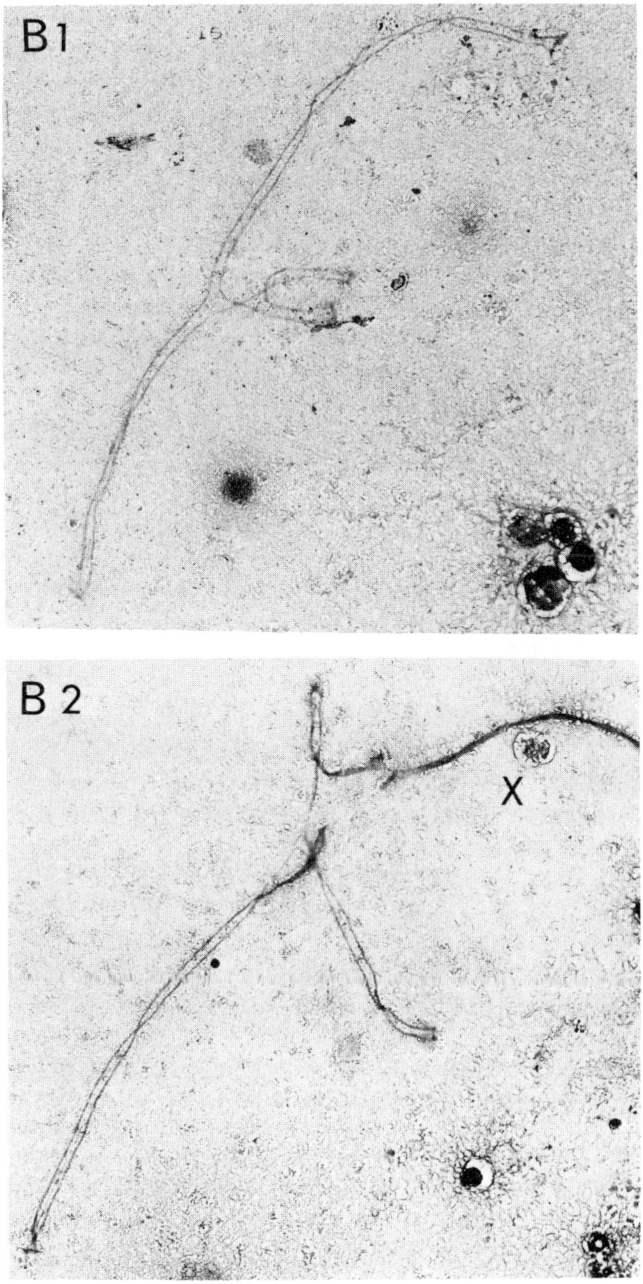

FIGURE 2. *(continued)*

formed that by chance orientation show 50% adjacent-2 segregation and 50% alternate segregation. This type of translocation, which ultimately is a centromere exchange between two nonhomologous chromosomes, is not picked up by a heritable translocation assay nor by scoring for multivalents during late diplotene–metaphase I of mutagen-treated male or female mice of the "all-acrocentric" karyotype. Crocker and Cattanach[31] used the MC stock of mice, which are homozygous for three metacentric chromosomes, for a study of translocation induction by irradiation at the spermatozoal level. They found that, when translocation induction involved at least one of the metacentric chromosomes, two breaks close to the centromere and leading to whole-arm exchanges could be picked up by G banding. Also, theoretically because the chromosome arm not involved in the translocation "extends" the short interstitial segment such that chiasma formation can follow, a multivalent is produced.

2. T(16;17)43H is an example of a mouse translocation where adjacent-2 segregation yields viable offspring in which the very short deletion (heterochromatin) for chromosome 16 is tolerated besides a duplication for the proximal 40% of chromosome 17, including the T region.[64]

The aneuploid segregation products appearing in the gametes after "3:1" segregation of the translocation complex at anaphase I lead to partially monosomic conceptuses and partially trisomic ones. Since primary monosomic or trisomic offspring usually are prenatal lethals (see Section 3.2), this condition can be picked up only in the offspring if one of the translocation chromosomes is smaller than the smallest autosomes, or alternatively, is genetically insignificant [for example, in T158H, T194H, T(5;12)31H, T(14;15)6Ca, and T(1;13)70H heterozygotes]. Usually, only the partially trisomic individuals (tertiary trisomics) are found. When the genetic content of the extra chromosome is not too important, the monosomic counterpart can also live into adulthood, as, for example, with T31H.

3.1.2. Quantitative Aspects of the Segregation Behavior of Translocation Heterozygotes

As stated earlier, the relative fertility of a translocation heterozygote in the mouse usually depends on factors influencing adjacent-2 and "3:1" segregation, since most translocations have at least one interstitial segment long enough to accommodate one chiasma. The only absolute way to distinguish between these two is to perform a cytological analysis of secondary meiocytes or young embryos. Usually, the first route is taken. In the normal mouse "all-acrocentric" karyotype, adjacent-2, can be measured only in the cell products of primary meiocytes characterized by an interstitial chiasma, provided the two attached translocated segments show enough difference in length. The interstitial chiasma will produce two chromosomes with homologous centromeres, but with chromatids that differ in absolute length. The case of T70H/+[33,100] is a nice example where such an analysis can be performed and is valid for almost all meiocytes concerned (see Figure 2). In contrast, T(2;8)26H[33] has only a chiasma frequency of 61.4% in the interstitial segments suitable for this type of analysis. Thus, cytological adjacent-2 estimates in secondary meiocytes only apply to this fraction of germinal cells. If secondary meiocytes and embryos do not yield aneuploid chromosome counts, then the adjacent-2 frequency can be estimated from relative fertility alone (by obtaining p from RF $= \frac{1}{2} - p$) or from the frequency of homozygotes for a recessive marker, located on an interstitial segment, after intercrossing of translocation heterozygotes (see refs. 47, 112, and 113 for the details of this method).

Jalbert *et al.*[74] studied 151 reciprocal translocations in the human. These translocations had all been picked up by an abnormal proband that could be classified as from adjacent-1, adjacent-2, or "3:1" origin. In this way, several rules were developed on the basis of which the segregation behavior of a reciprocal translocation could be predicted. Figure 1 gives the translocation and segregation types. It can be seen that in each case the prediction of segregation behavior coincides with the segregation type producing the smallest deficiency and not too great a duplication. In other words, the analysis is limited by the way the material was collected. Jalbert *et al.* propose that, this factor notwithstanding, long pairing segments have a greater potential for proper segregation than short pairing segments.

Figure 2 shows a number of pachytene pairing diagrams of mouse translocations that have been studied cytologically. The frequencies of adjacent-2 and "3:1" segregation found are given as well. It can be seen that the rules given by Jalbert *et al.* hold for the few cases studied in detail. A specific demonstration of this is the way the adjacent-2 frequency of T70H can be reduced by introducing an extra pairing segment (i.e., chromosome 11). For T70H itself, de Boer[34] found a coincidence of the frequency of late diakinesis–metaphase I primary spermatocytes that had a chiasma very proximal in the chromosome 13 interstitial segment (i.e., close to the centromere) and the adjacent-2 frequency in the same males. From this analysis, it was concluded that chains of III would show less adjacent-2 segregation. This prediction was verified in T70H translocation trisomics producing chains of III in excess.[35] Thus, in translocations between acrocentric chromosomes, a relation between the chiasma position in the interstitial segment and the segregation of the two centromeres involved was substantiated. This relation did not hold for T(1;11.13S)70H, +/+, Rb(11.13)4Bnr, however[95] (see Fig. 2). The extra pairing segment apparently offsets the effect of the single proximal chiasma.

For "3:1" segregation, the diagrams of Jalbert *et al.* (Figure 1) suggest that reciprocal translocations that give rise to aneuploid offspring are those characterized by a small translocation product that would not be bound by a chiasma to other chromosomes of the translocation complex at meiotic metaphase I. For mouse reciprocal translocations, the conclusion of "3:1" segregation usually is based on the recovery of tertiary trisomic offspring (see Section 3.1.1). For the following reason, it is often not possible to decide upon "3:1" segregation from chromosome counts of secondary spermatocytes and oocytes alone. Ideally, if all four chromosomes involved in the reciprocal translocation complex can be recognized in secondary spermatocytes and oocytes, than nondisjunction for other bivalents can be separated from "3:1" segregation. When, as with T70H and T(1;11.13S)70H, three of the four chromosomes are "marker" chromosomes in secondary spermatocytes, a fair assessment of the fraction of aneuploid meiocytes caused by "3:1" segregation can be established. For these two karyotypes, the fraction of "3:1" segregation is given along with the fraction of other bivalent nondisjunctions in Figure 2. For T26H (also included in Figure 2), these two cannot be separated. For all karyotypes studied, it is clear that the presence of a reciprocal translocation, at least in male carriers, increases the risk that the other bivalents will show first meiotic nondisjunction. This effect will cause an additional reduction of the relative fertility of translocation heterozygotes. A consequence of this observation is that in the formula $RF = \frac{1}{2} - p$, p also includes a contribution of nondisjunction for the other bivalents. Thus, the p-based estimate of adjacent-2 plus "3:1" segregations will be a slight overestimate.

3.1.3. Sex Differences of the Segregation Behavior of Reciprocal Translocations

In those cases where both sexes produce secondary meiocytes, the adjacent-2 frequencies are identical, but "3:1" segregation is very much increased in the female (see Table 2). In contrast to the findings of Oshimura and Sandberg[98] for T70H/+, Wauben-Penris and Prins[122] found no more "3:1" segregation in T70H/+ females after superovu-

TABLE 2. Sex Differences in Segregation Behavior for Two Reciprocal Translocations in the Mouse[a]

Translocation		Behavior for given mode of segregation[b]	
		Adjacent-2	"3:1"
T(14;15)6Ca/+[c]	♂	—	4.4%
	♀	—	22.2%
T(1;13)70H/+[d]	♂	31.5%	3.0%
	♀	30.1%	19.6%

[a] The estimates are based on cytological observations of secondary meiocytes. In line with these results, fertility comparisons between the sexes generally yield a lower score for females.(18,119)
[b] See Section 3.1.1.
[c] From Oshimura and Tagaki.(99) For the reasons outlined in Section 3.1.2, the adjacent-2 frequency cannot be cytologically estimated in T6Ca/+.
[d] From Oshimura and Sandberg.(98)

lation or after *in vitro* maturation of germinal vesicle dictyate oocytes until metaphase II. After normal ovulation the same stock produced more morphologically recognizable T70H tertiary trisomic offspring from T70H/+ females than from T70H/+ males.(33)

3.1.4. The Effect of Reciprocal Translocations on Gametogenesis

Both in man(27,28) and in the mouse(39,114) it is well known that reciprocal translocation heterozygosity can affect sperm production, up to the level of complete azoospermia. Because fertility in the mouse usually is assessed by the capability of the male carrier to impregnate females, sperm counts as low as 10% of normal go undetected because a normal male mouse produces a vast excess of spermatozoa.(41,111) Table 3 gives the relative sperm counts from the caput epididymis for a number of translocation heterozygotes. Available data for translocation homozygotes are also included. Also given is the percentage of abnormally shaped spermatozoa. It is clear that there is an inverse relationship between the sperm count and the percentage of spermatozoa with abnormal morphology in the epididymis. It can be seen that reciprocal translocation heterozygosity *per se* does not influence sperm output and sperm morphology.

A considerable amount of descriptive research has been carried

TABLE 3. Testis Weight and Relative Sperm Count and Morphology from the Caput Epididymis of Mouse Reciprocal Translocation Heterozygotes and Some Homozygotes[a]

Karyotype	Testis weight	Sperm count	Sperm with abnormal morphology
T26H/+[33]	—	99.2%	5.8%
T70H/+[33]	—	80.9%	4.3%
+/+[33]	—	100.0%	9.3%
T6Ca/+ H/M	89.1%[60]	60.1%[60]	11.0–93.0%[126]
T6Ca/+ L	67.8%[60]	20.6%[60]	
T6Ca/T6Ca	96.5%[60]	90.2%[60]	6.3%[126]
+/+	100.0%[60]	100.0%[60]	—
T43H/+[67]	53.0%	3.0%	—
T43H/T43H[67]	97.5%	87.1%	—
+/+[67]	100.0%	100.0%	—
T31H/+[10]	53.0%	8.9%	91.3%
+/+[10]	100.0%	100.0%	6.8%

[a] All data are given as percentages of normal.

out using male-sterile, reciprocal-translocation heterozygous males.[39] With the exception of the interstitial translocation (insertion) Is(7;1)40H, where the germ cells uniformly die during stage IV of the cycle of the seminiferous epithelium,[115] germ cell death usually occurs over stages from somewhere during pachytene until the stages of spermiogenesis. T(11:19)42H is extreme in that diakinesis of the first meiotic division is not reached, death occurring over a variety of pachytene substages.[43] On the other hand, T6Ca[60] and occasionally T31H[10] produce some carriers with sperm counts high enough to produce offspring. It is well known that fertility can be restored by a change in genetic background, as was shown by Forejt for T6Ca.[60]

When cell suspensions for the analysis of late diplotene–metaphase II chromosomes are prepared, the number of degenerating metaphase I's (DMIs) that can be seen is striking in all male-sterile translocations where this stage is reached.[101] De Boer and Speed,[40] studying meiosis in T70H translocation trisomics as a model, found that in primary spermatocytes prone to degenerate, bivalents spiralize precociously. Furthermore, these cells can be seen to persist into the stages I and II of the cycle of the seminiferous epithelium, when normally they are confined to stage XII. Thus, the total life span of degenerative metaphase is rather long, which adds to the impression that cell death in male-sterile, reciprocal-translocation heterozygotes preferentially occurs in metaphase I.

Hormone profiles of the male-sterile translocation T145H recently have been studied by Feldman *et al.*[57] In contrast to male-sterile syndromes in man, where FSH and LH levels are increased, these mice exhibited normal FSH and LH levels.

From experiments where male mice were irradiated in the post-meiotic stages and the sons and daughters from the irradiated stages were selected for sterility,[108] semisterility, and the production of 50% sterile sons,[81] it was concluded that the male-sterile reciprocal translocations were characterized by one breakpoint very proximal and one breakpoint very distal in the chromosomes concerned. Cacheiro *et al.*[17] came to the same conclusion when they also studied the sterile sons from postmeiotically treated (chemical mutagens, X rays) male mice. Thus, long and short "marker" translocation chromosomes are produced. Two pairing segments of the pachytene cross, one interstitial and one translocated segment, are so short that when no chiasma is formed in these (adjacent) segments, the small translocation product can be seen as a univalent in the first meiotic division (see Figures 1 and 2). Alternatively, if a chiasma is formed to bind this short chromosome, it usually is in the translocated segment, leaving the short interstitial segment containing the centric heterochromatin unpaired.

Forejt[59,61,63] has studied late diplotene–metaphase I stages in male-sterile translocations extensively and compared those with male fertiles. The most striking observation was that frequently there was an association between sex-chromosomal and autosomal material, preferentially the heterochromatic segment of the X and the heterochromatin of the short interstitial autosomal segment not occupied by a chiasma. The association is seen most frequently in late diplotene and less frequently in the stages approaching anaphase I. Forejt *et al.*[65] and de Boer *et al.*[42] extended these observations to the pachytene stages. The fraction of primary spermatocytes that have the axial elements of the proximal regions of mostly the X but also the Y chromosome close to the axial element of the unpaired interstitial segment of the translocation complex varied from 50 to 87.7%, depending on the translocation. When a univalent can already be seen at pachytene, this usually forms one mass with the sex vesicle chromatin.

Not much work has been done with regard to the reproductive capacity of mouse reciprocal-translocation homozygotes. The small numbers observed (see Table 3) do not show an effect of this condition on spermatogenesis.

It was long assumed that the fertility effects of reciprocal translocation heterozygosity are confined to the male. This assumption largely originated from the fact that meiosis in females heterozygous for a "male-sterile" reciprocal translocation had not been properly studied. One such report now available[84] shows that full homologous pairing is achieved during neither female nor male meiosis in Is40H (see also ref. 115). Studies of female reproduction of translocation heterozygotes for the two most severe male-sterile translocations, Is40H (which is an insertion) and T42H, indicate an effect on the ovary of both translocations. With Is40H/+ females,[115] as with XO females, reproductive life span seems to be reduced, indicating that there are fewer oocytes during adult life. With T42H/+,[88] the ovary weights of 3- to 5-days old carriers were lower. P. S. Burgoyne (personal communication) has found that the pattern of primary oocyte loss described for XO female mice (Section 2.2) also applies to T42H heterozygous females. By 4 days after birth, 35% have survived. Thus, the number of oogonia to enter meiosis could be less than normal in females carrying "male-sterile" reciprocal translocations. In the male, where the increase in the number of stem-cell spermatogonia continues until after birth, the effect of translocation heterozygosity on this process is not known. De Rooij[43] found that in adults the spermatogonial compartment of male-sterile translocations produced fewer primary spermatocytes. However, this result could have been produced also by tubular shrinkage due to the (partial) absence of the postmeiotic stages.

3.2. Robertsonian Translocations

Robertsonian translocations will not be treated too extensively here because there is no clear-cut evidence for their induction in the mouse by either chemical mutagens or ionizing irradiation (Searle, this volume, Chapter 3). Most of the work concerning heterozygosity for Robertsonian translocations has been carried out by Gropp and co-workers.[68,69] This section discusses their main findings.

Robertsonian translocations, like reciprocal translocations, exert their effect on fertility via two routes: (1) In heterozygous condition, they lead to the formation of aneuploid spermatozoa and oocytes. The aneuploidy concerns the acrocentric mouse chromosomes involved in the translocation. (2) Both homozygotes and heterozygotes are characterized by reduced testis weight and reduced sperm count.

Robertsonian translocations are derived principally from two sources:

1. They occur spontaneously in the wild species of *Mus musculus*, where from one to nine pairs of metacentrics can be found, depending on the geographic area.

2. Because of the introduction of single metacentric chromosomes from the wild into laboratory mouse stocks and strains, more *de novo* Robertsonian translocations are recovered in these as well.

The frequency of nondisjunction events for the acrocentric chromosomes taking part in the trivalent formed during first meiosis in a heterozygote depends on a number of factors: (1) the Robertsonian translocation concerned, (2) the sex of the carrier (the frequency for males is lower than that for females), and (3) the origin of the metacentric chromosome in the wild (identical arm combinations from different areas of Europe can, when crossed out to identical genetic backgrounds, have different nondisjunction frequencies). In addition, Robertsonian translocations that originated in laboratory-housed mice generally have low nondisjunction levels, although they are higher than for normal (acrocentric) bivalents. This effect is especially true for males, while in females the effect might be reduced. Finally, the two arms that participate within a particular Robertsonian translocation can have different nondisjunction frequencies.

When two animals heterozygous for the same Robertsonian translocation are crossed and one or both arms of the translocation are gene-tagged with recessive markers in one parent, homozygous recessive offspring occur, indicating that disomic sperm can fertilize monosomic eggs and vice versa, leading to euploid offspring. This method can be used to calculate frequencies of nondisjunction, although, due to the sex differences mentioned earlier between the levels of nondisjunction for the two acrocentric chromosomes concerned and because of the low frequencies of complementation usually seen, the method is one with wide margins.*

Much speculation has dealt with the origin of nondisjunction. Some results with gene-tagged acrocentrics suggest that the chances for nondisjunction increase when the crossover between the acrocentric and metacentric chromosomes is very close to the centromere.[21] Univalence, which might predispose to nondisjunction, generally is not seen in first meiotic metaphase. Some Robertsonian heterozygotes express crossover suppression in the vicinity of the centromere, while others do not. This might not be linked to the nondisjunction frequency

* This can be seen by solving the equation $(5p - \frac{1}{4})r^2 - 4pr + p$ for the 5% and g5% confidence limit values of p, where p = frequency of marked offspring. One of the two chromosome arms of the Robertsonian translocation has been marked and the nondisjunction values, r, of the two arms of the translocation have been assumed to be identical.[79]

TABLE 4. Some Male Mouse Pachytene Data and Male Fertility Characteristics of a Robertsonian Translocation Heterozygote of Feral Origin [Rb(11.13)4Bnr] and One of Laboratory Origin [Rb(9.19)163H]

	Rb4Bnr/+	Rb163H/+	+/+ Controls
Pachytene spermatocytes with unpaired chromosome segments[a]	26.7% (*n* = 60)	31.1% (*n* = 74)	—
Testis weight[b]	60.7%	68.9%	100.0%
Epididymal sperm count[b]	42.2%	51.2%	100.0%
Sperm morphology[b,c]	14.8%	16.8%	7.9%

[a] For both karyotypes, one male was studied. The number of cells inspected is indicated in parentheses. The unpaired segments exclusively concerned the proximal segments of the two acrocentric chromosomes that were involved in trivalent formation due to the Robertsonian translocation chromosome present. The impression was gained that pachytene spermatocytes of this type were found among all pachytene substages.

[b] Data collected by H. Korevaar as part of fulfillment of the requirement of the M.Sc. degree, Agricultural University, Wageningen, The Netherlands (1978). A portion of these data were published by Nijhoff and de Boer.[96] For methods see these authors.

[c] The percentage of spermatozoa with abnormal head shape.

observed.[22] Some research has been carried out with regard to meiotic pairing in Robertsonian translocation heterozygotes. We have compared meiotic pairing, testis weight, and sperm characteristics of the low-nondisjunction laboratory Robertsonian Rb(9.19)163H and the high-nondisjunction Rb(11.13)4Bnr, which is of *Mus poschiavinus* origin, when both were introduced into a Swiss random-bred genetic background. Table 4 gives the results, while Figure 3 presents electron photomicrographs taken after whole-mount spreading of pachytene spermatocytes on sucrose with pH 8.5 formalin fixation on grids afterward to show meiotic pairing.[90] Both trivalents sometimes exhibited nonhomologous pairing in the proximal regions of the acrocentrics. Frequently, these segments did not pair at all, and occasionally one of the two made contact with the unpaired elements of the sex vesicle. Sperm counts are reduced and the percentage of malformed spermatozoa increased in both Robertsonian heterozygotes. In Rb4Bnr, degenerated metaphase can be seen with an appreciable frequency (no attempt at quantification has been made).

Not only are reduced testis weight and diminished sperm count characteristic of Robertsonian translocation heterozygosity, homozygosity for the *poschiavinus*-type Robertsonians also has an affect on male gonad development and sperm output, as demonstrated by Cattanach and Moseley.[24] With regard to these homozygotes, another observation was that the overall percentage of male meiotic nondisjunction was increased. It is tempting to relate the increased number of germ cell deaths seen in Rb/+ karyotypes (at least when the Rb chromosome is of "wild-type" origin) with the high nondisjunction level encountered. Speed and de Boer,[120] in an autoradiographic study of Rb4Bnr/+ males, found that although the first diakinesis cells appearing after the labeling of the primary spermatocytes during the premeiotic S phase were not slower than in +/+ controls, a fraction of the primary spermatocytes was lagging behind during meiotic prophase. Nijhoff *et al.*[97] attempted to study meiotic segregation of Rb4Bnr/+ by following the first meiotic division in this karyotype after repopulation of the testis due to selective killing of spermatogonial stages during S phase by hydroxyurea. The results indicate that the first cells to reach meiosis from the reservoir of surviving A spermatogonia have somewhat lower nondisjunction rates than the subsequent spermatocytes.

3.3. Inversions

3.3.1. Paracentric Inversions

These are discovered both after treatment of premeiotic stages (irradiation of stem cell spermatogonia) and after treatment of post-meiotic stages (chemical mutagens, irradiation). Cytogenetically, they have been studied most extensively by means of whole-mount spreading of zygotene and pachytene stages.[93,102] Information regarding late diplotene–metaphase I and metaphase II chromosomes is more scanty. Also, no data on sperm counts and the quantitative histology of the stages of spermatogenesis are available. Based on what has been published or is otherwise available (M. J. Moses, personal communication), the following picture for male carriers can be reconstructed: in early pachytene, when all other bivalents have finished synapsis, mostly the proximal (and distal) segments of the inversion bivalent have not yet completed this process. Thickened axial elements for these segments show contact with the proximal segment of the X chromosome. When the inversion loop is resolved (synaptic adjustment) and unpaired chromosome segments are no longer seen, X-chromosome contact is maintained only infrequently. When the inversion spans most of the chromosome, synapsis starts within the inverted segment and the proximal and distal segments are opposite each other initially and are unsynapsed. When later pachytene stages of these inversions are inspected, nonhomologous pairing between the proximal–distal and distal–proximal segments is found to have occurred. It is not known whether in pachytene inversion bivalents of the former type (where the

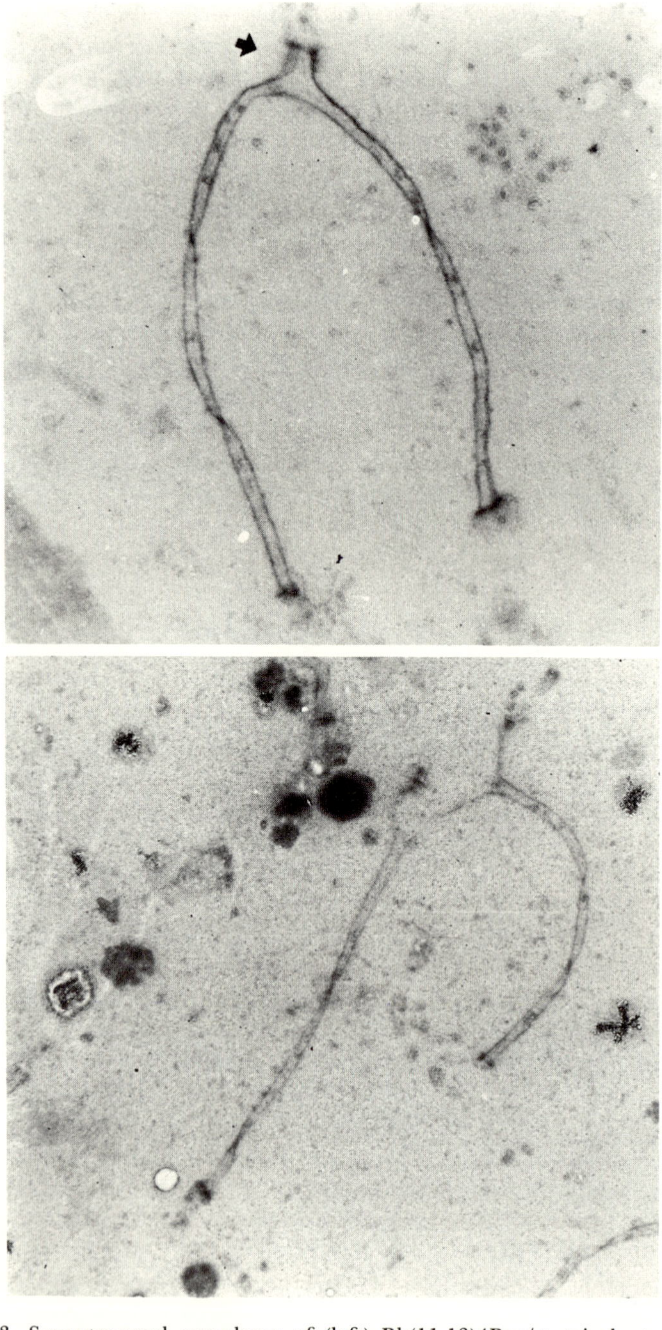

FIGURE 3. Synaptonemal complexes of (left) Rb(11.13)4Bnr/+ trivalents and (right) Rb(9.19)163H/+ trivalents from pachytene spermatocytes. Above: Meiotic pairing can be seen in the proximal regions of the two acrocentric chromosomes (arrow). Below: The

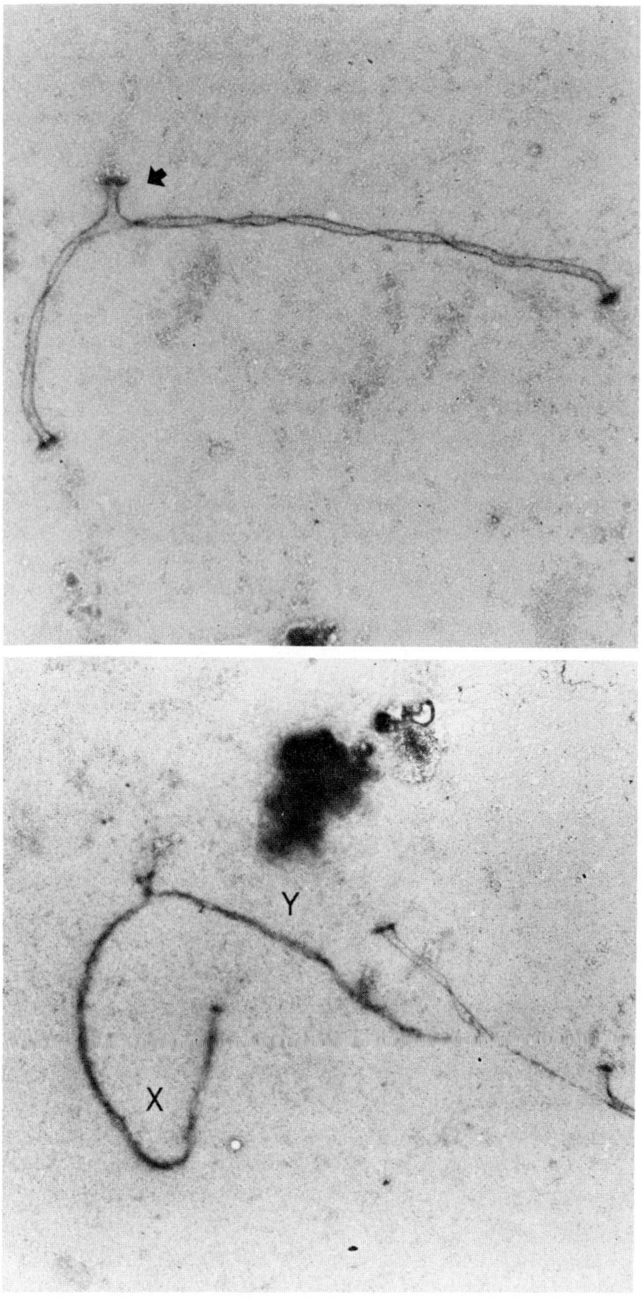

proximal regions of the acrocentric chromosomes and the centric regions of the translocation chromosome are unpaired. In the example of Rb163H, the unpaired proximal region of chromosome 9 makes contact with the Y chromosome.

proximal and distal segments are more prominent for the initiation of homologous pairing), pachytene spermatocyte death can follow from contact of autosomal material with X chromatin or whether this is a possibility in the second type (with the inversion covering most of the chromosome). Sperm counts, however, must be above the level of oligospermia (around 10% of normal[41,111]), since, according to T. H. Roderick (personal communication), no special fertility problems are encountered with male paracentric inversion heterozygotes. Evans and Ford[52] observed chiasmata within the inverted segment at late diplotene–metaphase I stages for the In(2)5Rk and In(5)9Rk paracentric inversions in the mouse. They found the resultant dicentric chromatids in "diploid" metaphase secondary spermatocytes, although, because a break can occur in any position of the chromatin stretch between the two centromeres at anaphase I, "unequal-armed" chromosomes (which have one more or less severe deletion chromatid) also have been encountered in diploid secondary spermatocytes. These cells can develop into diploid spermatozoa as in In9Rk.[72] Heterozygosity for this inversion also leads to double-headed spermatozoa.[72] Taking into account that there is selection at the uterotubal junction against spermatozoa with abnormal morphology and diploid spermatozoa, it is probable that the latter contribute little to the fertilizing population. The "haploid" secondary spermatocytes of In5Rk/+ and In9Rk/+ could also contain a dicentric chromosome or "unequal-armed" chromosomes. Ford *et al.*[58] have studied postimplantation development of In5Rk/+- and In9Rk/+- derived gametes after fusion with normal gametes. Cytological examination of meiosis had indicated that the anaphase bridges of In5Rk are less fragile than those of In9Rk. In line with this observation, although In9Rk has an anaphase bridge frequency of 74% (established after routine testis histology),[105] it produces only about 7% double-sized sperm.[72] Total pre- and postimplantation losses are considerable in In9Rk (42.5% of the number of corpora lutea) and negligible in In5Rk, an observation indicating that breaking anaphase bridges, through the production of unequal-size chromatids, result in 50% deletion-type sperm that are able to fertilize and have their effect expressed pre- or postimplantation. Thus, in the case of In9Rk, anaphase bridges lead to diploid spermatozoa and haploid balanced and unbalanced spermatozoa. The total production of spermatozoa will be high enough to guarantee high conception rates and fertilization percentages. Even when an anaphase bridge always leads to diploid spermatozoa and/or spermatocyte death, bridge frequencies must be of the order of 80% or more to make an effect on conception rate noticeable.

Little information is available concerning the reproductive behavior of female paracentric inversion heterozygotes. As in the male, the inversion loop produced in early pachytene seems to disappear through synaptic adjustment in later stages.[92] For In5Rk/+ females, Ford *et al.*[58] found more postimplantation death (25.6% in matings with +/+ males) than for the male (negligible; see above). Since the meiotic divisions and ovulation are coupled in the female, genetic embryonic death and the frequency and fate of anaphase I bridges should be directly related to each other. In male In5Rk carriers, the anaphase I bridge frequency is 19%.[105] Thus, the fertility reduction observed is in line with this figure (male and female bridge frequencies taken for convenience to be the same). A consequence of the anaphase I bridge in oocytes is seen in In(X)1H females that produce XO daughters (which received the X chromosome from their father) in about the same frequency as In/+ and +/+ daughters.[53] This points to the fact that during the first meiotic division, when an anaphase bridge is formed, it does not break, but instead the dicentric chromatid plus adjacent inversion X and normal X chromatids are expelled into the first polar body.

However, both dicentric and fragment can also be retained in the female pronucleus and give rise to interesting phenomena in the young embryo: replication and disjunction of the acentric fragment and the formation of $2n/4n$ mosaic embryos through coorientation at a cleavage division of the dicentric X chromosome.[51] The number of primary oocytes in the ovaries of In(X)1H females is reduced in a way comparable to XO females (about 65% survive; P. S. Burgoyne, personal communication).

Several inversion stocks are kept in the homozygous state. Since no quantitative fertility parameters have been published, they are taken generally to be of normal fertility. Hugenholtz and Bruce[72] reported that In9Rk homozygotes have 19.8% abnormal spermatozoa, which is higher than either controls or In9Rk/+ heterozygotes.

3.3.2. Pericentric Inversions

Because of the scarcity of mouse stocks with (sub)metacentric chromosomes in laboratories where the mouse is used to study transmission of induced genetic damage, not many pericentric inversions have been collected.[109] However, when such stocks are used, as by Crocker,[30] Davisson *et al.*,[32] and Nijhoff,[94] they can be discovered. Because of the few examples present in the mouse, no further treatment of this category of chromosome mutants is attempted here. For theoretical considerations with regard to the fertility of pericentric inversion heterozygotes, the reader is referred elsewhere.

3.4. Duplications and Deficiencies

The only tandem duplication known in the mouse, Dp(7)1R1, which comprises 24% of the length of chromosome 7 and involves the *Hbb-Mod2-c* loci,[107] has been studied during first meiotic prophase in the male by Poorman *et al.*[103] and Moses and Poorman.[91] In the first part of pachytene, a loop is formed that disappears during the later pachytene stages, and bivalent 7 can no longer be picked up by the characteristics of its lateral elements. No mention is made by these authors of any topographic relation between the duplication bivalent and the sex vesicle during meiotic prophase. Although there is growth retardation and extra preweaning loss among the carriers (60% loss up to day 10), fertility is retained. Russell and Maddux[106] mention a litter size of 5.5 for males, but no testis weights or relative sperm counts are reported. It seems that the effects on fertility of this chromosome mutant are contrary to all other types, in that the female is affected more than the male. Thus, litter size in the female is 3.5,[106] whereas sterility among females is more common with continuing inbreeding (L. B. Russell, personal communication). The physiological effects of the duplication probably are the reason for this observation rather than any "chromosome pairing" effect during meiosis. Homozygotes for this duplication are prenatal lethals.

Some of the many deficiencies for the *c* and *d-se* regions in the mouse that have been picked up in specific locus experiments designed to detect the effects of different types of ionizing irradiation and potential chemical mutagens,[109] such as c^{25H}, can be seen cytologically. No refined fertility analysis has been carried out in the female or in the male, but breeding results exclude any gross effect on this trait. Cacheiro[16] reported on a sterile male mouse carrying a deletion for chromosome 6, which was found after the treatment of postspermatogonial stages with triethylenemelamine (TEM). Thus, it seems that deficiencies can occur among the sterile descendants of mutagen-treated male mice.

Mice doubly heterozygous for small *c* deletions can be of abnormal fertility, which is taken as evidence that the two deficiencies do not entirely complement. Thus, c^{3H}/c^{6H} males exhibit increased germ cell death around the time of the meiotic divisions. Sperm motility, viability, and morphology were grossly abnormal.[76] In females showing partial complementation, the maintenance of pregnancy was disturbed and no fetus survived beyond mid-gestation. No cytogenetic meiotic analysis has been carried out on this karyotype. Because of the combination of male and female expression observed, no disturbed meiotic chromosome pairing as the origin of the affected fertility is anticipated. When small deletions do complement, no obvious effects on fertility have been recorded (L. B. Russell, personal communication).

4. Combinations of Chromosome Mutants within One Carrier

The general experience with most male carriers is that an accumulation of chromosome mutants within a single karyotype leads to a lower sperm production (up to the level of total azoospermia) and a higher fraction of genetically unbalanced spermatozoa (and secondary oocytes). There are exceptions to this rule. Examples of the above two categories will be given in the following sections.

4.1. Two Reciprocal Translocations That Have No Chromosomes in Common

Zwanenburg *et al.*(127) combined heterozygosity for two reciprocal translocations, T26H and T70H. While T26H has no effect on sperm production, T70H/+ is responsible for a reduction by 20% (see Table 3). Surprisingly, the double heterozygous males had a sperm count 60% of normal and 74% of T70H/+. Since no increase in the number of dead primary spermatocytes could be seen in these animals around the time of diakinesis metaphase I, the reduction of the output of spermatozoa is attributed to a reduction in the development of the stem cell spermatogonia.

Searle(108) performed a cytogenetic study on the sons generated after mutagenic treatment (X rays) of the father's postmeiotic stages. Of all single reciprocal translocations recovered after such treatment, 29% (n = 86) were sterile. Of the doubles, 54% (n = 13) were sterile, whereas all three triple translocation heterozygotes were sterile. In a similar experiment Cacheiro *et al.*(17) found three double translocation heterozygotes among 42 sterile sons, a fraction in line with that found by Searle (6/43).

4.2. Two Reciprocal Translocations That Have One Chromosome in Common

Rings or chains of six chromosomes that are regularly encountered at first meiotic metaphase in radiation experiments on irradiated stem cell spermatogonia will be of this type if they are transmitted to the progeny. Cacheiro *et al.*,[17] in the experiment cited in Section 4.1, found one such example. The male concerned was sterile. When generated through breeding experiments, as by Carter *et al.*,[19] the two translocations will be in repulsion. The one case mentioned by these authors, T(2;?)2/T(1;2)83, was sterile in all 14 males, but fertile (though with reduced litter size) in all 12 females tested.

Searle and Beechey[110] created artificially a number of triple translocation heterozygous males. They crossed homozygotes for two reciprocal translocations, T(10;18)18H and T(2;8)26H, with homozygotes for each of the following reciprocal translocation stocks: T(2;9)11H, T(2;4)13H, T(2;3)24H, T(7;18)50H, and T(1;13)70H. Thus, except for the cross with T(1;13)70H, meiocytes in the F_1 will be characterized by a potential hexavalent and a quadrivalent. Fertility was present in 15 of the 16 males tested, but testis weights were 25% and epididymal sperm counts 50% below normal.

4.3. Two Reciprocal Translocations That Have Two Chromosomes in Common

The earliest demonstration that male sterility can accompany this karyotype was provided by Carter *et al.*[19] for T(1;2)5/T(1;2)83. When the two breakpoints for each of the chromosomes are not too far apart, two heteromorphic bivalents will be formed, which by random orientation at first meiotic anaphase yield 50% duplication/deficiency gametes. At early pachytene, the two bivalents can be picked up by the formation of a loop. It is not yet known how big these loops must be for the generation of a quadrivalent. The best studied case is that of the combination of T(1;13)70H and T(1;13)1Wa. Chiasma formation in and segregation of the heteromorphic bivalents has been described by Wauben-Penris *et al.*[123] In the male, the meiotic analysis of this karyotype was confined to the minority of fertile carriers (though with a reduced sperm count): most male carriers are sterile. In both sexes. there are increased nondisjunction levels for both translocation bivalents and, surprisingly, the normal bivalents. For the small heteromorphic marker bivalent 1^{13}T70H;1^{13}T1Wa, the increased level of nondisjunction can be explained in terms of the frequency of univalent 1^{13} chromosomes at the first meiotic division. In contrast to T70H/T1Wa males, T(2;8)26H/T(2;8)2Wa males that also carry a small and a long heteromorphic bivalent are fertile in 75% of the cases, and occasionally sperm counts at the level of karyologically normal control mice are encountered. Inspection of the synaptonemal complexes at pachytene by electron microscope whole-mount spreading techniques suggests that meiotic pairing problems within the small heteromorphic bivalent are the prime cause for a reduction in sperm production with these karyotypes.

4.4. Combinations of Robertsonian Translocations That Have No Arms in Common

Several stocks that carry three or four metacentric chromosomes in homozygous condition have been created by Cattanach and coworkers.[23] These stocks, on a laboratory mouse genetic background, can carry metacentric chromosomes from feral origin alone or can contain a mixture of "laboratory" and feral metacentrics. The impression is that male fecundity is reduced when more Robertsonian translocations are combined into one carrier, especially if this concerns Robertsonian translocations from the "wild." In heterozygous condition, fertility effects are very apparent, due to multiple segregation disorders of the trivalents involved (both sexes, female affected more than male[124]) and to an effect on spermatogenesis in the male which also expresses itself in an increased frequency of morphologically abnormal sperm.[121]

4.5. Combinations of Two Robertsonian Translocations That Have One Chromosome in Common

The construction of these karyotypes has been very popular because of the fact that male (or female) carriers produce 25% gametes disomic for the arm present in both translocations and 25% gametes that are nullisomic for this arm. Of more concern in this discussion is that a considerable proportion (36%, $n = 42$[69]) of the combinations constructed between the various Robertsonian translocations that are of this type are male-sterile. At pachytene (electron microscope whole-mount observations) the proximal segments of the two acrocentric chromosomes participating in quadrivalent formation can usually be seen to be unpaired, and contact between these segments and X-chromosome chromatin during this stage can sometimes be seen.[68,125] In the case of Rb(16.17)7Bnr/Rb(8.17)1Iem, germ cell death follows abruptly during stage XII of the cycle of the seminiferous epithelium at the time of the meiotic divisions.[69] Forejt[62] noted that meiotic prophase was delayed in this karyotype. Primary spermatocytes that were ^{3}H-thymidine-labeled during premeiotic S phase reached late diplotene, diakinesis 18 hr later than control primary spermatocytes.

More elaborate structures combining Robertsonian translocations can be made by crossing feral mouse populations that are homozygous for different sets of Robertsonian translocations. This material and the offspring of crosses between laboratory mouse stocks containing feral mouse Robertsonian translocations have led to the insight that whenever the multivalent is a big ring in the male, the sperm production is relatively little affected (~ 33% in the case of an Rb1-9Rma/Rb10-18Rma karyotype that at first meiotic metaphase is a "super-ring" of 16 Robertsonian translocation chromosomes[69]). With chain formation at pachytene (with unpaired proximal chromosome segments at the ends) there usually is a complete breakdown of spermatogenesis.[69,125] However, in cases where a pachytene ring is formed, the many pre- and postimplantation lethals that originate from the aneuploid spermatozoa (or oocytes) that are the consequence of the segregation problems of long multivalent rings at first meiotic anaphase[69] will result in a very low litter size.

4.6. Combinations of Robertsonian and Reciprocal Translocations

No example is known to me where the effect on gamete production has been studied of the combination between a Robertsonian and reciprocal translocation where no homology exists. A few examples[6,82] are known where use of homology between one chromosome of both translocation types has made it possible to construct a double heterozygote (repulsion or coupling) that by backcrossing to the Robertsonian translocation can essentially be converted into a translocation between a metacentric and an acrocentric chromosome. This type of translocation was also discovered after irradiation of males that were homozygous for three Robertsonian translocations.[25] The most elaborate study of this kind was performed by Nijhoff[95] and a summary of the results is given in Table 5. By combining T(1;13)70H and Rb(11.13)4Bnr in the T70H, +/+, Rb4Bnr karyotype, the sperm production was lowered by more than could be expected on the basis of multiplying the relative sperm counts of T70H/+ and Rb4Bnr/+ (compared to +/+). When the Rb4Bnr segments are homozygous as in T(1;11.13S)70H, +/+, Rb4Bnr (see also Figure 2), the sperm count is somewhat higher, but not much. In both karyotypes, the frequency of abnormally shaped sperm is also higher. When segregation at first anaphase was studied (see Table 5) two facts became very apparent: "3:1" segregation of the multivalent with respect to the small 1^{13} marker chromosome is especially high in T70H, +/+, Rb4Bnr compared to T70H/+ (Figure 2). The other karyotype, homozygous for most of Rb4Bnr (but heterozygous for T70H), exhibits a reduction in adjacent-2 segregation

TABLE 5. Male Meiotic Segregation and Some Male Fertility Parameters of Combinations of Robertsonian and Reciprocal Translocations in the Mouse[a]

Meiotic segregation	Parameter for given karyotype: T(1;13)70H, +/+, Rb(11.13)4Bnr	Parameter for given karyotype: T(1;11.13S)70H, +/+, Rb(11.13)4Bnr
Adjacent-2	23.4%	12.3%
"3:1"	23.4%	10.9%
Nondisjunction in other bivalents	14.4%	23.2%
Testis weight	58.2%	65.6%
Sperm count	16.8%	27.8%
Sperm morphology (percent abnormal heads)	14.6%	14.0%

[a] All data are from Nijhoff.[95] Testis weights and epididymal sperm counts are given as percentages of +/+ controls. The control percentage of abnormal epididymal spermatozoa is 7.9%. The pachytene pairing diagram from T(1;11.13S)70H, +/+, Rb(11.13)4Bnr together with some of these data are given in Figure 2 also.

frequency. Thus, chiasmata on both sides of the 13 (and 11) centromere improve the chances for proper orientation (see also Section 3.1.2). Especially in this karyotype, but also in the double heterozygote karyotype, the nondisjunction frequency for the normal bivalents was sharply increased. This phenomenon, although expressed to a milder degree, seems to be a characteristic of mouse reciprocal translocations (see Figure 2). In "Robertsonian systems," nondisjunction is confined to the chromosomes participating in the translocation(s), whether it involves heterozygosity for a single translocation or two Robertsonian translocations with monobrachial homology. The relative fertility of the two combinations of reciprocal and Robertsonian translocations, as presented in Table 5, agreed exactly with the frequency of aneuploid and duplication/deficiency gametes that could be calculated from the second metaphase karyotypes. Thus, it seems to be the rule that chromosomally unbalanced spermatozoa function properly (see also Section 3.1.1).

A spectacular example of a combination between a Robertsonian and a reciprocal translocation where homology exists for both chromosomes has been presented by Forejt *et al.*[64] The male-sterile T(16;17)43H/+ was rendered fertile by introduction of Rb(16.17)7Bnr. Although homologous meiotic pairing between the long T43H marker chromosome and Rb(7)7Bnr must involve the formation of a loop, this probably has happened because an inhibition of meiotic pairing would probably have made this combination sterile.

4.7. The Combination of Two Paracentric Inversions within the Same Chromosome

One such example is known from the literature,[29] when In(1)1Rk was combined with In(1)12Rk. The inversion In(1)12Rk alone always has unpaired chromosome ends and X–Y contact before pairing adjustment during the second half of pachytene (M. Moses, personal communication). It is, as are most other paracentric inversion heterozygotes, fully fertile. When combined with In(1)1Rk or with In(1)24Rk (M. J. Moses, personal communication) sterility results. Chandley[29] attributed this to the fact that in whole-mount spreads, the complex chromosome 1 pairing structure and the XY bivalent were so often situated adjacent to each other. No comment was made with regard to the fertility of In1/In12 and In1/In24 females, but these probably produce oocytes.

5. Viable Unbalanced Progeny Obtained from Translocation Heterozygote Females and Males

Two categories can be distinguished here, aneuploid progeny that arise through "3:1" segregation of the translocation multivalent and 40-chromosome progeny that arise through adjacent-1 or adjacent-2 segregation of the multivalent. The "3:1" segregation products will be dealt with first. As in most mouse genetic backgrounds, primary trisomics are at best (chromosome 19) perinatal lethals, and primary monosomics usually die around implantation; the additional or missing chromosome segments must be small and/or genetically of little importance in order to be tolerated postnatally. Thus, translocations that yield aneuploid tertiary trisomic and tertiary monosomic progeny usually have their breakpoints at the ends of the chromosomes such that the small translocation product is shorter than chromosome 19; T194H, T70H, T6Ca, and T31H are examples of this class. Male meiotic pairing has been studied for T70H and T31H tertiary trisomics.[36] In both types, the small translocation chromosome was consistently found to behave like the sex chromosomes, forming one chromatin complex (an enlarged sex vesicle) during pachytene. In Ts(1^{13})70H tertiary trisomic sex vesicles, R. M. Speed (personal communication) found significant labeling of ^{3}H-uridine, indicating that some of the chromatin included (from the 1^{13}, X, and Y chromosomes) was transcriptionally active, something not seen in the sex vesicles of chromosomally normal males. The sperm counts of tertiary trisomic males are very low and sterility almost always follows. The Ts(1^{13})70H tertiary trisomics are usually fertile but oligospermic on a Swiss random-bred background, but sterile on the Harwell L III (linkage testing stock) background or after introduction of the Rb(11.13)4Bnr chromosome. Germ cells die preferentially at the first meiotic division and reduced numbers of secondary spermatocytes are seen. Usually there is some delay over the period of premeiotic S phase–diakinesis and possibly during anaphase I.[(120)] Most remarkable is the high frequency of spermatozoa with abnormal morphology.[(41)] Beechey *et al.*[(10)] compared the reproductive capacity of T(5;12)31H tertiary trisomic and monosomic descendants. Though both were sterile, sperm production in the latter (7% of normal) was higher than in the former (0.4%), a tendency reflected in the testis weights. Most spermatozoa of both types were abnormal. Apparently, a double dose of the T31H small translocation product can be tolerated in postnatal life. The one such male described was sterile despite the fact that 21 bivalents were usually formed at meiosis I and 21 chromosomes were present at meiosis II.[(9)] Female reproductive performance is impaired in tertiary trisomic and monosomic progeny and expresses itself as a reduction in litter size.[(10,81)] In Ts(1^{13})70H females, ovary weights often are reduced and the number of germinal vesicle primary oocytes that can be released by pricking the ovary is 29% of normal.[(7)] Thus, reproductive life span is expected to be shorter, although this has not yet been tested.

Females of the "male-sterile" reciprocal translocations T(10;13)199H and T(16;17)43H, through adjacent-2 segregation of the translocation multivalent at anaphase I, generate fertile euploid unbalanced males that are characterized by a small deletion and a relatively large duplication.[(48,67)] The 16, 17^{16}, 17, 17 segregation type male offspring of T43H/+ females had a sperm count of 46% from +/+ males and of 53% from T43H/T43H males.[(67)]

The recovery of aneuploid offspring from reciprocal translocation heterozygotes generally is much higher in the female carrier than in the male[(8,45)] (see also Table 2). For viable adjacent-1 and -2 segregation products, the yields probably are equal for both sexes, though in the cases reported above, they are only generated through the female, since the male carriers are sterile as a result of the translocation.

A special category yielding unbalanced viable progeny are X-autosome translocations; two such examples have been reported in detail by Searle *et al.*[(116)] Both T(X;4)37H/+ and T(X;11)38H/+ females give X^PO daughters (P = paternal) through "3:1" segregation. T(X;11)38H gives three more types of unbalanced viable progeny: $X^MX^{11}X^P$ females and $X^MX^{11}Y$ males (M = maternal) after "3:1" segregation and X^PX^{11} females after alternate/adjacent-1 segregation. Both female karyotypes can yield progeny or be sterile. One $X^MX^{11}Y$ male was sterile, but spermatogenesis persisted up to the meiotic stages. Thus, the gene(s) responsible for the deleterious effect of the second X chromosome on the spermatogonial multiplications (see Section 2.1) must be distal to the (proximally located) T38H breakpoint on the X chromosome.

6. Why is Chromosomal Infertility More Clearly Expressed in the Male Than in the Female? Consideration of X-Autosome Translocations

While X-autosome translocations and to a lesser extent Y-autosome translocations are perhaps the best known examples of translocations that cause sterility in the male, females with an X-autosome translocation have a normal oocyte production at ovulation. In a recent article concerning the X–A translocations T37H/+ and T38H/+, Searle *et al.*[(116)] give a list of all X–A translocations known with some of their characteristics. Apparently, with exceptions, germ cell death in male carriers of these translocations is at pachytene, which is earlier than in most male-sterile autosomal forms. Most X-autosome translocations involve chromosomes 4 and 7 and seem to have their breakpoints either distal or proximal to the centromere on the X chromosome. Thus, generally, one translocation chromosome is short and the other long, a situation that for autosome–autosome translocations would predispose to male sterility. Searle *et al.*[(116)] noted a parallel between the ultimate spermatogenic stage reached in X-autosome translocations and the localization of breakpoints. In agreement with autosome–autosome translocations, those with breakpoints very proximal and distal to the centromere lead to spermatocyte death earlier than when breakpoints are more intermediate as in T(X;16)16H. The importance of meiotic pairing in X-autosome rearrangements is even better illustrated in Cattanach's translocation, Is(In7;X)1Ct. Here, the unbalanced type II males can form a normal 7 bivalent in 90–95% of the diakinesis–metaphase I spermatocytes[(44)] (the chromosome 7 segment inserted in the X chromosome is present as a duplication). The epididymal sperm count in these males is 75% of normal, compared to 21% of normal in the balanced, type I males.[(87)] Disturbed X–Y meiotic pairing was not observed in the T(X;7)2R1, T(X;7)3R1, T(X;7)5R1, and T(X;7)6R1 translocations nor in Cattanach's translocation.[(2–4)] Thus, spermatogenesis breakdown in X-autosome translocations cannot be explained by the absence of pairing in the X–Y pairing region.

When the X-linked enzymes PGK-1, G6PD, and HGPRT were measured in late pachytene spermatocytes in control mice and in T(X;4)37H male-sterile mice, very low readings were obtained for +/+ mice but very high ones for T37H/+ mice.[70] Thus, it would seem that autosomal segments that are actively synthesizing hnRNA during male pachytene reactivate adjacently situated, senescent sex-chromosome DNA such that transcription takes place. In line with this result, sterility of c^{3H}/c^{6H} heterozygotes (see Section 3.4) can be corrected when the X chromosome carries an insertion for a chromosome 7 segment with the normal allele of the *c* locus as in type II males of Cattanach's translocation.[50] The generally unpaired status of this segment in these males (a conclusion that can be drawn from the pachytene pairing studies in males that are balanced for Cattanach's translocation[2]) apparently does not interfere with normal transcription.

X-autosome translocations are the strongest argument in favor of the X-reactivation male-sterility hypothesis as formulated by Lifschytz for *Drosophila*.[77] Forejt[61] has adopted this hypothesis for autosomal male-sterile chromosome mutations in the mouse, despite the fact that he often observed positive heteropycnosis for the unpaired autosomal (centromere plus heterochromatin-containing) segments that were in contact with the centric region of the X chromosome. This suggests that the inactivation is spreading from the sex chromosomes into the autosomal segment. Nevertheless, for the male-sterile autosomal chromosome mutations In(1)Rk/In(12)Rk, T31H/+, and Ts(5^{12})31H, elevated levels of PGK-1, G6PD, and HGPRT have been found as well.[70]

It is doubtful whether all chromosomal male sterility can be attributed to interference with sex-chromosomal activity during pachytene. The following arguments stress the importance of meiotic pairing *per se*.

1. The most stringent autosomal male-sterile chromosome abnormality, the insertion Is(7;1)40H,[115] is characterized by an abrupt cessation of spermatogenesis during the first half of pachytene at stage IV. It shows "contact" of unpaired autosomal material with the sex chromosomes in 40% of the pachytene cells during a stage where, for the inversion heterozygotes described in Section 3.3.1 and below, it is probably not lethal. Unpaired chromosome segments, however, are always present.

2. Unpaired chromosome segments plus concomitant "neighborship" of their axial elements and the axial elements of the X chromosome during the first half of pachytene is more the rule than the exception

in paracentric inversion heterozygotes where the inversion either spans most of the chromosome or is located at one end of the chromosome (see Section 3.3.1). Yet (non)homologous synaptic adjustment follows, sex vesicle contact stops, and spermatogenesis most likely continues without great cellular loss.

3. For the sex chromosomes themselves, pairing seems to be a prerequisite for meiotic progress in normal numbers (see Section 2.1).

4. Evidence is accumulating that females with chromosome abnormalities, such as XO, In(X)1H/+, T42H/+, and Ts(1^{13})70H, have reduced numbers of oocytes in the adult ovary. Late pachytene–early diplotene is the most likely meiotic stage for this reduction to occur (P. S. Burgoyne, personal communication). In all these mutants meiotic pairing problems can be inferred.

5. One more argument in favor of meiotic pairing as essential for meiotic progress is the high level of single-strand DNA nicking by endonucleases found in male-sterile chromosome mutants.[71] Previously the same group[71] had found that in lily meiocytes, effective meiotic pairing regulates the susceptibility of chromosomal DNA to nicking.

Thus, it seems that besides unwanted transcription for the sex chromosomes during the later stages of pachytene, unpaired chromosome segments that cannot be "neutralized" by pairing adjustment and/or nonhomologous pairing also lead to meiotic breakdown in the male. Often in male-sterile or oligospermic mutants, these two conditions cannot be separated within the same mutant.

ACKNOWLEDGMENTS

The typing of the manuscript by T. A. Makkes, E. van Liempt, and A. Sieswerda is gratefully acknowledged. Without the assistance of Prof. M. J. Moses, Dr. P. S. Burgoyne, Dr. A. G. Searle, Dr. L. B. Russell, and Dr. H. Winking, the sources covered by this chapter would have been less complete. I am grateful for their help and advice.

7. References

1. R. J. Andina, A study of X chromosome regulation during oogenesis in the mouse, *Exp. Cell Res. 111*, 211–218 (1978).
2. T. Ashley, Application of the spreading technique to structural heterozygotes, in: *Chromosomes Today* (M. D. Bennett, A. Gropp, and U. Wolf eds.), Vol. 8, pp. 80–89, Allen and Unwin, London (1984).

3. T. Ashley, L. B. Russell, and N. L. A. Cacheiro, Synaptonemal complex analysis of X-7 translocations in male mice I. R3 and R5 quadrivalents, *Chromosoma* (*Berl.*)*87*, 149–164 (1982).
4. T. Ashley, L. B. Russell, and N. L. A. Cacheiro, Synaptonemal complex analysis of X-7 translocations in male mice: R2 and R6 quadrivalents, *Chromosoma* (*Berl.*) *88*, 171–177 (1983).
5. C. V. Beechey, X–Y chromosome dissociation and sterility in the mouse, *Cytogenet. Cell Genet. 12*, 60–67 (1973).
6. C. V. Beechey, Private communication, *Mouse News Letter 63*, 14 (1980).
7. C. V. Beechey and P. de Boer, Private communication, *Mouse News Letter 54*, 53 (1976).
8. C. V. Beechey and A. G. Searle, Private communication, *Mouse News Letter 62*, 50 (1980).
9. C. V. Beechey and R. M. Speed, Private communication, *Mouse News Letter 64*, 56 (1981).
10. C. V. Beechey, M. Kirk, and A. G. Searle, A reciprocal translocation induced in an oocyte and affecting fertility in male mice, *Cytogenet. Cell Genet. 27*, 129–146 (1980).
11. P. S. Burgoyne and T. G. Baker, Oocyte depletion in XO mice and their XX sibs from 12 to 200 days *post partum, J. Reprod. Fertil. 61*, 207–212 (1981).
12. P. S. Burgoyne and T. G. Baker, The XO ovary—development and function, in: *Development and Function of Reproductive Organs* (Proceedings of the Vth Workshop on the Development and Function of the Reproductive Organs) (A. G. Byskov and H. Peters, eds.), pp. 122–128, Excerpta Medica, Amsterdam (1981).
13. P. S. Burgoyne and F. G. Biddle, Spermatocyte loss in XYY mice: Addendum to *Cytogenet. Cell Genet. 23*, 84–89, *Cyotogenet. Cell Genet. 28*, 143–144 (1980).
14. P. S. Burgoyne and J. D. Biggers, The consequences of X-dosage deficiency in the germ line: Impaired development *in vitro* of preimplantation embryos from XO mice, *Dev. Biol. 51*, 109–117 (1976).
15. P. S. Burgoyne, P. P. L. Tam, and E. P. Evans, Retarded development of XO conceptuses during early pregnancy in the mouse, *J. Reprod. Fertil. 68*, 387–393 (1983).
16. N. L. A. Cacheiro, Private communication, *Mouse News Letter 56*, 55 (1977).
17. N. L. A. Cacheiro, L. B. Russell, and M. S. Swartout, Translocations, the predominant cause of total sterility in sons of mice treated with mutagens, *Genetics 76*, 73–91 (1974).
18. T. C. Carter, M. F. Lyon, and R. J. S. Phillips, Gene-tagged chromosome translocations in eleven stocks of mice, *J. Genet. 53*, 154–166 (1955).
19. T. C. Carter, M. F. Lyon, and R. J. S. Phillips, Further genetic studies of eleven translocations in the mouse, *J. Genet. 54*, 462–473 (1956).
20. B. M. Cattanach, Sex reversal in the mouse and other mammals, in; *The Early Development of Mammals* (M. Balls and A. E. Wild, eds.), pp. 305–317, Cambridge University Press (1975).
21. B. M. Cattanach, I. Murray, and J. M. Tracey, Private communication, *Mouse News Letter 54*, 37–38 (1976).
22. B. M. Cattanach, Crossover suppression in mice heterozygous for tobacco mouse metacentrics, *Cytogenet. Cell Genet. 20*, 264–281 (1978).
23. B. M. Cattanach and J. Barlow, Private communication, *Mouse News Letter 67*, 18 (1982).
24. B. M. Cattanach and H. Moseley, Non-disjunction and reduced fertility caused by the tobacco mouse metacentric chromosomes, *Cytogenet. Cell Genet. 12*, 264–287 (1973).

25. B. M. Cattanach, I. Murray, and T. R. L. Bigger, Private communication, *Mouse News Letter 56*, 37–38 (1977).
26. B. M. Cattanach, A. J. M. Crocker, and C. Jones, Private communication, *Mouse News Letter, 63*, 15 (1980).
27. A. C. Chandley, The chromosomal basis of human infertility, *Br. Med. Bull. 35*, 181–186 (1979).
28. A. C. Chandley, The origin of chromosomal aberrations in man and their potential for survival and reproduction in the adult human population, *Ann. Genet. 24*, 5–11 (1981).
29. A. C. Chandley, A pachytene analysis of two male-fertile paracentric inversions in chromosome 1 of the mouse and in the male-sterile double heterozygote, *Chromosoma (Berl.) 85*, 127–135 (1982).
30. A. J. M. Crocker, Mutation Studies in Mice Carrying Robertsonian Translocation, Ph.D. Thesis, Council for National Academic Awards (1982).
31. A. J. M. Crocker and B. M. Cattanach, X-Ray induction of translocations in mice carrying metacentrics (Robertsonian fusions); detection of whole arm chromosome exchanges, *Mutat. Res. 91*, 353–357 (1981).
32. M. T. Davisson, P. A. Poorman, T. H. Roderick, and M. J. Moser, A pericentric inversion in the mouse. *Cytogenet. Cell Genet. 30*, 70–76 (1981).
33. P. de Boer, Male meiotic behaviour and male and female litter size in mice with the T(2;8)26H and T(1;13)70H reciprocal translocations, *Genet. Res.(Camb.) 27*, 369–387 (1976).
34. P. de Boer, Proximal chiasma localization within an interstitial chromosome segment, a likely correlate of adjacent 2 segregation of translocation caused multivalents in the mouse, *Environ. Health Perspect. 31*, 137–140 (1979).
35. P. de Boer and M. E. A. B. van Beek, Meiosis of T70H translocation trisomic male mice. I. Meiotic configuration and segregation, *Chromosoma (Berl.) 87*, 303–313 (1982).
36. P. de Boer and H. E. B. Branje, Association of the extra chromosome of tertiary trisomic male mice with the sex chromosomes during first meiotic prophase, and its significance for impairment of spermatogenesis, *Chromosoma (Berl.) 73*, 369–379 (1979).
37. P. de Boer and P. H. M. D. de Maar, A histological study of embryonic death caused by heterozygosity for the T26H reciprocal mouse translocation, *J. Embryol. Exp. Morphol. 35*, 595–606 (1976).
38. P. de Boer and J. H. Nijhoff, Incomplete sex chromosome pairing in oligospermic male hybrids of *Mus musculus* and *M. musculus molossinus* in relation to the source of the Y chromosome and the presence or absence of a reciprocal translocation, *J. Reprod. Fertil. 62*, 235–243 (1981).
39. P. de Boer and A. G. Searle, Summary and synthesis, Workshop on chromosomal aspects of male sterility in mammals, *J. Reprod. Fertil. 60*, 257–265 (1980).
40. P. de Boer and R. M. Speed, Meiosis of T70H translocation trisomic male mice. II. Meiotic rate, spermatocyte interactions and fertility, *Chromosoma (Berl.) 87*, 315–325 (1982).
41. P. de Boer, F. A. van der Hoeven, and J. A. P. Chardon, The production, morphology, karyotypes and transport of spermatozoa from tertiary trisomic mice and the consequences for egg fertilization, *J. Reprod. Fertil. 48*, 249–256 (1976).
42. P. de Boer, A. G. Searle, F. A. van der Hoeven, and C. V. Beechey, Spermatocyte meiotic pairing of the T(5;12)31H, T(6;12)32H, T(11;19)42H male-sterile and the T(1;13)70H male-fertile reciprocal translocations in the mouse using silver staining of synaptonemal complexes, *Cytogenet. Cell Genet. 27*, 202–203 (1980).

43. D. G. de Rooij, Spermatogenesis in mice heterozygous for a male-sterile translocation (T31H, T32H, T40H or T42H), *Cytogenet. Cell Genet. 27*, 210–211 (1981).
44. E. M. Eicher, X-Autosome translocations in the mouse: Total inactivation versus partial inactivation of the X chromosome, *Adv. Genet. 15*, 175–259 (1970).
45. E. M. Eicher, Translocation trisomic mice: Production by female but not male translocation carriers, *Science 180*, 81 (1973).
46. E. M. Eicher, Primary sex determining genes, in: *Prospects for Sexing Mammalian Sperm* (R. P. Amann and G. E. Seidel, eds.), pp. 121–135, Colorado Associated University Press, Boulder, Colorado (1982).
47. E. M. Eicher and M. C. Green, The T6 translocation in the mouse: Its use in trisomy mapping, centromere localization, and cytological identification of linkage group III, *Genetics 71*, 621–632 (1972).
48. E. M. Eicher and L. L. Washburn, Chromosomally unbalanced, viable, fertile mice, in: *47th Annual Report*, Jackson Laboratory, Bar Harbor, Maine, p. 34 (1976).
49. E. M. Eicher, S. J. Phillips, and L. L. Washburn, The use of molecular probes and chromosomal rearrangements to partition the mouse Y into functional regions, in: *Recombinant DNA and Medical Genetics* (Birth Defects Symposium XIII) (A. Messer and I. H. Porter, eds.), pp. 57–71, Academic Press, New York (1983).
50. R. P. Erickson, Cattanach's translocation (Is(7;X)Ct) corrects male sterility due to homozygosity for chromosome 7 deletions, *Genet. Res.* (*Camb.*) *43*, 35–41 (1984).
51. E. P. Evans and P. S. Burgoyne, The chromosome unbalance observed in early preimplantation embryos from mice heterozygous for the paracentric inversion of the X chromosome, In (X)1H, in: *Chromosomes Today* (M. D. Bennett, A. Gropp, and U. Wolf, eds.), Vol. 8, p. 305 (abstract), Allen and Unwin, London (1984).
52. E. P. Evans and C. E. Ford, Some cytological properties of paracentric inversions in the mouse, in: *Current Chromosome Research* (K. Jones and P. E. Brandham, eds.), pp. 216–217, Elsevier/North-Holland Biomedical Press, Amsterdam (1976).
53. E. P. Evans and R. J. S. Phillips, Inversion heterozygosity and the origin of XO daughters of *Bpa*/+ female mice, *Nature 256*, 40–41 (1975).
54. E. P. Evans, C. V. Beechey, and M. D. Burtenshaw, Meiosis and fertility in XYY mice, *Cytogenet. Cell Genet. 20*, 249–263 (1978).
55. E. P. Evans, M. D. Burtenshaw, and B. B. Brown, Meiosis in *Sxr* male mice II. Further absence of cytological evidence for a Y-autosome rearrangement in sex-reversed (*Sxr*) mice, *Chromosoma* (*Berl.*) *81*, 19–26 (1980).
56. E. P. Evans, M. D. Burtenshaw, and B. M. Cattanach, Meiotic crossing-over between the X and Y chromosomes of male mice carrying the sex-reversing (*Sxr*) factor, *Nature 300*, 443–445 (1982).
57. N. Feldman, M. André, J. F. Jarrige, and D. Boucher, Pituitary–gonadal axis in sterile male mice heterozygous for autosomal reciprocal translocation T145H, *J. Reprod. Fertil. 66*, 723–728 (1982).
58. C. E. Ford, E. P. Evans, and M. D. Burtenshaw, Meiosis and fertility in mice heterozygous for paracentric interversions, in: *Current Chromosome Research* (K. Jones and P. E. Brandham, eds.), pp. 123–132, Elsevier/North-Holland Biomedical Press, Amsterdam (1976).
59. J. Forejt, Non-random association between a specific autosome and the X chromosome in meiosis of the male mouse: Possible consequence of homologous centromeres' separation, *Cytogenet. Cell Genet. 13*, 369–383 (1974).
60. J. Forejt, Spermatogenic failure of translocation heterozygotes affected by H-2 linked gene in mouse, *Nature 260*, 143–145 (1976).
61. J. Forejt, Meiotic studies of translocations causing male sterility in the mouse. II.

Double heterozygotes for Robertsonian translocations, *Cytogenet. Cell Genet. 23*, 163–170 (1979).
62. J. Forejt, X–Y involvement in male sterility caused by autosome translocations—a hypothesis, in: *Genetic Control of Gamete Production and Function* (Serono Clinical Colloquia on Reproduction 3) (P. G. Crosignani, B. L. Rubin, and M. Fraccaro, eds.), pp. 135–151, Academic Press, London, and Grune and Stratton, New York (1982).
63. J. Forejt and S. Gregorová, Meiotic studies of translocations causing male sterility in the mouse I. Autosomal reciprocal translocations, *Cytogenet. Cell Genet. 19*, 159–179 (1977).
64. J. Forejt, J. Capková, and S. Gregorová, T(16;17)43H translocation as a tool in analysis of the proximal part of chromosome 17 (including T–t gene complex) of the mouse, *Genet. Res.* (*Camb.*) *35*, 165–177 (1980).
65. J. Forejt, S. Gregorová, and P. Goetz, XY pair associates with the synaptonemal complex of autosomal male-sterile translocations in pachytene spermatocytes of the mouse (*Mus musculus*), *Chromosoma* (*Berl.*) *82*, 41–53 (1981).
66. B. B. Gollapudi, O. P. Kamra, and S. R. Blecher, A search for a genetic basis for gonosomal univalency in mice, *Cytogenet. Cell Genet. 29*, 241–249 (1981).
67. S. Gregorová, V. S. Baranov, and J. Forejt, Partial trisomy (including T–t gene complex) of the chromosome 17 of the mouse. The effect on male fertility and the transmission to progeny, *Folia Biol.* (*Prague.*) *27*, 171–177 (1981).
68. A. Gropp and H. Winking, Robertsonian translocations: cytology, meiosis, segregation patterns and biological consequences of heterozygosity, in: *Biology of the House Mouse* (R. J. Berry, ed.), pp. 141–181, Academic Press, London, (1981).
69. A. Gropp, H. Winking, and C. Redi, Consequences of Robertsonian heterozygosity: segregation impairment of fertility versus male-limited sterility, in: *Genetic Control of Gamete Production and Function* (Serono Clinical Colloquia on Reproduction 3) (P. G. Crosignani, B. L. Rubin, and M. Fraccaro, eds.), pp. 115–134, Academic Press, London, and Grune and Stratton, New York (1982).
70. Y. Hotta and A. C. Chandley, Activities of X-linked enzymes in spermatocytes of mice rendered sterile by chromosomal alterations, *Gamete Res. 6*, 65–72 (1982).
71. Y. Hotta, A. C. Chandley, H. Stern, A. G. Searle, and C. V. Beechey, A disruption of pachytene DNA metabolism in male mice with chromosomally-derived sterility, *Chromosoma* (*Berl.*) *73*, 287–300 (1979).
72. A. D. Hugenholtz and W. R. Bruce, Sperm size abnormalities in homozygous and heterozygous In(5)9Rk mice, *Can. J. Genet. Cytol. 21*, 115–119 (1979).
73. H. T. Imai, Y. Matsuda, T. Shiroishi, and K. Moriwaki, High frequency of X–Y chromosome dissociation in primary spermatocytes of F_1 hybrids between Japanese wild mice (*Mus musculus molossinus*) and inbred laboratory mice, *Cytogenet. Cell Genet. 29*, 166–175 (1981).
74. P. Jalbert, B. Sele, and H. Jalbert, Reciprocal translocations: A way to predict the mode of imbalanced segregation by pachytene-diagram drawing. A study of 151 human translocations, *Hum. Genet. 55*, 209–222 (1980).
75. M. H. Kaufman, Non-random segregation during mammalian oogenesis, *Nature 238*, 465–466 (1972).
76. S. E. Lewis, H. A. Turchin, and T. E. Wojtowicz, Fertility studies of complementing genotypes at the albino locus of the mouse, *J. Reprod. Fertil. 53*, 197–202 (1978).
77. E. Lifschytz, X-chromosome inactivation: an essential feature of normal spermiogenesis in male heterogametic organisms, in: *The Genetics of the Spermatozoon* (R. A. Beatty and S. Gluecksohn-Waelsch, eds.), pp. 223–232, Beatty and Gluecksohn-Waelsch, Edinburgh (1972).

78. M. F. Lyon, Sex chromosome activity in germ cells, in: *Physiology and Genetics of Reproduction* (E. M. Coutinho and F. Fuchs, eds.), Part A, pp. 63–71, Plenum Press, New York (1974).
79. M. F. Lyon, The use of Robertsonian translocations for studies of nondisjunction, in: *Radiation-Induced Chromosome Damage in Man* (T. Ishihara and M. S. Sasaki eds.), pp. 327–346, Liss, New York (1983).
80. M. F. Lyon and S. G. Hawker, Reproductive life span in irradiated and unirradiated chromosomally XO mice, *Genet. Res.* (*Camb.*) *21*, 185–194 (1979).
81. M. F. Lyon and R. Meredith, Autosomal translocation causing male sterility and viable aneuploidy in the mouse, *Cytogenetics 5*, 335–354 (1966).
82. M. F. Lyon and I. M. Sayers, Private communication, *Mouse News Letter 58*, 44 (1978).
83. T. Magnuson and C. J. Epstein, Genetic control of very early mammalian development, *Biol. Rev. 56*, 369–408 (1981).
84. S. Mahadevaiah, U. Mittwoch, and M. J. Moses, Pachytene chromosomes in male and female mice heterozygous for the Is(7;1)40H insertion, *Chromosoma* (*Berl.*) *90*, 163–169 (1984).
85. Y. Matsuda, H. T. Imai, K. Moriwaki, K. Kondo, and F. Bonhomme, X–Y chromosome dissociation in wild derived *Mus musculus* subspecies, laboratory mice, and their F_1 hybrids, *Cytogenet. Cell Genet. 34*, 241–252 (1982).
86. Y. Matsuda, H. T. Imai, K. Moriwaki, and K. Kondo, Modes of inheritance of X–Y dissociation in inter-subspecies hybrids between BALB/c mice and *Mus musculus molossinus, Cytogenet. Cell Genet. 35*, 209–215 (1983).
87. M. L. Meistrich, W. Göhde, R. A. White, and J. Longtin, "Cytogenetic" studies of spermatids of mice carrying Cattanach's translocation by flow cytometry, *Chromosoma* (*Berl.*) *74*, 141–151 (1979).
88. U. Mittwoch, S. Mahadevaiah, and M. B. Olive, Retardation of ovarian growth in male-sterile mice carrying an autosomal translocation, *J. Med. Genet. 18*, 414–417 (1981).
89. T. Morris, The XO and OY constitution in the mouse, *Genet. Res.* (*Camb.*) *12*, 125–137 (1968).
90. M. J. Moses, Synaptonemal complex karyotyping in spermatocytes of the Chinese hamster (*Cricetulus griseus*) I. Morphology of the autosomal complement in spread preparations, *Chromosoma* (*Berl.*) *60*, 99–125 (1977).
91. M. J. Moses and P. A. Poorman, Synaptonemal complex analysis of mouse chromosomal rearrangements. II. Synaptic adjustment in a tandem duplication, *Chromsoma* (*Berl.*) *81*, 519–535 (1981).
92. M. J. Moses and P. A. Poorman, Synapsis, synaptic adjustment and DNA synthesis in mouse oocytes, in: *Chromosomes Today* (M. D. Bennett, A. Gropp, and U. Wolf, eds.), Vol. 8, pp. 90–103, Allen and Unwin, London (1984).
93. M. J. Moses, P. A. Poorman, T. H. Roderick, and M. T. Davisson, Synaptonemal complex analysis of mouse chromosomal rearrangements. IV. Synapsis and synaptic adjustment in two paracentric inversions, *Chromosoma* (*Berl.*) *84*, 457–474 (1982).
94. J. H. Nijhoff, Private communication, *Mouse News Letter 60*, 77 (1979).
95. J. H. Nijhoff, Meiotic Behaviour and Spermatogenesis in Male Mice Heterozygous for Translocation Types Also Occurring in Man, Ph.D. Thesis, Agricultural University, Wageningen, The Netherlands (1981).
96. J. H. Nijhoff and P. de Boer, A first exploration of a Robertsonian translocation heterozygote in the mouse for its usefulness in cytological evaluation of radiation-induced meiotic autosomal non-disjunction, *Mutat. Res. 61*, 77–86 (1979).
97. J. H. Nijhoff, P. de Boer, and M. C. A. Wessels-Kaalen, Private communication, *Mouse News Letter 71*, 52 (1984).

98. M. Oshimura and A. A. Sandberg, Sexual differences in meiotic disjunction of murine T(1;13)70H translocation heterozygotes, *Jpn. J. Genet. 53*, 215–218 (1978).
99. M. Oshimura and N. Takagi, Meiotic disjunction in T(14;15)6Ca heterozygotes and fate of chromosomally unbalanced gametes in embryonic development, *Cytogenet. Cell Genet. 15*, 1–16 (1975).
100. M. Oshimura, S. Sonta, and A. A. Sandberg, Cytogenetic evidence for and frequency of adjacent 2 disjunction in murine T(1;13)70H male heterozygotes, *Proc. Jpn. Acad. 52*, 199–202 (1976).
101. M. Oshimura, A. Tonomura, and N. Tanaka, Degenerative metaphase I's and meiotic configurations in male translocation carriers of mice, *Cytogenet. Cell Genet. 27*, 209 (1981).
102. P. A. Poorman, M. J. Moses, M. T. Davisson, and T. H. Roderick, Synaptonemal complex analysis of mouse chromosomal rearrangements. III. Cytogenetic observations on two paracentric inversions, *Chromosoma (Berl.) 83*, 419–429 (1981).
103. P. A. Poorman, M. J. Moses, L. B. Russell, and N. L. A. Cacheiro, Synaptonemal complex analysis of mouse chromosomal rearrangements. I. Cytogenetic observations on a tandem duplication, *Chromosoma (Berl.) 81*, 507–518 (1981).
104. M. Rapp, E. Therman, and C. Denniston, Non-pairing of the X and Y chromosomes in the spermatocytes of BDF_1 mice, *Cytogenet. Cell Genet. 19*, 85–93 (1977).
105. T. H. Roderick and N. L. Hawes, Nineteen paracentric chromosomal inversions in mice, *Genetics 76*, 109–117 (1974).
106. L. B. Russell, and S. Maddux, Private communication, *Mouse News Letter 56*, 54 (1977).
107. L. B. Russell, W. L. Russell, N. L. A. Cacheiro, C. M. Vaughan, R. A. Popp, and K. B. Jacobson, A tandem duplication in the mouse, *Genetics 80*, s71 (1975).
108. A. G. Searle, Nature and consequences of induced chromosome damage in mammals, *Genetic 78*, 173–186 (1974).
109. A. G. Searle, Chromosome variants, in: *Genetic Variants and Strains of the Laboratory Mouse* (M. C. Green, ed.), pp. 324–347, Gustav Fisher, Stuttgart (1981).
110. A. G. Searle and C. V. Beechey, Private communication, *Mouse News Letter 49*, 29 (1973).
111. A. G. Searle and C. V. Beechey, Sperm-count, egg fertilization and dominant lethality after X-irradiation of mice, *Mutat. Res. 22*, 63–72 (1974).
112. A. G. Searle and C. V. Beechey, Complementation studies with mouse translocations, *Cytogenet. Cell Genet. 20*, 282–303 (1978).
113. A. G. Searle, C. E. Ford, and C. V. Beechey, Meiotic disjunction in mouse translocations and the determination of centromere position, *Genet. Res. (Camb.) 18*, 215–235 (1971).
114. A. G. Searle, C. V. Beechey, and E. P. Evans, Meiotic effects in chromosomally derived male sterility of mice, *Ann. Biol. Anim. Biochim. Biophys. 18*, 391–398 (1978).
115. A. G. Searle, C. V. Beechey, P. de Boer, D. G. de Rooij, E. P. Evans, and M. Kirk, A male-sterile insertion in the mouse, *Cytogenet. Cell Genet. 36*, 617–626 (1983).
116. A. G. Searle, C. V. Beechey, E. P. Evans, and M. Kirk, Two new X-autosome translocations in the mouse, *Cytogenet. Cell. Genet. 35*, 279–292 (1983).
117. R. V. Short, Germ cell sex, in: *The Genetics of the Spermatozoon* (R. A. Beatty and S. Gluecksohn-Waelsch eds.), pp. 325–345, Beatty and Gluecksohn-Waelsch, Edinburgh (1972).
118. L. Singh and K. W. Jones, Sex reversal in the mouse (*Mus musculus*) is caused by a recurrent non-reciprocal crossover involving the X and an aberrant Y chromosome, *Cell 28*, 205–216 (1982).

119. G. D. Snell, An analysis of translocations in the mouse, *Genetics 31*, 157–180, (1946).
120. R. M. Speed and P. de Boer, Delayed meiotic development and correlated death of spermatocytes in male mice with chromosome abnormalities, *Cytogenet. Cell Genet. 35*, 257–262 (1983).
121. R. Stolla and A. Gropp, Variation of the DNA content of morphologically normal and abnormal spermatozoa in mice susceptible to irregular meiotic segregation. *J. Reprod. Fertil. 38*, 335–346 (1974).
122. P. J. J. Wauben-Penris and J.-B. Prins, Meiotic behaviour of alloxan-treated diabetic and nondiabetic T(1;13)70H/+ mice, *Hum. Genet. 63*, 268–273 (1983).
123. P. J. J. Wauben-Penris, F. A. van der Hoeven, and P. de Boer, Chiasma frequency and non-disjunction in heteromorphic bivalents: Meiotic behaviour in T(1;13)70H/T(1;13)1Wa mice as compared to T(1;13)70H/T(1;13)70H mice, *Cytogenet. Cell Genet. 36*, 547–553 (1983).
124. H. Winking and A. Gropp, Meiotic non-disjunction of metacentric heterozygotes in oocytes versus spermatocytes in: *Proceedings of the Serono Symposia* (P. G. Crosignani and D. R. Mishall eds.), Vol. 8, pp. 47–56, Academic Press, New York (1976).
125. H. Winking and R. Johanisson, Pattern of pachytene pairing in mouse hybrids with chain and ring multivalents, *Clin. Genet. 17*, 94 (1980).
126. A. J. Wyrobek, J. A. Heddle, and W. R. Bruce, Chromosomal abnormalities and the morphology of mouse sperma heads, *Can J. Genet. Cytol. 17*, 675–681 (1975).
127. T. S. B. Zwanenburg, P. de Boer and P. Stam, Clonal analysis of radiation induced translocations in stem-cell spermatogonia of normal and T70H translocation heterozygous mice, *Mutat. Res. 83*, 207–219 (1981).

CHAPTER 12

Mutagenesis and Plasmids

Kristien E. Mortelmans and Linda Dousman

1. Introduction

There are many naturally occurring plasmids that confer upon their host increased susceptibility to ultraviolet (UV) and/or chemical mutagenesis. This enhanced mutagenesis has been associated with an increased amount of cellular error-prone DNA repair. Other plasmid properties believed to involve cellular error-prone repair are ability to increase UV and chemical survival and ability to enhance spontaneous mutagenesis. Some plasmids have been shown to exhibit only a few of these effects, while other plasmids possess all of these properties, given the appropriate genetic DNA repair background of the host cell.

The first plasmid shown to enhance the frequency of UV-induced mutations in *Salmonella typhimurium* strain LT2 was the UV-protective colicin plasmid ColI.[25,26] The first resistance transfer factor (R plasmid) shown to enhance chemical mutagenesis in *S. typhimurium* LT2 was the UV-protective plasmid R205. This was reported by MacPhee in 1973,[41] whose findings prompted Ames and colleagues to analyze systematically a number of R plasmids for use in the Ames *Salmonella*/mammalian microsome assay. Because of its responsiveness to chemical mutagenesis, plasmid pKM101, a deletion derivative of plasmid R46, was introduced in several *Salmonella* strains, and has played a major role in the sensitivity

Kristien E. Mortelmans and Linda Dousman • Microbial Genetics Department, SRI International, Menlo Park, California 94025.

of the Ames assay.[3,46] Recently, the pKM101 *muc* genes (mutagenesis-enhancing genes) have been cloned in *Escherichia coli,*[59] and have been successfully introduced into a *Saccharomyces cerevisiae* strain with three auxotrophic markers. The presence of the *muc* genes in this strain has been shown to enhance the reversion frequency of all three markers.[62]

In this chapter, we first give a short overview of bacterial plasmids. We then discuss our research efforts with 33 R plasmids in an attempt to find a plasmid(s) superior to pKM101 in enhancing chemical mutagenesis. Finally, we discuss the mechanism(s) of the plasmid-mediated effects by reviewing the literature. No attempt is made to review bacterial DNA repair processes, since there are a number of excellent articles on this subject.[22,28,63,94]

2. Historical Overview of Plasmids

Plasmids are nonessential extrachromosomal genetic elements that may be added to or lost from bacteria without lethal effects. They are circular double-stranded DNA molecules, which can exist in three forms: (1) covalently closed, circular duplex (supercoiled), (2) open circular duplex (one nick in one strand), (3) linear duplex. Under certain conditions, plasmids may integrate into the host chromosome. Some, but not all, plasmids mediate their own transfer by conjugation. Plasmids range in size from 1.5 to over 200 million daltons, and the number of plasmid copies per bacterium varies from one to 30.

2.1. Colicin Plasmids

Colicin plasmids, carried by some strains of the Enterobacteriaceae, code for a range of bactericidal agents (colicins), and are probably proteinlike in nature because they are destroyed by proteolytic enzymes. Different colicins are designated by capital letters (B_1, E_2, E_2, I, K, V) and may be distinguished experimentally by their diffusibility and host specificity. The colicins absorb to specific receptors of the bacterial cell walls. *Salmonella* strains, whether or not they produce colicins, are naturally resistant to them. The *S. typhimurium* strain LT2 is naturally noncolicinogenic, but can easily be rendered colicinogenic by mixed growth with *E. coli, Shigella,* or other *Salmonella* strains that carry the colicin plasmid.[16]

2.2. Resistance Transfer Plasmids (R Plasmids)

R plasmids can confer resistance to one or more antibacterial drugs on their hosts. The first R plasmids were discovered in Japan, when after World War II various sulfonamide antibiotics were used for the treatment of bacillary dysentery. Initially, these drugs were very effective in treating the disease. However, *Shigella* strains resistant to the sulfonamides were encountered for the first time about 1949. The newer antibiotics, such as streptomycin, chloramphenicol, and tetracycline, were found to be of therapeutic value only for 4 years. *Shigella* isolates resistant to tetracycline or streptomycin were discovered in 1963,[91] and the first multiply resistant pathogenic bacteria emerged in 1965.[49] The transferable nature of the resistance determinants was not known until 1960, when Akiba *et al.*[1] demonstrated this transfer *in vitro*. R plasmids with one or multiple resistance traits have since then been found in many parts of the world. Very often, plasmid-determined antibiotic resistance emerges within a few weeks or months after the introduction of a novel antibacterial drug. Such plasmids are prevalent in areas where the use of antibiotics is common, but they have also been found in antibiotic-free places.[8] Needless to say, the therapeutic effectiveness of many antibiotics has been severely curtailed by the presence of R plasmids. Resistance determinants can be transferred widely among the Enterobacteriaceae and, in some instances, also to *Vibrio cholera, Pasteurella pestis*, or *Serratia marcescens*.[47]

2.3. Plasmid Compatibility

Plasmids have been classified into several groups according to their compatibility. Plasmids belonging to the same incompatibility group are unable to coexist stably in a single cell. Compatible plasmids are placed in different incompatibility groups. At least 20 plasmid incompatibility groups have been identified. Incompatibility may occur between pairs of plasmids that are nearly isogenic (e.g., F′-*lac* and F′-*gal*) or between pairs that have a close interrelationship, but are not isogenic (e.g., ColV2-K94 and F; F and ColV3-K30; ColB2-K77 and 222/R).[7] Plasmid incompatibility occurs even in crosses between isogenic bacteria and therefore cannot result from DNA restriction. It has been suggested that incompatibility of plasmids may indicate a common evolutionary origin.[70] Plasmid incompatibility is a phenomenon different from "entry exclusion," which is a change on the surface of plasmid-containing cells. This change results in the inability of the DNA of a closely related plasmid to infect a plasmid-bearing cell by conjugative route, but not by transduction.[56]

2.4. Plasmid Replication

Two distinct classes of plasmids have been identified on the basis of the number of copies that are present in the host cell. Some plasmids are under "stringent" control, and replicate at the same rate as the chromosome. Only one or two copies of such plasmids are present in the cell at any time.[31] Other plasmids are under a "relaxed-regulation" control and produce ten or more copies per cell.[66] The relaxed-regulation plasmids appear to make three times as many copies when the cells are in stationary phase as when the host cells are in logarithmic phase. In addition, the control of plasmid replication appears to be size-related.[7] In general, only smaller plasmids show relaxed regulation, which may imply that controlling genes are lacking. The large plasmids may code for these genes, which could be involved in the regulation of the initiation of plasmid replication. Some plasmids are dependent upon host function(s) for their replication and maintenance. Chromosomal mutations resulting in temperature sensitivity for R-plasmid maintenance have been reported.[57,92] Kingsburg and Helinski in 1971[30] showed that the maintenance of the small ColEl plasmids is dependent on the DNA polymerase I activity of the host.

2.5. Plasmid Size

Plasmids belonging to the same incompatibility group have several genetic similarities, such as frequency and kinetics of transfer. Molecular studies of R-plasmid incompatibility groups have shown that the lengths of plasmids within a group are very similar.[20,21] R plasmids of group N (to which R46, parent plasmid of pKM101, belongs) are smaller than all other conjugative plasmids studied, the length varying between 13 and 16 μm, with a molecular weight between 27.3 and 32.9 million daltons.

2.6. DNA Damage and Plasmids

Howarth[25] showed that some colicin plasmids reduce the susceptibility of wild-type cells of *S. typhimurium* to the lethal effects of UV irradiation. She also discovered that the UV-protective ColIb plasmid increased the rate of UV-induced mutations.[26] Drabble and Stocker,[12] Marsh and Smith,[45] and Siccardi[71] reported the UV-protective ability of some R plasmids in *S. typhimurium* or *E. coli*. In addition, Siccardi[71] found that some conjugative R plasmids increased the sensitivity of *E. coli* to the bactericidal effect of UV irradiation. The sensitivity was considerably affected by some pre- and postirradiation treatment. Involvement of plasmid gene(s) in one or more DNA repair pathway was implicated in these plasmid-mediated effects.

Mortelmans and Stocker[52,53] found that the UV-protective plasmid

R46, which is the same as the R Brighton of Drabble and Stocker,[12] provides UV protection in *S. typhimurium* strains of wild-type UV sensitivity and in *polA* and *uvrB*, but not in *recA* mutants. A similar dependence of UV-protecting ability on the host *recA* function has been observed for both R46 and other plasmids in *E. coli* and *S. typhimurium*.[42,76] The presence of plasmid R46 was further shown to enhance the UV-induced mutation frequency in all classes of histidine-dependent base-pair substitution mutants, e.g., missense and nonsense (UAG, UAA, and UGA). Enhanced UV mutagenesis was also observed in *S. typhimurium* histidine-dependent frameshift mutants. The increase in the UVınduced mutation frequency was observed in wild-type, *polA*, and *uvrB* strains, but not in *recA* mutants. In *E. coli*, the UV-protective effect of R46 was shown to be dependent on the *recA lexA* genotype, and, as was shown in *Salmonella*, not on the *uvrA* or *polA* genotype.[76,84]

In 1972, MacPhee[40] discovered that the UV-protective plasmid R-Utrecht (R205) increased the survival of wild-type *S. typhimurium* exposed to γ-irradiation, to thymine deprivation, and to *N*-methyl-*N'*-nitro-*N'*nitrosoguanidine (MNNG) and methyl methanesulfonate (MMS). In 1973, he reported that plasmid R205 enhanced the mutagenic effect of MMS in *Salmonella hisG46*.[41] This report led to the systematic analysis of a number of plasmids by Ames and colleagues for use in the *Salmonella*/mammalian microsome assay. Among the plasmids they examined were R46 and some of its incomplete derivatives, which Mortelmans and Stocker[54] had isolated. One of the derivatives, pKM101, proved to be superior in making some of the tester strains respond to chemical mutagens previously not detectable in the system. This led to the incorporation in 1975 of plasmid pKM101 into two of the five standard tester strains.[46]

2.7. Spontaneous Mutagenesis and Plasmids

Mortelmans and Stocker[52–54] reported that plasmid R46 enhances not only UV-induced mutagenesis in *Salmonella hisG46*, but also the frequency of spontaneous mutation (mutator effect). About a tenfold increase in the spontaneous reversion of several *his* UAG, UAA, and UAG mutants was observed. Three of five *his* missense mutants showed a similar increase in the spontaneous mutation frequency. In contrast, no mutator effect was observed in several frameshift mutants of wild-type UV sensitivity. However, R46 exhibited the mutator effect in *uvrB* frameshift mutants with the frameshift *hisC3076* allele. About 31 spontaneous revertants were obtained with *hisG46*(R46) versus about 340 with *hisG46* (*uvrB* R46).[52] The high spontaneous background associated with *Salmonella* strain TA100, about 100–150 colonies per plate versus about 15–20 for TA1535, is attributable to the presence of plasmid pKM101.

3. Survey of Thirty-Three R Plasmids

The sensitivity of the Ames *Salmonella* mutagenicity assay was significantly improved by the introduction of plasmid pKM101 in some of the tester strains.(46) However, the test still does not detect certain classes of carcinogens or suspected carcinogens, such 1,2-dimethylhydrazine, safrole, and dieldrin. In an attempt to improve the sensitivity of the Ames assay, we screened 33 R plasmids in *Salmonella* strains TA1535, TA1537, and TA1538 (K. E. Mortelmans and L. Dousman, unpublished results). It was hoped that plasmid(s) would be found superior to pKM101 in enhancing chemical mutagenesis, or supplementing pKM101 by making one or more tester strains responsive to known or suspected carcinogens not detected by the test in its present form. Two sets of chemicals were used. One set consisted of three chemicals that elicit mutagenic activity in the Ames assay, namely methylmethanesulfonate (MMS), 4-nitroquinoline-1-oxide (4NQO), and ICR-191. The effect of 16 plasmids on chemical mutagenesis and survival was studied in strains TA1535, TA1537, and TA1538, and were compared with plasmid pKM101. The second set of chemicals contained seven known or suspected carcinogens not detected as mutagens in the Ames *Salmonella* assay. These chemicals are: acetamide, amitrole, carbon tetrachloride, 1,2-dimethylhydrazine, safrole, thiourea, and urethane. Most of the 33 plasmids were tested in strains TA1535, TA1537, and TA1538 with one or more of these seven chemicals in the presence of an Aroclor 1254-induced rat-liver S9 activation system. In order to gain additional information about the mechanism(s) of plasmid-mediated protection and mutagenesis, all 33 plasmids were examined in strains TA1535, TA1537, and TA1538 for the spontaneous mutator effect, and for UV-protective and UV-mutagenesis-enhancing ability. The experimental procedures and results are presented below.

3.1. Materials and Methods

3.1.1. Plasmids and Their Transfer

A total of 33 R plasmids were selected for analysis in strains TA1535 (*hisG46, uvrB-bio*$_{\Delta}$, *rfa*), TA1537 (*hisC3076, uvrB-bio*$_{\Delta}$, *rfa*), and TA1538 (*hisD3052, uvrB-bio*$_{\Delta}$, *rfa*). The plasmids represent a wide range of characteristics, including various incompatibility groups, antibiotic resistant markers, metal resistant markers, molecular weights, and deletion plasmids. Table 1 lists the plasmids' characteristics, in addition to the bacterial host that served as donor in the conjugation experiments, and the source. The presence of the plasmids in the donor strains was analyzed prior to each conjugation experiment. For transfer of the plasmids, donor and recipient bacteria were inoculated in nutrient broth in the ratio 1:10 and incubated, unshaken, at 37°C for about 18 hr. A loop full of the mixture was then streaked out on defined medium selective for the recipient strain with an appropriate antibiotic: ampicillin, 25 μg/ml; chloramphenicol, 25 μg/ml; kanamycin, 100 μg/ml; streptomycin, 800 μg/ml; sulfathiazole, 100 μg/ml; or tetracycline, 25 μg/ml. All antibiotics were obtained from Sigma Chemical Company, except tetracycline, which was obtained from Calbiochem. Colonies picked from selection plates were purified by single-colony reisolation at least twice on the selective medium. Purified clones were tested for each of the resistances of the donor.

3.1.2. Media

The nutrient broth used was "Oxoid" nutrient extract broth #2 (CM67). The nutrient agar medium consisted of a peptone-enriched yeast complete medium. The defined medium was Vogel–Bonner[3] with 1% glucose (hereafter called GM medium) and supplemented when appropriate with required amino acids or other growth factors. A 0.2 M sodium phosphate buffer, pH 7.4, was used to dilute the cells, or to adjust the incubation mixture in the absence of the S9 mix in the chemical protection and chemical mutagenesis preincubation assay with the three direct-acting mutagens. Top agar consisted of 0.6% Difco agar and 0.6% sodium chloride, and was, unless stated otherwise, supplemented with biotin and histidine at a final concentration of 0.5 mM each.

3.1.3. Spontaneous Mutator Effect

Overnight nutrient broth cultures containing approximately 10^9 cells/ml were used. A 0.1-ml aliquot was added to 2.0 ml of molten top agar (43°C), which was vortexed and poured onto GM medium plates. After the top agar had hardened, the plates were incubated for 48 hr at 37°C, at which time the spontaneous revertant colonies were scored.

TABLE 1. Bacterial R Plasmids

Plasmid	Incompatibility group[a]	Source[b]	Donor strain[c]	Expected antibiotic resistance[d]	Observed antibiotic resistance[d]
RA1	A	PRC	J53-1	Sul, Tc	Sul, Tc
R16	B	DB	DB196	Amp, Sm, Sul, Tc	Amp, Sm, Sul, Tc
R724	B	PRC	J53	Chl, Sm, Sul, Tc	Chl, Sm, Sul, Tc
RpIP55	C	PRC	J53 F^-	Amp, Chl, Sul	Amp, Chl, Sul, Km, Sm
R1	FII	DB	DB196	Amp, Chl, Km, Sm, Sul	Amp, Chl, Km, Sm, Sul
R1-16	FII	PRC	J53	Km, Nm	Km, Nm^{nt}
R6	FII	BS	KM409	Chl, Km, Sm, Sul, Tc	Chl, Km, Sm, Sul^{nt}, Tc
R6-Tc	FII	BS	KM343	Tc	Sm, Tc
R62	Iα	DB	DB202	Sm, Sul, Tc	Sm, Sul, Tc
R64	Iα	PRC	J53 F^-	Sm, Tc	Sm, Km, Tc^s
R391	J	PRC	J53	Km, Nm, Hg	Km, Nm^{nt}, Hg^{nt}
R387	K	PRC	J53	Chl, Sm	Chl, Sm
R471a	L	PRC	J53 F^-	Amp, Chl, Km	Amp, Chl^s, Km
R446b	M	PRC	J53 F^-	Sm, Tc	Km, Sm, Tc^s
RN3	N	PRC	J53	Sm, Sul, Tc	Sm, Sul, Tc
R48	N	BS	KM336	Amp, Sm, Sul, Tc	Amp, Sm, Sul^{nt}, Tc^s
R45	N	BS	KM335	Amp, Sul, Tc	Amp, Sul^{nt}, Tc^s
R205	N	BS	KM337	Amp, Sul, Tc	Amp, Sul^{nt}, Tc

R269	N	PRC	SL3810	Amp, Sm	Amp, Sm, Tc
R313N	N	PRC	J53	Tc	Tc
R348N	N	BS	SL1654	Sm, Tc	Sm, Tc
R390	N	PRC	SL3810	Amp, Chl, Sm, Sul, Tc	Amp, Chl, Sm, Sul^{nt}, Tc^{s}
R390-4	N	DB	DB198	Amp	Amp
R390-8	N	DB	DB199	Amp	Amp
RP4	P	DB	DB192	Amp, Km, Tc	Amp, Km, Tc
R478	S	DB	DB193	Chl, Km, Tc	Chl, Km, Tc
Rtsl	T	DB	DB190	Km	Km
RSa	W	DB	DB191	Chl, Km, Sm, Sul	Chl, Km, Sm, Sul
PJD47	n.d.	JD	*E. coli (met A)*	Sm, Tc	Sm, Tc
PJD48	n.d.	JD	*E. coli (met A)*	Amp, Sm, Tc	Amp, Sm, Tc
R3916-CDPH	n.d.	CDPH	KM341	Amp, Sm, Tc	Amp, Sm, Tc
R7842-CDPH	n.d.	CDPH	KM329	Km, Sm, Sul, Tc	Km, Sm, Sul^{nt}, Tc
R9095-CDPH	n.d.	CDPH	KM339	Sm, Sul, Tc	Sm, Sul^{nt}, Tc
pKM101	N	BA	TA100	Amp	Amp

[a] n.d., Not determined.

[b] PRC, Plasmid Reference Center, Stanford University; DB, Dr. D. Brunner, Illinois State University; BS, Dr. B. A. D. Stocker, Stanford University; BA, Dr. B. Ames, University of California, Berkeley; JD, Dr. J. Donch, Palo Alto Research Foundation; CDPH, Dr. M. Bissett, California Department of Public Health, Berkeley.

[c] J53: *E. coli met F63*, *pro-22*, *nal*, (λ); KM strains: *S. typhimurium* LT2 *hisG46*; DB strains: *S. typhimurium hisG46 trp1011::Tn10*; SL1654: *S. typhimurium met trp* Sm^{r} (chromosomal marker); SL3810: *S. typhimurium* LT2 *pyrE125 rfa-738*.

[d] Dose levels for resistance (μg/ml): ampicillin (Amp), 25; chloramphenicol (Chl), 25; kanamycin (Km), 100; streptomycin (Sm), 800; sulfathiazole (Sul), 100; tetracycline (Tc), 25; mercury (Hg) and neomycin (Nm) resistance were not tested; nt, not tested; s, sensitive.

3.1.4. UV Protection and UV Mutagenesis

For the UV protection experiments, the overnight cultures containing approximately 10^9 cells/ml were serially diluted to about 10^3 cells/ml. Volumes of 0.1 ml were spread onto yeast complete plates, which were then irradiated with a Gelman lamp (#51438, 253.7 nm). The doses used were 0, 0.3, and 0.9 J/m^2. The plates were immediately wrapped in aluminum foil to prevent photoreactivation from occurring. They were then incubated for 24 hr at 37°C. The colony count was scored for each dose level, and percent survival was calculated as the number of surviving colonies per UV dose divided by the total number of colonies on the control (nonirradiated) plates. The UV-mutagenesis experiments were performed in parallel with the UV-protection experiments with the same overnight cultures. Aliquots of 0.1 ml undiluted overnight culture (approximately 10^8 cells) were added to 2.0 ml of molten (43°C) top agar. After vortexing, the top agar was poured onto GM plates. When the top agar had hardened, the plates were exposed to the same levels of UV irradiation used in the UV-protection experiments. The plates were wrapped in aluminum foil and incubated at 37°C. The histidine-independent revertant colonies in the thin film of "background" growth were counted after 2 days of incubation. The number of spontaneously induced revertant colonies is directly proportional to the reversion frequency per bacterium per generation. Therefore, the number of induced colonies is obtained by subtracting the number of revertants appearing on the control (nonirradiated) plates from the total (adjusted for killing) number appearing on the irradiated plates. The ratio P^+/P^- (Table 2) represents the ratio of induced revertants using data transformed to account for the killing effect.

3.1.5. Chemical Protection and Chemical Mutagenesis

Chemical protection data were obtained by using the Ames preincubation assay. Overnight cultures, grown to about 10^9 cells/ml, were serially diluted to 10^3 cells/ml. A 0.1-ml aliquot of the cultures was added to 0.05 ml of a particular dose level of the test chemical (or solvent for the negative controls) and 0.5 ml of buffer (plasmid-mediated protection was examined only with the direct-acting mutagens). This mixture was vortexed and incubated at room temperature for 20 min, at which time 2.0 ml of molten (43°C) top agar was added. The mixture was then vortexed and poured onto yeast complete plates. The surviving colonies were scored after 24 hr of incubation at 37°C. Percent survival was calculated as the number of colonies on the mutagen-treated plates divided by the number of cells on the untreated control plates. Since the number of cells used in the Ames chemical mutagenicity assay has been shown to affect the toxicity results quantitatively,[83] it should be kept in mind that our calculated percent survival is a measure of relative survival only. The chemically induced mutagenicity experiments were performed in parallel with the chemical survival experiments. A 0.1-ml aliquot of the undiluted overnight culture (about 10^8 cells) was added to 0.05 ml of chemical (or solvent for the negative controls) and 0.5 ml of S9 liver mix or buffer. The mixture was vortexed and preincubated at room temperature for 20 min, at which tim 2.0 ml of molten (43°C) top agar containing 0.05 mM biotin and histidine was added. The mixture was then vortexed and poured on GM plates, which were incubated at 37°C for 48 hr, at which time revertant colonies were scored. Because the percent survival is only a relative measure of toxicity, the data were not adjusted for the killing effect. Therefore the P^+/P^- ratio was not calculated.

3.1.6. Chemicals

The effect of the different plasmids on chemical survival and mutagenesis was examined with two sets of chemicals. One set consisted of the following chemicals presently not detected as mutagens in the Ames assay: acetamide (Aldrich), amitrole (Aldrich), carbon tetrachloride (Mallinckrodt), 1,2-dimethylhydrazine (Aldrich), safrole (Pfaltz and Bauer), thiourea (National Cancer Institute), and urethane (Pfaltz and Bauer). These chemicals were tested only in the presence of an Aroclor 1254-induced Sprague-Dawley rat-liver S9 mix (50 μl S9 fraction). The other set consisted of the following three Ames *Salmonella* direct-acting mutagens: ICR-191 (Polyscience Corp.), methyl methanesulfonate (Aldrich), and 4-nitroquinoline-1-oxide (Starks).

3.1.7. Evaluation of Plasmid-Protective Ability

The following criteria for UV-protective ability (UVP) were used: nonprotective, when the difference in survival between the P^+ and P^- strains was $\pm 10\%$; protective, when the survival of the P^+ strain was more than 10% higher than that of the P^- strain; sensitizing, when the survival of the P^+ strain was more than 10% lower than that of the P^- strain; more protective, when the survival level of the P^+ strain was higher than the isogenic pKM101-carrying strain at more than one dose

TABLE 2. Protective and Mutagenesis-Enhancing Effects of 33 Plasmids in *Salmonella* TA1535, TA1537, and TA1538 As Compared with pKM101

Plasmid and host	Mutator effect[a]	Percent UV survival[b]			UV mutagenesis[c]			Percent chemical survival[d]			Chemical mutagenesis[e]	
		P^+	P^-	UVP	P^+	P^-	P^+/P^-	P^+	P^-	CP	P^+	P^-
No plasmid												
TA1535	14 ± 8											
TA1537	10 ± 6											
TA1538	15 ± 7											
RA1												
TA1535	14 ± 5	45	33	(+)	57 (13)	43 (8)	0.92	50	51	(−)	30 (13)	17 (14)
TA1537	8 ± 5	98	54	(+)	24 (7)	28 (9)	0.49	27	12	(+)	73 (17)	83 (2)
TA1538	11 ± 5	100	61	(+)	7 (4)	9 (11)	*f*	9.0	7.5	(−)	Toxic	Toxic
R16												
TA1535	99 ± 8	67	59	(+)	613 (109)	34 (9)	18		NT		NT	
TA1537	24 ± 7	54	68	(s)	905 (18)	38 (4)	33		NT		NT	
TA1538	11 ± 3	100	62	(++)	15 (14)	15 (12)	0.20		NT		NT	
R724												
TA1535	18 ± 6	30	40	(−)	140 (21)	53 (22)	5.1		NT		NT	
TA1537	20 ± 8	6.0	46	(s)	22 (14)	59 (7)	1.2		NT		NT	
TA1538	21 ± 3	28	53	(s)	20 (23)	28 (28)	*f*		NT		NT	
RpIP55												
TA1535	16 ± 8	47	59	(s)	47 (13)	34 (9)	1.4		NT		NT	
TA1537	9 ± 4	68	59	(−)	83 (16)	79 (8)	.83		NT		NT	
TA1538	17 ± 7	72	40	(+)	18 (20)	15 (13)	0.00		NT		NT	
R1												
TA1535	11 ± 3	42	59	(s)	55 (13)	34 (9)	2.4		NT		NT	
TA1537	8 ± 6	54	72	(s)	98 (7)	92 (6)	1.4		NT		NT	
TA1538	22 ± 8	84	62	(+)	53 (60)	15 (14)	*f*		NT		NT	

R1-16																
TA1535	16 ± 8	79	62	(+)	29	(15)	81	(34)	0.24		NT			NT		
TA1537	10 ± 4	54	72	(s)	63	(7)	36	(7)	2.6		NT			NT		
TA1538	13 ± 3	48	52	(−)	6	(10)	14	(13)	0.00		NT			NT		
R6																
TA1535	14 ± 5	20	53	(s)	89	(13)	72	(9)	3.2	68	79	(s)	13	(9)	14	(10)
TA1537	11 ± 5	45	55	(−)	97	(13)	99	(29)	1.0	8	21	(s)	189	(14)	182	(17)
TA1538	10 ± 1	28	48	(s)	12	(10)	12	(90)	*f*	82	73	(+)	414	(11)	792	(10)
R6-Tc																
TA1535	8 ± 3	50	53	(−)	95	(9)	72	(9)	1.4	91	79	(+)	14	(6)	14	(10)
TA1537	11 ± 4	44	55	(−)	64	(14)	99	(29)	0.90	11	21	(−)	203	(12)	182	(17)
TA1538	10 ± 3	60	48	(+)	15	(17)	12	(20)	*f*	77	73	(−)	396	(13)	792	(10)
R62																
TA1535	123 ± 33	74	62	(+)	963	(130)	81	(34)	14		NT			NT		
TA1537	47 ± 12	80	72	(+)	1632	(37)	36	(7)	50		NT			NT		
TA1538	33 ± 8	100	54	(+)	107	(33)	22	(25)	*f*		NT			NT		
R64																
TA1535	11 ± 6	44	59	(s)	12	(3)	34	(9)	0.33		NT			NT		
TA1537	13 ± 4	84	68	(+)	758	(16)	38	(4)	18		NT			NT		
TA1538	17 ± 7	86	62	(+)	44	(14)	15	(12)	3.2		NT			NT		
R391																
TA1535	22 ± 5	62	60	(−)	764	(29)	99	(9)	7.9	48	51	(−)	108	(20)	18	(12)
TA1537	10 ± 4	90	48	(+)	362	(10)	56	(13)	4.4	46	21	(+)	316	(16)	182	(17)
TA1538	14 ± 5	69	32	(+)	39	(18)	17	(14)	3.2	31	22	(+)	99	(11)	675	(15)
R387																
TA1535	20 ± 8	36	43	(−)	52	(31)	54	(22)	0.78		NT			NT		
TA1537	13 ± 4	49	48	(−)	69	(12)	95	(9)	0.65		NT			NT		
TA1538	27 ± 12	84	53	(+)	20	(26)	28	(28)	*f*		NT			NT		
R471a																
TA1535	23 ± 6	100	33	(+)	1411	(33)	43	(8)	13	95	52	(++)	82	(23)	17	(14)
TA1537	27 ± 12	80	75	(+)	1652	(19)	59	(7)	30	32	12	(+)	793	(20)	83	(20)
TA1538	16 ± 4	100	61	(+)	13	(8)	9	(11)	*f*	16	7.5	(−)	Toxic		Toxic	

(*continued*)

TABLE 2. *(continued)*

Plasmid and host	Mutator effect[a]	Percent UV survival[b]			UV mutagenesis[c]			Percent chemical survival[d]			Chemical mutagenesis[e]	
		P+	P−	UVP	P+	P−	P+/P−	P+	P−	CP	P+	P−
R446b												
TA1535	43 ± 15	55	33	(+)	1120 (53)	190 (11)	3.6	66	51	(+)	327 (34)	18 (12)
TA1537	11 ± 3	100	48	(++)	918 (11)	56 (13)	10	46	21	(+)	301 (16)	182 (17)
TA1538	53 ± 13	86	27	(+)	988 (68)	15 (7)	36	73	22	(+)	131 (46)	675 (15)
RN3												
TA1535	86 ± 10	91	33	(+)	956 (83)	43 (8)	9.1	71	51	(+)	905 (82)	17 (14)
TA1537	16 ± 9	100	54	(+)	522 (19)	28 (9)	14	71	20	(+)	712 (11)	46 (4)
TA1538	22 ± 12	74	61	(+)	46 (12)	9 (11)	[f]	0.52	7.5	(s)	Toxic	Toxic
R48												
TA1535	66 ± 16	61	33	(+)	561 (74)	54 (13)	6.4		NT		NT	
TA1537	15 ± 4	100	67	(++)	1352 (19)	36 (5)	29		NT		NT	
TA1538	13 ± 4	100	48	(++)	25 (9)	8 (8)	[f]		NT		NT	
R45												
TA1535	89 ± 17	100	59	(+)	495 (98)	34 (9)	9.9		NT		NT	
TA1537	19 ± 17	83	68	(+)	40 (15)	38 (4)	0.60		NT		NT	
TA1538	113 ± 37	70	62	(+)	645 (165)	15 (12)	137		NT		NT	
R205												
TA1535	17 ± 3	89	60	(+)	76 (14)	99 (9)	0.47	79	51	(+)	26 (18)	18 (12)
TA1537	26 ± 11	89	56	(+)	1182 (40)	68 (15)	13	77	23	(+)	789 (35)	93 (7)
TA1538	83 ± 13	100	32	(+)	1478 (100)	17 (14)	147	42	22	(+)	880 (64)	675 (15)
R269												
TA1535	60 ± 8	51	30	(+)	1490 (65)	51 (18)	25	91	73	(+)	550 (40)	19 (10)
TA1537	21 ± 7	100	67	(+)	908 (26)	36 (5)	19	53	25	(+)	734 (11)	129 (6)
TA1538	15 ± 6	73	48	(+)	44 (7)	8 (8)	[f]	57	13	(+)	846 (16)	440 (38)

R313N																
TA1535	118 ± 30	93	62	(+)	935	(117)	81	(34)	12		NT		NT			
TA1537	23 ± 3	87	70	(+)	1059	(25)	118	(11)	11		NT		NT			
TA1538	26 ± 6	78	54	(+)	50	(23)	22	(25)	*f*		NT		NT			
R348N																
TA1535	130 ± 27	89	62	(+)	139	(113)	81	(34)	12		NT		NT			
TA1537	45 ± 15	92	70	(+)	1563	(58)	118	(11)	15		NT		NT			
TA1538	32 ± 14	92	54	(+)	90	(46)	22	(25)	*f*		NT		NT			
R390																
TA1535	71 ± 18	93	50	(+)	589	(60)	51	(18)	5.2	91	73	(+)	495	(44)	17	(10)
TA1537	21 ± 8	72	67	(−)	763	(16)	36	(5)	22	30	25	(−)	606	(18)	129	(6)
TA1538	17 ± 7	61	48	(+)	27	(11)	8	(8)	*f*	35	13	(+)	515	(20)	440	(38)
R390-4																
TA1535	126 ± 6	80	43	(+)	715	(127)	54	(22)	9.9		NT		NT			
TA1537	21 ± 6	48	46	(−)	43	(25)	59	(7)	0.80		NT		NT			
TA1538	22 ± 6	88	32	(−)	37	(17)	16	(14)	11		NT		NT			
R390-8																
TA1535	85 ± 19	61	43	(+)	751	(97)	54	(22)	14		NT		NT			
TA1537	26 ± 19	47	46	(−)	933	(29)	59	(7)	17		NT		NT			
TA1538	42 ± 27	85	53	(+)	940	(52)	28	(28)	*f*		NT		NT			
RP4																
TA1535	12 ± 4	53	33	(+)	71	(17)	43	(8)	0.96	51	51	(−)	18	(9)	17	(14)
TA1537	10 ± 6	49	75	(s)	50	(12)	59	(7)	1.1	38	12	(+)	82	(6)	83	(63)
TA1538	19 ± 5	100	62	(+)	0	(0)	7	(12)	*f*	32	7.5	(+)	Toxic		Toxic	
R478																
TA1535	21 ± 2	64	43	(+)	32	(19)	54	(22)	0.86		NT		NT			
TA1537	14 ± 1	48	46	(−)	79	(14)	59	(7)	1.2		NT		NT			
TA1538	16 ± 5	77	53	(+)	15	(22)	28	(28)	*f*		NT		NT			
Rts1																
TA1535	12 ± 4	53	33	(+)	108	(12)	43	(8)	1.7	52	51	(−)	84	(17)	17	(14)
TA1537	11 ± 3	81	54	(+)	414	(5)	28	(9)	14	17	12	(+)	647	(10)	83	(20)
TA1538	5 ± 3	100	61	(++)	1	(4)	9	(11)	*f*	97	7.5	(+)	Toxic		Toxic	

(continued)

TABLE 2. *(continued)*

Plasmid and host	Mutator effect[a]	Percent UV survival[b]			UV mutagenesis[c]			Percent chemical survival[d]			Chemical mutagenesis[e]	
		P^+	P^-	UVP	P^+	P^-	P^+/P^-	P^+	P^-	CP	P^+	P^-
RSa												
TA1535	17 ± 4	13	38	(s)	56 (20)	55 (11)	2.4	50	51	(−)	21 (13)	17 (4)
TA1537	10 ± 5	64	76	(s)	35 (9)	28 (9)	1.2	45	12	(+)	61 (6)	83 (20)
TA1538	3 ± 1	100	62	(++)	2 (2)	9 (11)	*f*	90	7.5	(+)	Toxic	Toxic
PJD47												
T1535	12 ± 5	56	59	(−)	62 (19)	34 (9)	1.8		NT		NT	
TA1537	11 ± 4	67	68	(−)	51 (12)	38 (4)	1.2		NT		NT	
TA1538	14 ± 6	52	88	(s)	89 (21)	13 (10)	3.3		NT		NT	
PJD48												
TA1535	17 ± 9	39	40	(−)	38 (21)	53 (22)	0.57		NT		NT	
TA1537	20 ± 9	67	46	(+)	68 (28)	59 (7)	0.65		NT		NT	
TA1538	85 ± 27	65	53	(+)	460 (128)	28 (28)	*f*		NT		NT	
R3916-CDPH												
TA1535	49 ± 7	62	30	(+)	547 (49)	51 (18)	7.3	58	50	(−)	403 (46)	14 (12)
TA1537	18 ± 7	74	61	(+)	599 (22)	25 (9)	30	52	25	(+)	277 (12)	129 (6)
TA1538	67 ± 14	30	48	(s)	857 (84)	8 (8)	*f*	34	13	(+)	737 (79)	440 (16)

R7842-CDPH																
TA1535	57 ± 7	31	30	(−)	448	(58)	51	(18)	11	55	50	(−)	457	(54)	14	(12)
TA1537	12 ± 4	100	61	(++)	832	(18)	25	(9)	31	70	25	(+)	459	(6)	129	(6)
TA1538	15 ± 6	45	48	(−)	36	(13)	8	(8)	*f*	69	13	(+)	844	(826)	440	(16)
R9095-CDPH																
TA1535	45 ± 8	38	30	(−)	530	(40)	51	(18)	12	72	73	(−)	303	(31)	19	(10)
TA1537	11 ± 5	84	67	(+)	891	(14)	36	(5)	23	91	25	(+)	279	(6)	129	(6)
TA1538	15 ± 7	74	48	(−)	20	(8)	8	(8)	*f*	65	13	(+)	417	(19)	440	(10)
pKM101																
TA1535	123 ± 25	100	40	(+)	767	(119)	22	(15)	65	71	51	(+)	1425	(153)	23	(10)
TA1537	34 ± 12	82	54	(+)	1611	(41)	28	(19)	113	100	21	(+)	950	(40)	93	(7)
TA1538	25 ± 10	100	53	(+)	107	(31)	28	(28)	*f*	97	22	(+)	1070	(20)	675	(15)

[a] The number of spontaneous revertants per plate from at least three independent experiments.

[b] The data presented are for UV survival at 0.3 J/m^2; UVP, UV-protective ability. (−), Nonprotective; (+), protective; (s), sensitizing; (++), more protective than pKM101 (see Section 3.1 for evaluation criteria).

[c] The total number of histidine-independent colonies per irradiated plate is given first, followed by the number of spontaneous revertants (parentheses); P^+/P^- is the ratio of UV-induced revertants adjusted for the killing effect.

[d] The data presented are for the chemical survival at the following concentrations: 1500 μg/plate MMS (TA1535); 0.1 μg/plate 4NQO (TA1537); 20 μg/plate ICR-191 (TA1538); CP, chemical-protective ability. (−), Nonprotective; (+), protective; (s), sensitizing; (++), more protective than pKM101 (see Section 3.1 for evaluation criteria). NT, Not tested.

[e] The total number of histidine-independent colonies per treated plate is given first, followed by the number of spontaneous revertants (parentheses). The data presented were obtained at the following concentrations: 1500 μg/plate MMS (TA1535); 0.5 μg/plate 4NQO (TA1537); 50 μg/plate ICR-191 (TA1538). NT, Not tested.

[f] The net P^+/P^- ratio could not be calculated, since no net UV-induced mutants were obtained with the P^- strains after subtraction of the spontaneous background.

level. The same criteria for evaluating the chemical-protective ability of the plasmids was used as for the UV-protective ability.

3.2. Results

3.2.1. Mutator Effect

All 33 plasmids were screened for their ability to enhance spontaneous mutagenesis (mutator effect) in *Salmonella* strains TA1535, TA1537, and TA1538. Table 2 lists the average number of spontaneous revertants obtained in at least three independent experiments. The greatest variation in mutator effect was observed in the base-pair substitution mutant strain TA1535. Four plasmids (R62, R313N, R348N, and R390-4) enhanced the induction of spontaneous mutants about nine-fold, an effect similar to that observed with pKM101. Five plasmids (R16, RN3, R45, R390, and R390-8) were found to enhance spontaneous mutagenesis about sixfold, while six plasmids (R446b, R48, R269, R3916-CDPH, R7842-CDPH, and R9095-CDPH) exhibited a mutator effect that increased the spontaneous induction of histidine revertants by about fourfold. The remainder of the plasmids housed in TA1535 had no effect or a very small effect on spontaneous mutagenesis. In general, very little, if any, plasmid-mediated mutator effect was observed in the frameshift mutant strain TA1537, with the exception of plasmids R62 and R348N, which enhanced the spontaneous mutation level about fivefold, compared to a 3.4-fold enhancement observed with plasmid pKM101. A mutator effect was observed in host strain TA1538 with the following plasmids: R45, 7.5-fold; R205 and PJD48, 5.5-fold; R3916-CDPH, 4.5-fold; R446b and R390-8, about threefold; and R62, twofold. A less than twofold increase or no increase was observed in the number of revertant colonies of host strain TA1538 with the remainder of the plasmids. A slight, less than twofold, mutator effect was obtained with plasmid pKM101.

3.2.2. UV Protection and UV Mutagenesis

The 33 plasmids were tested for their ability to protect the bacterial hosts (TA1535, TA1537, TA1538) against the lethal effects of UV radiation (Table 2). The pKM101-carrying strains were always tested in parallel with the other plasmid-carrying strains for comparative purposes. Nine plasmids provided UV protection in all three host strains, but to a lesser extent than pKM101. These plasmids are: RA1, R62, R471a, RN3, R45, R205, R269, R313N, and R348N. The remaining

24 plasmids exhibited one of the following characteristics: (1) protective in one or two strains and nonprotective in the other host(s): plasmids R387, R6-Tc, R390, R390-4, R390-8, R391, R478, PJD48, and R9095-CDPH; (2) protective in all three strains, but more protective than pKM101 in one or two strains: plasmids R446b, R48, and Rtsl; (3) nonprotective in one or two strains, sensitizing in other strain(s): plasmids R724, R6, and PJD47; (4) protective in two strains, and sensitizing in one strain: plasmids R64, RP4, and R3916-CDPH; (5) sensitizing in TA1537 and TA1535 and protective to a greater extent than pKM101 in TA1538: plasmid RSa; (6) sensitizing in TA1535 and TA1537 and protective in TA1538: plasmid R1; (7) sensitizing, protective, and nonprotective in one each of the three strains: plasmids R1-16 and RpIP55; (8) nonprotective in strains TA1535 and TA1538 and protective to a greater extent than pKM101 in TA1537: plasmid R7842-CDPH; (9) protective in TA1535, sensitizing in TA1537, and protective to a greater extent than pKM101 in TA1538: plasmid R16.

Varying degrees of enhanced UV-induced mutagenesis were observed with the 33 plasmids examined in the three host strains (Table 2). In many cases the extent (P^+/P^- ratio) of plasmid-mediated UV-induced mutagenesis could not be calculated for host strains TA1538 because there were no net UV-induced mutants when the spontaneous background was subtracted. With the two other strains, correlation between the UV-protective ability of the plasmids and their ability to increase the number of UV-induced revertant colonies was not always evident. When such correlation existed, the extent of UV protection did not, however, necessarily correlate with the extent of enhanced mutagenesis. For instance, a few UV-protective plasmids did not enhance UV mutagenesis: e.g., RA1 in TA1535 and TA1537, R45 in TA1537, and R205 in TA1535. Several non-UV-protective plasmids enhanced UV mutagenesis: e.g., R391 in TA1535, R724 in TA1535, and R390 in TA1537. Enhanced UV mutagenesis was observed with several sensitizing plasmids: e.g., R6 in TA1535, PJD47 in TA1538, and R16 in TA1537.

Compared to plasmid pKM101 directly, several plasmids enhanced UV-induced mutagenesis in one or more tester strains to the same extent. A few plasmids enhanced UV mutagenesis to a greater extent than pKM101. However, the P^+/P^- ratio was never greater than the pKM101/P^- ratio. The data presented for all plasmids, including pKM101, are from one experiment. The variation observed in the number of UV-induced P^- revertants can be explained by the fact that not all plasmids were tested on the same day; slight variations in the assay procedures, e.g., the use of log-phase cells versus early or late stationary phase cells, or slight differences in the UV fluence used, could account for this variation.

3.2.3. Chemical Protection and Mutagenesis with MMS, 4NQO, and ICR-191

Only 16 plasmids were tested for their effect on chemical protection and mutagenesis. The three known mutagens used were MMS (TA1535), 4NQO (TA1537), and ICR-191 (TA1538). The following concentrations (μg/plate) were used: MMS, 100, 500, 1500, and 2500; 4NQO, 0.1, 0.3, 0.5, and 1.0; ICR-191, 5, 10, 20, and 50. The results obtained with these direct-acting mutagens at one concentration are presented in Table 2.

3.2.3a. Chemical Protection. Three plasmids provided protection to a lesser extent or to the same extent as pKM101 in all three strains. These plasmids are R446b, R205, and R269. Seven plasmids provided protection in two strains and were nonprotective in the third strain: R391, RP4, Rtsl, RSa, R3916-CDPH, R7842-CDPH, and R9095-CDPH. Plasmid R6 was protective in TA1538, but sensitized strains TA1535 and TA1537; plasmid RN3 protected TA1535 and TA1537, but sensitized strain TA1538; plasmid R390 was protective in TA1535 and TA1538 and nonprotective in TA1537; plasmids RA1 and R6-Tc were nonprotective in two strains and protective in the other strain; plasmid R471a was protective in TA1537, nonprotective in TA1538, and more protective than pKM101 in TA1535.

3.2.3b. Chemical Mutagenesis. As explained in Section 3.1, the chemical mutagenesis results could not be adjusted for the killing effect because the calculated percent survival is only a relative measure of toxicity. Based on the net number of induced revertants per plate, pKM101 exhibited the greatest enhancing effect in all three strains. Plasmid R205, the first plasmid shown to enhance chemical mutagenesis[41] in *S. typhimurium hisG46*, did not significantly enhance MMS-induced mutagenesis in TA1535, which also carries the *hisG46* allele. In contrast, enhancement by plasmid R205 was observed almost to the same extent as that observed with pKM101 in TA1537 and TA1538. Plasmid RN3 was almost as good as pKM101 in enhancing MMS- and 4NQO-induced mutagenesis in TA1535 and TA1537, respectively. However, this plasmid sensitized strain TA1538 to ICR-191, and no induced revertants were obtained. The remaining plasmids exhibited either no enhancement or enhancement to a lesser extent than that observed with plasmids RN3 and pKM101.

3.2.4. Chemical Mutagenesis with *Salmonella* "Nonmutagenic" Chemicals

The following seven chemicals, reported nonmutagenic in the Ames *Salmonella* assay, but suspected of being carcinogenic, were tested for mutagenicity with 23 plasmids in host strains TA1535, TA1537, and TA1538 in the presence of an Aroclor 1254-induced Sprague-Dawley rat liver S9 mix. The dose range used is given in parentheses: acetamide (5, 50, 500, 5000 μg/plate); amitrole (5, 50, 500, 5000 μg/plate); carbon tetrachloride (8, 80, 800, 8000 μg/plate); safrole (1, 10, 100, 1000 μg/plate); thiourea (5, 50, 500, 5000 μg/plate); urethane (10, 100, 1000, 10,000 μg/plate); 1,2-dimethylhydrazine (4, 40, 400, 4000 μg/plate). The 23 plasmids tested are: RA1, RN3, RP4, Rtsl, RSa, R1, R45, PJD47, PJD48, R48, RpIP55, R64, R205, R269, R387, R390, R390-8, R391, R446b, R471a, R3916-CDPH, R7842-CDPH, and R9095-CDPH. The seven chemicals listed above were also tested in strain TA1535 with plasmid R6 and in strains TA1535 and TA1537 with plasmid R16. The effect of plasmid R390-4 on mutagenesis in the three strains was examined with acetamide, amitrole, and carbon tetrachloride only. The remaining plasmids were tested in one or two strains with at least one chemical. Plasmids R62, R313N, and R348N in strain TA1535 were tested with acetamide, amitrole, safrole, and thiourea; in strain TA1537 these plasmids were tested with acetamide, safrole, thiourea, and urethane; and in strain TA1538 they were tested with amitrole and safrole. The remaining plasmids, R1-16, R6-Tc, R478, and R724, were not tested with any of the above chemicals. No mutagenic activity was observed using the plasmid-carrying strains with any of the chemicals under the assay conditions. The strains with and without pKM101 were always used as controls.

3.3. Discussion

Of all the plasmids examined, it appears that plasmid pKM101 remains the most appropriate one for use in the *Salmonella* assay. None of the plasmids could enhance chemically induced mutagenesis to a greater extent than pKM101, or be responsive to carcinogens that are not detected as mutagens with the present system. This survey of plasmids also showed that the genetic background of the bacterial host influences the plasmid-mediated mutator effect and their effects on UV and chemical survival and mutagenesis. An example is given in Figure 1. No UV killing was observed at a dose level of 0.9 J/m^2 with strain TA1538 carrying plasmid RSa, while plasmid pKM101 increased

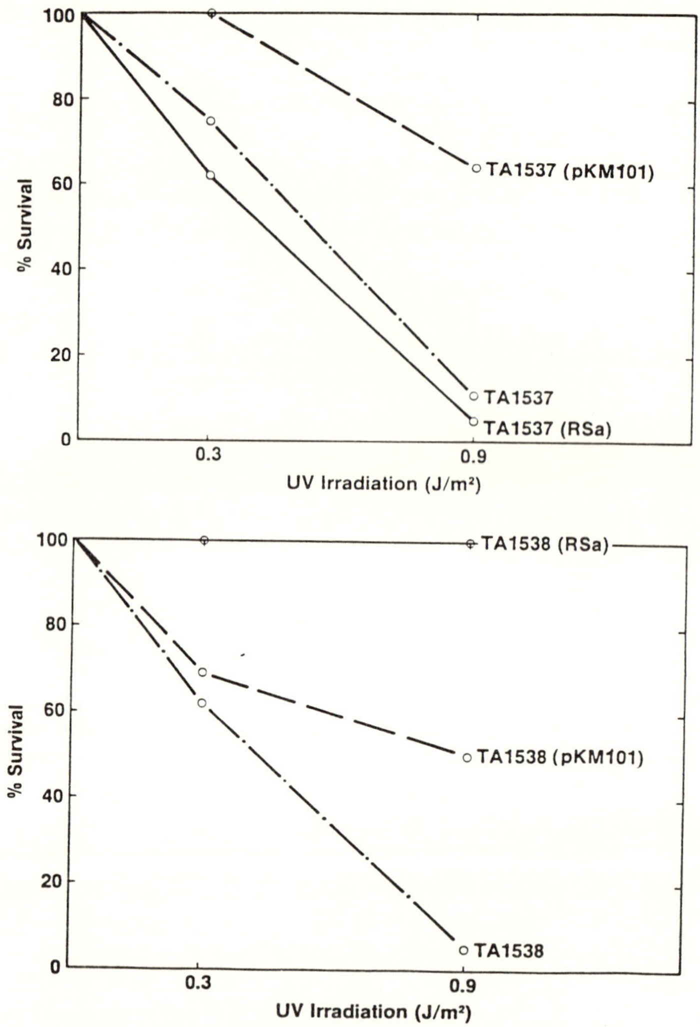

FIGURE 1. Effect of plasmids pKM101 and RSa on UV survival in *Salmonella* TA1537 and TA1538.

the survival of TA1538 to about 50%, compared to about 7% for TA1538 without a plasmid. In contrast, at a dose level of 0.9 J/m^2, plasmid RSa sensitized strain TA1537, while plasmid pKM101 increased the survival to about 65%. This finding is not too surprising, since the *Salmonella* tester strains have been independently isolated with genetic manipulations at different stages during the course of their construction.[3] For instance, the *hisG46* mutation present in TA1535 was obtained spontaneously,[2] while the *hisC3076* mutation in TA1537[58] and the *hisD3052* mutation in TA1538[27] were obtained after chemical treatment with ICR364-OH.

We know that the *uvrB* deletion extends into the nearby biotin gene. However, a precise genetic analysis of each of the *uvrB-bio* deletions in the different *Salmonella* strains is not available. Thus, it is quite possible that the deletion mutations extent into or affect different regions of the chromosome needed for the plasmid-mediated effects. Similarly, the *rfa* mutations have not been mapped, and their spontaneous occurrence may have been associated with other unknown chromosomal events that affect plasmid functions. Of course, the interpretation of the results obtained with the 33 plasmids would have been made easier if the research had been performed with isogenic *Salmonella* strains. The use of such strains would also have helped in determining whether just one mechanism or different mechanisms are involved in the plasmid functions. The issue will be discussed in greater detail in the following section on the mechanisms of plasmid-mediated effects.

4. Mechanisms of Plasmid-Mediated Effects on Mutagenesis and Survival

Since the discovery that certain plasmids can change the survival rate of their hosts after UV irradiation as well as enhance UV-induced mutagenesis,[25,26,52] DNA repair has been implicated in these plasmid-mediated phenomena. Of special interest is the question of whether these plasmid functions are dependent on bacterial DNA repair systems. The discovery that some plasmids, in addition to their UV effects, enhance the spontaneous and chemical induction of mutations as well as chemical survival prompted speculations about one versus more than one mechanism being involved in all these plasmid-mediated effects. Several studies using genetic, biochemical, or molecular approaches were conducted in attempts to answer these questions.

4.1. Genetic Studies

Marsh and Smith[45] examined 50 R plasmids in *E. coli* K12 for ability to protect against the lethal effects of UV irradiation. Two were found that increased the host survival by about 60%. Dependence of the plasmid on the host's photoreactivation DNA repair mechanism was excluded, since light exposure did not affect the results. It was also concluded that there was no difference in the number of UV-induced dimers between plasmid-containing and plasmid-free *E. coli* cells.

Mortelmans[52] tested 18 R plasmids and three colicin plasmids in *S. typhimurium hisG46* for three plasmid properties: mutator effect, UV protection, and enhanced UV mutagenesis. Table 3 presents the results. Seven of the R plasmids were from the incompatibility group N, two were from the FII incompatibility group, and nine had not been assigned an incompatibility group. Six of these were obtained from the California Department of Public Health and had been isolated from human, chicken, calf, or bovine sources. Of the seven R plasmids belonging to incompatibility group N, five (R45, R48, R205, RN3, and R447a) showed all three effects, while two provided UV protection and enhanced UV mutagenesis, but failed to enhance spontaneous mutagenesis (R269 and R390). One plasmid, R6, incompatibility group FII, exhibited all three effects. However, its derivative, R6-Tc, exhibited UV protection and enhanced UV mutagenesis only. Three of the remaining nine R plasmids failed to exhibit any of the three effects: R-Peru, R-South Africa, and R-Singapore. Two R plasmids obtained from the California Department of Public Health showed the three effects: R3916-CDPH and R9095-CDPH; two plasmids, R3729-CDPH and R11576-CDPH, did not exhibit any of the three effects. Plasmid R7842-CDPH was UV-protective and enhanced spontaneous mutagenesis, but was not tested for UV-mutagenesis-enhancing ability. Plasmid R11687-CDPH was not UV-protective and did not enhance spontaneous mutagenesis. Its ability to enhance UV mutability was not tested. The three colicin plasmids, ColIa-53, ColIb-P9, and ColIa-596, enhanced spontaneous and UV-induced mutagenesis, and were found to be UV-protective.

Of the 21 plasmids tested in *S. typhimurium hisG46* by Mortelmans,[52] we also tested 11 in TA1535 for the mutator effect and UV-protective and mutagenesis-enhancing ability. Qualitatively, similar results were obtained with four plasmids: R45, R48, R3916-CDPH, and RN3; these plasmids exhibited all three effects. One plasmid, R7842-CDPH, was not UV-protective, but exhibited the mutator effect and UV-mutagenesis-enhancing effect in TA1535 and showed the mutator and UV-protective property in *hisG46*; the UV-mutagenesis-enhancing effect of this plasmid was not tested in *hisG46*. The remaining six plasmids exhibited properties in TA1535, which carries the same *his* allele as *S. typhimurium hisG46*, different from those observed in *hisG46*. For in-

TABLE 3. Effect of 18 R Plasmids and Three Colicin Plasmids on UV Survival and on Rates of Spontaneous and UV-Induced His$^+$ Reversion in *S. typhimurium hisG46*[a]

Plasmid[b]	Percent survival (6 J/m^2)	Number of revertants/plate	
		Spontaneous	At (6 J/m^2)
None	32	2.3	13.0
R45 (N)	95	45.3	275.0
R48 (N)	93	36.6	234.6
R205 (N)	90	32.6	203.3
R6 (FII)	85	19	155.0
R6-Tc (FII)	97	4.3	124.0
RN3 (N)	92	19.6	163.0
R447a (N)	90	15.3	170.0
R269 (N)	90	3.6	123.6
R390 (N)	85	4.6	151.6
R3916-CDPH (human isolate)	90	14.0	171.3
R7842-CDPH (bovine isolate)	94	25.3	NT[c]
R9095-CDPH (chicken isolate)	87	16.0	157.6
R-Peru	32	2.0	19.0
R-South Africa	29	3.3	24.3
R-Singapore	28	3.3	14.6
R3729-CDPH (calf isolate)	40	2.3	19.0
R11576-CDPH (human isolate)	35	2.6	17.0
R11687-CDPH (human isolate)	42	4.3	NT[c]
ColIa-53	95	9.6	245.6
ColIb-P9	90	11.3	215.0
ColIa-596	82	11.0	177

[a] Adapted from the data of Mortelmans.[52]
[b] Plasmid incompatibility groups, when known, are given in parentheses.
[c] NT, Not tested.

stance, plasmid R205 exhibited all three effects in *hisG46*, yet showed only the UV-protective effect in TA1535. Plasmid R6 also showed the three effects in *hisG46*, but was able to enhance UV mutagenesis only slightly in TA1535. In addition, plasmid R6 sensitized TA1535 to UV irradiation. Plasmid R6-Tc did not exhibit the UV-protective ability in TA1535, that was observed in *hisG46*, but was still able to enhance UV-induced mutagenesis slightly; no mutator effect was observed in either TA1535 and *hisG46*. Plasmids R269 and R390 enhanced spontaneous mutagenesis in TA1535, but failed to do so in *hisG46*. They exhibited the UV-protective effect and UV-mutagenesis-enhancing effect in both *hisG46* and TA1535. Plasmid R9095-CDPH showed all three effects in *hisG46*, but exhibited only the mutator and UV-mutagenesis-enhancing effect in TA1535. These results clearly demonstrate that the genetic background of the host strain influences the expression of plasmid genes. These results also point to the complexity involved in studying plasmid-mediated mutagenesis and protective effects.

Siccardi(71) examined 31 R plasmids for their UV-protective effect in *E. coli*. Fifteen of the 31 plasmids provided UV protection, 11 had no or little effect, and five caused increased UV sensitivity. Molina *et al.*(50) tested 72 plasmids in *E. coli* WP2 for their effect on MMS-induced Trp^+ reversion. Most of the plasmids were also tested for their effect on UV survival. They found a good correlation between ability to enhance chemical mutagenesis and UV protection ability. In addition, plasmids from the same incompatibility group were generally found to behave in a similar fashion. One plasmid, R391, was found to sensitize its host against the lethal effect of UV radiation. This observation was also made by Magarian,(44) who extensively studied plasmid R391 in *E. coli*. She found R391 to enhance UV-induced mutagenesis, despite its UV-sensitizing property. A similar observation was made by Upton and Pinney(79) with plasmid R391, and in the present work with several other plasmids surveyed in *S. typhimurium* (Table 2). Magarian found that the presence of both R391 and pKM101 in *E. coli* resulted in decreased UV sensitivity when compared to the same strain carrying R391 only, but resulted in increased UV sensitivity when compared to the strain with pKM101. Upton and Pinney(79) reported similar results with R391 and R46 housed together in *E. coli* TK501 *umuC* mutants. The R391 phenotypes studied by Magarian and by Upton and Pinney have been compared with those of an *E. coli tif*-1 subclone isolated by Witkin and co-workers.(95) They observed enhanced spontaneous and UV-induced mutagenesis in addition to increased UV sensitivity in this *tif*-1/R391 subclone. The allele implicated in these effects was mapped in or very close to the *recA* region. A few other UV-sensitizing plasmids in *E. coli* have been reported.(6,50,61,81)

In 1980, Pinney(61) tested one representative plasmid from each of 19 incompatibility groups and four plasmids from incompatibility group N for effects on UV mutagenesis and UV survival in *E. coli*. No significant enhancement of UV-induced mutagenesis and UV protection was observed with plasmids from 12 groups. Plasmids from the remaining eight incompatibility groups showed a pronounced UV mutator activity. Six of these were found to be UV-protective, one was found to be non-UV-protective, and one increased bacterial sensitivity to UV irradiation.

Todd and Glickman[75] reported the UV-protective property of plasmid pKM101 in *E. coli* to be *recL*-dependent, and confirmed its *uvrE* dependence and *uvrD* independence.[19] They also showed that the UV-mutagenesis enhancement property of pKM101, unlike its UV-protective property, was *uvrE*-independent, and confirmed its *uvrD* and *recL* independence.[81]

Fowler and colleagues[14,15] examined the mutational specificity of pKM101 (spontaneous and UV-induced) in *E. coli* in an attempt to resolve the question of whether one or several mechanisms are involved in the UV-induced and spontaneous mutagenesis. Plasmid pKM101 was found to have the same mutational specificity, in that it enhanced mutations at A:T base pairs, especially transversions, in both instances. These results suggest the involvement of a similar mechanism in pKM101-mediated enhanced spontaneous and UV-induced mutagenesis. Spanggord[72] reported that two plasmids, R269 and R390, appear to be very similar to pKM101 when housed in *E. coli.* In addition to the mutator effect, the two plasmids confer UV protection and enhance UV-induced mutagenesis. These two plasmids have also been shown to have the same spontaneous and UV-induced mutational specificity as pKM101, i.e., preferential enhanced transversion of A:T basepairs.

The spontaneous mutator effect as well as the UV-protective and UV-enhancement properties have all been shown to be independent of the excision repair pathway in *E. coli* and *S. typhimurium.*[45,53,76] In addition, Walker[84] reported that these three properties in *E. coli* were also *uvrB, polA, recB, recC,* and *recF* independent, but completely dependent on the *recA lexA* genotype. Except for one study,[71] all plasmids examined for their UV effects in *E. coli, S. typhimurium,* and *Pseudomonas aeruginosa* mutants were found to require a functional *recA* gene.[45,53,71,76,82,84] These findings suggest that the plasmid effects are dependent on the SOS repair system.[63,95] As far as the one exception[71] is concerned, it is conceivable that a recombination-deficient *recA* mutant was used that was still able to perform SOS repair.[65,94] Additional support for the *recA* dependence of plasmid-mediated effects in mutagenesis has come from studies with plasmids R46, R205, and pKM101. In addition to enhancing UV mutagenesis, these plasmids were also shown to be dependent on the *recA* function of the host cell to enhance chemically induced mutagenesis.[42,46,53,80,81]

Support for the dependence on the SOS repair pathway has come from studies with pKM101 in *E. coli tif* mutants.[84] An increase in the number of *E. coli*/pKM101 *tif* mutants was observed to a greater than additive amount at 42°C. The *tif* mutation is believed to result in the induction of SOS functions at 42°C without any DNA damage.[93] Therefore, the SOS process is activated and mutations are induced in the absence of any DNA damage. If indeed the mutator effect of plasmid pKM101 is independent of the SOS repair system, then one would expect only an additive mutagenesis effect in the *tif*/pKM101 strain. Walker interpreted the multiplicative increase to be due to interaction between the plasmid and cellular SOS repair. It has been speculated that on a regulatory level, plasmids interfere with the inducible nature of the SOS error-prone repair pathway either by making the induction process constitutive or making it more responsive to DNA damage. Evidence against constitutive SOS repair in plasmid-carrying strains has come from phage UV reactivation (Weigle-reactivation) experiments.[84] Plasmid pKM101-carrying *E. coli* exposed to a low level of UV radiation prior to phage infection reactivated more UV-irradiate phage than the nonirradiated plasmid-carrying cells. Goze and Devoret[19] also concluded that SOS repair remains inducible in bacteria carrying DNA repair-enhancing plasmids, since they found other cellular SOS activities not to be affected in a constitutive way by the presence of pKM101.

4.2. Biochemical Studies

It has been postulated that the plasmid-mediated DNA repair effects come about by interaction of a plasmid-coded enzyme(s) with one or more of the bacterial repair systems, which changes their rate-limiting step. The following plasmid products have been proposed: (1) a nonspecific DNA ligase, (2) an error-prone DNA polymerase, (3) a protein that modifies an error-free DNA polymerase into an error-prone polymerase, (4) a new nuclease, or (5) a DNA-binding protein that stabilizes DNA lesions. Several plasmid-coded enzymes have been reported. MacPhee[43] reported that *S. typhimurium polA* mutants carrying plasmid R205 had about 25% of the wild-type DNA polymerase I activity, compared to no activity in the plasmid-free isogenic bacteria. Lehrbach *et al.*,[36] working with plasmid pMG2 in *Pseudomonas aeruginosa*, also detected DNA polymerase I-like activity in *polI*-deficient host strains. However, Kronish and Walker[33] failed to detect such activity with plasmids R205 and pKM101 in several *E. coli* K12 *polA* mutants. Similarly, Stokes and Krishnapillai[73] working with several UV-protecting plasmids in *P. aeruginosa*, could not detect any plasmid-mediated

DNA polymerase activity. Failure to demonstrate *polI*-like activity was also reported by Upton and Pinney,[78] who compared DNA polymerase activity in *polA* and *polA*$^+$ *E. coli* strains carrying plasmids R205, R46, or pKM101, even though the cell extracts were prepared from UV-irradiated (SOS-induced) cultures. From these studies it appears that plasmid DNA polymerase *per se* does not play a role in plasmid-mediated UV protection.

A plasmid pKM101-coded endonuclease was reported by Lackey *et al.*[34] However, there is no evidence that this enzyme is involved in the enhanced mutagenesis and survival effects, since mutant pKM101 plasmids that have lost the ability to protect against or enhance UV mutagenesis still code for this endonuclease.[85] Little and Hanawalt[37] examined plasmid pKM101 for the ability to induce the *recA* gene product. Their study indicated that the plasmid does not spontaneously induce this gene product.

4.3. Molecular Studies

Most molecular studies to elucidate the mechanisms involved in plasmid-mediated enhanced mutagenesis and protection have been performed by Walker and colleagues with plasmid pKM101 in *E. coli*.[17,35,59,60,68,84,85,90] These studies will be summarized here.

Plasmid pKM101 is a 35.4-kilobase plasmid belonging to the N incompatibility group.[35] It is a deletion derivative of the clinically isolated parent plasmid R46. It was obtained through a series of consecutive transduction and conjugation experiments in *S. typhimurium hisG46*.[54] The deletion extents over a 13.8-kilobase region of plasmid R46, which carries the antibiotic resistance marker against streptomycin, sulfonamide, and tetracycline. Plasmid pKM101 retained the remaining ampicillin resistance marker.

It has been shown that plasmid pKM101 codes for two proteins analogous to the cellular *umuC* proteins. The *E. coli umuC* mutants are deficient in DNA repair and are nonmutable.[28] The *umuC* DNA region has been cloned and has been shown to code for two gene products, those of *umuC* and *umuD*, with molecular weights of 45,000 and 16,000, respectively.[59] Introduction of plasmid pKM101 in *umuC* cells restored the DNA repair and mutagenesis activities.[90] It was concluded that pKM101 carries a DNA region analogous to the cellular *umuC* region.[68] Further studies have indicated that the *umuD* and *umuC* gene-like products coded for by plasmid pKM101 increase the extent of error-prone DNA repair, which results in increased UV survival and mutagenesis. The *umuC*-like DNA region has been located on the plasmid by using the transposon Tn5, and is referred to as the *muc* region (mutagenesis: UV and chemical). Subcloning of the *muc* region in the plasmid vector pBR322 revealed the presence of two protein products.[59] Complementation analyses determined that the two protein products were coded for by two genes, namely *mucA* and *mucB*; their molecular weights are 16,000 and 45,000, respectively. Langer *et al.*[35] showed that the *muc* region of pKM101 is bounded by inverted repeated sequences. It is postulated that the *mucA* and *mucB* genes have a chromosomal origin, and that their transfer to the plasmid occurred via a transposon. However, from amino acid sequencing studies, a 35% homology was estimated between *E. coli umuCD* and pKM101 *mucAB* gene products.[89] Several other plasmids (N3, R45, and R48) have been found to contain DNA regions homologuous to the *muc* region of pKM101.[89] Functional *mucA* and *mucB* gene products are necessary for the plasmid to exert its mutagenesis and protective effects.

Evidence is accumulating that, like the *umu* genes, the *mucA* and *mucB* genes may be directly regulated by the *recA* and *lexA* gene products.[17] One model, postulated by Walker and co-workers,[60] is that the *lexA* acts as a repressor of the *muc* DNA region, and induction requires the cleavage of the *lexA* protein by the *recA* protease. Support in favor for this model comes from the work of Ciesla,[5] who introduced plasmid pKM101 into a *tif–umuC* strain, which codes for an altered *recA* protease. The *umuC* mutation eliminates the spontaneous mutagenesis in *tif*-1 mutants when the temperature is raised to 42°C. However, spontaneous mutations are induced at 42°C when pKM101 is introduced in this mutant. Evidence for pKM101 dependence on the *lexA* gene comes from studies by Venturini and Monte-Bragadin[80] and Walker.[84]

In addition to the well-studied plasmid pKM101, a considerable number of UV-protecting plasmids have been shown to suppress the deficiencies in mutagenesis and DNA repair of *umuC* mutants after UV irradiation. Kopylov *et al.*[32] reported that the colicin-producing plasmid ColIb-P9 restore UV mutagenesis and UV resistance in *E. coli umuC* mutants. Comparing the effects of pKM101 and ColIb-P9 on colicin synthesis (ICS E1),[74] these researchers reported that the presence of pKM101 increases the levels of colE1 in mitomycin C-treated host cells, whereas a decrease was seen with plasmid ColIb-P9. In addition, pKM101, unlike ColIb-P9, reduced the survival of the *E. coli* cells after mitomycin C treatment. In the case of UV irradiation, however, the two plasmids caused a reduction of ICS E1 to about the same extent. These results suggest that the difference in the effects of the two plasmids on ColE1 synthesis is specifically associated with the type of

DNA damage induced by mitomycin C. The two plasmids most likely have similar effects on the repair of UV damage, but different effects on the repair of mitomycin C-induced DNA damage, which can be acted upon by two different repair pathways involving either the *recF* or *recBC* pathways of SOS induction.[11]

Dowden *et al.*[11] studied UV protection and mutagenesis in *E. coli* carrying the I group plasmid TP110 by *in vivo* fusion studies in which β-galactosidase levels were measured. It was shown that the UV-protection and mutation properties of plasmid TP110 were inducible functions under control of the *rec/lex* system, and that the plasmid suppressed *umuC* mutations just like plasmid pKM101. Because of the widespread occurrence of UV-protection and mutagenesis genes on bacterial plasmids, it has been suggested that these functions are related and mediated through transposable elements. A comparison of the restriction sites in the *muc* region of pKM101 and the *imp* (I mutagenesis and protection) of TP110 revealed almost no similarities. Therefore, the *muc* genes of pKM101/R46 and the *imp* genes of TP110 are unlikely to be related.

4.4. One versus Several Mechanisms

Notwithstanding the apparent complexity of plasmid-mediated mutagenicity and protection, a tremendous amount of insight has been obtained over the past few years into how plasmids interact with the genetic makeup of the host to exert their effects. Because of this interaction, one can speculate that no simple straightforward mechanism will emerge that can account for all the plasmid-associated functions. Following is a brief overview of recent findings that might shed some light on the mechanism(s) of plasmid-mediated effects.

Evidence is rapidly accumulating that all plasmid-encoded as well as chromsome-encoded SOS genes are repressed by the chromosomal *lexA* protein, which is essential in the initiation of the SOS repair process. Therefore, understanding the mechanism(s) of plasmid-mediated effects in response to UV and chemically induced DNA damage, as well as plasmid effects on the spontaneous mutagenesis, requires a knowledge of how the bacterial inducible SOS repair process operates. Such processes have been studied extensively in *E. coli* and have recently been reviewed in depth by Walker.[89] An attempt to summarize this excellent review would be beyond the scope of this chapter. However, whenever possible we will integrate aspects of the review that will help explain plasmid-mediated phenomena.

A complicating factor in resolving the issue of one versus several mechanisms is the fact that the genetic background of the host may influence the plasmid-mediated activities. For instance, comparing plasmid effects in *E. coli* K12 and *S. typhimurium* LT2 is already complicated by the fact that *S. typhimurium* LT2 is believed to be deficient in *umuDC* function relative to *E. coli.*[87,88] A number of bacteria are nonmutable with UV and in that respect resemble *E. coli umuD* and *umuC* mutants. Examples of such bacteria are *Haemophilus influenzae,*[29,55] *Proteus mirabilis,*[23] and *Streptococcus pneumoniae.*[18] When plasmid R46, parent of pKM101, is introduced in *P. mirabilis,* UV mutagenesis is induced.[23] Even comparing plasmid effects in different *S. typhimurium* LT2 strains, as was done in this chapter (Section 3), revealed that the presumed differences in the host's genetic background influenced plasmid-mediated effects.

Another important factor complicating the genetic analysis of plasmid-mediated effects is that the plasmid *umuDC* gene analogues carried by many UV-protective plasmids can be differentially expressed, regulated, and/or modified by other gene products, host- or plasmid-encoded. For instance, Elledge and Walker[13] have shown that the *lexA* protein acts as a direct repressor of the pKM101 *mucB* gene. A slight alteration in the *lexA* protein could therefore increase or decrease its effectiveness as a repressor, resulting in increased or decreased levels of the *muc* gene product, respectively. Regulation of the *muc* genes could also be affected by plasmid-associated genes that negatively regulate their expression, as has been suggested for plasmid R46, parent of pKM101.[10] The unusual characteristics of plasmid R391, which increases UV-induced mutagenesis but sensitizes its host against UV,[44,79] may also result from the activity of plasmid-coded *muc* proteins. Upton and Pinney[79] have suggested that competition for DNA lesions between the host *umuC* and R391 *muc* gene products could conceivably lead to an overall reduction in repair efficiency in $umuC^+$ cells. This could occur if the R391 *mucA* or *mucB* gene product was altered in such a way that error-prone repair leading to mutagenesis could proceed normally, but with the activity of an essential later event reduced. A similar situation may exist in the *E. coli* cells harboring either R391/pKM101[44] or R391/R46,[79] with the presumed altered R391 *muc* gene product greatly interfering with the UV-protective pathway of R46 or pKM101, but not with their UV-induced mutagenesis pathway.

Evidence is accumulating that DNA not damaged by exogenous agents may be acted upon by the SOS processing system. When an *E. coli recA441* mutant is grown at 42°C in the presence of adenine, a condition which causes the induction of many of the SOS responses, a significant increase in certain types of spontaneous reverse mutations is observed. The presence of pKM101 in the *recA441* strain enhances this increased spontaneous rate at 42°C.[5,9,84] Since not all types of mutations are spontaneously reversed by shifting *recA441* mutants to 42°C, it has been suggested that there are hotspots on the DNA, not damaged by exogenous agents, that facilitate the action of the SOS processing system. This hypothesis has been substantiated by the observation that pKM101/R46 enhances the spontaneous reversion frequency of certain types of mutations but not others.[14,52,53] A possible explanation for this enhanced spontaneous reversion rate is that it is the result of untargeted mutagenesis, which occurs at random sites, whether or not lesions are present. This is in contrast to targeted mutagenesis, which is site-specific and occurs in response to UV and chemically induced DNA damage.[48,89] Insertion of transposable elements in *S. typhimurium* LT2 has been shown by Shanabruch *et al.*[69] to generate spontaneous mutators. This raises intriguing questions about the possible role of transposons in plasmid mediated enhanced spontaneous mutagenesis, since at least in the case with plasmid pKM101, the *mucA* and *mucB* genes may have been introduced via a transposon.

Sargentini and Smith[67] have analyzed the spontaneous mutations rate of a number of different *E. coli* DNA repair mutants grown under different conditions. The following conclusions were reached: (1) the spontaneous mutation rate is reduced in *umuD, recA*(Def), or *lexA*(Ind^-) mutants because there is only a basal level of SOS processing activity operating in uninduced cells, and (2) the spontaneous mutation rate is raised in *uvrA* and *uvrB* mutants, because of the action of the basal level of the SOS repair pathway. The presence of an operating SOS repair pathway in uninduced cells can be explained by the fact that the SOS-inducing signal is largely dependent on the activation of the *recA* gene product, which is required for the cleavage of the *lexA* protein.[24,38,39] The basal level of the *recA* protein in an uninduced cell has been estimated to consist of between 800 and 1200 molecules. It has been shown that synthesis of the *recA* protein is not required for the cleavage of the *lexA* protein. Instead, it is a conformational change of the *recA* protein that is necessary to impart its proteolytic activity. Thus, synthesis/overproduction of the *recA* protein is not sufficient for inducing the SOS signal.

Upton and Pinney[79] have suggested that not all UV-protective plamids are regulated by the classical SOS pathway proposed by Witkin.[94] Their hypothesis is based on the observation that some muc^+ plasmids increase the spontaneous (uninduced) mutagenesis,[14,73] and plasmids R46[77] and pKM101[84,86] possess both constitutive and inducible DNA repair functions. The constitutive DNA repair capacity of R46 and pKM101 was deduced from observations that plasmids R46[82] and pKM101[19] increased UV survival and restored UV-induced mutagenesis in *E. coli* with the *lexB30* allele, which is a mutation in *recA* that prevents induction of SOS functions, but that leaves other *recA* protein functions unaffected.[64] These findings also suggest that for plasmid-mediated DNA repair, the *recA* gene product provides an additional function besides its involvement in the cleavage of the *lexA* protein repressor during induction of the SOS repair. It is therefore possible that plasmid R46/pKM101-enhanced UV- and MMS-induced mutagenesis in certain *E. coli lexA* mutants[51,79,82,84] is due to interference of the mutant *lexA* protein with the host *recA* protein synthesis, and not to the direct repression of the plasmid *muc* genes by the *lexA* gene product.

The muc^+ plasmids have recently been divided in two groups depending on (1) whether their *muc* genes are directly controlled by the *lexA* protein, or (2) whether the expression of *muc* genes is independent of the *lexA* gene product, except for the control of the *recA* protein synthesis.[4] Plasmids belonging to the first group are expected to have little or no effect on the spontaneous mutation frequency, while plasmids in the second group are expected to enhance spontaneous mutagenesis significantly. Babudri and Monti-Bragadin[4] provided support for this model when they studied plasmid-mediated MMS-induced mutagenesis in *E. coli lexA* mutants. It is speculated that the two groups of plasmids arose through mutation or error during transposition of the chromosomal *umuCD* genes into their ancestral plasmids.[79]

From all the available evidence on plasmid-mediated effects on bacterial survival and spontaneous and induced mutagenesis, it appears unlikely that one simple mechanism is involved in all the effects. It is most likely that branched pathways are available, which may include salvage pathways, and which are used depending on, e.g., the type of DNA damage, growth conditions, ability of the host to repair the DNA, and genetic makeup of both the host and the plasmid.

Recently, the host range of the pKM101 *muc* genes has been effectively broadened when the cloned genes introduced in the yeast organism *Saccharomyces cerivisiae* were shown to enhance chemical mutagenesis.[62] With the advent of recombinant DNA technology, it is quite possible that plasmid *muc* genes will also be transferred to mammalian cells. Assuming stable maintenance of these genes, the sensitivity of *in vitro* mammalian mutagenicity tests could possibly be affected as much as bacterial mutagenicity tests have been.

ACKNOWLEDGMENTS

We would like to thank Dr. B. N. Ames, Dr. M. Bisset, Dr. B. Brunner, Dr. J. Donch, and Dr. B. A. D. Stocker for the generous gift of strains and plasmids used in the survey of 33 R plasmids, which was supported by NIH grant 5 R23 ES02004-03. We also appreciate Dr. E. Lederberg's assistance in sending us plasmids from the Plasmid Reference Center, and the expert technical assistance of Cindy Falk.

5. References

1. T. Akiba, K. Kayama, Y. Ishiki, S. Kemura, and T. Fukushima, On the mechanism of the development of multiple drug-resistant clones of *Shigella, Jpn. J. Microbiol. 4,* 219–227 (1960).
2. B. N. Ames, The detection of chemical mutagens with enteric bacteria, in: *Chemical Mutagens: Principles and Methods for Their Detection,* Vol. 1, pp. 267–282 (A. Hollaender, ed.), Plenum Press, New York (1971).
3. B. N. Ames, J. McCann, and E. Yamasaki, Methods for detecting carcinogens and mutagens with the *Salmonella*/mammalian-microsome mutagenicity test, *Mutat. Res. 31,* 347–364 (1975).
4. N. Babudri and C. Monti-Bragadin, Restoration of mutability in nonmutable *Echerichia coli* carrying different plasmids, *Mol. Gen. Genet. 155,* 287–290 (1977).
5. Z. Ciesla, Plasmid pKM101-mediated mutagenesis in *Escherichia coli* is inducible, *Mol. Gen. Genet. 186,* 298–300 (1982).
6. L. S. Chernin and V. S. Mikoyan, Effects of plasmids on chromosome metabolism in bacteria, *Plasmid 6,* 119–140 (1981).
7. R. C. Clowes, Molecular structure of bacterial plasmids, *Bacteriol. Rev. 36,* 361–405 (1972).
8. C. E. Davis and J. Arrandan, The evolution of R factors. A study of a "preantibiotic" community in Borneo, *N. Engl. J. Med. 282,* 117–122 (1970).
9. O. P. Doubleday, M. H. L. Green, and B. S. Bridges, Spontaneous and ultraviolet-induced mutation in *Escherichea coli*: Interaction between plasmid and *tif-1* mutator efforts, *J. Gen. Microbiol. 101,* 163–166 (1977).
10. S. B. Dowden and P. Strike, R46-derived recombinant plasmids affecting DNA repair and mutation in *E. coli, Mol. Gen. Genet. 186,* 140–144 (1982).
11. S. B. Dowden, J. S. Glazebrook, and P. Strike, UV inducible UV protection and mutation functions on the I group plasmid TP110, *Mol. Gen. Genet. 193,* 316–321 (1984).
12. W. T. Drabble and B. A. D. Stocker,, R (transmissible drug-resistance) factors in *Salmonella typhimurium*: Pattern of transduction by phage P22 and ultraviolet-protection effect, *J. Gen. Microbiol. 53,* 109–123 (1968).
13. S. J. Elledge and G. C. Walker, The *muc* genes of pKM101 are induced by DNA damage, *J. Bacteriol. 155,* 1306–1315 (1983).
14. R. G. Fowler, L. D. McGinty, and K. E. Mortelmans, Spontaneous mutational specificity of drug resistance plasmid pKM101 in *Escherichia coli, J. Bacteriol. 140,* 929–937 (1979).
15. R. G. Fowler, L. D. McGinty, and K. E. Mortelmans, Mutational specificity of

ultraviolet light in *Escherichia coli* with and without the R-plasmid pKM101, *Genetics 99,* 25–40 (1981).

16. P. Fredericq, Colicins, *Annu. Rev. Microbiol. 11,* 7–22 (1957).
17. A. Gabb, C. J. Kenyon, and G. C. Walker, Inducibility of a gene product required for UV and chemical mutagenesis, *Proc. Natl. Acad. Sci. USA 78,* 5749–5753 (1981).
18. A. Gasc, M. N. Sicard, J. P. Claverys, and A. M. Sicard, Lack of SOS repair in *Streptococcus pneumoniae, Mutat. Res. 70,* 157–165 (1980).
19. A. Goze and R. Devoret, Repair promoted by plasmid pKM101 is different from SOS repair, *Mutat. Res. 61,* 163–179 (1979).
20. N. D. F. Grindley, J. N. Grindley, and E. S. Anderson, R factor compatibility groups, *Mol. Gen. Genet. 119,* 287–297 (1972).
21. N. D. F. Grindley, G. O. Humphreys, and E. S. Anderson, Molecular studies of R factor compatibility groups, *J. Bacteriol. 115,* 387–383 (1973).
22. P. C. Hanawalt and R. B. Setlow (eds.), *Molecular Mechanisms for the Repair of DNA,* Plenum Press, New York (1975).
23. J. Hofemeister, H. Kohler, and V. D. Filippov, DNA repair in *Proteus mirabilis.* VI. Plasmid (R46-) mediated recovery and UV mutagenesis, *Mol. Gen. Genet. 176,* 265–273 (1979).
24. T. Horii, T. Ogawa, T. Nakatani, T. Hase, H. Matsubara, and H. Ogawa, Regulation of SOS functions: Purification of *E. coli LexA* protein and determination of its specific site cleaved by the *RecA* protein, *Cell 27,* 515–522 (1981).
25. S. Howarth, Resistance to the bactericidal effect of ultraviolet radiation conferred on enterobacteria by the colicin factor ColI, *Mutat. Res. 40,* 43–55 (1965).
26. S. Howarth, Increase in frequency of ultraviolet-induced mutation brought about by the colicin factor ColI in *Salmonella typhimurium, Mutat. Res. 3,* 129–134 (1966).
27. K. Isono and J. Yourno, Chemical carcinogens as frameshift mutagens: *Salmonella* DNA sequence sensitive to mutagenesis by polycyclic carcinogens, *Proc. Natl. Acad. Sci. USA 71,* 1612–1617 (1974).
28. T. Kato and Y. Shinoura, Isolation and characterization of mutants of *Escherichia coli* deficient in induction of mutations by ultraviolet light, *Mol. Gen. Genet. 156,* 121–131 (1977).
29. R. F. Kimball, M. E. Boling, and S. W. Perdue, Evidence that inducible error-prone repair is absent in *Hemophilus influenzae* Rd, with a discussion of the relation of error-prone repair of alkylating agent damage, *Mutat. Res. 44,* 183–196 (1977).
30. D. T. Kingsbury and D. R. Helinski, DNA polymerase as a requirement for the maintenance of the bacterial plasmid colicinogenic factor E1, *Biochem. Biophys. Res. Commun. 41,* 1538–1544 (1971).
31. P. Kontomichalou, M. Mitani, and R. C. Clowes, Circular R-factor molecules controlling penicillinase synthesis, replicating in *Escherichia coli* under either relaxed or stringent control, *J. Bacteriol. 104,* 34–44 (1970).
32. V. M. Kopylov, I. A. Khmel, I. P. Vorobjeva, V. A. Lipoasova, and M. N. Kolot, Effect of the plasmid ColIb-P9 on cellular processes related to DNA repair, *Mol. Gen. Genet. 193,* 520–524 (1984).
33. J. Kronish and G. C. Walker, The effects of the ultraviolet protecting plasmids pKM101 and R205 on DNA polymerase I activity in *Escherichia coli* K12, *Mutat. Res. 60,* 135–142 (1979).
34. D. Lackey, G. C. Walker, T. Keng, and S. Linn, Characterization of an endonuclease associated with the drug resistance plasmid pKM101, *J. Bacteriol. 131,* 583–588 (1977).
35. P. J. Langer, W. G. Shanabruch, and G. C. Walker, Functional organization of plasmid pKM101, *J. Bacteriol. 145,* 1310–1316 (1981).

36. P. Lehrbach, A. H. C. Kung, and B. T. O. Lee, Loss of ultraviolet light protection and enhanced ultraviolet light-induced mutability in *Pseudomonas aeruginosa* carrying mutant R plasmids, *J. Gen. Microbiol. 101,* 135–141 (1977).
37. J. W. Little and P. C. Hanawalt, Induction of protein X in *Escherichia coli* by UV irradiation, *Mol. Gen. Genet. 150,* 237–248 (1977).
38. J. W. Little, S. H. Edmiston, L. Z. Pacelli, and D. W. Mount, Cleavage of the *Escherichia coli lexA* protein by the *recA* protease, *Proc. Natl. Acad. Sci. USA 77,* 3225–3229 (1980).
39. J. W. Little, D. W. Mount, and C. R. Yanisch-Perron, Purified *lexA* protein is a repressor of the *recA* and *lexA* genes, *Proc. Natl. Acad. Sci. USA 78,* 4199–4203 (1981).
40. D. G. MacPhee, Effect of an R factor on resistance of *Salmonella typhimurium* to radiation and chemical treatment, *Mutat. Res. 14,* 450–453 (1972).
41. D. C. MacPhee, *Salmonella typhimurium hisG46* (R-Utrecht): Possible use in screening mutagens and carcinogens, *Appl. Microbiol. 26,* 1004–1005 (1973).
42. D. G. MacPhee, Effect of *rec* mutations on the ultraviolet protecting and mutation-enhancing properties of the plasmid R-Utrecht in *Salmonella typhimurium, Mutat. Res. 19,* 357–359 (1973).
43. D. G. MacPhee, DNA polymerase activity determined by the ultraviolet-protecting plasmid R-Utrecht, *Nature 251,* 432–434 (1974).
44. M. S. Magarian, The Effect of R Plasmid R391 on UV-sensitivity, Spontaneous and UV-Induced Mutagenesis in *Escherichia coli,* M.A. Thesis, San Jose State University, San Jose, California (1983).
45. E. B. Marsh, Jr., and D. H. Smith, R factors improving survival of *E. coli* K-12 after ultraviolet irradiation, *J. Bacteriol. 100,* 128–139 (1969).
46. J. McCann, N. E. Spingarn, J. Kobori, and B. N. Ames, Detection of carcinogens as mutagens: Bacterial tester strains with R factor plasmids, *Proc. Natl. Acad. Sci. USA 72,* 979–983 (1975).
47. E. Meynell, G. G. Meynell, and N. Datta, Phylogenetic relationships of drug-resistance factors and other transmissible bacterial plasmids, *Bacteriol. Rev. 32,* 55–83 (1968).
48. J. H. Miller, Carcinogens induce targeted mutations in *Escherichia coli, Cell 31,* 5–7 (1982).
49. S. Mitsuhashi, Transmissible drug-resistance factor R, *Gunma J. Med. Sci. 13,* 169 (1965).
50. A. M. Molina, N. Badudri, M. Tamaro, S. Venturini, and C. Monti-Bragadin, *Enterobacteriaceae* plasmids enhancing chemical mutagenesis and their distribution among incompatibility groups, *FEMS Lett. 5,* 33–37 (1979).
51. C. Monti-Bragadin, N. Babudri, and L. Samer, Expression of the plasmid pKM101-determined DNA repair system in *recA*$^-$ and *lex*$^-$ strains of *Escherichia coli, Mol. Gen. Genet. 145,* 303–306 (1976).
52. K. E. Mortelmans, Effect of R Plasmids on *Salmonella typhimurium:* UV-Killing, UV-Mutagenesis, and Spontaneous Mutability, Ph.D. Thesis, Stanford University, Stanford, California (1975).
53. K. E. Mortelmans and B. A. D. Stocker, Ultraviolet light protection, enhancement of ultraviolet light mutagenesis and mutator effect of plasmid R46 in *Salmonella typhimurium, J. Bacteriol. 128,* 271–282 (1976).
54. K. E. Mortelmans and B. A. D. Stocker, Segregation of the mutator property of plasmid R46 from its ultraviolet protecting property, *Mol. Gen. Genet. 167,* 317–327 (1979).
55. N. K. Notani and J. E. Setlow, Inducible repair system in *Haemophilus influenzae, J. Bacteriol. 143,* 516–519 (1980).
56. R. P. Novick, Extrachromosomal inheritance in bacteria, *Bacteriol. Rev. 33,* 210–234 (1969).

57. R. P. Novick and D. Bouanchaud, Extrachromosomal nature of drug resistance in *Staphylococcus aureus, Ann. N. Y. Acad. Sci. 182,* 279–294 (1971).
58. N. S. Oeschger and P. E. Hartman, ICR-induced frameshift mutations in the histidine operon of *Salmonella, J. Bacteriol. 101,* 490 (1970).
59. K. L. Perry and G. C. Walker, Identification of plasmid (pKM101)-coded proteins involved in mutagenesis and UV resistance, *Nature 300,* 278–281 (1982).
60. K. L. Perry, S. J. Elledge, M. R. Lichtman, and G. C. Walker, Plasmid mediated enhancement of chemical mutagenesis, in: *Environmental Mutagens and Carcinogens* (T. Sugimura, S. Kondo, and H. Takebe, eds.), pp. 113–120, University of Tokyo Press, Tokyo, and Liss, New York (1982).
61. R. J. Pinney, Distribution among incompatibility groups of plasmids that confer UV mutability and UV resistance, *Mutat. Res. 72,* 155–159 (1980).
62. A. A. Potter, E. R. Nestmann, and V. N. Iyer, Transfer of the pKM101-associated *muc* genes to *Sacharomyces cerevisiae* (abstract), *Environ. Mutagen. 6(3)*, 444 (1984).
63. M. Radman, SOS repair hypothesis: Phenomenology of an inducible DNA repair which is accompanied by mutagenesis, in: *Molecular Mechanisms for the Repair of DNA* (P. C. Hanawalt and R. B. Setlow, eds.), p. 283, Plenum Press, New York (1975).
64. J. W. Roberts and C. W. Roberts, Two mutations that alter the regulatory activity of *E. coli recA* protein, *Nature 290,* 422–424 (1981).
65. J. W. Roberts, C. W. Roberts, and N. L. Craig, *Escherichia coli recA* gene product inactivates phage λ repressor, *Proc. Natl. Acad. Sci. USA 75,* 4714–4718 (1978).
66. R. Rowns, Replication of a bacterial episome under relaxed control, *J. Mol. Biol. 44,* 387–402 (1969).
67. N. J. Sargentini and K. C. Smith, Much of spontaneous mutagenesis in *Escherichia coli* is due to error-prone DNA repair: Implications for spontaneous carcinogenesis, *Carcinogenesis 2,* 863–872 (1981).
68. W. G. Shanabruch and G. C. Walker, Localization of the plasmid (pKM101) gene(s) involved in *recA*$^+$ *lexA*$^+$-dependent mutagenesis, *Mol. Gen. Genet. 179,* 289–297 (1980).
69. W. G. Shanabruch, I. Behlau, and G. C. Walker, Spontaneous mutators of *Salmonella typhimurium* LT2 generated by insertion of transposable elements, *J. Bact. 147,* 827–835 (1981).
70. P. A. Sharp, S. N. Cohen, and N. Davidson, Electron microscope heteroduplex studies of sequence relations among plasmids of *Escherichia coli.* II. Structure of drug resistance (R) factors and F factors, *J. Mol. Biol. 75,* 235–255 (1973).
71. A. G. Siccardi, Effect of R factors and other plasmids on ultraviolet susceptibility and host cell reactivation property of *E. coli, J. Bacteriol. 100,* 337–356 (1969).
72. D. L. Spanggord, The Effects of the R Plasmids R269, R390, and RSa on UV-Sensitivity, Spontaneous and UV-Induced Mutagenesis in *Escherichia coli,* M.A. Thesis, San Jose State University, San Jose, California, (1982).
73. H. W. Stokes and V. Krishnapillai, Prevalence of *Pseudomonas aeruginosa* FP plasmids which enhance spontaneous and UV-induced mutagenesis, *Mutat. Res. 50,* 19–28 (1978).
74. E. S. Tessman, C. A. Gritzmacher, and P. K. Peterson, Derepression of colicin E1 synthesis in the constructive *tif* mutant strain (*spr tif sfi*) and in a *tif sfi* mutant strain of *Escherichia coli* K-12, *J. Bacteriol 135,* 29–38 (1978).
75. P. A. Todd and B. W. Glickman, UV protection and mutagenesis in *uvrD, uvrE,* and *recL* strains of *Escherichia coli* carrying the pKM101 plasmid, *Mutat. Res. 62,* 451–457 (1979).
76. D. J. Tweats, M. L. Thompson, R. J. Pinney, and J. T. Smith, R factor mediated resistance to ultraviolet light in strains of *Escherichia coli* deficient in known repair functions, *J. Gen. Microbiol. 93,* 103–110 (1976).

77. C. Upton, Plasmid-Mediated DNA Repair and Mutagenesis in *Escherichia coli,* Ph.D. Thesis, School of Pharmacy, University of London, London, England (1982).
78. C. Upton and R. J. Pinney, Absence of ultraviolet-inducible DNA polymerase I-like activity in *Escherichia coli* strains harbouring R plasmids, *J. Gen. Microbiol. 125,* 131–137 (1981).
79. C. Upton and R. J. Pinney, Expression of eight unrelated Muc^+ plasmids in eleven DNA repair-deficient *E. coli* strains, *Mutat. Res. 112,* 261–273 (1983).
80. S. Venturini and C. Monti-Bragadin, Improvement in the sensitivity of DNA polymerase I-deficient *E. coli* for detecting mutagens and carcinogens, *Appl. Environ. Microbiol. 36,* 794–797 (1978).
81. S. Venturini and C. Monti-Bragadin, R plasmid mediated enhancement of mutagenesis in strains of *Escherichia coli* deficient in known repair functions, *Mutat. Res. 50,* 1–8 (1978).
82. N. S. Waleh and B. A. D. Stocker, Effect of host *lex, recA, recF,* and *uvrD* genotypes on the ultraviolet light protecting and related properties of plasmid R46 in *Escherichia coli, J. Bacteriol. 137,* 830–838 (1979).
83. N. S. Waleh, S. J. Rapport, and K. Mortelmans, Development of a toxicity test to be coupled to the Ames *Salmonella* assay and the method of construction of the required strains, *Mutat. Res. 97,* 247–256 (1982).
84. G. C. Walker, Plasmid (pKM101)-mediated enhancement of repair and mutagenesis: Dependence on chromosomal genes in *Escherichia coli, Mol. Gen. Genet. 152,* 93–103 (1977).
85. G. C. Walker, Isolation and characterization of mutants of the plasmid pKM101 deficient in their ability to enhance mutagenesis and repair, *J. Bacteriol. 133,* 1203–1211 (1978).
86. G. C. Walker, Mutagenesis and repair-enhancing activities associated with the plasmid pKM101, *Cold Spring Harbor Symp. Quant. Biol. 43,* 893–896 (1979).
87. G. C. Walker, Molecular principles underlying the Ames *Salmonella*–microsome test: Elements and design of short term test, in: *In Vitro Toxicology Testing of Environmental Agents: Current and Future Possibilities* (A. R. Kobler, T. K. Wang, L. D. Grant, R. S. DeWoskink, and T. J. Hughes, eds.), pp. 15–39, Plenum Press, New York (1983).
88. G. C. Walker, Regulation and function of cellular gene products involved in UV and chemical mutagenesis in *E. coli,* in: *Induced Mutagenesis: Molecular Mechanisms and Their Implications for Environmental Protection* (C. W. Lawrence, ed.), pp. 181–198, Plenum Press, New York (1983).
89. G. C. Walker, Mutagenesis and inducible responses to deoxyribonucleic acid damage in *Escherichia coli, Microbiol. Rev. 48,* 60–93 (1984).
90. G. C. Walker and P. P. Dobson, Mutagenesis and repair deficiencies of *Escherichia coli umuC* mutants are suppressed by the plasmid pKM101, *Mol. Gen. Genet. 172,* 17–24 (1979).
91. T. Watanabe, Infective heredity of multiple drug resistance in bacteria, *Bacteriol. Rev. 27,* 87–115 (1963).
92. N. S. Willet, The genetics of transmissible plasmids, *Annu. Rev. Genet. 6,* 257–268 (1972).
93. E. M. Witkin, Thermal enhancement of ultraviolet mutability in a *tif-1 uvrA* derivative of *Escherichia coli* B/r: Evidence that ultraviolet mutagenesis depends on an inducible function, *Proc. Natl. Acad. Sci. USA 71,* 1930–1934 (1974).
94. E. M. Witkin, Ultraviolet mutagenesis and inducible DNA repair in *Escherichia coli, Bacteriol. Rev. 40,* 869–907 (1976).
95. E. M. Witkin, J. O. McCall, M. R. Volkert, and I. E. Wermundsen, Constitutive expression of SOS functions and modulation of mutagenesis resulting from resolution of genetic instability at or near the *recA* locus of *Escherichia coli, Mol. Gen. Genet. 185,* 43–50 (1982).

CHAPTER 13

Mutation in Somatic Cells as Determined by Electrophoretic Analysis of Mutagen-Exposed Chinese Hamster Ovary Cells

Michael J. Siciliano, Raymond L. Stallings, Ronald M. Humphrey, and Gerald M. Adair

1. Introduction

1.1. Characterization of Electrophoretic Mutations in Natural Populations

After Ingram[17] demonstrated that the electrophoretic differences in inherited forms of hemoglobin were due to specific substitutions of amino acids of one charge for those of another, electrophoresis has been a useful procedure in identifying genetically variant forms of proteins. The procedure was made available to the study of a wide

Michael J. Siciliano and Raymond L. Stallings • Department of Genetics, University of Texas System Cancer Center, M. D. Anderson Hospital and Tumor Institute at Houston, Houston, Texas 77030. **Ronald M. Humphrey and Gerald M. Adair** • Science Park Research Division, University of Texas System Cancer Center, Smithville, Texas 78957.

range of proteins (enzymes) as a result of the application of histochemical staining to gels after electrophoresis.[16,21] It became practical to screen hosts of organisms and populations for mutant alleles that directed the production of electrophoretically variant proteins at a wide range of enzyme loci (any locus that produced an enzyme for which a histochemical stain was available). Such detected electrophoretic variants were proven to be due to the presence of mutant alleles by family or breeding studies, population frequencies, and electrophoretic patterns in heterozygotes consistent with the subunit structures of the enzymes (see Harris[12] for review).

In family or breeding studies, mutant and wild-type alleles were shown to segregate according to Mendelian expectations, producing single-banded wild-type or mutant patterns in appropriate homozygotes and multiple-banded heterozygous patterns. In population studies, the frequencies of homozygous and heterozygous electrophoretic patterns were shown to be consistent with the frequencies of mutant and wild-type alleles according to Hardy–Weinberg expectations. Electrophoretic patterns in heterozygotes were shown to be multiple-banded, with the number of bands a function of the number of polypeptide subunits making up the enzyme, since the products of wild-type and mutant alleles (electrophoretically variant) were produced in the same cell. If the enzyme was monomeric (composed of only a single polypeptide), both electrophoretically separable polypeptides were produced in the same cell of heterozygotes, resulting in two-banded patterns. If the enzyme was dimeric (composed of two polypeptides), three electrophoretically resolvable enzyme molecules were possible: aa (homopolymer made up of wild-type subunits), bb (homopolymer made up of mutant-type subunits), and ab (heteropolymer, having mobility between aa and bb, with one wild-type and one mutant subunit). Furthermore, since the subunits produced in the same cell combined randomly into multimeric molecules according to binominal expectations,[19,27] twice as many ab molecules were formed than either homopolymer, resulting in a three-banded pattern with band intensity ratios of 1:2:1. By the same principles, heterozygotes for trimeric enzymes produced four-banded patterns (aaa, aab, abb, and bbb) with band intensity ratios of 1:3:3:1, and heterozygotes for tetrameric enzymes produced five-banded patterns (aaaa, aaab, aabb, abbb, and bbbb) with activity ratios of 1:4:6:4:1. While variations in the activity of wild-type or mutant polypeptides or occurrence of "secondary" isozymes due to posttranslational modification (real or artifactual) of any of the primary components may alter such patterns, heterozygotes for a wide variety of enzymes

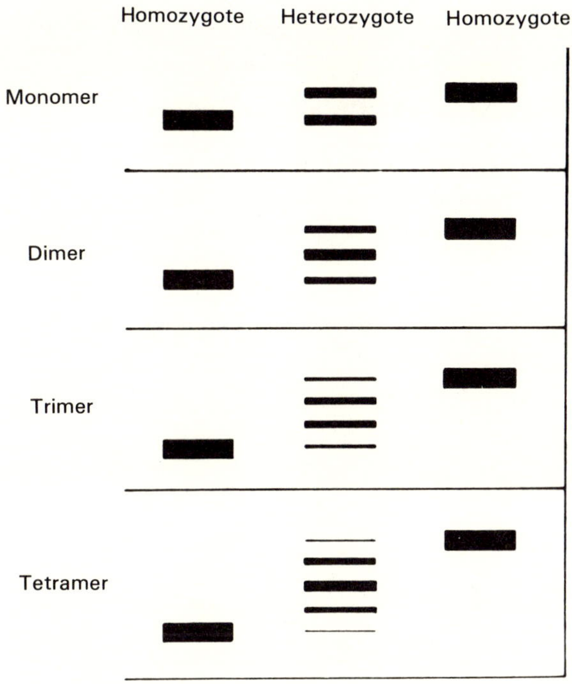

FIGURE 1. Diagramatic representations of typical isozyme patterns seen in heterozygotes for enzymes with monomeric, dimeric, trimeric, and tetrameric quaternary structures (from Harris[12]). Since CHO cells produce only one electrophoretic form for all the loci in the electrophoretic screen, patterns for all loci products (irrespective of the quaternary structure of the enzyme or ploidy of the locus) may be represented by the left column. Electrophoretic shift mutations at functionally diploid loci would be expected to produce multiple-band patterns typical of heterozygotes consistent with the subunit structures of the enzymes (middle column). Shift mutations at haploid loci would produce only a single modified protein (right column).

have been examined and generally fit the situations described above and illustrated diagramatically in Figure 1.[12]

Having been so well characterized, the naturally induced electrophoretic shift mutations have since been used as markers for a wide range of genetic applications: population analysis, gene mapping and linkage, cellular differentiation, X-chromosome inactivation and dosage compensation, evolution, paternity evaluation, and genetic signature identification of cell lines. Here, we describe our efforts on the isolation of electrophoretic shift mutations in mutagen-exposed Chinese hamster ovary (CHO) cells.

1.2. Rationale for Isolation of Electrophoretically Detectable Mutations

1.2.1. In Mammalian Somatic Cells in General

The introduction of mutations associated with specific modified proteins as markers into cultured mammalian cells by the isolation of induced electrophoretic shift mutations is an attractive prospect. However, the identification of such mutations would appear to be a considerably more rigorous procedure compared to the identification of the more classical "selectable" mutations. For both approaches, mutagen induction is identical. Cells are treated with mutagen to some predetermined level of cytotoxicity and allowed to go through two rounds of cell division in order to segregate mutant DNA strands into daughter cells. The problem then focuses down to identifying cells with mutations. For selective systems, cells are simply replated and allowed to grow a certain number of days under normal (permissive) conditions to allow expression of the mutant phenotype (expression time). They are then selected under conditions (physical and/or chemical) designed to allow the survival of only cells that had become mutant at specific loci (auxotrophic, drug-resistant, or temperative-sensitive; for a review of these procedures, see refs. 34 and 44). For electrophoretically detectable mutations, following strand segregation, each cell must be cloned and each clone expanded and electrophoretically analyzed for mutations at ~40 enzyme loci. Clearly, the selective approach will give a much quicker, much less labor-intensive production of mutation information than searching unselected clones for electrophoretic shift mutations.

However, as indicated in Section 1.1, and as further verified by Henning and Yanofsky[14] in analyzing the electrophoretic characteristics of tryptophane synthetase mutants, the electrophoretic system will provide specific information on nucleotide substitution mutations (point mutations). In addition to point mutations, mutants isolated via selective systems may arise through a number of mechanisms: e.g., chromosome loss and/or deletion, frameshift, or epigenetic effects. For experiments designed to determine the mutagenic potential of physical and chemical agents, selective systems, if they are shown to be truly efficient in isolating mutants, should be adequate. However, for mechanism studies, where information on the induction of point mutations is important, the electrophoretic approach may be the most efficient. It has been suggested that selective systems for mutations at essential loci—ouabain resistance[5] and diphtheria toxin resistance[22]—can only be point mutations because a partially functional protein is necessary to avoid lethality. However, the electrophoretic approach should give a more meaningful approximation of such events, since electrophoretically informative point mutations that have little or no effect on protein function and may take place anywhere in the protein molecule are detectable. Furthermore, since the electrophoretic approach is set up to screen for mutations at ~40 enzyme loci/clone, sampling error due to screening only one or a few loci by selection when loci may be differentially susceptible to mutagens[7,11] is avoided.

The ability to incorporate 40 or more loci into the screen is clearly a function of the codominant nature of electrophoretic shift mutations. Since, aside from the one popular dominant selectable marker, ouabain resistance, most selectable mutations are recessive, target loci are limited to hemizygous loci, either sex-linked loci such as *HPRT* (6-thioguanine resistance) or autosomal loci that have been rendered hemizygous (e.g., *TK* in mouse L5178Y cells).[8] Therefore, the codominant nature of electrophoretic shift mutations at known protein loci is a unique advantage in searching for mutations in diploid mammalian cells.

Perhaps the most important attribute of the electrophoretic approach is its ostensible liability: the lack of the ability to select for mutants. Following mutagen treatment, cells are cloned in maximum happiness medium. A truer view of the effects of mutagen divorced from further physical or chemical insult is unattainable. Further, there is no worry about nonmutant cells (false positives) slipping through selection because of incomplete selective stringency or membrane phenomena, or mutant cells being killed off because of metabolic cooperation, inappropriate expression time, or overly stringent selection conditions.

Therefore, in addition to providing genetic markers at many protein loci, we might expect induced electrophoretic shift mutations to give a very different perspective on the impact of induced mutations on the genetic variability of mammalian somatic cells.

1.2.2. In CHO Cells

CHO cells have been used in a wide variety of somatic cell genetic experiments because of their excellent adaptation to tissue culture conditions, high plating efficiency, and stable karyotype with a low number of readily identifiable chromosomes. In addition to their user-friendliness, we have used CHO cells in our experiments in order to test an interesting hypothesis advanced to explain mutation frequencies at recessive, selectable, autosomal loci in these cells. For a great many such loci,[34] mutations have been isolated at frequencies comparable to hemizygous loci such as *HPRT* ($>10^{-6}$). Since the total DNA content of CHO cells is close to that of diploid normal Chinese hamster cells and most of the chromosomal bands can be identified,[9] most autosomal loci should be represented twice and therefore autosomal recessive mutations should be recovered much more rarely. In an attempt to resolve this question, and citing the observations made by Deaven and Peterson[9] that the pattern of late-replicating regions in CHO chromosomes differed from the pattern in diploid Chinese hamster cells, Siminovitch[34] suggested that the extensive chromosomal rearrangements that occurred during CHO evolution resulted in changes in the behavior of the genome such that the cells are functionally hemizygous for a large number of genes. While it was recognized that this hemizygosity probably did not extend over the entire genome, invoking this concept to explain the easy isolation of phenotypically recessive mutations at such a large number of loci (30–36 as determined by complementation studies) implied that CHO cells are effectively haploid at most genetic loci.

Since we have demonstrated that CHO cells are not electrophoretically heterozygous at any of the ~40 enzyme loci in our electrophoretic screen,[37] the proportion of the CHO genome that is functionally hemizygous should be readily identified by the isolation of electrophoretic shift mutations. Loci at which mutations result in heterozygous patterns (multiple-band shifts; Figure 1, middle column) must be functionally diploid. Loci that are functionally hemizygous or haploid should give rise to shifted homozygous patterns (single-band shifts; Figure 1, right column). We might also expect to recover nulls or mutations for loss of enzyme activity much more frequently at loci shown to be haploid (by the above criteria) than those shown to be diploid. Since we have also had the opportunity to assign many of the loci studied in the mutation experiments to both euploid Chinese hamster chromosomes and to the normal and rearranged chromosomes in CHO cells, the genetic basis of the ploidy findings can be verified.

Finally, comparisons of the frequency of nulls obtained at haploid loci with the frequency of colonies obtained at selectable haploid loci will give us an opportunity to evaluate the efficiency of selective systems.

2. Methods of Procedure

2.1. Isolation of Clones Exposed to Mutagens

All experiments were conducted using the CHO cell line. CHO cells have a modal chromosome number of 21,[45] an average generation time of 12 hr, and a plating efficiency of 70–90%.[15] Cells were cloned and all experiments were conducted with aliquots of cells derived from a cloned population. Cells (1×10^6) were plated into a $60mm^2$ Petri dish and incubated overnight. They were then exposed to varying doses of mutagen—ultraviolet light (UV), ethyl methane sulfonate (EMS), or ethylnitrosourea (ENU)—and the percentage of survivors was determined.[15] Incubation was continued in complete medium until the population had divided twice. The cells were then trypsinized and resuspended in medium at a dilution of one surviving cell per 0.25 ml. Each well of a 96-well Falcon microtest plate then received 0.25 ml of culture. After 7–9 days of incubation, wells containing a single clone were picked (with trypsin) and each clone was expanded to $(40–50) \times 10^6$ cells. This same procedure was followed for untreated cells to produce control clones. Cells of a clone were trypsinized out of bottles and $(2–3) \times 10^6$ cells were resuspended in medium and incubated until electrophoretic analysis of the remainder was completed. If a variant was detected in a clone, this sample was regrown for confirmation and then subcloned for heritability determination.

2.2. Electrophoretic Detection of Mutations at Enzyme Loci

The remainder of the cells of a clone was homogenized. Cleared homogenates—up to 30 at a time—were subjected to vertical starch gel electrophoresis. A gel was sliced into as many as ten slabs so that each could be histochemically stained for a different enzyme. Stained slices (zymograms) were then photographed and fixed. Staining was done for 33 "housekeeping" enzymes. Because many of the enzymes studied have multiple forms (isozymes) coded by separate genetic loci,[20] the products of 44 isozyme loci were studied for induced variation. For various reasons, every clone was not necessarily screened for the products of all 44 loci. Each clone was examined for the products of at least 30 loci and >95% of all clones were studied for 39–44 loci.

The following seven gel buffer systems were used to resolve the products of the 44 isozyme loci (see Table 1 for the list of isozyme loci in the screen and the buffer systems most appropriate for their resolution):

1. Electrode: 0.13 M Tris, 0.043 M citrate, pH 7.0. Gel: 40 ml of electrode buffer diluted up to 600 ml with water; three drops of mercaptoethanol added to the molten gel prior to degassing.
2. Same as above, but 40 mg NAD also added to the molten gel prior to degassing.

TABLE 1. Isozyme Loci in the Mutation Screen and Buffer Systems Used to Resolve Their Products

Isozyme and locus[a]	Buffer system[b]
Acid phosphatase 1 (ACP1)	1
Acid phosphatase 2 (ACP2)	1
Acid phosphatase 3 (ACP3)	1
Adenosine deaminase (ADA)	1, 5
Adenylate kinase 2 (AK2)	3, 7
Enolase 1 (ENO1)	5
Esterase D (ESD)	1, 5
Esterase 1 (ES1)	1, 5
Esterase 2 (ES2)	1, 5
β-*Glucoronidase (GUSB)*	5
Glucose-6-phosphate dehydrogenase (G6PD)	6
Glucose phosphate isomerase (GPI)	1
α-*Glucosidase (GAA)*	5
Glutamate oxaloacetate transaminase 1 (GOT1)	1, 4
Glutamate oxaloacetate transaminase 2 (GOT2)	1, 4
Glutathione reductase (GSR)	5, 6
Glyceraldehyde-3-phosphate dehydrogenase (GAPD)	2
Glyoxalase 1 (GLO1)	4, 5
Hexokinase 1 (HK1)	6, 7
Hexokinase 2 (HK2)	6, 7
Hexosaminadase A (HAA)	5
Hexosaminadase B (HAB)	1
Inosine triphosphatase (ITPA)	5
Isocitrate dehydrogenase 1 (IDH1)	6
Isocitrate dehydrogenase 2 (IDH2)	3
Lactate dehydrogenase A (LDHA)	1, 2
Malate dehydrogenase 1 (MDH1)	1, 2
Malate dehydrogenase 2 (MDH2)	1, 2
Malic enzyme (ME1)	6
Mannose phosphate isomerae (MPI)	1, 3
Nucleoside phosphorylase (NP)	1, 5
Peptidase A (PEPA)	5
Peptidase B (PEPB)	5
Peptidase C (PEPC)	5
Peptidase D (PEPD)	5
Peptidase S (PEPS)	5
Phosphoglucomutase 1 (PGM1)	1, 3
Phosphoglucomutase 2 (PGM2)	1, 3
Phosphoglucomutase 3 (PGM3)	4
6-Phosphogluconate dehydrogenase (PGD)	6
Phosphoglycerate kinase (PGK)	7
Phosphoglyceric acid mutase (PGAM)	7
Superoxide dismutase 1 (SOD1)	3, 5
Triosephosphate isomerase (TPI)	1

[a] Abbreviations follow the recommendations for human gene nomenclature,[28] since human–Chinese hamster homologies are readily identifiable.[37]
[b] Numbers refer to buffer system described in Section 2.2.

3. Same as above, except 40 mg NADP instead of NAD added.
4. Electrode: 0.1 M Tris, 0.1 M maleic acid, 0.01 M EDTA, 0.01 M $MgCl_2$, adjusted to pH 7.4 with 40% NaOH. Gel: one part tank buffer diluted with 14 parts water; three drops of mercaptoethanol added to molten gel before degassing.
5. Electrode: 0.5 M Tris, 0.016 M EDTA, 0.65 M borate, pH 8.0. Gel: 60 ml electrode buffer diluted up to 600 ml with water; three drops of mercaptoethanol added prior to degassing.
6. Same as above, but 40 mg NADP also added to the molten gel prior to degassing.
7. Same as above, except 40 mg ATP instead of NADP added.

Recipes for the histochemical stains are contained in Siciliano and Shaw[29] and Harris and Hopkinson.[13] Methods for homogenization, starch gel electrophoresis, slicing, photography, and fixing as carried out in our laboratory have been published.[29]

2.3. Mutant Criteria

Confidence that shift variants represented true mutations in these experiments was relatively high because variants were identified as having a modification in a specific protein. However, after a cell extract containing a putative mutant gene product had been run several times to verify that the variant was not an artifact of a particular electrophoretic run, the clone was regrown and retested to determine that the variant was not an artifact of sample preparation. Finally, the clone was subcloned with the recovery of the variant.

3. Results and Discussion

3.1. Loci at Which Mutations Were Obtained

Mutations were recovered at all 44 loci in the screen. These were shown to be mutagen-induced, with a optimum level of mutagen treatment resulting in 10–20% survival of the cells.[32] Mutations were also shown to be heritable. In our system, heritability was determined by subcloning a clone shown to have a variant phenotype. Ten to 20 subclones were produced by the limiting dilution technique. Each subclone was then examined electrophoretically for the variant phenotype. Usually, all subclones showed the variation, indicating a true mutation. Occasionally, only a subset of the subclones showed the

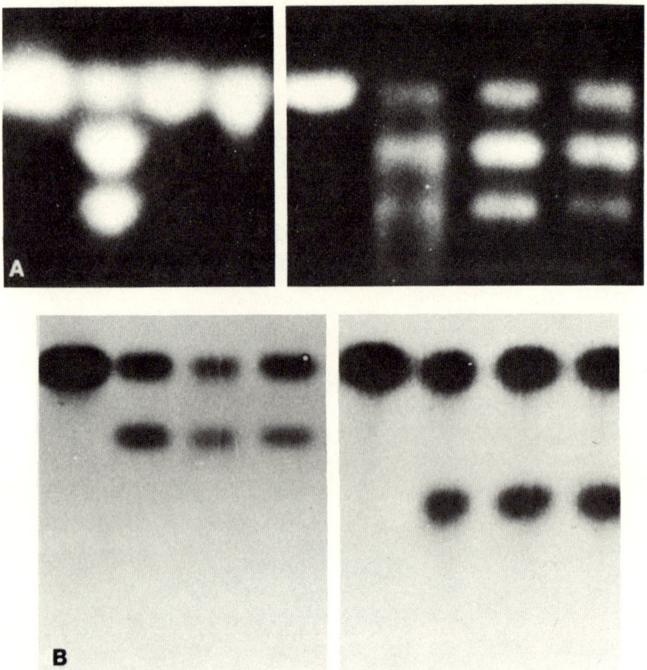

FIGURE 2. Zymograms demonstrating multiple-banded electrophoretic shift mutations induced in CHO cells at loci specifying enzymes representative of the four basic levels of quaternary structure. For each zymogram, the anodal end is toward the top and the origin is at the bottom. (A) Left panel shows a clone variant for *ESD* with three unaffected controls. The three-banded pattern in the variant is consistent with heterozygosity of *ESD*, which specifies a dimeric enzyme. Heritability, and therefore the true mutational nature of the variant, is demonstrated in the left panel, showing three, three-banded subclones of the original variant with a single-banded control. (B) Three subclones each (left and

variation. Such a result may have been due to an artifact in the original cloning: more than a single cell may have been deposited in a cloning well following exposure to mutagen. Another reason for such a result may have been the delayed fixation of the mutant phenotype, with wild-type and mutant cells segregating after cloning. In either case, the true mutational nature of the variant was verified by further subcloning of one of the variant subclones and recovering 100% variants. Representative variants at all loci met these criteria and are therefore considered the products of mutational events. Examples of the subclone analysis are shown in Figure 2.

Table 2 lists the loci at which mutations were obtained according to the three major classes that the electrophoretic systems is designed

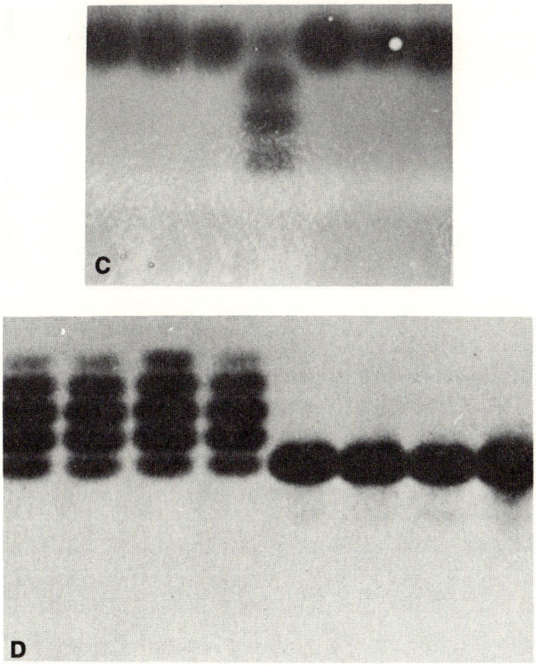

right panels) of two different *ADA* mutants. Each panel contains a single-banded control to the left. The two-banded patterns in the mutant subclones are consistent with heterozygosity in the monomeric *ADA* products. (C) Four-banded shift mutation for *NP* (center channel). The pattern is consistent with heterozygosity for the trimeric *NP* product. The six other samples are controls. (D) Four subclones of a five-banded shift mutation for *LDH* (left side) and four controls (right side). The *LDH* product is a well-known tetramer and the patterns in the subclones are therefore consistent with induced heterozygosity at that locus.

to detect: multiple-band shifts, indicating the induction of heterozygosity, single-band shifts, indicating point mutation at possibly hemizygous loci, and nulls. In all cases of induced heterozygosity, the multiple-banded patterns were consistent with the subunit structures of the enzymes. (Examples of all classes of mutations are shown in Figures 2 and 3.)

3.2. Functional Ploidy in CHO Cells and Its Effect on Mutation Frequency

As indicated in Section 1.2.2, multiple-band shift mutations should be limited to autosomal loci, which are functionally diploid in CHO

TABLE 2. Loci at Which Mutations Were Obtained

Multiple-band shifts		Single-band shifts	Nulls
ACP1	*LDHA*	*ACP2*	*ACP2*
ACP3	*MDH1*	*ES2*	*ES2*
ADA	*ME1*	*GOT2*	*GOT2*
AK2	*MPI*	*G6PD*	*G6PD*
ENO1	*NP*	*GPI*	*GPI*
ES1	*PEPA*	*HAA*	*HAA*
ESD	*PEPB*	*HK2*	*HK2*
GAA	*PEPC*	*IDH2*	*IDH2*
GAPD	*PEPS*	*MDH2*	*MDH2*
GLO1	*PGD*	*PEPD*	*PEPD*
GOT1	*PGM1*	*PGK*	*PGK*
GSR	*PGM2*	*PGAM*	*HAB*
GUSB	*PGM3*		*ADA*
HK1	*SOD1*		*ENO1*
IDH1	*TPI*		*GAA*
ITPA			

cells, while single-band shift mutations should be restricted to hemizygous loci, either sex-linked or autosomal (rendered hemizygous by chromosomal changes known to have taken place in CHO cells[9,48]). There should be a preference for null mutations to have taken place at loci having only a single functional copy of the structural gene

+

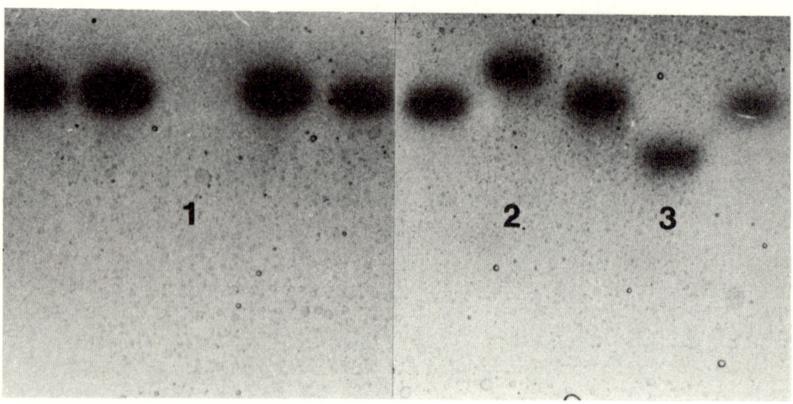

O

FIGURE 3. Zymograms demonstrating two different single-banded shift mutations (samples 2 and 3) and a null mutation (sample 1) for *PEPD*.

(hemizygous). These points are supported by inspection of the loci that fall into the three variant classes (Table 2). No loci that produced multiple-band shift mutations also produced single-band shift mutations. Also, of the 12 loci at which single-band shift mutations were recovered, 11 also produced null mutations, indicating the hemizygosity of those loci. Of the remaining loci that produced null mutations, *HAB* did not produce any shift mutations, so it may be hemizygous, but is uninformative in this analysis. Only three null mutations, one each at *ADA*, *ENO1* and *GAA*, were inconsistent with the remainder of the observations, since several multiple-band shift mutations were recovered at these loci. The nulls at *ADA*, *ENO1*, and *GAA* may have been regulatory mutations, or in clones at which one of the alleles may have been lost and the second mutated to null, or in clones at which two null mutations had taken place. The great bulk of the data are consistent with the idea that loci producing multiple-banded mutations are dizygous, whereas those producing single-band shifts (with the possible exception of *PGAM*) and nulls (with the possible exceptions of *HAB* and the likely exceptions of *ADA*, *ENO1*, and *GAA*) are hemizygous.

This hypothesis is supported by our gene mapping studies. Normal euploid Chinese hamster spleen or fibroblast cells, as well as CHO cells, were polyethylene glycol-fused with mouse C11D cells. Interspecific hybrid clones segregated hamster genetic material. In a series of such studies[1–4,25,26,31,38–43,46,47] summarized in Siciliano *et al.*,[33] Chinese hamster isozyme loci were assigned to chromosomal elements. The loci *ACP1, ADA, AK2, ENO1, ESD, GAA, GAPD, GLO1, GSR, ITPA, LDHA, ME1, MPI, NP, PEPB, PEPS, PGD, PGM1, PGM2, PGM3*, and *TPI* (21 of the 31 loci that produced multiple-band shift mutations) mapped onto Chinese hamster autosomes in euploid cells. In CHO cells, these loci were shown to have two alleles each mapping onto specific CHO chromosomes (Table 3). The remaining loci producing multiple-band shift mutations have not yet been mapped. The loci *G6PD* and *PGK* (two of the loci producing single-band shift and null mutations) were shown to be functionally hemizygous due to their X-linkage in both euploid and CHO cells.[10,26,47] The loci *GPI*, *IDH2*, and *PEPD* (three of the remaining seven loci producing single-band shift and null mutations) were assigned to Chinese hamster autosomes, but alleles for each mapped onto only single genetic elements in CHO cells (Table 3).[1,31] Remaining loci that produced single-band shift or null mutations have yet to be mapped. Therefore, the data confirm the genetic basis of the multiple-band shift, single-band shift, and null mutations observed in CHO cells and indicate that the mutations observed were at the target isozyme loci.

TABLE 3. Chromosomal Assignment in CHO Cells of Loci at Which Mutations Were Obtained[a]

Loci producing multiple-band shifts	Chromosome(s) assigned	Loci producing single-band shifts and nulls	Chromosome assigned
ACP1	7, Z8	*G6PD*	X
ADA	Z8, Z9	*PGK*	X
AK2	2, Z2	*IDH2*	Z3
ENO1	2, Z2	*PEPD*	9
ESd	1, Z6	*GPI*	9
GAA	Z3, Z4		
GAPD	8, 8		
GLO1	1, Z1		
GSR	1, Z1		
ITPA	Z8, Z9		
LDHA	Z3, Z4		
ME1	Z5, Z7		
MPI	Z5, Z7		
NP	1, Z1		
PEPB	1, Z1		
PEPS	1, Z1		
PGD	2, Z2		
PGM1	2, Z2		
PGM2	1, Z1		
PGM3	Z5, Z7		
TPI	8, 8		

[a] Chromosomes in CHO cells that are morphologically identical to normal Chinese hamster chromosomes are designated accordingly; e.g., 1, 2, . . ., X. CHO chromosomes that differ morphologically from euploid Chinese hamster chromosomes because of deletions, inversions, and translocations that have occurred during the evolution of the cell line[9] are known as chromosomes of the Z group and are designated by the prefix Z. A review discussing the genetic origin of CHO Z-group chromosomes as determined by G banding and gene mapping analysis is contained in Siciliano *et al.*[33]

From these data it is possible to evaluate the extent of functional haploidy in CHO cells and relate that to the question of mutation frequency in these cells. In an earlier report,[30] in which shift mutations at only ten loci were available for analysis, nine were shown to be diploid and only one (*MDH2*) haploid. It was then our conclusion that CHO cells appeared to be only as functionalliy hemizygous as one would expect of a slightly hypodiploid cell line derived from an organism in which the haploid number of chromosomes was 11 and that functional hemizygosity[34] as a general theory to explain the expression of recessive mutations in CHO was unlikely. The updated information summarized here supports that conclusion. As seen from Table 2, we have shown

31 loci to be clearly functionally diploid (the 31 giving rise to multiple-band shifts, indicating induced heterozygosity). Eleven loci (26.8% of the total number of loci whose ploidy could be unambiguously established by this approach) are clearly hemizygous, since they produced single-band shift *and* null mutations. The extent of hemizygosity is more than our original data indicated, but still is insufficient to suggest that CHO cells are functionally haploid as a general phenomenon.

With the identification of more hemizygous loci (both electrophoretically detectable as well as selectable) and as the mapping studies clarify the chromosomal basis for their hemizygosity in CHO cells, a picture is beginning to emerge which suggests that hemizygosity is clustered in the CHO genome into blocks or regions in which the activity of homologous alleles has been lost by a variety of phenomena.[4] The hemizygous loci producing single-band shift and null mutations, five of which have been mapped (Table 3), can be considered along with other drug-resistant and isozyme loci (those for *hypoxanthine-phosphoribosyl transferase, HPRT; emetin resistance, EMT; chromate resistance, CHR; leucyl-tRNA synthetase, LARS; diptheria toxin resistance, DTS; galactose-1 phosphate uridyl transferase, GALT; methylglyoxalbisguanyl hydrazone, MBG*), which have been mapped and shown to be hemizygous in CHO cells. The loci *HPRT*, *G6PD* and *PGK* are all on the X chromosome,[10,26,47] which is present in only one copy in CHO cells.[9] The loci *EMT*, *CHR*, *LARS*, *DTS*, and *GALT* are all on chromosome 2q,[6,25,40,46] the hemizygosity of which can be directly attributed to physical deletion of a large portion of the 2q arm giving rise to the Z2 chromosome in CHO cells.[9] The hemizygosity of *GPI* and *PEPD* is clearly a function of their location on chromosome 9, which is monosomic in CHO cells.[31] The loci *IDH2* and *MBG* are linked on the Z3 chromosome.[4] However, as opposed to the other hemizygous loci, there does not appear to have been any cytologically detectable loss of chromosomal material in the vicinity of the inactive allele.[1,3] As can be seen from Figure 4, the presence of the active *IDH2* allele on the CHO chromosome Z3 is a consequence of the position of the locus on the Chinese hamster chromosome 3 and the origin of the Z3 as a result of a reciprocal translocation between the portion of the chromosome 3 bearing *IDH2* (along with *LDHA* and *GAA*) and chromosome 4.[3] The inactive *IDH2* allele therefore must be a consequence of the inversion that took place in the homologous hamster chromosome 3 that gave rise to the Z4 chromosome in CHO cells (Figure 4). Thus, the *IDH2* locus (and possibly the *MBG* locus) may be *functionally*, and not *physically*, hemizygous in CHO cells. Close proximity of the *IDH2* locus to the inversion breakpoint on Chinese hamster chromosome 3q that gave

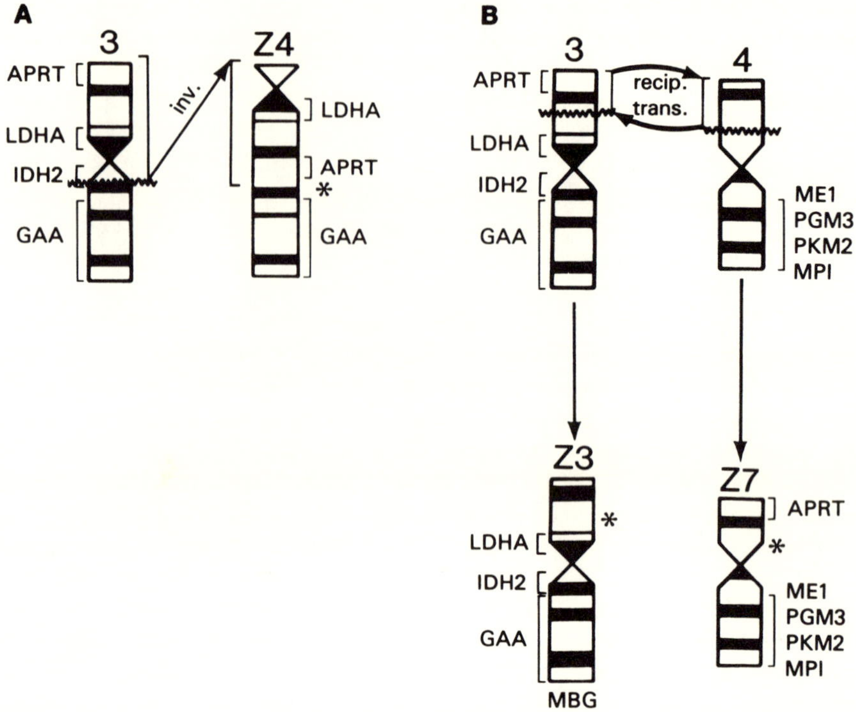

FIGURE 4. Cytological alterations to the two Chinese hamster chromosome 3's that gave rise to chromosomes Z4, Z3, and Z7 in the evolution of CHO cells (from Siciliano *et al.*[33]). The asterisks indicate rejoining points. (A) The inversion in one chromosome 3, which resulted in the Z4. (B) The reciprocal translocation between the other chromosome 3 and one of the chromosome 4's to produce the Z3 and Z7 chromosomes in CHO cells.[9]

rise to the Z4 chromosome may have resulted in deletion of the gene or its inactivation by some type of position effect, leaving the Z3 allele as the single functional *idh2* gene in CHO.[1,3] Proximity of the Z4 *adenine phosphoribosyl transferase* (*APRT*) allele to the same breakpoint region[1–3] appears to render the *APRT* gene highly susceptible to deletion[2,35,36] or inactivation events[36] that give rise to $APRT^{-/+}$ heterozygotes at high frequencies in unmutagenized CHO cell populations. Confirmation of the physical or functional nature of hemizygosity for *MBG* and *IDH2* in CHO may require cloned gene probes.

We conclude that there is appreciable haploidy in the CHO genome (~25%). However, for the most part the haploidy tends to be clustered into specific regions cytologically identifiable by monosomies or deletions. Localized *functional* haploidy due to position effects appears a possibility, but is not likely to be the basis for high frequencies for selectable, recessive mutations at a large number of loci throughout the genome. This question will be dealt with further as we consider our mutation frequency information.

3.3. Mutation Frequency in Mammalian Somatic Cells

As was suggested in Section 1, the multiple-locus screen for induced, electrophoretically detected mutations in mammalian cells was expected to give a more accurate view of these events than was possible using the more classical selective systems. Since our UV dose–response data indicated that the most effective mutagen treatment level was one that resulted in 10–20% survival of the cells,[32] we looked at the results of the three mutagens used at that dose point. EMS proved to be the most effective mutagen in the system. This is consistent with the results O'Niell *et al.*[24] obtained by screening CHO cells for mutations selected by 6TG. However, the frequency at which we obtained mutations with EMS, 5.21×10^{-3}/locus screened, is the highest ever obtained in a generalized mutation screen (since ~40 loci were screened/clone, one of every five clones picked had a mutation!). The results with ENU were not much lower, $f = 3.70 \times 10^{-3}$. Mutation frequency results with UV (1×10^{-3}) were significantly lower than for either alkylating agent, but still significantly higher than in the controls (0.06×10^{-3}).[32] Clearly, the impact of mutation on the induction of mammalian somatic cell variability is greater than one would have predicted on the basis of data from selective systems.

We believe one of the reasons for this finding is differential locus susceptibility to mutagen. The multiple-locus approach used here was designed to test the impact of that phenomenon. For instance, in the UV experiments[32] three loci were candidates for such differentially susceptible mutagenesis, since more than two mutations were recovered at them in the irradiated groups. These included four *MDH2* shifts, three *IDH2* nulls, and five *HK2* nulls, a total of 12 mutations out of 2838 loci screened (*MDH2, IDH2*, and *HK2* were each screened in all 946 UV-exposed clones). The fact that 12 of the 28 mutations obtained in the irradiated groups were at only three of the mean 41 loci screened/clone indicates that loci may be heterogeneous with respect to their susceptibility to UV mutagenesis. This was consistent with the *in vivo* germ-line results of Johnson and Lewis,[18] where six of the nine mutations were seen at only three of the 21 loci in their screen. For EMS, we recovered mutations at five loci (*PEPB*, *MDH2*, *HK2*, *PGM1*, and *PGM2*) in excess of 1%, while for ENU, *MPI*, *PEPB*, *ADA*, *MDH2*, *HK2*, *PEPD*, and *IDH2* mutations were recovered at frequencies of ~1% and above (details to be presented elsewhere). Obviously, some loci appear to be more generally susceptible to mutagenesis than others, e.g., *MDH2*, *IDH2*, and *HK2*. Perhaps *PEPB* should also be in this category, and the recovery of only one such mutation in the UV studies(32) may be a reflection of the much lower mutagenic potential of that mutagen compared to the alkylating agents.

Of much greater interest are the results that indicate loci exquisitely sensitive to one mutagen, but not to others. For instance, the high frequency of *PGM1* and *PGM2* mutations after exposure to EMS was not obtained with UV and ENU, and the high frequency of *MPI*, *ADA*, and *PEPD* mutations after exposure to ENU was not obtained after exposure to UV or EMS. The molecular bases of these differences has yet to be determined. However, for the present, the results indicate the need for caution in interpreting results on the mutagenic potential of agents where such potential is based on tests in which just one or a few loci were available for evaluation.

This information on differential locus susceptibility to mutagens has allowed us to generate an alternative hypothesis to explain the high frequency ($>10^{-6}$) of selectable, recessive, autosomal mutations in CHO cells. Data from three loci (*PEPD*, *HK2*, and *IDH2*) shown to be hemizygous in CHO cells suggest that the induced mutation frequency of null alleles at certain loci exposed to certain mutagens may be high enough to account for the induction of recessive phenotypes at functionally diploid loci. Three *PEPD* null mutations (e.g., Figure 3) were obtained following analysis of 527 EMS-exposed clones.(31) The frequency (f) of EMS-induced null mutations at the haploid *PEPD* locus was therefore 3/527, or 5.69×10^{-3}. Also, as indicated above, three null mutations for *IDH2* and five nulls for *HK2*(32) were isolated from 946 UV-exposed clones ($f = 3.17 \times 10^{-3}$ and 5.28×10^{-3}, respectively). Therefore, the frequency of null mutations isolated at these loci following UV and EMS exposure averaged 4.71×10^{-3}.

If these loci were diploid in these cells and if systems existed to select for nulls at them after EMS or UV exposure, one would expect the frequency of their recovery to be not less than the square of the frequency of induction of a null allele, i.e., $(4.71 \times 10^{-3})^2$, or 2.21×10^{-5}. This is a frequency within the range obtained for drug-resistant mutants at putatively functionally haploid autosomal loci in CHO cells.(34) We therefore suggest that the frequencies obtained for recessive drug-resistant mutants in CHO cells may be due not only to haploidy caused by deletions and monosomy, but also to great sensitivity of certain loci to particular mutagens.

It is also of interest to note that the frequency of electrophoretically detectable null mutations at these haploid loci is two to several orders of magnitude higher than have been obtained for selectable dominant mutations (Oua^r) or recessive mutations selected at haploid loci in CHO cells after mutagen treatment.[11] While these results may be attributable to differential locus sensitivity, they may also reflect a much higher efficiency for the nonselective electrophoretic approach in identifying mutations than has been possible for selective systems.

In addition to being able to overcome problems associated with selection and differential locus sensitivity, another major reason for our frequency results must be related to the subtle nature of the mutations that the system is capable of detecting—amino acid substitutions occurring anywhere in an enzyme molecule and which do not necessarily effect gene function. A fascinating aspect of this is that only one-third of amino acid substitutions result in charge changes that have the potential of being electrophoretically informative.[23] Also, due to the redundancy of the genetic code, not all possible nucleotide substitutions result in an amino acid substitution of any kind. Taking these facts together with our EMS results at 10–20% survival, where the frequency of induced electrophoretically detectable mutations was 5.21×10^{-3}/locus screened, it would appear that >1% of all loci might be expected to have a point mutation (nucleotide substitution) somewhere in their coding region following such exposure.

We therefore conclude, from the various aspects of our results using the electrophoretic approach, that the effect of chemical mutagens on the mutational load of somatic cells may be considerably more formidable than previously imagined.

4. Summary

The formal genetics of electrophoretic shift mutations occurring in natural populations established the genetic basis of mutant isozyme patterns at enzyme loci following electrophoresis.

Such mutations were induced at 44 isozyme loci in CHO cells by cloning cells following exposure to UV, EMS, or ENU, then expanding the clones, and analyzing them by starch gel electrophoresis and histochemical staining. Mutations were verified by subclone, subunit structure, and gene mapping analyses.

Multiple-band shift mutants, indicating induced heterozygosity, identified ~75% of the CHO loci to be functionally diploid. Single-band shift and null mutations at the remaining loci suggested they were functionally haploid.

Gene mapping studies verified the ploidy of the loci in CHO cells and indicated that haploid loci are clustered into regions in which the activity of homologous alleles had been lost due to chromosome rearrangement and loss.

At the most effective mutagen dose (10–20% survival) mutation frequencies after exposure to EMS, ENU, and UV were, respectively, 5.21×10^{-3}, 3.70×10^{-3}, and 1.04×10^{-3} per locus.

The multiple-locus nature of the screen revealed that the high frequency of mutation obtained was due in large part to differential locus susceptibility. It was further determined that loci varied in their susceptibility to different mutagens.

The frequency of null mutants obtained at certain haploid loci served as a model to account for the high frequency ($>10^{-6}$) of selectable recessive mutations in somatic cells and indicated that selective systems focused on just a few loci and dependent on stringent physical and/or chemical conditions to isolate mutations grossly underestimate mutation frequencies.

The electrophoretically detectable mutation frequencies obtained and the known limitations of the approach in detecting point mutations suggest that >1% of all loci may be expected to contain at least one nucleotide substitution following chemical mutagen exposure.

ACKNOWLEDGMENTS

This research was supported in part by NIH grants CA4484, CA28711, CA34797, and CA34936 and a gift from the Exxon Corporation. We also acknowledge Kenneth D. Muller for the creation of the Professorship in Tumor Genetics awarded to M. J. S.

5. References

1. G. M. Adair, R. L. Stallings, K. K. Friend, and M. J. Siciliano, Gene mapping and linkage analysis in Chinese hamster: Assignment of the genes for APRT, LDHA, IDH2, and GAA to chromosome 3, *Somat. Cell Genet. 9*, 477–487 (1983).
2. G. M. Adair, R. L. Stallings, R. S. Nairn, and M. J. Siciliano, High frequency structural deletion as the basis for functional hemizygosity of the *aprt* locus in CHO cells, *Proc. Natl. Acad. Sci. USA 80*, 5961–5964 (1983).
3. G. M. Adair, R. L. Stallings, and M. J. Siciliano, Chromosomal rearrangements and

gene expression in CHO cells: Mapping of alleles for eight enzyme loci on CHO chromosomes Z3, Z4, Z5, and Z7, *Somat. Cell Mol. Genet 10*, 283–295 (1984).
4. G. M. Adair and M. J. Siciliano, Linkage of the *MBG* locus to another functionally hemizygous gene locus (*IDH2*) on chromosome Z3 in Chinese hamster ovary (CHO) cells, *Mol. Cell. Biol. 5*, 109–113 (1985).
5. R. M. Baker, D. M. Brunette, R. Mankovitz, L. H. Thompson, G. F. Whitmore, L. Siminovich, and J. E. Till, Ouabain-resistant mutants of mouse and hamster cells in culture, *Cell 1*, 9–21 (1974).
6. C. E. Campbell and R. G. Worton, Linkage of genetic markers *EMT* and *CHR* in Chinese hamster cells, *Somat. Cell Genet. 6*, 215–224 (1980).
7. J. H. Carver, G. M. Adair, and D. L. Wandres, Mutagenicity testing in mammalian cells. II. Validation of multiple drug resistant markers having practical application for screening potential mutagens, *Mutat. Res. 72*, 203–230 (1980).
8. D. Clive, K. O Johnson, J. F. S. Spector, A. G. Batson, and M. M. M. Brown, Validation and characterization of the L5178Y/TK$^{+/-}$ mouse lymphoma mutagen assay system, *Mutat. Res. 59*, 61–108 (1979).
9. L. L. Deaven and D. F. Peterson, The chromosomes of CHO, an aneuploid Chinese hamster cell line: G-band, C-band, and autoradiographic analysis, *Chromosoma 41*, 129–144 (1973).
10. S. A. Farrell and R. G. Worton, Chromosome loss is responsible for segregation at the *HPRT* locus in Chinese hamster cell hybrids, *Somat. Cell Genet. 3*, 539–551 (1977).
11. R. S. Gupta and B. Singh, Mutagenic responses of five independent genetic loci in CHO cells to a variety of mutagens. Development and characteristics of a mutagen screening system based on selection for multiple drug-resistant markers, *Mutat. Res. 94*, 449–466 (1982).
12. H. Harris, *The Principles of Human Biochemical Genetics*, pp. 1–45, Elsevier/North-Holland and Biomedical Press, New York (1980).
13. H. Harris and D. A. Hopkinson, *Handbook of Enzyme Electrophoresis in Human Genetics*, North-Holland, Oxford.
14. U. Henning and C. Yanofsky, An electrophoretic study of mutually altered A proteins of the tryptophan, synthetase of *Escherichia coli*, *J. Mol. Biol. 6*, 16–21 (1963).
15. R. M. Humphrey, B. A. Sedita, and R. E. Meyn, Recovery of Chinese hamster cells from ultraviolet damage, *Int. J. Radiat. Biol. 18*, 61–69 (1970).
16. R. L. Hunter and C. L. Markert, Histochemical demonstration of enzymes separated by zone electrophoresis in starch gels, *Science 125*, 1294–1295 (1957).
17. V. M. Ingram, Gene mutations in human haemoglobin: The chemical difference between normal and sickle cell haemoglobin, *Nature 180*, 326–328 (1957).
18. F. M. Johnson and S. E. Lewis, Electrophoretically detected germinal mutations induced in the mouse by ethylnitrosourea, *Proc. Natl. Acad. Sci. USA 78*, 3138–3141 (1981).
19. C. L. Markert, Lactate dehydrogenase isozymes: Dissociation and recombination of subunits, *Science 140*, 1329–1330 (1963).
20. C. L. Markert, The molecular basis for isozymes, *Ann. N. Y. Acad. Sci. 151*, 14–40 (1968).
21. C. L. Markert and F. Moller, Multiple forms of enzymes: Tissue, ontogenetic and species specific patterns, *Proc. Natl. Acad. Sci. USA 45*, 753–763 (1959).
22. T. Moehring and J. Moehring, Selection and characterization of cells resistant to diphtheria toxin, *Cell 11*, 447–454 (1977).
23. M. Nei, *Molecular Population Genetics and Evolution*, pp. 26–28, American Elsevier, New York (1975).

24. J. P. O'Neill, D. B. Couch, R. Machanoff, J. R. San Sebastian, P. A. Brimer, and A. W. Hsie, A quantitative assay of mutation induction at the hypoxanthine-guanine phosphoribosyl transferase locus in Chinese hamster ovary cells (CHO/HGPRT system): Utilization with a variety of mutagenic agents, *Mutat. Res. 45*, 103–109 (1977).
25. M. Roberts and F. H. Ruddle, The chinese hamster gene map: Assignment of four genes (*DTS, PGM2, 6PGD, ENO1*) to chromosome 2, *Exp. Cell Res. 127*, 47–54 (1980).
26. M. J. Rosenstraus and L. A. Chasin, Isolation of mammalian cell mutants deficient in glucose-6-phosphate: Linkage to hypoxanthine phosphoribosyl transferase, *Proc. Natl. Acad. Sci. USA 72*, 493–497 (1975).
27. C. R. Shaw, The use of genetic variants in the analysis of isozyme structure, *Brookhaven Symp. Biol. 17*, 117–129 (1964).
28. T. B. Shows, C. A. Alper, D. Bootsma *et al.*, International system for human gene nomenclature, *Cytogenet. Cell Genet. 25*, 96–116 (1979).
29. M. J. Siciliano and C. R. Shaw, Separation and visualization of enzymes on gels, in: *Chromatographic and Electrophoretic Techniques*, (D. Smith, ed.), Vol. II, pp. 185–209, Heinemann, London (1976).
30. M. J. Siciliano, J. Siciliano, and R. M. Humphrey, Electrophoretic shift mutants in Chinese hamster ovary cells: Evidence for genetic diploidy, *Proc. Natl. Acad. Sci. USA 75*, 1919–1923 (1978).
31. M. J. Siciliano, R. L. Stallings, G. M. Adair, R. M. Humphrey, and J. Siciliano, Provisional assignment of *TPI* and *GPI/PEPD* to Chinese hamster autosomes 8 and 9: A cytogenetic basis for functional haploidy of an autosomal linkage group in CHO cells, *Cytogenet. Cell Genet. 35*, 15–20 (1983).
32. M. J. Siciliano, B. F. White, and R. M. Humphrey, Electrophoretically detectable mutations induced in CHO cells by varying doses of ultraviolet radiation, *Mutat. Res. 107*, 167–176 (1983).
33. M. J. Siciliano, R. L. Stallings, G. M. Adair, and R. M. Humphrey, The genetic map of the Chinese hamster and the genetic consequences of chromosomal rearrangement in CHO cells, in: *Molecular Cell Genetics: The Chinese Hamster Cell* (M. M. Gottesman, ed.), pp. 95–135, Wiley, New York, 1985.
34. L. Siminovitch, On the nature of hereditable variation in cultured somatic cells, *Cell 7*, 1–11 (1976).
35. A. E. Simon and M. W. Taylor, High-frequency mutation at the adenine phosphoribosyl transferase locus in Chinese hamster ovary cells due to deletion of the gene, *Proc. Natl. Acad. Sci. USA 80*, 810–814 (1983).
36. A. E. Simon, M. W. Taylor, W. E. C. Bradley, and L. H. Thompson, Model involving gene inactivation in the generation of autosomal recessive mutants in mammalian cells in culture, *Mol. Cell. Biol. 2*, 1126–1133 (1982).
37. R. L. Stallings and M. J. Siciliano, Genetic polymorphisms and gene expression variations at enzyme loci in Chinese hamster cell lines, *Somat. Cell Genet. 7*, 295–306 (1981).
38. R. L. Stallings and M. J. Siciliano, Confirmational, provisional and/or regional assignment of fifteen enzyme loci onto Chinese hamster autosomes 1, 2, and 7, *Somat. Cell Genet. 7*, 683–698 (1981).
39. R. L. Stallings and M. J. Siciliano, Assignment of *ADA*, *ITPA*, *AK1*, and *AK2* to Chinese hamster chromosomes: Genetic and structural evidence for the conservation of mammalian autosomal synteny, *J. Hered. 73*, 399–404 (1982).
40. R. L. Stallings, M. J. Siciliano, G. M. Adair, and R. M. Humphrey, Regional assignment of Chinese hamster *DTS* and *GALT* loci and further identification of hemizygous and dizygous chromosome 2 loci in CHO cells, *Somat. Cell Genet. 8*, 413–422 (1982).

41. R. L. Stallings, G. M. Adair, J. Siciliano, J. Greenspan, and M. J. Siciliano, Genetic effects of chromosomal rearrangements in CHO cells: Expression and chromosomal assignment of loci for TK, GALK, ACP1, ADA and ITPA, *Mol. Cell. Genet. 3*, 1967–1974 (1983).
42. R. L. Stallings, G. M. Adair, J. Lin, and M. J. Siciliano, Expression and regional assignment of Chinese hamster *esD* and rRNA genes associated with a translocation giving rise to chromosomes Z1 and Z6 in CHO cells, *Cytogenet. Cell Genet. 38*, 132–137 (1984).
43. R. L. Stallings, G. M. Adair, and M. J. Siciliano, Provisional assignment of *MPI, PKM2, PGM3*, and *ME1* to Chinese hamster chromosome 4, *Somat. Cell Mol. Genet. 10*, 109–111 (1984).
44. L. H. Thompson and R. M. Baker, Isolation of mutants of cultured mammalian cells, in: *Methods of Cell Biology* (D. M. Prescott, ed.), Vol. VI, pp. 209–281, Academic Press, New York (1973).
45. J. H. Tijo and T. T. Puck, Genetics of somatic mammalian cells, *J. Exp. Med. 108*, 259–268 (1958).
46. J J. Wasmuth and L.-Y. Chu, Linkage in cultured Chinese hamster cells of two genes, *EMTB* and *LEUS*, involved in protein synthesis and isolation of cell lines with mutations in three linked genes, *J. Cell Biol. 87*, 697–702 (1980).
47. A. Westerveld, R. P. L. S. Visser, M. A. Freeke, and D. Bottsma, Evidence for linkage of 3-phosphoglycerate kinase, hypoxanthine-guanine-phosphoribosyl transferase, and glucose 6-phosphate dehydrogenase loci in Chinese hamster cells studied by using a relationship between gene multiplicty and enzyme activity, *Biochem. Genet. 7*, 33–40 (1972).
48. R. G. Worton, C. C. Ho, and C. Duff, Chromosome stability in CHO cells, *Somat. Cell Genet. 3*, 27–45 (1977).

Index